Springer Textbooks in Earth Sciences, Geography and Environment

The Springer Textbooks series publishes a broad portfolio of textbooks on Earth Sciences, Geography and Environmental Science. Springer textbooks provide comprehensive introductions as well as in-depth knowledge for advanced studies. A clear, reader-friendly layout and features such as end-of-chapter summaries, work examples, exercises, and glossaries help the reader to access the subject. Springer textbooks are essential for students, researchers and applied scientists.

Jun-Yi Guo

Physical Geodesy

A Theoretical Introduction

 Springer

Jun-Yi Guo
Division of Geodetic Science, School of
Earth Sciences
The Ohio State University
Columbus, OH, USA

ISSN 2510-1307   ISSN 2510-1315   (electronic)
Springer Textbooks in Earth Sciences, Geography and Environment
ISBN 978-3-031-23322-7   ISBN 978-3-031-23320-3   (eBook)
https://doi.org/10.1007/978-3-031-23320-3

Preface

This book is intended as a comprehensive but accessible introduction to *Physical Geodesy*, one of the largest subdisciplines of *Geodesy* dealing with the theory, determination, and representation of the Earth's external gravity field, which significantly overlaps adjacent disciplines such as geophysics and orbital mechanics. The objectives are twofold: (1) to serve as a first course textbook addressing the contemporary understanding of *Physical Geodesy*; (2) to expand students' knowledge and skills in those areas of mathematics essential to this subdiscipline, preparing them for more advanced study and research in this field. Students are required to have a good understanding of single and multivariable calculus and some elementary knowledge of ordinary differential equations and matrix algebra. Some knowledge of the least squares method is desirable, but not necessary. This is a mathematical background similar to that acquired by most physical science or engineering undergraduates during their freshman and sophomore years. Beyond such a requirement, this book is self-sufficient.

The style of this book is to build the reader's knowledge and understanding "from the ground up," with little reference to other textbooks or the professional scientific literature. Mathematical formulations are developed from the practical point of view of applied physics. Rigorous understanding of the physics is emphasized, but the derivation of formulae may appear to be less rigorous from the point of view of a pure mathematician. One design characteristic of this book is the inclusion of many of the intermediate steps of mathematical derivations usually omitted in other textbooks. By discouraging gaps of mathematical understanding, I hope to promote a more complete physical understanding of the material. For the same reason, this text expresses important mathematical results in a form that fully evokes their physical meaning. Whenever practicable, it avoids the use of terse symbolic references to component sub-expressions defined previously, and elsewhere, which generates more compact representations, but does so at the cost of conceptual transparency. I prefer to spell out the details of each important result "locally," placing less burden on students' memory and minimizing the need to double back to re-establish the meaning of some forgotten and therefore abstract mathematical token. I hope this book draws a reasonable compromise between providing an easily read tutorial for novices, while retaining lasting value as a reference book for more experienced geodesists.

To keep the book to a reasonable size, while maintaining its objectives and style, the topics covered—and excluded—have been chosen with some care. The emphasis is on the static component of the gravity field, as most books entitled *Physical Geodesy* do. Temporal variations, such as the tidal phenomena, are not a focus of this book. Nor is satellite gravimetry, though the static gravity field determined using satellite missions such as GRACE and GOCE is incorporated into the narrative, so as to build up a more complete understanding of the contemporary determination of the gravity field. Relevant potential theory and spherical harmonics are treated at some length so as to familiarize students with the "machinery" of these powerful mathematical tools, as well as their various applications to the gravity field. Stokes' and Molodensky's classical boundary value theories using gravity anomaly as data are systematically presented and are extended to cover other less celebrated boundary value theories using various other types of data obtained using contemporary space geodesy techniques. Gravity reduction is systematically formulated based on not only the conventional flat Earth model but also the spherical Earth model for higher accuracy. The internal flattening of the Earth is treated in more detail than is usual in introductory geodesy texts, so as to properly develop the geophysics. An emphasis on the practicalities of computation pervades the entire book.

This book is the culmination of a long journey, both geographical and conceptual. My first effort along these lines produced a comprehensive set of lecture notes, written in Chinese, while teaching at the Wuhan Technical University of Surveying and Mapping (which merged into Wuhan University in 2001). These notes, published in 1994, benefited from the advice of many of my teachers, especially Professors Jin-Sheng Ning, Ze-Lin Guan, and Ding-Bo Cao. Through that experience, I secured a reasonably thorough understanding of *physical geodesy*, especially in the underlying mathematics, and I have been following the development of this field ever since, while contributing to it as an active researcher. I have now spent many years working as a researcher, and also as a teacher during the last three years, in the Geodetic Science program at The Ohio State University. I gradually realized that there were not many English language textbooks in *physical geodesy* that attempted to teach this subject "from the ground up," and none using the approach that I have adopted. So, I decided to rewrite my lecture notes in English, and comprehensively update them, to reflect my improved understanding of classical topics and methodologies, and the tremendous development of this field since the early 1990s. I have benefitted from my long and fruitful association with Professor C. K. Shum, who has constantly supported, encouraged, and assisted me in my scientific endeavors. Professor Michael Bevis has provided me with unflagging support and encouragement too: in particular, he pushed me to begin writing this long-imagined textbook, and to finish it! I thank all these generous people, as well as many others, colleagues and students, my esteemed fellow travelers, who have provided me with so many helpful comments, corrections, challenges, and suggestions over the years.

Columbus, OH, USA   Jun-Yi Guo
October 2022

Contents

List of Figures

List of Tables

The Earth's Gravity Field and Some of Its Properties

1

Abstract

This chapter introduces the definitions of the most general concepts in the book: gravitational attraction, centrifugal acceleration, gravity, potential, equipotential surface, geoid, and dynamic and orthometric heights with a discussion on the orthometric height system and datum. The gravitational potential and attraction of some simple configurations are formulated. Some fundamental properties of the gravitational potential and attraction are proven, most notably their continuities across surfaces and Laplace's and Poisson's equations of the potential. Finally, the curvatures of equipotential surfaces and plumb lines are formulated.

1.1 Gravitational Attraction, Centrifugal Acceleration, and Gravity

1.1.1 Gravitational Attraction

Gravitational Attraction of a Point Mass

According to Newton's law of gravitation, any two objects attract each other, i.e., each object attracts the other toward it. The force of attraction is customarily referred to as force of gravitational attraction between masses. If the sizes of the objects are much smaller than the distance between them, the force of gravitational attraction between them is proportional to the masses of both objects and is inversely proportional to the square of the distance between them.

As shown in Fig. 1.1, we assume that two point masses (which could be understood as infinitely small particles), m and m', are located at M and P, respectively. We use (ξ, η, ζ) and (x, y, z) to denote their coordinates, and use $\overrightarrow{l}$ and l to represent the relative position vector of P with respect to M and its length,

Fig. 1.1 The relative
position vector

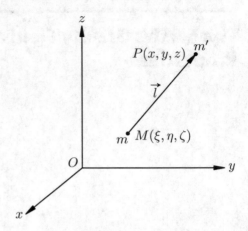

respectively. We then have

$$\vec{l} = (x - \xi)\hat{e}_x + (y - \eta)\hat{e}_y + (z - \zeta)\hat{e}_z , \qquad (1.1)$$

where $\hat{e}_x, \hat{e}_y, \hat{e}_z$ are the unit vectors along the respective Cartesian axes, and

$$l = [(x - \xi)^2 + (y - \eta)^2 + (z - \zeta)^2]^{1/2} . \qquad (1.2)$$

The magnitude of the force of gravitational attraction between the two masses is

$$F = G\frac{mm'}{l^2} , \qquad (1.3)$$

where G is the universal gravitational constant, which can only be determined by experiments, $G = 6.6742 \times 10^{-11}\text{m}^3 \cdot \text{kg}^{-1} \cdot \text{s}^{-2} = 6.6742 \times 10^{-8}\text{cm}^3 \cdot \text{g}^{-1} \cdot \text{s}^{-2}$ being its most up-to-date determination. In other words, each point mass attracts the other toward it with the force F.

For theoretical formulation, a vector form of the force is more convenient. The force of gravitational attraction exerted by m on m' can be expressed as

$$\vec{F} = -G\frac{mm'}{l^2}\hat{e}_l = -G\frac{mm'}{l^3}\vec{l} , \qquad (1.4)$$

where $\hat{e}_l = \vec{l}/l$ is the unit vector along the direction of $\vec{l}$. The negative sign implies that $\vec{F}$ is opposite to $\vec{l}$ in direction. The force of gravitational attraction exerted by m' on m is opposite to $\vec{F}$ with the same magnitude, i.e., the same norm.

In (1.4), we call m attracting mass and m' attracted mass. When the attracted mass is a unit point mass, i.e., $m' = 1$, (1.4) becomes

$$\overrightarrow{F} = -G\frac{m}{l^3}\overrightarrow{l}\,, \tag{1.5}$$

which is the value of the gravitational acceleration of m' under the action of the force of gravitational attraction of m. This is equivalent to redefine the notation $\overrightarrow{F}$ as $-G[(mm'/l^3)\,\overrightarrow{l}\,]/m'$.

From now on, we always assume the attracted mass to be a unit point mass and refer $\overrightarrow{F}$ as the gravitational attraction, which is also briefly referred to as attraction, of m at the point P, i.e., $\overrightarrow{F}$ is considered as a function of the coordinates of P where the attracted mass $m' = 1$ is located. Strictly speaking, the gravitational attraction $\overrightarrow{F}$ thus defined is the intensity of the force of gravitational attraction and has the unit of acceleration. By substituting (1.1) into (1.5), we obtain the 3 components of the gravitational attraction,

$$F_x = -G\frac{m}{l^3}(x - \xi)\,, \qquad F_y = -G\frac{m}{l^3}(y - \eta)\,, \qquad F_z = -G\frac{m}{l^3}(z - \zeta)\,. \tag{1.6}$$

With such a convention, the gravitational attraction is the gravitational acceleration of a point mass at P due to the gravitational attraction of the point mass m.

Gravitational Attraction of a System of Point Masses

If the attracting point mass is not limited to 1, but there are n, with masses m_1, $m_2, \ldots, m_n$. We denote the relative position vectors of the unit attracted mass with respect to the point masses as $\overrightarrow{l}_1, \overrightarrow{l}_2, \ldots, \overrightarrow{l}_n$, and their norms as $l_1, l_2, \cdots,$ and l_n. The gravitational attraction of this system of point masses at the location of the unit attracted mass is then

$$\overrightarrow{F} = -G\sum_{i=1}^{n}\frac{m_i}{l_i^3}\overrightarrow{l}_i\,. \tag{1.7}$$

This implies that the gravitational attraction satisfies the law of vector addition, which should be understood as a fact proven by experiments. By substituting (1.1) into (1.7), we obtain the 3 components of the gravitational attraction:

$$F_x = -G\sum_{i=1}^{n}\frac{m_i}{l_i^3}(x-\xi_i)\,, \qquad F_y = -G\sum_{i=1}^{n}\frac{m_i}{l_i^3}(y-\eta_i)\,, \qquad F_z = -G\sum_{i=1}^{n}\frac{m_i}{l_i^3}(z-\zeta_i)\,. \tag{1.8}$$

Gravitational Attraction of a Material Surface

Now we assume that the attracting mass is condensed on a surface S as shown in Fig. 1.2, called material surface. On this surface, we denote the coordinates of any

Fig. 1.2 Gravitational attraction of a material surface

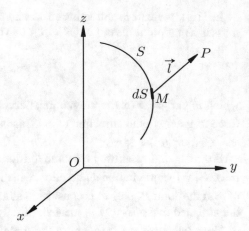

point M as (ξ, η, ζ). The surface mass density μ is a function of the point M, or its coordinates ξ, η and ζ. The mass of an infinitesimal surface element dS at the point M behaves as a point mass μdS; therefore, its gravitational attraction at the point P of coordinate (x, y, z) is $d\overrightarrow{F} = -G[(\mu dS)/l^3]\,\overrightarrow{l}$, where $\overrightarrow{l}$ is the relative position vector of P with respect to M, and l its norm, as expressed in (1.1) and (1.2). The gravitational attraction of the material surface at the point P is then the integral

$$\overrightarrow{F} = -G \int_S \frac{\mu}{l^3}\,\overrightarrow{l}\,dS. \tag{1.9}$$

It is to be noted that $\overrightarrow{F}$ is a function of x, y and z; the variables of integration are ξ, η and ζ; and x, y and z are treated as constant parameters within the integral. By substituting (1.1) and (1.2) into (1.9), we obtain the 3 components of the gravitational attraction,

$$F_x = -G \int_S \frac{\mu}{l^3}(x - \xi)dS, \; F_y = -G \int_S \frac{\mu}{l^3}(y - \eta)dS, \; F_z = -G \int_S \frac{\mu}{l^3}(z - \zeta)dS. \tag{1.10}$$

Gravitational Attraction of a Solid Body

In general, the attracting mass is not concentrated at a point or a surface, but fills a spatial region, and is referred to as a solid body. When the size of the solid body is much smaller than the dimension of the problem studied, it may be approximately considered as a point mass. When the solid body is a thin shell, and the thickness of the shell is much smaller than the dimension of the problem studied, it may be approximately considered as a material surface. For example, when the gravitational

Fig. 1.3 Gravitational
acceleration of a solid body

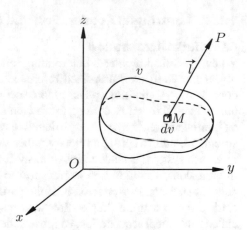

attraction on the Earth's surface due to the Earth and all celestial bodies is studied,
the Sun, the Moon, and other celestial bodies may be considered as point masses
(the dimension of the problem is the distance between the Earth and the celestial
bodies), but the Earth must be considered as a solid body. When the gravitational
attraction of the Earth system (all masses enclosed from the Earth's center to the top
of the atmosphere) at satellite height is studied, the variable part of ocean surface
height and variation of land water storage may be considered as a material surface.

In Fig. 1.3, we assume the solid body occupies the spatial region v. Within the
body, we denote the coordinates of any point M as (ξ, η, ζ). The volume mass
density ρ is a function of the point M, or its coordinates ξ, η and ζ. The mass of
an infinitesimal volume element dv at the point M behaves as a point mass ρdv;
therefore, its gravitational attraction at the point P of coordinate (x, y, z) is $d\overrightarrow{F} =
-G[(\rho dv)/l^3]\overrightarrow{l}$, where $\overrightarrow{l}$ is the relative position vector of P with respect to M,
and l its norm, as expressed in (1.1) and (1.2). The gravitational attraction of the
solid body at the point P is then the integral

$$\overrightarrow{F} = -G \int_v \frac{\rho}{l^3} \overrightarrow{l} \, dv. \tag{1.11}$$

Like the case of material surface: $\overrightarrow{F}$ is a function of x, y, and z; the variables of
integration are ξ, η and ζ; and x, y and z are treated as constant parameters within
the integral. Substitute (1.1) and (1.2) into (1.11). We obtain the 3 components of
the gravitational attraction,

$$F_x = -G \int_v \frac{\rho}{l^3}(x-\xi)dv, \quad F_y = -G \int_v \frac{\rho}{l^3}(y-\eta)dv, \quad F_z = -G \int_v \frac{\rho}{l^3}(z-\zeta)dv. \tag{1.12}$$

1.1.2   Centrifugal Acceleration and Gravity

Centrifugal Acceleration

When dynamical problems in a rotating reference system are studied, like how an object moves in the Earth-fixed reference system, the centrifugal force must be considered, which is not a physical force originated from the action of other objects, but an inertial force. According to Newton's first law of motion, when observed in an inertial reference system, a particle free of any force maintains a motion along a straight line with a constant velocity. When the particle is held standstill in a rotating reference system, it tends to move away from the rotation axis for maintaining a motion along a straight line with respect to the inertial space, as if a force, called centrifugal force, is exerted on it. If the particle is suddenly released free, it would start to move with an acceleration with respect to the rotating reference system, called centrifugal acceleration, which is understood as the consequence of the action of the centrifugal force. Like gravitational attraction, we understand the centrifugal acceleration as the centrifugal force exerted on a unit point mass. In the literature of geodesy, the term *centrifugal force* is also sometimes used to indicate the centrifugal acceleration.

As shown in Fig. 1.4, we assume that the reference system $Oxyz$ rotates with a constant rotation rate Ω around the z-axis with respect to the inertial space. The direction of rotation and that of the z-axis follow the right-hand rule. The vector $\vec{\Omega} = \Omega \hat{e}_z$ is called angular velocity. The magnitude of the centrifugal acceleration at the point P of coordinate (x, y, z) is

$$F_c = \Omega^2 (x^2 + y^2)^{1/2} = \Omega^2 r \sin\theta, \tag{1.13}$$

which is proportional to the distance to the rotation axis. Its direction is perpendicular to the rotation axis and points outward. In vector form, the centrifugal

Fig. 1.4 Rotating reference system and centrifugal acceleration

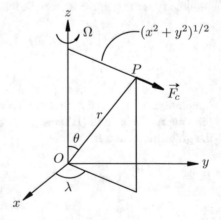

acceleration is

$$\vec{F}_c = \Omega^2(x\hat{e}_x + y\hat{e}_y). \tag{1.14}$$

In the above two formulae, the SI unit of Ω is radian per second. As radian is dimensionless, the dimension of Ω is s^{-1}. Similar to the case of gravitational attraction, the centrifugal acceleration is the intensity of the centrifugal force, and therefore, we also refer to it as centrifugal force.

Gravity

In the general sense, the gravity is the free fall acceleration of a particle in the Earth-fixed reference system not subjecting to electromagnetic force. More strictly speaking, it is the acceleration when the velocity of the particle with respect to the Earth-fixed reference system vanishes. When the velocity does not vanish, another inertial force—the Coriolis force—is also present. The temporal variation of the Earth's angular velocity brings yet another inertial force—the Poincaré or Euler force, which is extremely small, and is completely negligible in the context of this book. For these inertial forces, readers are referred to textbooks of classical mechanics.

The same as the gravitational attraction and centrifugal acceleration, we also understand the gravity as a force exerted on a unit point mass, which is, strictly speaking, the intensity of gravity.

In the context of this book, we consider only the effects of the Earth's mass and rotation. By definition, the gravity $\vec{g}$ is the sum of the gravitational attraction and centrifugal acceleration,

$$\vec{g} = \vec{F} + \vec{F}_c, \tag{1.15}$$

where $\vec{F}$ is the gravitational attraction of the Earth, and $\vec{F}_c$ is the centrifugal acceleration corresponding to the instantaneous angular velocity of the Earth.

Furthermore, we restrict the gravity to be the time-invariable part of the Earth's gravitational attraction and centrifugal acceleration. Therefore, the mass distribution within the Earth is assumed to be time-invariable, and the angular velocity of the Earth-fixed reference system is chosen to be a kind of the Earth's mean angular velocity assumed to represent the Earth's long term time-invariable rotation. The meaning of the Earth's mean angular velocity is of twofold: it includes the definitions of a mean rotation rate of the Earth with respect to the inertial space, and a mean pole defining the mean axis of the Earth rotation as seen in the Earth-fixed reference system. Here we understand the pole as the intersection point of the Earth's rotation axis going through the center of mass of the Earth system (all the masses enclosed from the Earth's center to the top of the atmosphere) with the Earth's surface. Its position varies with time, which is referred to as polar motion, and the Earth's rotation rate also varies with time. Choosing a mean of them to define the Earth-fixed reference system ensures the centrifugal acceleration to be time-invariable within the context of this book.

The gravity measured using a gravimeter is the instantaneous value of gravity including all possible effects, which is not the time-invariable gravity as defined in the previous two paragraphs. Hence, there is a need to correct the measured value by removing the effect of extraterrestrial celestial bodies and the time-variable part of the gravity of the Earth. The corrections may include the gravitational effect of the Sun, the Moon, and other planets, i.e., the tidal generating force, the gravitational effect of mass redistribution within the Earth system, particularly in the hydrosphere–cryosphere–atmosphere system, the effect of the Earth rotation variation including rotation rate variation and polar motion on the centrifugal acceleration, and the gravitational effect of the Earth's deformation caused by these forcings. Whether a correction is applied depends on the accuracy of measurement and the magnitude of the correction.

From now on, we implicitly understand the gravitational attraction, centrifugal acceleration, and gravity as time-invariable. According to the definition of the mean pole of the Earth rotation, the origin of the Earth-fixed reference system should be located on the center of mass of the Earth system as precisely as possible.

Unit

The gravitational attraction, centrifugal acceleration, and gravity are all understood as a force exerted on a unit point mass. Consequently, their values are identical to the respective accelerations. Therefore, the unit of acceleration is adopted for them. The SI unit is $m \cdot s^{-2}$. In geodesy, $cm \cdot s^{-2}$ is traditionally used as standard unit and is named as Gal to honor the Italian astronomer Galileo. Smaller units include miliGal denoted as mGal, microGal denoted as μGal, ..., to represent 10^{-3}Gal, 10^{-6}Gal, ...

The Concept of Field

Gravity field is often mentioned. Here the word *field* is understood in the sense of physics. In a space domain v, if a physical quantity has a value at each point, which may also be a function of time, we say that the physical quantity is a field in v. This quantity may be a scalar, vector, or tensor and is named scalar field, vector field, or tensor field. For example, temperature, density, etc., are scalar fields, and force, magnetic flux density, etc., are vector fields. The fields associated with gravitational attraction, centrifugal acceleration, and gravity are referred to as gravitational field, centrifugal field, and gravity field, respectively.

A field is an actual existence. For example, for any mass, an associated gravitational field exists whether an attracted mass is placed in it or not. When a mass is placed in a gravitational field, the mass is subject to the action of gravitational attraction.

1.2 Gravitational Potential, Centrifugal Potential, and Gravity Potential

1.2.1 Definition of Potential of a Vector

For a vector field

$$\vec{f} = f_x \hat{e}_x + f_y \hat{e}_y + f_z \hat{e}_z \,, \tag{1.16}$$

if there exists a scalar field ϕ such that

$$f_x = \frac{\partial \phi}{\partial x} \,, \quad f_y = \frac{\partial \phi}{\partial y} \,, \quad f_z = \frac{\partial \phi}{\partial z} \,, \tag{1.17}$$

which can be written in vector form as

$$\vec{f} = \frac{\partial \phi}{\partial x} \hat{e}_x + \frac{\partial \phi}{\partial y} \hat{e}_y + \frac{\partial \phi}{\partial z} \hat{e}_z = \nabla \phi \,, \tag{1.18}$$

the vector $\vec{f}$ is said to have a potential ϕ. The quantity $\nabla \phi$ is called gradient of ϕ, where $\nabla = \hat{e}_x (\partial/\partial x) + \hat{e}_y (\partial/\partial y) + \hat{e}_z (\partial/\partial z)$ is called Nabla operator. Therefore, the vector is the gradient of its potential, if the potential exists.[1] A vector or scalar field will also be briefly referred to as vector or scalar.

Furthermore, the directional derivative of the potential along any direction is equal to the projection of the vector along this direction. In the coordinate system $Oxyz$, a unit vector $\hat{e}_l$ can be expressed as

$$\hat{e}_l = \cos(\hat{e}_l, \hat{e}_x)\hat{e}_x + \cos(\hat{e}_l, \hat{e}_y)\hat{e}_y + \cos(\hat{e}_l, \hat{e}_z)\hat{e}_z \,, \tag{1.19}$$

where $(\hat{e}_l, \hat{e}_x)$, $(\hat{e}_l, \hat{e}_y)$, and $(\hat{e}_l, \hat{e}_z)$ represent the angle between $\hat{e}_l$ and the 3 axes, respectively. The directional derivative of ϕ along $\hat{e}_l$, denoted as $\partial \phi/\partial l$, has the property

$$\frac{\partial \phi}{\partial l} = \frac{\partial \phi}{\partial x} \cos(\hat{e}_l, \hat{e}_x) + \frac{\partial \phi}{\partial y} \cos(\hat{e}_l, \hat{e}_y) + \frac{\partial \phi}{\partial z} \cos(\hat{e}_l, \hat{e}_z) \,. \tag{1.20}$$

Hence, according to (1.16)–(1.18), we have

$$\frac{\partial \phi}{\partial l} = \nabla \phi \cdot \hat{e}_l = \vec{f} \cdot \hat{e}_l = f_l \,. \tag{1.21}$$

---

[1] In physics, the relation $\vec{f} = -\nabla \phi$ is used to define the potential of a vector. In this book, we follow the convention in geodesy without the minus sign. However, the minus sign may also be present in the literature of geodesy and geophysics. Hence, there is always a need to verify or clarify how the potential is defined.

Based on the property (1.21), we can infer that $\nabla\phi$ is normal to the surface $\phi = C$, where C is a constant: As ϕ is constant on the surface, for any vector tangential to the surface, $\hat{e}_l$, we have $\partial\phi/\partial l = 0$, meaning that $\nabla\phi$ is perpendicular to all vectors tangential to the surface, thus being normal to the surface.

It will be verified that the gravitational attraction does have a potential, and consequently, the directional derivative of the gravitational potential is equal to the projection of the gravitational attraction along the direction. The same is for the centrifugal acceleration and gravity.

1.2.2 Gravitational Potential

The gravitational potential of a point mass is

$$V = G\frac{m}{l}, \tag{1.22}$$

where l is defined in (1.2) based on Fig. 1.1. The coordinate of the point mass m is (ξ, η, ζ), and the gravitational potential V is understood as a function of (x, y, z).

In order to prove the correctness, we only need to verify that V satisfies the definition of potential, which is straightforward:

$$
\begin{cases}
\dfrac{\partial V}{\partial x} = Gm\dfrac{\partial}{\partial x}\left(\dfrac{1}{l}\right) = -G\dfrac{m}{l^3}(x - \xi) = F_x, \\[2mm]
\dfrac{\partial V}{\partial y} = Gm\dfrac{\partial}{\partial y}\left(\dfrac{1}{l}\right) = -G\dfrac{m}{l^3}(y - \eta) = F_y, \\[2mm]
\dfrac{\partial V}{\partial z} = Gm\dfrac{\partial}{\partial z}\left(\dfrac{1}{l}\right) = -G\dfrac{m}{l^3}(z - \zeta) = F_z.
\end{cases}
\tag{1.23}
$$

It can be verified that the gravitational potential of a system of point masses is

$$V = G\sum_{i=1}^{n}\frac{m_i}{l_i}, \tag{1.24}$$

and those of a material surface and a solid body are

$$V = G\int_S \frac{\mu}{l}dS \tag{1.25}$$

and

$$V = G\int_v \frac{\rho}{l}dv, \tag{1.26}$$

respectively.

In order to verify the correctness of the gravitational potential for a material surface or a solid body, we only have to use the facts that ξ, η and ζ are the variables of integration, and the partial derivative with respect to x, y and z can be moved to within the integral sign. For example, for the case of a material surface,

$$
\begin{cases}
\dfrac{\partial V}{\partial x} = G \displaystyle\int_S \mu \dfrac{\partial}{\partial x}\left(\dfrac{1}{l}\right) dS = -G \displaystyle\int_S \dfrac{\mu}{l^3}(x-\xi)dS = F_x\,, \\[3mm]
\dfrac{\partial V}{\partial y} = G \displaystyle\int_S \mu \dfrac{\partial}{\partial y}\left(\dfrac{1}{l}\right) dS = -G \displaystyle\int_S \dfrac{\mu}{l^3}(y-\eta)dS = F_y\,, \\[3mm]
\dfrac{\partial V}{\partial z} = G \displaystyle\int_S \mu \dfrac{\partial}{\partial z}\left(\dfrac{1}{l}\right) dS = -G \displaystyle\int_S \dfrac{\mu}{l^3}(z-\zeta)dS = F_z\,.
\end{cases} \tag{1.27}
$$

It can be inferred from (1.24) to (1.26) that the gravitational potential of various masses can be added together to obtain the total gravitational potential of all the masses (integration is also an addition), which is a consequence of the fact that the gravitational attraction obeys the law of vector addition.

We emphasize that the gravitational potential is defined according to the expression of gravitational attraction. If we add a constant into the expression of the gravitational potential, the gravitational attraction remains unaffected. Therefore, if (1.17) or (1.18) is the only relation to be satisfied, the definition of gravitational potential chosen is preferred, but not unique.

By definition, the SI unit of gravitational potential is $\mathrm{m}^2 \cdot \mathrm{s}^{-2}$.

1.2.3 Gravitational Potential of a Dipole or a Double Layer

Gravitational Potential of a Dipole

As shown in Fig. 1.5, we assume two point masses, m and $-m$, are located at the points M' and M, respectively. The total gravitational potential of these two masses

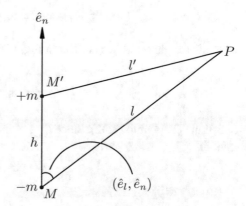

Fig. 1.5 Gravitational potential of a dipole

at the point P is

$$V = G\frac{m}{l'} - G\frac{m}{l} = Gm\left(\frac{1}{l'} - \frac{1}{l}\right),$$ (1.28)

where l' and l are the distances from M' and M to P, respectively. Denote the distance between M and M' as h, and the direction from M to M' as $\hat{e}_n$. The quantity $1/l'$ can then be expressed using $1/l$ as a Taylor series:

$$\frac{1}{l'} = \frac{1}{l} + \frac{\partial}{\partial n}\left(\frac{1}{l}\right)h + \frac{1}{l}\frac{\partial^2}{\partial n^2}\left(\frac{1}{l}\right)h^2 + \cdots.$$ (1.29)

Assume $m \to \infty$ as $h \to 0$, so that

$$M = \lim_{h\to 0} mh$$ (1.30)

remains finite. We obtain, by substituting (1.29) into (1.28) and then setting the limit $h \to 0$,

$$V = GM\frac{\partial}{\partial n}\left(\frac{1}{l}\right).$$ (1.31)

This is the gravitational potential of a dipole, and M is referred to as dipole moment.

The expression of the gravitational potential of a dipole (1.31) can be further simplified. According to the definition of directional derivative, we have

$$\frac{\partial l}{\partial n} = \lim_{h\to 0}\frac{l' - l}{h}.$$ (1.32)

As

$$l'^2 = l^2 + h^2 - 2lh\cos(\hat{e}_l, \hat{e}_n),$$ (1.33)

where the angle $(\hat{e}_l, \hat{e}_n)$ is defined in Fig. 1.5, we have

$$l' - l = \frac{h^2 - 2lh\cos(\hat{e}_l, \hat{e}_n)}{l' + l}.$$ (1.34)

Substitute (1.34) into (1.32). Since $l' \to l$ as $h \to 0$, we obtain

$$\frac{\partial l}{\partial n} = -\cos(\hat{e}_l, \hat{e}_n).$$ (1.35)

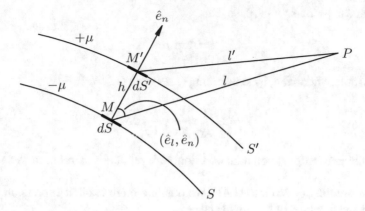

Fig. 1.6 Gravitational potential of a double layer

So we have

$$\frac{\partial}{\partial n}\left(\frac{1}{l}\right) = -\frac{1}{l^2}\frac{\partial l}{\partial n} = \frac{\cos(\hat{e}_l, \hat{e}_n)}{l^2}. \tag{1.36}$$

Substitute this formula into (1.31). We obtain a more explicit expression of the gravitational potential of a dipole,

$$V = GM\frac{\cos(\hat{e}_l, \hat{e}_n)}{l^2}. \tag{1.37}$$

Gravitational Potential of a Double Layer

As shown in Fig. 1.6, we assume that two material surfaces S' and S of densities μ and $-\mu$, respectively, are very closely located; the distance between them, h, is infinitesimally small. Draw a straight line perpendicular to S' at the point M', which intersects S at the point M. We denote the direction from M to M' as $\hat{e}_n$. The total gravitational potential of these two material surfaces is

$$V = G\int_{S'}\frac{\mu}{l'}dS' - G\int_S\frac{\mu}{l}dS, \tag{1.38}$$

where dS' and dS are infinitesimal surface elements on S' and S at the points M' and M, respectively, and l' and l are the distances from M' and M to P, respectively. Set $h \to 0$. As S' and S tend to overlap, (1.38) becomes

$$V = G\int_S\left[\lim_{h\to 0}\mu\left(\frac{1}{l'}-\frac{1}{l}\right)\right]dS. \tag{1.39}$$

We assume $\mu \to \infty$ as $h \to 0$, so that

$$v = \lim_{h \to 0} \mu h \tag{1.40}$$

remains finite. We obtain, by substituting (1.29) into (1.39),

$$V = G \int_S v \frac{\partial}{\partial n} \left(\frac{1}{l} \right) dS. \tag{1.41}$$

This is the gravitational potential of a double layer, and v is referred to as dipole density.

By substituting (1.36) into (1.41), we obtain a more explicit expression for the gravitational potential of the double layer,

$$V = G \int_S v \frac{\cos(\hat{e}_l, \hat{e}_n)}{l^2} dS, \tag{1.42}$$

where the angle $(\hat{e}_l, \hat{e}_n)$ is defined in Fig. 1.6.

Since the mass of material we are concerned is always positive, the gravitational potentials of a dipole or a double layer are mostly mathematically meaningful. However, in electromagnetism, the potentials of a dipole and a double layer do have real physical meaning.

As compared to a double layer, a material surface is also referred to as a single layer.

1.2.4   Centrifugal Potential and Gravity Potential

It can be readily verified that centrifugal acceleration also has a potential, called centrifugal potential, which is

$$Q = \frac{1}{2}\Omega^2(x^2 + y^2) = \frac{1}{2}\Omega^2 r^2 \sin^2 \theta, \tag{1.43}$$

where all symbols are defined in Fig. 1.4.

The gravity of the Earth is the sum of the gravitational attraction of the Earth and the centrifugal acceleration. Therefore, the gravity potential of the Earth is the sum of the gravitational potential of the Earth and the centrifugal potential,

$$W = V + Q. \tag{1.44}$$

A more concrete expression of it is

$$W = G \int_v \frac{\rho}{l} dv + \frac{1}{2}\Omega^2 r^2 \sin^2 \theta, \tag{1.45}$$

where v is the space domain representing the body of the Earth, ρ the density of the Earth, and Ω the rotation rate of the Earth. We denote the point where W is to be computed as P, as shown in Fig. 1.4. The other symbols in (1.45) are as follows: l is the distance between the location of dv and the point P; r is the distance between the origin of the coordinate system and the point P; and θ the angle between the rotation axis of the Earth and the direction from the origin of the coordinate system to the point P.

The same as gravitational potential, the SI unit of both centrifugal and gravity potentials is $m^2 \cdot s^{-2}$.

1.3   Equipotential Surface of Gravity, Geoid, and Orthometric Height

1.3.1   Equipotential Surface of Gravity

An equipotential surface of gravity is defined such that the values of the gravity potential are identical at all points on it. According to the definition, the equation of an equipotential surface of gravity is

$$W = \alpha,\tag{1.46}$$

where α is a constant. When α takes a series values, we obtain a series of equipotential surface of gravity. The values of the gravity potential are identical on the same equipotential surface of gravity, but different on different equipotential surfaces of gravity.

Based on the discussion below (1.18), the gravity $\vec{g} = \nabla W$ is perpendicular to the equipotential surface of gravity $W = \alpha$. In the rest of this subsection, this property is to be discussed in more detail.

At any point, the gravity is perpendicular to the equipotential surface of gravity passing through this point. To demonstrate this, we choose a direction $\hat{e}_t$ that is tangential to the equipotential surface of gravity. As the gravity potential is constant on the equipotential surface of gravity, the directional derivative of the gravity potential along the direction $\hat{e}_t$, $\partial W/\partial t$, vanishes. According to the definition of the gravity potential, this directional derivative is the projection of the gravity along the direction $\hat{e}_t$. Hence, we have

$$g_t = \frac{\partial W}{\partial t} = 0.\tag{1.47}$$

This implies that the gravity is perpendicular to any direction tangential to the equipotential surface of gravity, meaning that the gravity must be perpendicular to the equipotential surface of gravity.

Adjacent equipotential surfaces of gravity are not necessarily parallel, and they cannot intersect or be tangential to each other. The distance between two infinitesimally close equipotential surfaces of gravity is inversely proportional to the gravity. To demonstrate these, we choose a direction $\hat{e}_h$ that is perpendicular to the equipotential surface of gravity. By definition, $\hat{e}_h$ is either along or opposite to the direction of gravity. Here we choose it to be opposite to the direction of gravity, i.e., the gravity points downward, while $\hat{e}_h$ is chosen to point upward. We express the gravity as $\vec{g} = -g\hat{e}_h$ to ensure the magnitude g to be positive. The directional derivative of the gravity potential along the direction $\hat{e}_h$, $\partial W / \partial h$, is then $-g$. We have

$$g = -\frac{\partial W}{\partial h}. \tag{1.48}$$

The distance between two adjacent equipotential surfaces of gravity with a difference of gravity potential dW is then

$$dh = -\frac{dW}{g}. \tag{1.49}$$

The conclusions we are seeking follow from this equation and the fact that, on an equipotential surface of gravity, g is finite, but it is not necessarily constant.

The above are for the Earth's gravity field. We will see that, in some special instances, the gravitational potential may be constant in a domain, such as the inside of a homogeneous spherical shell, where by definition, the gravitational attraction vanishes.

1.3.2 Geoid, Dynamic, and Orthometric Heights

First of all, we imagine an ideal state that all the oceans are at standstill. In this case, the gravity on the ocean surface must be perpendicular to the surface, as the water would flow if this is not true. Hence, the ocean surface at standstill coincides with an equipotential surface of gravity. This particular equipotential surface of gravity is referred to as geoid.

In reality, the oceans are not at standstill. The flow of ocean water not only causes changes of gravity and ocean surface, but also renders the mean ocean surface to deviate from an equipotential surface of gravity. We assume that we know the time-invariable gravity and the mean ocean surface after correcting the time-variable part and define the geoid as the equipotential surface of gravity chosen to be the nearest possible to the mean ocean surface following a prescribed convention. In this way, we may define the geoid to be the equipotential surface of gravity through the mean ocean surface at a particular tide gauge, the equipotential surface of gravity through the average of the mean ocean surface at a number of tide gauges, or even the equipotential surface of gravity that best fits the global mean ocean surface determined by satellite altimetry.

Without relying on gravity potential that is not something we can feel, the geoid could be depicted as a closed surface satisfying the following conditions: (1) the gravity is perpendicular to it everywhere; (2) it is chosen to be the nearest possible to the mean ocean surface according to a prescribed standard as convention.

The geoid is a very important concept in the theory of the Earth's shape and gravity field; it is considered as a physical surface representing the shape of the Earth. One of the main focus of this book is to present the fundamental knowledge for the determination of the geoid.

By definition, the geoid is a level surface; the surface of the oceans would coincide with it if the oceans were at standstill. Therefore, it represents an idealized sea level and thus serves as an ideal reference surface to measure height—the height of any point on it is zero. For any point around the Earth's surface, the distance from the point to the geoid is defined as the orthometric height, which is also referred to as height above sea level. When the point is above the geoid, the height is positive; when the point is below the geoid, the height is negative. The geoid used to define the height is referred to as vertical datum.

In surveying, height is traditionally determined using spirit leveling. The instrument used is called level. The main component of a level is a telescope turning on an axis strictly perpendicular to its line of sight. While doing observations, a level is leveled with its turning axis pointing to the direction of gravity, so that the line of sight turns in the horizontal plane. When two identically graduated poles adjusted to stand along the direction of gravity are read at the line of sight, the difference between the readings is the height difference between the bottoms of the two poles. Hence, what a level measures is the local distance between two equipotential surfaces of gravity. As shown in Fig. 1.7, the height difference H_i could be understood as that measured by a level by reading two poles standing at A_{i-1} and A_i, which is the distance between the equipotential surface of gravity passing through A_{i-1} and that passing through A_i. Now we assume that A is on the geoid where the orthometric height is 0. The height of P obtained by leveling

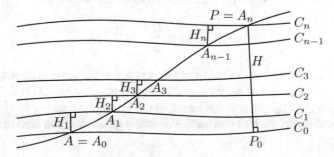

Fig. 1.7 Determination of height by spirit leveling

through the specific path is then

$$H_L = \sum_{i=1}^{n} H_i \,, \tag{1.50}$$

where the subscript L indicates leveling.

As the distance between two equipotential surfaces of gravity is different at different locations, the height of a point determined by leveling through different paths may be different; therefore, it cannot be used as a standard measure of height and must be converted into something else that is unique. The simplest intuition is that the height of an equipotential surface of gravity be identical everywhere, which obeys strictly the life experience that water flows from higher places to lower places. The quantity that is unique for an equipotential surface of gravity is the gravity potential on it. Hence, a quantity named potential number C is defined for any point P, which is the difference between the gravity potential on the geoid, $W_A = W_0$, and that at P, W_P,

$$C = W_0 - W_P = \sum_{i=1}^{n} (W_{i-1} - W_i) = \sum_{i=1}^{n} g_{i-1} H_i = \int_{A}^{P} g \, dH_L \,, \tag{1.51}$$

where W_i is the gravity potential on the equipotential surface of gravity C_i as shown in Fig. 1.7. The third identity is inferred using (1.49), and the last identity states the fact that every step of leveling is infinitesimally small as compared to the distance between A and P.

Although C is unique, and it is the most important parameter to define height, it is not a length, while a height should be. Hence, it must be divided by a gravity to be converted into a length. Different type of height is defined following the choice of different kind of gravity. A straightforward choice is to use a fixed value of gravity, such as the average gravity $\bar{g}$ in the region where the height system is used, or chosen otherwise. This type of height is referred to as dynamic height,[2]

$$H_D = \frac{C}{\bar{g}} \,. \tag{1.52}$$

The height thus defined obeys the intuition that water flows from higher places to lower places. However, this height is not the length of a line explicitly defined.

The orthometric height H is the length of the line $P_0 P$ as shown in Fig. 1.7, and it is measured along the line perpendicular to the geoid. We can also obtain C by

---

[2] In the Great Lakes region, a dynamic height system, called International Great Lakes Datum (IGLD), is defined using the normal gravity at the latitude $45°$, which is about the latitude at the central point of the region. The term *normal gravity* will be explained in Chap. 4.

integrating g along P_0P,

$$C = \int_{P_0}^{P} g(H')dH' = H\left(\frac{1}{H}\int_{P_0}^{P} g(H')dH'\right),$$ (1.53)

where H' is the height above the point P_0. Define

$$g_m = \frac{1}{H}\int_{P_0}^{P} g(H')dH' = \frac{1}{H}\int_{0}^{H} g(H')dH',$$ (1.54)

which is the average of g over P_0P. We have

$$H = \frac{C}{g_m}.$$ (1.55)

According to this formula, the key to compute orthometric height from levelling result is to compute g_m. It can already be realized that computing g_m exactly is impossible, as that requires to know g between the ground and the geoid, which cannot be measured. We mention that a more accurate definition of the orthometric height is based on the plumb line, which is a curve whose tangent at any point is in the direction of the gravity. In this case, P_0P is along the plumb line, and the orthometric height is the length of the line P_0P, which is no longer a straight line segment.

From the formulation, we see that measurements of gravity are indispensable in the determination of height. An approximate formulation to practically compute H will be provided in Chap. 6.

If the geoid defining the vertical datum is chosen to be the equipotential surface of gravity through the mean ocean surface at a particular tide gauge, it is then the equipotential surface of gravity through a reading on the gauge. In this case, a benchmark could be constructed on the shore nearby, where the orthometric height can be determined by spirit leveling to the gauge. This benchmark, together with its orthometric height, can be understood as the materialized origin of the vertical datum; it tells how far the geoid is below the benchmark and is independent of any model of the geoid. When the geoid is defined to be the equipotential surface of gravity through the average of the mean ocean surface at a number of tide gauges, or the equipotential surface of gravity that best fits the mean ocean surface globally, a model of the geoid is required to link the observations at different locations. Therefore, a benchmark serving as the materialized origin of the vertical datum independent of the model of the geoid is no longer possible. These discussions could be understood more thoroughly after learning Chap. 5.

When the orthometric height is determined by spirit leveling, the vertical datum is provided as a leveling network composed of permanent benchmarks, which is referred to as vertical control network. The geoid is located at the orthometric height below the benchmarks.

1.4   Gravitational Potential and Attraction of Some Simple Configurations

In the previous sections, we have introduced some concepts representing the gravity field. The computation of the gravitational potential and attraction is complicated, as it requires the evaluation of integrals. In this section, we analytically evaluate the gravitational potential and attraction for some regular configurations, first for illustrative purpose for better understanding the concepts, and second for later usage in theoretical formulations.

1.4.1   Homogeneous Spherical Surface

In Fig. 1.8, we show a spherical shell extremely thin so that it can be considered as a material surface. We use R to denote the radius of the spherical surface S and assume its mass density μ to be constant.

For convenience of formulation, we use a spherical coordinate system with origin on the center of the sphere O and polar axis along OP, P being the point where the gravitational potential and attraction are to be computed. The gravitational potential of the shell at P is

$$V = G \int_S \frac{\mu}{l} dS = G\mu \int_S \frac{1}{l} dS, \tag{1.56}$$

where dS is an infinitesimal surface element at M on the sphere, and l is the distance between M and P. In the spherical coordinate system $r\psi\lambda$, where λ is not shown in Fig. 1.8, we have $dS = R^2 \sin\psi \, d\psi \, d\lambda$. We can thus further develop (1.56) to

$$V = G\mu \int_0^{2\pi} d\lambda \int_0^\pi \frac{R^2 \sin\psi}{l} d\psi = 2\pi G\mu \int_0^\pi \frac{R^2 \sin\psi}{l} d\psi . \tag{1.57}$$

Fig. 1.8 Gravitational potential of a homogeneous spherical material surface

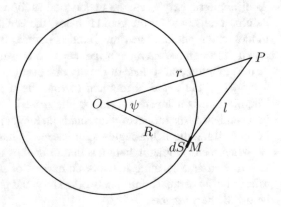

According to the cosine theorem, we have

$$l^2 = R^2 + r^2 - 2Rr \cos \psi . \tag{1.58}$$

This is a relation between l and ψ, where both R and r are constant. By taking the differential at both sides, we obtain

$$l \, dl = Rr \sin \psi \, d\psi . \tag{1.59}$$

It follows from this relation that

$$\frac{R^2 \sin \psi}{l} d\psi = \frac{R}{r} dl . \tag{1.60}$$

The above derivations are valid whenever P is in the exterior or interior of the shell. By substituting (1.60) into (1.57), we can convert the integral of ψ into an integral of l, which can be readily evaluated analytically. However, the limits of the integral of l are different for the two cases when P is in the exterior or interior of the shell. Hence, these two cases have to be treated separately hereafter.

We use V_E and V_I to denote the gravitational potential in the exterior and interior, respectively, and simply name them as external potential and internal potential. When P is in the exterior, it follows from Fig. 1.8 that $\psi = 0$ corresponds to $l = r - R$ and $\psi = \pi$ to $l = r + R$. Hence, we obtain, by substituting (1.60) into (1.57),

$$V_E = 2\pi G \mu \int_{r-R}^{r+R} \frac{R}{r} dl = 4\pi G \mu \frac{R^2}{r} . \tag{1.61}$$

When P is in the interior, it follows from Fig. 1.8 that $\psi = 0$ corresponds to $l = R - r$, and $\psi = \pi$ to $l = R + r$. Hence, we obtain, by substituting (1.60) into (1.57),

$$V_I = 2\pi G \mu \int_{R-r}^{R+r} \frac{R}{r} dl = 4\pi G \mu R , \tag{1.62}$$

which is constant. The mass of the shell is

$$M = 4\pi R^2 \mu . \tag{1.63}$$

By substituting it into (1.61) and (1.62), we obtain

$$V_E = G \frac{M}{r} , \tag{1.64}$$

$$V_I = G \frac{M}{R} . \tag{1.65}$$

Based on the symmetry of the problem, for both the exterior and interior, the gravitational attraction $\vec{F}$ must be along the radial direction $\hat{e}_r$; therefore, we have $\vec{F} = F\hat{e}_r$, where $F = dV/dr$. We denote F as F_E and F_I for the exterior and interior, respectively. It follows that

$$F_E = \frac{dV_E}{dr} = -G\frac{M}{r^2}, \tag{1.66}$$

$$F_I = \frac{dV_I}{dr} = 0. \tag{1.67}$$

As F_E is defined to be the magnitude in the direction of $\hat{e}_r$, the negative sign at the right-hand side of (1.66) implies that the direction of the force is opposite to $\hat{e}_r$ and therefore points to the center of the spherical surface.

It can be inferred from the above four equations that, for a homogeneous spherical surface, the gravitational potential is continuous, the gravitational attraction is discontinuous across the surface, the gravitational potential and attraction in the exterior are identical to those of a particle of the same mass located at the center of the sphere, the gravitational potential in the interior is constant, and the gravitational attraction in the interior vanishes.

1.4.2 Homogeneous Sphere

We denote the radius of the sphere as R and the constant mass density as ρ. In order to compute the gravitational potential, we divide the sphere into a series spherical shells, which are all extremely thin and can be treated as spherical material surfaces. We use R' and dR' to denote the radius and the thickness of such a spherical shell. The density of a material surface representing the thin spherical shell, μ, is then

$$\mu = \rho dR'. \tag{1.68}$$

The gravitational potential of the sphere is the sum of that of all the thin spherical shells.

If the point P is in the exterior of the sphere, as shown in Fig. 1.9, it is in the exterior of all thin spherical shells. Hence, we obtain according to (1.61) and (1.68),

$$V_E = 4\pi G\rho \int_0^R \frac{R'^2}{r}dR' = \frac{4}{3}\pi G\rho\frac{R^3}{r}. \tag{1.69}$$

The mass of the sphere is

$$M = \frac{4}{3}\pi R^3 \rho. \tag{1.70}$$

Fig. 1.9 Gravitational potential of a homogeneous sphere

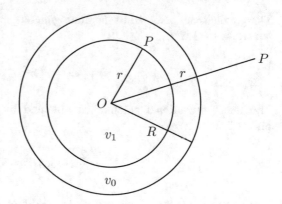

So, (1.69) can be simplified to

$$V_E = G\frac{M}{r}.$$ (1.71)

The same as in the case of a homogeneous spherical surface, based on the symmetry of the problem, the gravitational attraction $\vec{F}_E$ must be along the radial direction $\hat{e}_r$; therefore, we have $\vec{F}_E = F_E\hat{e}_r$, where $F_E = dV_E/dr$. It follows that

$$F_E = \frac{dV_E}{dr} = -G\frac{M}{r^2}.$$ (1.72)

The negative sign at the right-hand side also implies that the gravitational attraction points to the center of the sphere. It can be inferred from the two formulae above that, for a homogeneous sphere, the gravitational potential and attraction outside the sphere are the same as those of a particle of the same mass located at the center of the sphere.

If the point P is in the interior of the sphere, also as shown in Fig. 1.9, we separate the sphere into two parts using the sphere of radius r, where P is located on. The outer part is a homogeneous spherical shell with inner and outer radii r and R, respectively, and is denoted as v_0. The inner part is a homogeneous sphere of radius r and is denoted as v_1. Evidently, the total gravitational potential is the sum of those of v_0 and v_1,

$$V_I = V_0 + V_1.$$ (1.73)

The point P is in the interior of all the thin spherical shells inside the homogeneous spherical shell v_0. Therefore, according to (1.62) and (1.68), we obtain the gravitational potential of v_0 at P to be

$$V_0 = 4\pi G\rho \int_r^R R'dR' = 2\pi G\rho(R^2 - r^2).$$ (1.74)

The gravitational potential of the homogeneous sphere v_1 can be readily obtained according to (1.69),

$$V_1 = \frac{4}{3}\pi G r^2 \rho .$$ (1.75)

The total gravitational potential is obtained by substituting (1.74) and (1.75) into (1.73),

$$V_I = \frac{2}{3}\pi G \rho (3R^2 - r^2) .$$ (1.76)

Again, based on the symmetry of the problem, the gravitational attraction $\overrightarrow{F_I}$ must be along the radial direction $\hat{e}_r$; therefore, we have $\overrightarrow{F_I} = F_I \hat{e}_r$, where $F_I = dV_I/dr$. It follows that

$$F_I = \frac{dV_I}{dr} = -\frac{4}{3}\pi G r \rho = -G\frac{M_r}{r^2} ,$$ (1.77)

where M_r is the mass enclosed within the sphere of radius r, i.e.,

$$M_r = \frac{4}{3}\pi r^3 \rho .$$ (1.78)

The negative sign in (1.77) also implies that the gravitational attraction points toward the center of the sphere.

According to (1.69), (1.72), (1.76), and (1.77), we see $V_E = V_I$ and $F_E = F_I$ on the surface of the sphere, implying that the gravitational potential and attraction are both continuous across the surface of the sphere.

1.4.3   Homogeneous Disk

We use S, a, and μ to denote the disk, its radius, and density, respectively. As shown in Fig. 1.10, we use a cylindrical coordinate system $Ozr\alpha$ in the derivation. We only compute the gravitational potential and attraction on the symmetry axis of the disk.

By definition, the gravitational potential at a point P on the symmetry axis as shown in Fig. 1.10 is

$$V = G\int_S \frac{\mu}{l}dS = G\mu\int_S \frac{1}{l}dS ,$$ (1.79)

where dS is an infinitesimal surface element at the point M on the disk, and l is the distance between M and P. In the cylindrical coordinate system $Ozr\alpha$, $dS =$

Fig. 1.10 Gravitational potential of a homogeneous disk

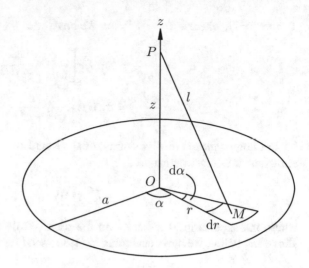

$rdrd\alpha$. So we have

$$V = G\mu \int_0^{2\pi} d\alpha \int_0^a \frac{r}{l} dr = 2\pi G\mu \int_0^a \frac{r}{l} dr .$$ (1.80)

According to Fig. 1.10, $l = (r^2 + z^2)^{1/2}$. Hence,

$$V = \pi G\mu \int_0^a \frac{dr^2}{(r^2 + z^2)^{1/2}} dr = 2\pi G\mu \left[(r^2 + z^2)^{1/2} \right]_0^a$$

$$= 2\pi G\mu \left[(a^2 + z^2)^{1/2} - |z| \right] .$$ (1.81)

When the point P is above the disk, $z > 0$, we use V^+ to express V. We have

$$V^+ = 2\pi G\mu \left[(a^2 + z^2)^{1/2} - z \right] .$$ (1.82)

When the point P is below the disk, $z < 0$, we use V^- to express V. We have

$$V^- = 2\pi G\mu \left[(a^2 + z^2)^{1/2} + z \right] .$$ (1.83)

Evidently, $\lim_{z \to 0} V^+ = \lim_{z \to 0} V^-$, i.e., the gravitational potential on the symmetry axis is continuous across the disk.

According to the symmetry of the problem, the gravitational attraction is along the z-axis. We can express the gravitational attraction above and below the disk as

$\vec{F}^{\pm} = F^{\pm}\hat{e}_z$, where $F^{\pm} = dV^{\pm}/dz$. We have

$$F^+ = \frac{dV^+}{dz} = 2\pi G\mu \left[\frac{z}{(a^2 + z^2)^{1/2}} - 1 \right], \tag{1.84}$$

$$F^- = \frac{dV^-}{dz} = 2\pi G\mu \left[\frac{z}{(a^2 + z^2)^{1/2}} + 1 \right]. \tag{1.85}$$

As the homogeneous disc is symmetrical around the center O, the gravitational attraction at O must vanish, i.e.,

$$F^0 = 0, \tag{1.86}$$

where the superscript 0 indicates on the disk. While approaching the disk from above and below, we have, according to (1.84) and (1.85),

$$\lim_{z \to 0} F^+ = -2\pi G\mu, \tag{1.87}$$

$$\lim_{z \to 0} F^- = 2\pi G\mu. \tag{1.88}$$

These formulae imply that the gravitational attraction of a homogeneous disc on the symmetry axis is discontinuous across the disk; there is a jump of $4\pi G\mu$. Both above and below, the gravitational attraction points to the disk with identical magnitude, thus having opposite signs at the two sides.

1.4.4   Homogeneous Cylinder

Finally, we consider a homogeneous cylinder as shown in Fig. 1.11. Its radius, height, and density are denoted as a, h, and ρ, respectively. As in the case of a homogeneous disk, we also compute only the gravitational potential and attraction on the symmetry axis.

We divide the cylinder into a series of extremely thin layers, so that every layer can be treated as a homogeneous disk. The surface density of a layer, μ, is related to the volume density of the cylinder, ρ, in the form of

$$\mu = \rho dz', \tag{1.89}$$

where z' is the distance from the center of the thin layer to the point P where the gravitational potential is to be computed, and dz' is the thickness of the layer.

In the derivation, only one coordinate z is required to express the position of the point P as shown in Fig. 1.11. The origin O is at the center of the top surface, and the z-axis points above.

Fig. 1.11 Gravitational potential of a homogeneous cylinder

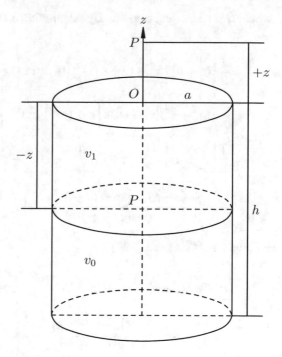

First, we assume that the point P is above or on the top of the cylinder; thus $z \geq 0$. According to (1.82) and (1.89), we have

$$V_E = 2\pi G\rho \int_{z}^{z+h} \left[(a^2 + z'^2)^{1/2} - z' \right] dz'$$

$$= 2\pi G\rho \left\{ \int_{z}^{z+h} (a^2 + z'^2)^{1/2} dz' - \frac{1}{2} \left[(z+h)^2 - z^2 \right] \right\} . \qquad (1.90)$$

The integral

$$I = \int_{z}^{z+h} (a^2 + z'^2)^{1/2} dz' \qquad (1.91)$$

must be evaluated. From a table of indefinite integrals, we can find

$$\int (a^2 + x^2)^{1/2} dx = \frac{1}{2} \left\{ x(a^2 + x^2)^{1/2} + a^2 \ln \left[x + (a^2 + x^2)^{1/2} \right] \right\} + C , \qquad (1.92)$$

where C is an integral constant. By substituting this integral formula into (1.91), we obtain

$$I = \frac{1}{2}\left\{(z+h)\left[a^2+(z+h)^2\right]^{1/2} + a^2\ln\left\{(z+h)+\left[a^2+(z+h)^2\right]^{1/2}\right\}\right.$$

$$\left. - z(a^2+z^2)^{1/2} - a^2\ln\left[z+(a^2+z^2)^{1/2}\right]\right\}$$

$$= \frac{1}{2}\left\{(z+h)\left[a^2+(z+h)^2\right]^{1/2} - z(a^2+z^2)^{1/2}\right.$$

$$\left. + a^2\ln\frac{(z+h)+\left[a^2+(z+h)^2\right]^{1/2}}{z+(a^2+z^2)^{1/2}}\right\}. \tag{1.93}$$

Finally, we obtain from (1.90)

$$V_E = \pi G\rho\left\{(z+h)\left\{\left[a^2+(z+h)^2\right]^{1/2} - (z+h)\right\} - z\left[(a^2+z^2)^{1/2} - z\right]\right.$$

$$\left. + a^2\ln\frac{(z+h)+\left[a^2+(z+h)^2\right]^{1/2}}{z+(a^2+z^2)^{1/2}}\right\}. \tag{1.94}$$

Due to symmetry, the gravitational attraction must be along the direction of the z-axis. We have $\vec{F_E} = F_E\hat{e}_z$, where $F_E = dV_E/dz$. After some differential and algebraic operations, we obtain

$$F_E = \frac{dV_E}{dz} = -2\pi G\rho\left\{(a^2+z^2)^{1/2} + h - \left[a^2+(z+h)^2\right]^{1/2}\right\}. \tag{1.95}$$

The derivation of this formula is left to readers.

Now we assume that the point P is within the cylinder, where its coordinate z is negative, i.e., $z < 0$. The result can be inferred from the result for the case when P is on or above the cylinder. We divide the cylinder into two, one below the point P denoted as v_0, and one above denoted as v_1. The total gravitational potential at P is the sum of those of v_0 and v_1 at the same point,

$$V_I = V_0 + V_1. \tag{1.96}$$

For v_0, it is a homogeneous cylinder of radius a and height $z + h$, where we emphasize that z is the coordinate of the point P and is negative. The point P is on its top surface. Hence, we need to replace z and h in (1.94) by 0 and $z + h$, respectively, to obtain V_0. We have

$$V_0 = \pi G\rho \left\{ (z+h) \left\{ \left[a^2 + (z+h)^2\right]^{1/2} - (z+h) \right\} \right. $$
$$\left. + a^2 \ln \frac{(z+h) + \left[a^2 + (z+h)^2\right]^{1/2}}{a} \right\}. \tag{1.97}$$

This is the gravitational potential of a homogeneous cylinder of radius a, height $z+h$, and density ρ at the center of the top surface. For v_1, its gravitational potentials at P and O are evidently identical. Hence, V_1 is the gravitational potential of a homogeneous cylinder of radius a and height $-z$ on the top. Consequently, it can be obtained by replacing $z + h$ by $-z$ in (1.97). We obtain

$$V_1 = \pi G\rho \left\{ -z \left[(a^2 + z^2)^{1/2} + z \right] + a^2 \ln \frac{-z + (a^2 + z^2)^{1/2}}{a} \right\}. \tag{1.98}$$

Finally, we obtain V_I by substituting the above two formulae into (1.96),

$$V_I = \pi G\rho \left\{ (z+h) \left\{ \left[a^2 + (z+h)^2\right]^{1/2} - (z+h) \right\} - z \left[(a^2 + z^2)^{1/2} + z \right] \right.$$
$$\left. + a^2 \ln \frac{\left\{ (z+h) + \left[a^2 + (z+h)^2\right]^{1/2} \right\} \left[-z + (a^2 + z^2)^{1/2} \right]}{a^2} \right\}. \tag{1.99}$$

The gravitational attraction can be obtained similarly. For v_0, its gravitational attraction at P can be obtained by replacing z and h by 0 and $z + h$ in (1.95), respectively. We obtain

$$F_0 = -2\pi G\rho \left\{ (a + (z+h)) - \left[a^2 + (z+h)^2\right]^{1/2} \right\}. \tag{1.100}$$

For v_1, as the point P is under it, we need to replace $z + h$ by z in (1.100) and then reverse the sign to obtain the gravitational attraction. We have

$$F_1 = 2\pi G\rho \left\{ (a - z - (a^2 + z^2)^{1/2} \right\}. \tag{1.101}$$

The total gravitational attraction of the cylinder at the point P on the symmetry axis in the interior is then the sum of F_0 and F_1. We obtain

$$F_I = -2\pi G\rho \left\{ (a^2 + z^2)^{1/2} + (2z + h) - \left[a^2 + (z+h)^2\right]^{1/2} \right\}. \tag{1.102}$$

We note that $z < 0$ here and that, as for the case of F_E, $\overrightarrow{F_I} = F_I \hat{e}_z$ is implicitly assumed.

When approaching the top surface from both above and inside, we have

$$\lim_{z \to 0} V_E = \lim_{z \to 0} V_I = \pi G \rho \left\{ h \left[(a^2 + h^2)^{1/2} - h \right] + a^2 \ln \frac{h + (a^2 + h^2)^{1/2}}{a} \right\},$$
(1.103)

$$\lim_{z \to 0} F_E = \lim_{z \to 0} F_I = -2\pi G \rho \left[a + h - (a^2 + h^2)^{1/2} \right].$$
(1.104)

These imply that both the gravitational potential and attraction of a homogeneous cylinder are continuous on the symmetry axis.

1.5   Some Properties of the Gravitational Potential and Attraction

1.5.1   Regularity at Infinity

If the attracting mass occupies a finite space, the gravitational potential is regular at infinity, i.e., it obeys the regularity properties

$$\lim_{r \to \infty} V = 0,$$
(1.105)

$$\lim_{r \to \infty} r \frac{\partial V}{\partial r} = 0,$$
(1.106)

$$\lim_{r \to \infty} rV \text{ is finite},$$
(1.107)

$$\lim_{r \to \infty} r^2 \frac{\partial V}{\partial r} \text{ is finite},$$
(1.108)

where r is the distance to a fixed point within the attracting mass. The first two of them follow straightforwardly from the last two. Therefore, we only prove the last two.

We take a solid body as an example to prove these relations. With notations as shown in Fig. 1.12, the gravitational potential at the point P is

$$V = G \int_v \frac{\rho}{l} dv,$$
(1.109)

where v represents the body, ρ its density, and l is related to r as follows:

$$l^2 = r^2 + r'^2 - 2rr' \cos \psi.$$
(1.110)

Fig. 1.12 Limits while approaching infinity

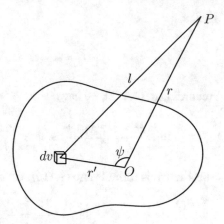

Multiplying both sides of (1.109) with r, we obtain

$$rV = G \int_v \rho \frac{r}{l} dv.$$

(1.111)

As the body is assumed to occupy a finite space, r' is finite. Hence, we obtain from (1.110)

$$\lim_{r \to \infty} \frac{r}{l} = \lim_{r \to \infty} \frac{1}{\left[1 + (r'/r)^2 - 2(r'/r) \cos \psi\right]^{1/2}} = 1.$$

(1.112)

Substitute it into (1.111), we obtain

$$\lim_{r \to \infty} rV = G \lim_{r \to \infty} \int_v \rho \frac{r}{l} dv = G \int_v \rho dv = GM,$$

(1.113)

where M is the total attracting mass. This completes the proof of (1.107).

According to (1.109), we have

$$r^2 \frac{\partial V}{\partial r} = G \int_v \rho r^2 \frac{\partial}{\partial r} \left(\frac{1}{l} \right) dv.$$

(1.114)

It can be obtained from (1.110) that

$$\frac{\partial}{\partial r} \left(\frac{1}{l} \right) = -\frac{1}{l^2} \frac{\partial l}{\partial r} = -\frac{r - r' \cos \psi}{l^3}.$$

(1.115)

Hence,

$$r^2 \frac{\partial V}{\partial r} = -G \int_v \rho \frac{r^3 - r^2 r' \cos \psi}{l^3} dv. \tag{1.116}$$

According to (1.112), we have

$$\lim_{r \to \infty} \frac{r^3 - r^2 r' \cos \psi}{l^3} = \lim_{r \to \infty} \left(\frac{r}{l}\right)^3 \left(1 - \frac{r'}{r} \cos \psi\right) = 1. \tag{1.117}$$

Substitute this formula into (1.116). We obtain

$$\lim_{r \to \infty} r^2 \frac{\partial V}{\partial r} = -G \int_v \rho dv = -GM. \tag{1.118}$$

This complete the proof of (1.108).

We have intentionally written (1.107) and (1.108) in a less concrete form for later reference as regularity for similar but different functions.

The proof for a system of point masses or material surface is similar.

1.5.2   Material Surface: Continuity of Potential and Discontinuity of Attraction

For a material surface, the gravitational potential is finite and continuous everywhere, but the gravitational attraction is discontinuous across the surface. We have already discussed a homogeneous spherical surface in detail and a homogeneous disc along the symmetry axis. Here we discuss the general case.

First of all, we formulate the directional derivative of the gravitational potential of a material surface. The gravitational potential is defined in (1.25) which we rewrite here for convenience of reference,

$$V = G \int_S \frac{\mu}{l} dS, \tag{1.119}$$

where μ is the surface mass density, and other symbols are defined in Fig. 1.13. Its directional derivative with respect to a direction $\hat{e}_t$ is

$$\frac{\partial V}{\partial t} = G \int_S \mu \frac{\partial}{\partial t} \left(\frac{1}{l}\right) dS = -G \int_S \frac{\mu}{l^2} \frac{\partial l}{\partial t} dS. \tag{1.120}$$

According to Fig. 1.13,

$$(l + \Delta l)^2 = l^2 + \Delta t^2 + 2l \Delta t \cos(\hat{e}_l, \hat{e}_t). \tag{1.121}$$

Fig. 1.13 Directional
derivative of the gravitational
potential of a material surface

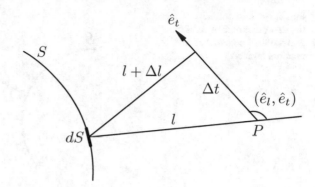

Expand the square at the left-hand side. We obtain

$$2l\Delta l + \Delta l^2 = \Delta t^2 + 2l\Delta t \cos(\hat{e}_l, \hat{e}_t). \tag{1.122}$$

According to the definition of directional derivative, we obtain from this equation

$$\frac{\partial l}{\partial t} = \lim_{\Delta t \to 0} \frac{\Delta l}{\Delta t} = \cos(\hat{e}_l, \hat{e}_t). \tag{1.123}$$

We finally obtain, by substituting this formula into (1.120),

$$\frac{\partial V}{\partial t} = -G \int_S \mu \frac{\cos(\hat{e}_l, \hat{e}_t)}{l^2} dS. \tag{1.124}$$

We see that the expression of the directional derivative of the gravitational potential of a material surface, (1.124), is similar to the expression of the gravitational potential of a double layer, (1.42). However, they are different, particularly, the meaning of $\hat{e}_n$ in (1.35) and (1.42) is different from the meaning of $\hat{e}_t$ in (1.123) and (1.124). We will discuss this again later while discussing the discontinuity of the gravitational potential of a double layer.

First, it can be readily inferred from (1.119) and (1.124) that the gravitational potential and its first order derivative are finite and continuous if the point P is not on the surface.

Second, the gravitational potential is finite on, and continuous across, the surface. Here we provide a proof only for such a point P on the surface where the surface is smooth and the mass density is continuous. We divide the surface into two parts. One part is an infinitesimal disk with P as its center, which can be considered as a homogeneous disk, and the other part is the rest. We denote the surface without the infinitesimal disk as S_0, and the infinitesimal disk as S_1. The total gravitational potential is then the sum of those of S_0 and S_1. As P is not on S_0, the gravitational potential of S_0 is finite and continuous at P. As S_1 can be considered as a homogeneous disk, and P is at its center, it has been shown in the last section

Fig. 1.14 Discontinuity of
the first order derivative of the
gravitational potential of a
material surface

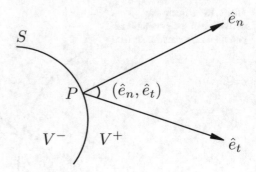

that its gravitational potential is finite and continuous across the surface at P. The
properties we are attempting to prove follow from those for S_0 and S_1. From this
proof, we see that results of the formulations of simple configurations may be used
as a tool to study the general case.

Finally, we formulate the jump of the first order derivative of the gravitational
potential across the surface. As in the previous paragraph, we still consider only
a point P on the surface where the surface is smooth and the mass density is
continuous and also divide the surface into the same two parts, S_0 and S_1. As P is
not on S_0, the first order derivative of the gravitational potential of S_0 is continuous
at P. Hence, the jump of the first order derivative of the gravitational potential of
the whole surface is purely related to that of S_1.

As shown in Fig. 1.14, we draw a normal vector of the surface $\hat{e}_n$. We denote
the values of the gravitational potential at the normal vector's "point to" and "point
from" sides of the surface as V^+ and V^-, respectively, and supplementally denote
the value on the surface as V^0. The gravitational attraction of S_1 is along the
direction of $\hat{e}_n$, i.e., $\overrightarrow{F}_1 = F_1\hat{e}_n$, and the same superscripts are used for F_1 to
represent its values at the two sides of or on the surface. According to the results of
the previous section, (1.86) to (1.88), we have

$$F_1^+ = \frac{\partial V_1^+}{\partial n} = -2\pi G\mu\,, \quad F_1^- = \frac{\partial V_1^-}{\partial n} = 2\pi G\mu\,, \quad F_1^0 = \frac{\partial V_1^0}{\partial n} = 0\,,$$

$$(1.125)$$

where the first identity of every equation states the fact that the gravitational
attraction is the directional derivative of the gravitational potential. For any direction
$\hat{e}_t$ as shown in Fig. 1.14, the directional derivative of the gravitational potential along
it is the projection of the gravitational attraction along it. We therefore have

$$\begin{cases} \dfrac{\partial V_1^+}{\partial t} = F_1^+ \cos(\hat{e}_n, \hat{e}_t) = -2\pi G\mu \cos(\hat{e}_n, \hat{e}_t)\,, \\[2mm] \dfrac{\partial V_1^-}{\partial t} = F_1^- \cos(\hat{e}_n, \hat{e}_t) = 2\pi G\mu \cos(\hat{e}_n, \hat{e}_t)\,, \\[2mm] \dfrac{\partial V_1^0}{\partial t} = F_1^0 \cos(\hat{e}_n, \hat{e}_t) = 0\,. \end{cases} \qquad (1.126)$$

The first order derivative of the gravitational potential of S_0 is continuous at P, i.e.,

$$\frac{\partial V_0^+}{\partial t} = \frac{\partial V_0^-}{\partial t} = \frac{\partial V_0^0}{\partial t} = \frac{\partial V_0}{\partial t}, \tag{1.127}$$

where the term $\partial V_0/\partial t$ could be used to represent all other terms in the identities. We finally obtain

$$\begin{cases} \dfrac{\partial V^+}{\partial t} = \dfrac{\partial V_0^+}{\partial t} + \dfrac{\partial V_1^+}{\partial t} = \dfrac{\partial V_0}{\partial t} - 2\pi G\mu \cos(\hat{e}_n, \hat{e}_t), \\[2mm] \dfrac{\partial V^-}{\partial t} = \dfrac{\partial V_0^-}{\partial t} + \dfrac{\partial V_1^-}{\partial t} = \dfrac{\partial V_0}{\partial t} + 2\pi G\mu \cos(\hat{e}_n, \hat{e}_t), \\[2mm] \dfrac{\partial V^0}{\partial t} = \dfrac{\partial V_0^0}{\partial t} + \dfrac{\partial V_1^0}{\partial t} = \dfrac{\partial V_0}{\partial t}, \end{cases} \tag{1.128}$$

where according to the last equation, the term $\partial V_0/\partial t$ in the first two equations can be replaced by $\partial V^0/\partial t$. The following relations can be obtained by simple combinations of the 3 equations in (1.128),

$$\frac{\partial V^0}{\partial t} = \frac{1}{2}\left(\frac{\partial V^+}{\partial t} + \frac{\partial V^-}{\partial t}\right), \tag{1.129}$$

$$\frac{\partial V^+}{\partial t} - \frac{\partial V^-}{\partial t} = -4\pi G\mu \cos(\hat{e}_n, \hat{e}_t). \tag{1.130}$$

The former implies that the directional derivative of the gravitational potential of a material surface on the surface is equal to the average of the limits of those at the two sides; the later implies that the directional derivative of the gravitational potential of a material surface has a jump of $4\pi G\mu \cos(\hat{e}_n, \hat{e}_t)$ across the surface.

1.5.3   Double Layer: Discontinuity of Potential

The gravitational potential of a double layer is given by (1.42), which is copied to here for convenience of reference,

$$V = G\int_S v\frac{\cos(\hat{e}_l, \hat{e}_n)}{l^2} dS, \tag{1.131}$$

where v is the dipole density, and other symbols are defined in Fig. 1.6. Although its form is similar to the directional derivative of the gravitational potential of a material surface (1.124), their physical meanings are completely different. The meanings of the symbol $\hat{e}_l$ in both formulae are identical. However, the symbol $\hat{e}_t$ in (1.124) represents a fixed direction independent of the variables of integration, but the symbol $\hat{e}_n$ in (1.131) is the normal vector of the surface depending on the variables of integration.

We also discuss only a point P on the double layer where the surface S is smooth and the dipole moment v is continuous. We again can select an infinitesimal disk with P at the center, which can be treated as a homogeneous double layer disk. As this disk is taken as a plane, its normal vector $\hat{e}_n$ is then invariant, thus being independent of the variables of integration, and its double layer potential would behave in the same way as the directional derivative of the gravitational potential of a material surface expressed by (1.124), where $\hat{e}_t$ is independent of the variables of integration. The only difference is that a negative sign is present at the right-hand side of (1.124). Therefore, a derivation similar to the derivation of (1.128) can be made to obtain

$$V^+ = V^0 + 2\pi G v , \quad V^- = V^0 - 2\pi G v . \tag{1.132}$$

1.5.4   Solid Body: Continuity of Potential and Attraction

We have already discussed these properties for the case of a homogeneous sphere in detail and for a homogeneous cylinder along the symmetry axis. Here we provide an outline of the proof for the general case. The gravitational potential of a solid body is defined in (1.26), which is copied to here,

$$V = G \int_v \frac{\rho}{l} dv. \tag{1.133}$$

Similar to the case of a material surface, the directional derivative is

$$\frac{\partial V}{\partial t} = -G \int_v \rho \frac{\cos(\hat{e}_l, \hat{e}_t)}{l^2} dv , \tag{1.134}$$

which is the projection of the gravitational attraction along the direction $\hat{e}_t$. Evidently, both V and $\partial V / \partial t$ are finite and continuous outside the body. For a point P inside the body where the density is continuous, we can separate the body into an infinitesimal sphere with P inside, which can be treated as homogeneous, and the rest of the body. The gravitational potential and attraction of both parts are continuous. Therefore, their sums are continuous. For a point P on the surface of the body, we only discuss the case when the density around the point P is continuous, and the surface of the body around P is smooth. We can then divide the body into two parts, one part being an infinitesimal cylinder with the top on the surface of the body centered by the point P, and the other part being the rest. In this case, the infinitesimal cylinder can be treated as homogeneous, and its gravitational potential and attraction are all continuous across the surface at the point P. The continuity of the gravitational potential and attraction at a point P on a surface of discontinuous density inside the solid body can be inferred similarly by defining an infinitesimal cylinder at each side of the surface with the point P on the center of the top/bottom surfaces. The detail of the proof is left to readers.

1.5.5 Laplace's Equation of the Gravitational Potential Outside the Attracting Mass

We provide a proof for a solid body only. The proofs for a system of point masses or a material surface are similar.

According to (1.2) and the definition of the gravitational potential of a solid body (1.26) or (1.133), we can readily obtain

$$
\begin{cases}
\dfrac{\partial^2 V}{\partial x^2} = G \displaystyle\int_v \rho \dfrac{\partial^2}{\partial x^2}\left(\dfrac{1}{l}\right) dv = -G \displaystyle\int_v \rho \left[\dfrac{1}{l^3} - 3\dfrac{(x-\xi)^2}{l^5}\right] dv\,, \\[4mm]
\dfrac{\partial^2 V}{\partial y^2} = G \displaystyle\int_v \rho \dfrac{\partial^2}{\partial y^2}\left(\dfrac{1}{l}\right) dv = -G \displaystyle\int_v \rho \left[\dfrac{1}{l^3} - 3\dfrac{(y-\eta)^2}{l^5}\right] dv\,, \\[4mm]
\dfrac{\partial^2 V}{\partial z^2} = G \displaystyle\int_v \rho \dfrac{\partial^2}{\partial z^2}\left(\dfrac{1}{l}\right) dv = -G \displaystyle\int_v \rho \left[\dfrac{1}{l^3} - 3\dfrac{(z-\zeta)^2}{l^5}\right] dv\,.
\end{cases}
\tag{1.135}
$$

Outside the body, the integrands of all these three integrals are finite. Hence, all the integrals are integrable. By adding them together, we obtain

$$
\Delta V = \frac{\partial^2 V}{\partial x^2} + \frac{\partial^2 V}{\partial y^2} + \frac{\partial^2 V}{\partial z^2} = 0\,.
\tag{1.136}
$$

This equation is called Laplace's equation, where the operator $\Delta = \partial^2/\partial x^2 + \partial^2/\partial y^2 + \partial^2 \partial z^2$ is called Laplace operator, or more frequently, Laplacian.

A function satisfying the Laplace's equation is called harmonic function. Therefore, the gravitational potential outside the attracting mass is a harmonic function.

Based on definition, we can obtain, for the centrifugal potential,

$$
\Delta Q = \frac{\partial^2 Q}{\partial x^2} + \frac{\partial^2 Q}{\partial y^2} + \frac{\partial^2 Q}{\partial z^2} = 2\Omega^2\,,
\tag{1.137}
$$

which is called Poisson's equation.

Outside the Earth, the Earth's gravity potential satisfies the Poisson's equation

$$
\Delta W = \Delta V + \Delta Q = 2\Omega^2\,.
\tag{1.138}
$$

1.5.6   Poisson's Equation of the Gravitational Potential Inside a Solid Body

When the point P is inside the solid body, the integrands in the three integrals in (1.135) tend to be infinitely large at P. To avoid this difficulty, we divide the body into two parts. One part is an infinitesimal sphere with the point P in the inside, and the other part is the rest. We use v_0 to denote the body without the infinitesimal sphere and v_1 the infinitesimal sphere. The total gravitational potential is the sum of those of v_0 and v_1,

$$V = V_0 + V_1 \, . \tag{1.139}$$

The point P is outside v_0. Hence,

$$\Delta V_0 = 0 \, . \tag{1.140}$$

According to (1.76), we write the gravitational potential of v_1 as

$$V_1 = \frac{2}{3} \pi G \rho (3R^2 - r^2) \, , \tag{1.141}$$

where R is the radius and r is the distance of P to the center of the sphere. We use (ξ, η, ζ) to denote the coordinate of the center of the sphere and (x, y, z) that of the point P. We have

$$r^2 = (x - \xi)^2 + (y - \eta)^2 + (z - \zeta)^2 \, . \tag{1.142}$$

We obtain from the two equations above,

$$\begin{cases} \dfrac{\partial^2 V_1}{\partial x^2} = \dfrac{2}{3} \pi G \rho \dfrac{\partial^2}{\partial x^2} (3R^2 - r^2) = -\dfrac{4}{3} \pi G \rho \, , \\[2mm] \dfrac{\partial^2 V_1}{\partial y^2} = \dfrac{2}{3} \pi G \rho \dfrac{\partial^2}{\partial y^2} (3R^2 - r^2) = -\dfrac{4}{3} \pi G \rho \, , \\[2mm] \dfrac{\partial^2 V_1}{\partial z^2} = \dfrac{2}{3} \pi G \rho \dfrac{\partial^2}{\partial z^2} (3R^2 - r^2) = -\dfrac{4}{3} \pi G \rho \, . \end{cases} \tag{1.143}$$

Adding the three equations together yields

$$\Delta V_1 = -4\pi G \rho \, . \tag{1.144}$$

Finally, we obtain, according to (1.139), (1.140), and (1.144),

$$\Delta V = \Delta V_0 + \Delta V_1 = -4\pi G \rho \, , \tag{1.145}$$

which is a Poisson's equation. By definition, we already know that (1.26) is its solution, as the quantity itself is the solution of its equation. However, the solution is not unique, for which the reason is left to readers to figure out.

Inside the Earth, the Earth's gravity potential satisfies the following Poisson's equation:

$$\Delta W = \Delta V + \Delta Q = -4\pi G\rho + 2\Omega^2 . \tag{1.146}$$

1.6 Curvature of Equipotential Surfaces of Gravity and Plumb Lines

1.6.1 Definition of Curvature

The curvature is a measure of the intensity of curving, which is expressed as the change of direction of the tangent of the curve per unit length of the curve.

We start by considering the plane curve

$$z = z(x) . \tag{1.147}$$

As shown in Fig. 1.15, let P be a point on the curve with coordinate (x, z), P' a neighboring point on the curve with coordinate $(x + \Delta x, z + \Delta z)$, and PQ and $P'Q'$ are tangential to the curve. Accurate to the first order, the relation between Δz and Δx is

$$\Delta z = \frac{dz}{dx}\Delta x . \tag{1.148}$$

According to the geometrical property of derivative,

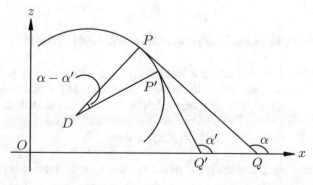

Fig. 1.15 Definition of curvature

$$\tan\alpha = \frac{dz}{dx}.$$ (1.149)

We then have

$$\tan\alpha' = \tan\alpha + \frac{d}{dx}(\tan\alpha)\Delta x = \frac{dz}{dx} + \frac{d^2z}{dx^2}\Delta x.$$ (1.150)

Accurate to the first order of Δx, we have

$$\tan(\alpha - \alpha') = \frac{\tan\alpha - \tan\alpha'}{1 + \tan\alpha\tan\alpha'}$$

$$= -\frac{(d^2z/dx^2)\Delta x}{1 + (dz/dx)[dz/dx + (d^2z/dx^2)\Delta x]}$$

$$= -\frac{(d^2z/dx^2)\Delta x}{1 + (dz/dx)^2}.$$ (1.151)

As the length of PP' is $(\Delta x^2 + \Delta z^2)^{1/2} = (1 + dz/dx)^{1/2}\Delta x$, the curvature κ and radius of curvature ρ are defined as

$$\kappa = \frac{1}{\rho} = \lim_{\Delta x \to 0}\frac{\alpha - \alpha'}{PP'} = \lim_{\Delta x \to 0}\frac{\tan(\alpha - \alpha')}{(\Delta x^2 + \Delta z^2)^{1/2}} = -\frac{d^2z/dx^2}{[1 + (dz/dx)^2]^{3/2}},$$ (1.152)

where $\alpha - \alpha'$ is the change of direction of the tangent through the length PP'.

A special case we will make use is when the tangent of the curve at P is parallel to the x-axis. In this case, $dz/dx = 0$, and we have simply

$$\kappa = \frac{1}{\rho} = -\frac{d^2z}{dx^2}.$$ (1.153)

1.6.2   Curvature of an Equipotential Surface of Gravity

As shown in Fig. 1.16, we set up a local coordinate system $Oxyz$ with the plane Oxy tangential to the equipotential surface of gravity, and z points upward. The curvature at the point O is to be computed. The equation of the surface is

$$W(x, y, z) = W_0.$$ (1.154)

The curvature of the intersection line of the surface with the plane Oxz is

$$K_x = -\frac{\partial^2z}{\partial x^2}.$$ (1.155)

Fig. 1.16 Curvature of an equipotential surface of gravity

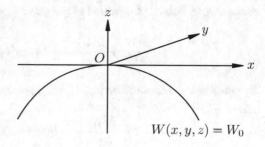

$$W(x, y, z) = W_0$$

Take the derivative of both sides of (1.154) with respect to x two times. By writing the explicit form of the equation of the surface as $z = z(x, y)$, we obtain successively

$$\frac{\partial W}{\partial x} + \frac{\partial W}{\partial z}\frac{\partial z}{\partial x} = 0, \tag{1.156}$$

$$\frac{\partial^2 W}{\partial x^2} + 2\frac{\partial^2 W}{\partial x \partial z}\frac{\partial z}{\partial x} + \frac{\partial^2 W}{\partial z^2}\left(\frac{\partial z}{\partial x}\right)^2 + \frac{\partial W}{\partial z}\frac{\partial^2 z}{\partial x^2} = 0. \tag{1.157}$$

As the x-axis is tangential to the curve, $\partial z/\partial x = 0$. We obtain from (1.157)

$$\frac{\partial^2 z}{\partial x^2} = -\left(\frac{\partial W}{\partial z}\right)^{-1}\frac{\partial^2 W}{\partial x^2}. \tag{1.158}$$

The gravity is in opposite direction of the z-axis. Hence, the absolute magnitude of the gravity is

$$g = -\frac{\partial W}{\partial z}. \tag{1.159}$$

We obtain finally

$$K_x = -\frac{1}{g}\frac{\partial^2 W}{\partial x^2}. \tag{1.160}$$

Similarly,

$$K_y = -\frac{1}{g}\frac{\partial^2 W}{\partial y^2}. \tag{1.161}$$

Define an average curvature

$$J = \frac{1}{2}(K_x + K_y) = -\frac{1}{2g}\left(\frac{\partial^2 W}{\partial x^2} + \frac{\partial^2 W}{\partial y^2}\right). \tag{1.162}$$

According to Poisson's equation of the gravity potential (1.146), we have

$$\frac{\partial^2 W}{\partial x^2} + \frac{\partial^2 W}{\partial y^2} = -4\pi G\rho + 2\Omega^2 - \frac{\partial^2 W}{\partial z^2}. \tag{1.163}$$

By substituting this equation into (1.162), we obtain

$$-\frac{\partial^2 W}{\partial z^2} = -2gJ + 4\pi G\rho - 2\Omega^2. \tag{1.164}$$

This equation can be finally written as, using (1.159),

$$\frac{\partial g}{\partial z} = -2gJ + 4\pi G\rho - 2\Omega^2. \tag{1.165}$$

This formula is referred to as Bruns' formula, which links a physical quantity, i.e., the rate of change of the gravity in the vertical direction, to a geometrical quantity, i.e., the average curvature of the equipotential surface. It can be inferred from this formula that the average curvature J is independent of the direction of the axes x and y.

1.6.3   Curvature of a Plumb Line

By definition, a plumb line is a curve such that its tangent at every point is in the direction of the gravity. We write an infinitesimal increment of position vector on the line as

$$d\vec{r} = dx\hat{e}_x + dy\hat{e}_y + dz\hat{e}_z. \tag{1.166}$$

It is in the same direction as the gravity,

$$\vec{g} = \frac{\partial W}{\partial x}\hat{e}_x + \frac{\partial W}{\partial y}\hat{e}_y + \frac{\partial W}{\partial z}\hat{e}_z. \tag{1.167}$$

Therefore, the components of $d\vec{r}$ and $\vec{g}$ are proportional:

$$\frac{dx}{\partial W/\partial x} = \frac{dy}{\partial W/\partial y} = \frac{dz}{\partial W/\partial z}. \tag{1.168}$$

This is the equation of the plumb line.

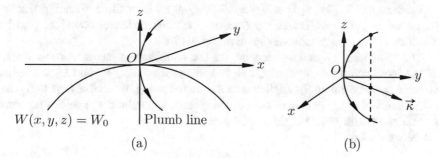

Fig. 1.17 Projection of a plumb line in the Oxz plane (**a**) and definition of the curvature vector (**b**)

The equation of the projection of the plumb line in the plane Oxz can be obtained from (1.168) to be

$$\frac{dx}{dz} = \frac{\partial W}{\partial x} \bigg/ \frac{\partial W}{\partial z}, \tag{1.169}$$

as shown in Fig. 1.17a. We formulate the curvature of the plumb line at O, where the plumb line is tangential to the z-axis. According to (1.153), with the notations x and z swapped, we have

$$\kappa_x = \frac{d^2 x}{dz^2} = \left(\frac{\partial W}{\partial z}\right)^{-2} \left[\frac{\partial W}{\partial z}\left(\frac{\partial^2 W}{\partial z \partial x} + \frac{\partial^2 W}{\partial x^2}\frac{dx}{dz}\right) - \frac{\partial W}{\partial x}\left(\frac{\partial^2 W}{\partial x \partial z}\frac{dx}{dz} + \frac{\partial^2 W}{\partial z^2}\right)\right]. \tag{1.170}$$

As the plane Oxy is tangential to an equipotential surface at O, we have $\partial W/\partial x = \partial W/\partial y = 0$ at O. It can be solved from (1.169) and (1.170) that

$$\kappa_x = \left(\frac{\partial W}{\partial z}\right)^{-1} \frac{\partial^2 W}{\partial z \partial x}. \tag{1.171}$$

Finally,

$$\kappa_x = \frac{1}{g}\frac{\partial g}{\partial x}. \tag{1.172}$$

Similarly,

$$\kappa_y = \frac{1}{g}\frac{\partial g}{\partial y}. \tag{1.173}$$

Evidently, if $\partial g/\partial x > 0$ or $\partial g/\partial y > 0$, κ_x or κ_y is positive, meaning that the projection of the plumb line in the Oxz or Oyz plane concave toward the positive direction of the x- or y-axis, as shown in Fig. 1.17b.

The quantities κ_x and κ_y depend on the coordinate system. As the choice of the local coordinate system $Oxyz$ is somewhat at will, κ_x and κ_y are hence somewhat inconvenient for applications. In the following, we define a quantity that is independent of the choice of coordinate system, and κ_x and κ_y can be computed from it and the directions of the x- and y-axes.

Define a vector

$$\vec{\kappa} = \kappa_x \hat{e}_x + \kappa_y \hat{e}_y, \tag{1.174}$$

as shown in Fig. 1.17b. We use κ and α to denote its norm and the angle between it and the x-axis. We have

$$\kappa = (\kappa_x^2 + \kappa_y^2)^{1/2}; \tag{1.175}$$

$$\cos\alpha = \frac{\kappa_x}{\kappa}, \quad \sin\alpha = \frac{\kappa_y}{\kappa}. \tag{1.176}$$

The components κ_x and κ_y can be computed according to

$$\kappa_x = \kappa \cos\alpha, \quad \kappa_y = \kappa \sin\alpha. \tag{1.177}$$

We will show that $\vec{\kappa}$ is independent of the choice of coordinate system, and hence, so is κ.

In the local coordinate system $Oxyz$, according to (1.165), (1.172), and (1.173), we can express the gradient of g in the following form:

$$\nabla g = g\kappa_x \hat{e}_x + g\kappa_y \hat{e}_y + (-2gJ + 4\pi G\rho - 2\Omega^2)\hat{e}_z. \tag{1.178}$$

The gradient is related to the directional derivative in the form of

$$\frac{\partial g}{\partial t} = \nabla g \cdot \hat{e}_t, \tag{1.179}$$

which was already applied to the potential. We see that the norm of ∇g is the maximum of $\partial g/\partial t$, and its direction is the corresponding direction of $\hat{e}_t$. Hence, ∇g is independent of the coordinate system $Oxyz$. This is a fundamental property of gradient. Furthermore, the choice of z is not random; its direction is opposite to the direction of gravity. Hence, the projection of ∇g along the z-axis is independent of the x- and y-axes. Based on these discussions, we infer from (1.178) that the vector $g\kappa_x \hat{e}_x + g\kappa_y \hat{e}_y$ is independent of the choice of the x- and y-axes. Finally, as g is independent of coordinates, we conclude that the vector $\kappa_x \hat{e}_x + \kappa_y \hat{e}_y$ is independent of coordinates and so is its norm κ.

It follows from (1.174) to (1.177) that κ is the maximum that κ_x or κ_y may attain for all possible choices of the coordinate system Oxy. If we define a coordinate system Otz in the osculating plane (approximately speaking, the plane that includes a small portion of the curve in it) of the plumb line, $\kappa = \kappa_t$. Therefore, we follow (1.175) to define κ as the curvature,

$$\kappa = \frac{1}{g} \left[\left(\frac{\partial g}{\partial x} \right)^2 + \left(\frac{\partial g}{\partial y} \right)^2 \right]^{1/2}, \tag{1.180}$$

which is independent of the directions of x and y.

Further reading

Blakely, R.J. (1996). *Potential theory in gravity and magnetic applications* (Chapter 1). Cambridge University Press.
MacMillan, W. D. (1958). *The theory of the potential* (Chapter 1). Dover Publications Incorporated.

Elementary Potential Theory

2

Abstract

This chapter is an introduction to potential theory. Readers are assumed to be familiar with Gauss' divergence theorem, which is taught in multivariable calculus. Green's three identities for both the internal and external domains of a bounding surface are proven. These are then applied to the Earth's gravity field to prove, notably, Gauss' law of gravity, the expression of the Earth's external gravitational potential as that of a surface mass layer, and Stokes' theorem. Boundary value problems of an external domain are summarized with proofs of the uniqueness of their solutions. The solution of boundary value problems using the method of Green's function is introduced, and the solution of Dirichlet's boundary value problem outside a sphere, i.e., the Poisson integral, is derived. In the last section, the gradient of the gravitational attraction in the spherical coordinate system is formulated, which readers not prepared to learn gravity gradient can skip.

2.1   Green's Identities

2.1.1   Green's First and Second Identities for an Internal Domain

Green's identities have very important applications in potential theory. They are equivalent to the as famous Gauss' theorem, for which most textbooks of multivariable calculus provide a derivation. Here we cite Gauss' theorem without proof and derive Green's identities from it.

Let the closed three-dimensional domain v_i be connected, i.e., any two points within it can be connected with a broken line completely within it, and its boundary S be piecewise smooth. Let the functions $P(x, y, z)$, $Q(x, y, z)$, and $R(x, y, z)$ and their first order derivatives be all continuous over v_i. The Gauss' theorem

$$\int_{v_i} \left(\frac{\partial P}{\partial x} + \frac{\partial Q}{\partial y} + \frac{\partial R}{\partial z} \right) dv$$

$$= \int_S \left[P \cos(\hat{e}_n, \hat{e}_x) + Q \cos(\hat{e}_n, \hat{e}_y) + R \cos(\hat{e}_n, \hat{e}_z) \right] dS \tag{2.1}$$

holds, where $\hat{e}_n$ is the normal of the surface S pointing to the outside of the domain v_i. The expression of the vector $\hat{e}_n$ in terms of $\hat{e}_x$, $\hat{e}_y$, and $\hat{e}_z$ is

$$\hat{e}_n = \cos(\hat{e}_n, \hat{e}_x)\hat{e}_x + \cos(\hat{e}_n, \hat{e}_y)\hat{e}_y + \cos(\hat{e}_n, \hat{e}_z)\hat{e}_z. \tag{2.2}$$

Define a vector-valued function

$$\vec{A}(x, y, z) = P(x, y, z)\hat{e}_x + Q(x, y, z)\hat{e}_y + R(x, y, z)\hat{e}_z, \tag{2.3}$$

and denote

$$\nabla \cdot \vec{A} = \frac{\partial P}{\partial x} + \frac{\partial Q}{\partial y} + \frac{\partial R}{\partial z}, \tag{2.4}$$

which is called divergence of the vector $\vec{A}$. We can write Gauss' theorem in the brief form

$$\int_{v_i} \nabla \cdot \vec{A} \, dv = \int_S \vec{A} \cdot \hat{e}_n dS. \tag{2.5}$$

For this reason, it is also called divergence theorem. It is a relation between a surface integral over S and a volume integral over v_i, which is the closed domain inside S.

In the following, we derive Green's identities from Gauss' theorem.

In (2.5), we set

$$\vec{A} = U\frac{\partial V}{\partial x}\hat{e}_x + U\frac{\partial V}{\partial y}\hat{e}_y + U\frac{\partial V}{\partial z}\hat{e}_z, \tag{2.6}$$

which can be briefly written in vector form as

$$\vec{A} = U\nabla V. \tag{2.7}$$

We can obtain

$$\nabla \cdot \vec{A} = \frac{\partial}{\partial x}\left(U\frac{\partial V}{\partial x} \right) + \frac{\partial}{\partial y}\left(U\frac{\partial V}{\partial y} \right) + \frac{\partial}{\partial z}\left(U\frac{\partial V}{\partial z} \right)$$

$$= U\left(\frac{\partial^2 V}{\partial x^2} + \frac{\partial^2 V}{\partial y^2} + \frac{\partial^2 V}{\partial z^2} \right) + \frac{\partial U}{\partial x}\frac{\partial V}{\partial x} + \frac{\partial U}{\partial y}\frac{\partial V}{\partial y} + \frac{\partial U}{\partial z}\frac{\partial V}{\partial z}, \tag{2.8}$$

which can be briefly written in vector form as

$$\nabla \cdot \vec{A} = U\Delta V + \nabla U \cdot \nabla V. \tag{2.9}$$

The substitution of (2.6) and (2.9) into (2.5) yields

$$\int_{v_i} (U\Delta V + \nabla U \cdot \nabla V)dv = \int_S U\nabla V \cdot \hat{e}_n dS. \tag{2.10}$$

According to the property of directional derivative

$$\frac{\partial V}{\partial n} = \frac{\partial V}{\partial x}\cos(\hat{e}_n, \hat{e}_x) + \frac{\partial V}{\partial y}\cos(\hat{e}_n, \hat{e}_y) + \frac{\partial V}{\partial z}\cos(\hat{e}_n, \hat{e}_z) = \nabla V \cdot \hat{e}_n, \tag{2.11}$$

we can finally write (2.10) as

$$\int_{v_i} (U\Delta V + \nabla U \cdot \nabla V)dv = \int_S U\frac{\partial V}{\partial n}dS, \tag{2.12}$$

which is referred to as Green's first identity for an internal domain.

Switching the position of U and V in Green's first identity (2.12), we obtain

$$\int_{v_i} (V\Delta U + \nabla V \cdot \nabla U)dv = \int_S V\frac{\partial U}{\partial n}dS. \tag{2.13}$$

Subtract (2.13) from (2.12). We obtain

$$\int_{v_i} (U\Delta V - V\Delta U)dv = \int_S \left(U\frac{\partial V}{\partial n} - V\frac{\partial U}{\partial n}\right)dS, \tag{2.14}$$

which is referred to as Green's second identity for an internal domain.

In Green's both identities, the functions U and V and their derivatives up to the order concerned must be continuous over v_i.

2.1.2 Green's First and Second Identities for an External Domain

In the previous subsection, the domain v_i is the interior of the surface S. Now we consider a domain v_e that is the exterior of the surface S.

In order to derive Green's identities for the external domain v_e, we first apply the Green's identities to a domain v_e', as shown in Fig. 2.1, which has an inner boundary S and an outer boundary Σ that is a sphere of radius R enclosing S completely inside. We then set the limit $R \rightarrow \infty$ to expand v_e' to v_e.

Fig. 2.1 Green's identities
for an external domain

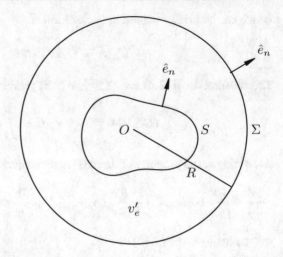

The application of Green's first identity for an internal domain, (2.12), to the domain v'_e with boundaries S and Σ yields

$$\int_{v'_e} (U\Delta V + \nabla U \cdot \nabla V)dv = -\int_S U\frac{\partial V}{\partial n}dS + \int_\Sigma U\frac{\partial V}{\partial n}d\Sigma . \quad (2.15)$$

There are two surface integrals at the right-hand side, as the domain has two distinct boundaries. The negative sign of the first surface integral is due to the fact that the normal vector of S toward outside points to the inside of v'_e, as shown in Fig. 2.1.

We first analyze the second surface integral at the right-hand side of (2.15). We denote the distance to the center of the sphere Σ as r and define a spherical coordinate system $Or\theta\lambda$. The unit outward normal vector of Σ, $\hat{e}_n$, is then the same as $\hat{e}_r$, and the surface element is $d\Sigma = R^2 \sin\theta d\theta d\lambda$. Hence,

$$\int_\Sigma U\frac{\partial V}{\partial n}d\Sigma = \int_0^{2\pi} d\lambda \int_0^\theta \left(Ur^2\frac{\partial V}{\partial r} \right)_{r=R} \sin\theta d\theta . \quad (2.16)$$

We make the following assumption:

$$\lim_{r\to\infty} U = 0 , \quad (2.17)$$

$$\lim_{r\to\infty} r^2\frac{\partial V}{\partial r} \quad \text{is finite} . \quad (2.18)$$

We can then obtain

$$\lim_{R\to\infty} \int_\Sigma U\frac{\partial V}{\partial n}d\Sigma = \lim_{R\to\infty} \int_0^{2\pi} d\lambda \int_0^\theta \left(Ur^2\frac{\partial V}{\partial r} \right)_{r=R} \sin\theta d\theta = 0 . \quad (2.19)$$

Finally, we obtain by setting $R \to \infty$ in (2.15)

$$\int_{v_e} (U\Delta V + \nabla U \cdot \nabla V)dv = -\int_S U\frac{\partial V}{\partial n}dS.\qquad(2.20)$$

This is Green's first identity for an external domain, where U and V must satisfy (2.17) and (2.18).

Similar to the case of an internal domain, Green's second identity for an external domain can be obtained to be

$$\int_{v_e} (U\Delta V - V\Delta U)dv = -\int_S \left(U\frac{\partial V}{\partial n} - V\frac{\partial U}{\partial n}\right)dS,\qquad(2.21)$$

where, besides (2.17) and (2.18), U and V should satisfy

$$\lim_{r\to\infty} V = 0,\qquad(2.22)$$

$$\lim_{r\to\infty} r^2\frac{\partial U}{\partial r} \quad \text{is finite}.\qquad(2.23)$$

In Green's both identities, U and V must also satisfy the continuity requirements for themselves and their derivatives up to the order concerned. The negative sign for the surface integral at the right-hand side is due to the fact that the outward normal of the surface S points to the interior of v_e.

We mention that the conditions (2.17), (2.18) and (2.22), (2.23) are the regularity properties of the gravitational potential at infinity expressed in (1.105)–(1.108).

2.1.3 Green's Third Identity

We will derive the identity for the case of an internal domain only and write out the result for an external domain directly.

The basic equation is Green's second identity (2.14). We first define some necessary symbols. We define two points: P is a fixed point, and M is the point where the volume element dv is located, and denote the distance between them as l. Hence, l depends on the position of M. In (2.14), we set $U = 1/l$. We can then obtain three identities when P is outside, inside, and on S. We mention again that $U = 1/l$ and V are functions of the position of M.

When P is Outside
As shown in Fig. 2.2a, when P is outside S, $U = 1/l$ is finite and harmonic in v_i. Hence, it follows right away that (2.14) becomes

$$\int_{v_i} \frac{1}{l}\Delta V dv = \int_S \left[\frac{1}{l}\frac{\partial V}{\partial n} - V\frac{\partial}{\partial n}\left(\frac{1}{l}\right)\right]dS.\qquad(2.24)$$

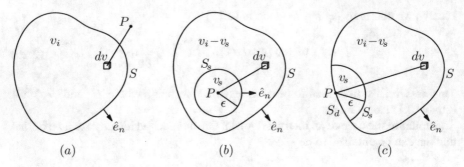

Fig. 2.2 Three cases of Green's third identity

When P Is Inside

As shown in Fig. 2.2b, when P is inside S, $U = 1/l$ and $\Delta U = \Delta(1/l)$ tend to infinitely large when M approaches P. Hence, (2.14) cannot be applied as is to the domain v_i and its surface S. To solve this problem, we define an infinitesimal sphere v_s of radius ϵ with P as its center, and denote its surface as S_s. We can then apply (2.14) to the closed domain $v_i - v_s$ and its surfaces S and S_s. We obtain

$$\int_{v_i - v_s} \frac{1}{l} \Delta V dv = \int_S \left[\frac{1}{l} \frac{\partial V}{\partial n} - V \frac{\partial}{\partial n} \left(\frac{1}{l} \right) \right] dS - \int_{S_s} \left[\frac{1}{l} \frac{\partial V}{\partial n} - V \frac{\partial}{\partial n} \left(\frac{1}{l} \right) \right] dS .$$

(2.25)

The limit $\epsilon \to 0$ in (2.25) is to be studied, for which we define a spherical coordinate system with P as origin, $Pr\theta\lambda$. We first study the second surface integral at the right-hand side and then the volume integral at the left-hand side.

On the spherical surface S_s, $r = \epsilon$, $dS = \epsilon^2 \sin\theta d\theta d\lambda$, and as defined, $l = r$ and $\hat{e}_n = \hat{e}_r$; thus $[\partial/\partial n(1/l)]_{S_s} = [\partial/\partial r(1/r)]_{r=\epsilon} = -(1/\epsilon^2)$. We have

$$\int_{S_s} \left[\frac{1}{l} \frac{\partial V}{\partial n} - V \frac{\partial}{\partial n} \left(\frac{1}{l} \right) \right] dS = \int_0^{2\pi} d\lambda \int_0^\pi \left[\epsilon \frac{\partial V}{\partial n} + V \right] \sin\theta d\theta .$$

(2.26)

In this equation, we set $\epsilon \to 0$. We assume that the first order derivatives of V are continuous and finite within v_i, including certainly the point P. Then, $\partial V/\partial n$ is finite, and $\epsilon(\partial V/\partial n) \to 0$ when $\epsilon \to 0$. We obtain

$$\lim_{\epsilon \to 0} \int_{S_s} \left[\frac{1}{l} \frac{\partial V}{\partial n} - V \frac{\partial}{\partial n} \left(\frac{1}{l} \right) \right] dS = \lim_{\epsilon \to 0} \int_0^{2\pi} d\lambda \int_0^\pi V \sin\theta d\theta = 4\pi V(P) ,$$

(2.27)

where $V(P)$ is the value of V at P. The second identity in the above formula is derived from the fact that, as V is assumed continuous, it can be considered as a constant on the surface S_s when $\epsilon \to 0$, which is $V(P)$.

In the domain v_i, we assume that ΔV is finite. As the volume element is $dv = r^2 \sin\theta dr d\theta d\lambda$, we obtain, remembering $l = r$,

$$\lim_{\epsilon \to 0} \int_{v_s} \frac{1}{l} \Delta V dv = \lim_{\epsilon \to 0} \int_0^{2\pi} d\lambda \int_0^{\pi} d\theta \int_0^{\epsilon} r \Delta V \sin\theta dr = 0, \qquad (2.28)$$

which leads to

$$\lim_{\epsilon \to 0} \int_{v_i - v_s} \frac{1}{l} \Delta V dv = \int_{v_i} \frac{1}{l} \Delta V dv. \qquad (2.29)$$

Finally, we obtain, by setting $\epsilon \to 0$ in (2.25),

$$\int_{v_i} \frac{1}{l} \Delta V dv = \int_S \left[\frac{1}{l} \frac{\partial V}{\partial n} - V \frac{\partial}{\partial n} \left(\frac{1}{l} \right) \right] dS - 4\pi V(P). \qquad (2.30)$$

When P Is on the Surface

As shown in Fig. 2.2c, when P is on S, the same problem arises as in the case when P is inside S: $U = 1/l$ and $\Delta U = \Delta(1/l)$ tend to infinitely large when M approaches P, and thus (2.14) cannot be applied as is to the domain v_i and its surface S. A similar technique is used to solve this problem. Assume the surface is smooth around P, so that an infinitesimal disk S_d of radius ϵ can be defined on S with its center on P. We also define an infinitesimal half sphere v_s of radius ϵ with P as its center and denote its surface as S_s. We can then apply (2.14) to the closed domain $v_i - v_s$ and its surfaces $S - S_d$ and S_s. We obtain

$$\int_{v_i - v_s} \frac{1}{l} \Delta V dv = \int_{S - S_d} \left[\frac{1}{l} \frac{\partial V}{\partial n} - V \frac{\partial}{\partial n} \left(\frac{1}{l} \right) \right] dS$$

$$- \int_{S_s} \left[\frac{1}{l} \frac{\partial V}{\partial n} - V \frac{\partial}{\partial n} \left(\frac{1}{l} \right) \right] dS. \qquad (2.31)$$

Similar to the case when P is inside v_i, we can show that

$$\lim_{\epsilon \to 0} \int_{S_s} \left[\frac{1}{l} \frac{\partial V}{\partial n} - V \frac{\partial}{\partial n} \left(\frac{1}{l} \right) \right] dS = 2\pi V(P), \qquad (2.32)$$

$$\lim_{\epsilon \to 0} \int_{v_i - v_s} \frac{1}{l} \Delta V dv = \int_{v_i} \frac{1}{l} \Delta V dv. \qquad (2.33)$$

In (2.31), the first surface integral at the right-hand side also needs some further treatment. We define a polar coordinate system over S_d with origin at P, $Pr\alpha$. We see that $\hat{e}_n \perp \hat{e}_r$ and $l = r$. Hence, $[\partial/\partial n(1/l)]_{S_d} = [\partial/\partial n(1/r)]_{S_d} = 0$. Assume

$\partial V/\partial n$ to be finite. We obtain, as $dS = r\,dr\,d\alpha$ on S_d,

$$\lim_{\epsilon \to 0} \int_{S_d} \left[\frac{1}{l} \frac{\partial V}{\partial n} - V \frac{\partial}{\partial n} \left(\frac{1}{l} \right) \right] dS = \lim_{\epsilon \to 0} \int_0^{2\pi} \int_0^{\epsilon} \frac{\partial V}{\partial n} dr = 0, \qquad (2.34)$$

which leads to

$$\lim_{\epsilon \to 0} \int_{S-S_d} \left[\frac{1}{l} \frac{\partial V}{\partial n} - V \frac{\partial}{\partial n} \left(\frac{1}{l} \right) \right] dS = \int_S \left[\frac{1}{l} \frac{\partial V}{\partial n} - V \frac{\partial}{\partial n} \left(\frac{1}{l} \right) \right] dS. \qquad (2.35)$$

Finally, we obtain, by setting $\epsilon \to 0$ in (2.31),

$$\int_{v_i} \frac{1}{l} \Delta V dv = \int_S \left[\frac{1}{l} \frac{\partial V}{\partial n} - V \frac{\partial}{\partial n} \left(\frac{1}{l} \right) \right] dS - 2\pi V(P). \qquad (2.36)$$

Summary

We now gather the three identities together. For the case where the domain v_i is the inside of S, we define

$$p_i = \begin{cases} 0, & P \text{ outside } S; \\ 4\pi, & P \text{ inside } S; \\ 2\pi, & P \text{ on } S. \end{cases} \qquad (2.37)$$

We write the identities (2.24), (2.30), and (2.36) together as

$$\int_{v_i} \frac{1}{l} \Delta V dv = \int_S \left[\frac{1}{l} \frac{\partial V}{\partial n} - V \frac{\partial}{\partial n} \left(\frac{1}{l} \right) \right] dS - p_i V(P), \qquad (2.38)$$

which is referred to as Green's third identity for an internal domain.

For the case where the domain v_e is the outside of S, we write out the result without proof. We define

$$p_e = \begin{cases} 0, & P \text{ inside } S; \\ 4\pi, & P \text{ outside } S; \\ 2\pi, & P \text{ on } S. \end{cases} \qquad (2.39)$$

We write the identities as

$$\int_{v_e} \frac{1}{l} \Delta V dv = -\int_S \left[\frac{1}{l} \frac{\partial V}{\partial n} - V \frac{\partial}{\partial n} \left(\frac{1}{l} \right) \right] dS - p_e V(P), \qquad (2.40)$$

which is referred to as Green's third identity for an external domain. We mention that, the same as in Green's second identity for an external domain (2.21), in the last formula, $U = 1/l$ satisfies the condition (2.17) and (2.23), and V must satisfy (2.18) and (2.22).

2.2   Some Applications of Green's Identities

2.2.1   Gauss' Law of Gravity

Set $U = 1$ in the Green's second identity for an internal domain, i.e., (2.14), and assume V to be the gravitational potential of a solid body. We obtain Gauss' theorem for the gravitational potential,

$$\int_S \frac{\partial V}{\partial n} dS = \int_{v_i} \Delta V dv. \tag{2.41}$$

When the domain v_i is completely outside the solid body, the gravitational potential satisfies Laplace's equation $\Delta V = 0$. We obtain

$$\int_S \frac{\partial V}{\partial n} dS = \int_{v_i} \Delta V dv = 0. \tag{2.42}$$

When the domain v_i is completely inside the solid body, the gravitational potential satisfies Poisson's equation $\Delta V = -4\pi G\rho$, where ρ is the density. We obtain

$$\int_S \frac{\partial V}{\partial n} dS = \int_{v_i} \Delta V dv = -4\pi G \int_{v_i} \rho dv = -4\pi G M_i, \tag{2.43}$$

where M_i is the mass enclosed in the domain v_i. This formula is referred to as Gauss' law of gravity.

A more general form of (2.43) can be derived using the formula

$$\int_S \frac{\partial}{\partial n}\left(\frac{1}{l}\right) dS = \begin{cases} 0, & P \text{ outside } S; \\ -4\pi, & P \text{ inside } S; \\ -2\pi, & P \text{ on } S, \end{cases} \tag{2.44}$$

which is obtained by setting $V = 1$ in (2.38). We assume that a point mass m is located at P. By multiplying both sides of (2.44) by Gm, we obtain an equation satisfied by the gravitational potential of the mass, $V = Gm/l$,

$$\int_S \frac{\partial V}{\partial n} dS = \begin{cases} 0, & P \text{ outside } S; \\ -4\pi Gm, & P \text{ inside } S; \\ -2\pi Gm, & P \text{ on } S. \end{cases} \tag{2.45}$$

This equation can be readily generalized to a system of point masses, which can be written as

$$\int_S \frac{\partial V}{\partial n} dS = -4\pi G M_i - 2\pi G M_S, \tag{2.46}$$

where M_i is the total mass within S, and M_S the total mass on S. Although the gravitational potential of a system of point masses is a sum, while that of a material surface or a body is an integral, but as an integral is indeed a sum by its nature, (2.46) can be generalized as is to a complicated system of masses composed of point masses as well as material surfaces and solid bodies.

2.2.2   Determination of the Earth's Mass Using Gravity

Set $U = 1$ and v_i to be the Earth in Green's second identity for an internal domain, i.e., (2.14), and replace V by the gravity potential W. As W satisfies the Poisson's equation $\Delta W = -4\pi G\rho + 2\Omega^2$, where ρ is the density, and Ω the rotation rate, we obtain

$$\int_S \frac{\partial W}{\partial n} dS = \int_{v_i} \Delta W dv = -4\pi GM + 2\Omega^2 v_E , \qquad (2.47)$$

where M is the Earth's mass, and v_E is the Earth's volume. With

$$g_n = -\frac{\partial W}{\partial n}, \qquad (2.48)$$

we obtain

$$M = \frac{1}{4\pi G} \int_S g_n dS + \frac{\Omega^2 v_E}{2\pi G} . \qquad (2.49)$$

We make a particular comment on the physical meaning of g_n defined in (2.48): it is the projection of the gravity to the normal direction of the Earth's surface pointing to the interior of the Earth.

In fact, v_i could be any domain enclosing the whole Earth inside. The differences in the result are that v_E becomes the volume of v_i, and g_n the projection of the gravity along the normal of S pointing to the Earth.

2.2.3   Expression of the External Gravitational Potential as Surface Integrals

In Green's third identity for an external domain, i.e., (2.40), we assume that V is the gravitational potential of masses confined within the boundary S, thus satisfying Laplace's equation. We assume that P is outside the boundary S; hence, $p_e = 4\pi$. We obtain the gravitational potential at P to be

$$V = -\frac{1}{4\pi} \int_S \frac{1}{l} \frac{\partial V}{\partial n} dS + \frac{1}{4\pi} \int_S V \frac{\partial}{\partial n} \left(\frac{1}{l}\right) dS . \qquad (2.50)$$

We note that, as the gravitational potential V is regular at infinity, it satisfies the requirement of (2.40). This formula is referred to as Green's theorem of equivalent layers, which implies that the external gravitational potential can be expressed as the sum of the gravitational potentials of a single layer (i.e., a material surface) of surface density $-(4\pi/G)(\partial V/\partial n)$ and a double layer of dipole density $(4\pi/G)V$ at a boundary enclosing all masses inside.

If S is an equipotential surface, V is constant on S, and (2.50) can be written as

$$V = -\frac{1}{4\pi} \int_S \frac{1}{l} \frac{\partial V}{\partial n} dS + \frac{V}{4\pi} \int_S \frac{\partial}{\partial n}\left(\frac{1}{l}\right) dS. \qquad (2.51)$$

The second surface integral at the right-hand side is given in (2.44), which vanishes when P is outside S. Hence, we have

$$V = -\frac{1}{4\pi} \int_S \frac{1}{l} \frac{\partial V}{\partial n} dS, \qquad (2.52)$$

which is referred to as Chasles' theorem. This formula implies that the external gravitational potential can be expressed as that of a single layer on an equipotential surface S, provided that the masses are all enclosed within S.

We can derive a formula similar to (2.52) without the requirement that S is an equipotential surface with the help of Green's second identity for an internal domain (2.14). To the author's knowledge, this proof is provided only in Guan and Ning (1981). We use l to denote the distance to the point P located outside S. In (2.14), both U and V are functions defined in v_i, which is the domain within S. We assume U to be a harmonic function, and $V = 1/l$, which is also a harmonic function. This leads (2.14) to

$$0 = \int_S \left[U \frac{\partial}{\partial n}\left(\frac{1}{l}\right) - \frac{1}{l} \frac{\partial U}{\partial n} \right] dS. \qquad (2.53)$$

As this equation is derived using Green's second identity for an internal domain, the boundary is approached from the inside, while it is approached from the outside in (2.50). However, on the boundary S, the distance l in both equations is the same. Divide both sides of this equation by 4π, and then subtract it from (2.50). We obtain

$$V = -\frac{1}{4\pi} \int_S \frac{1}{l} \left[\left(\frac{\partial V}{\partial n}\right)_e - \left(\frac{\partial U}{\partial n}\right)_i \right] dS + \frac{1}{4\pi} \int_S (V-U) \frac{\partial}{\partial n}\left(\frac{1}{l}\right) dS. \qquad (2.54)$$

We mention again that V is defined outside and on S and U inside and on S. Therefore, $(\partial V/\partial n)_e$ is a limit from the outside as in (2.50), and $(\partial U/\partial n)_i$ from the inside as in (2.53). Since U is arbitrary, we define U such that $V - U = C$ is constant on the surface S. In this case, the second term in the above equation vanishes according to (2.44), and we obtain

$$V = -\frac{1}{4\pi} \int_S \frac{1}{l} \left[\left(\frac{\partial V}{\partial n} \right)_e - \left(\frac{\partial U}{\partial n} \right)_i \right] dS = G \int_S \frac{\mu}{l} dS, \qquad (2.55)$$

where

$$\mu = -\frac{1}{4\pi G} \left[\left(\frac{\partial V}{\partial n} \right)_e - \left(\frac{\partial U}{\partial n} \right)_i \right] \qquad (2.56)$$

is an equivalent surface density, noting that $(\partial U / \partial n)_i$ is based on an internal solution with $U = V - C$ as boundary value on the surface S. The main point of (2.55) is that the gravitational potential of masses completely enclosed within S can be expressed as that of a material surface on S.

2.2.4   Stokes' Theorem

Consider an attracting body of total mass M that is rotating with a constant angular velocity $\vec{\Omega}$. Let S be an equipotential surface of gravity that is known and encloses all attracting masses inside. The Stokes' theorem states that the gravity potential and attraction on and outside the surface S are uniquely determined by M, $\vec{\Omega}$, and S.

To prove Stokes' theorem, we consider two different mass distributions, which produce two different gravitational potentials V_1 and V_2. The centrifugal potential is the same for both cases and is denoted as Q. The gravity potentials for the two cases are then $W_1 = V_1 + Q$ and $W_2 = V_2 + Q$. As S remains an equipotential surface of gravity for both cases, both W_1 and W_2 are constant on it, which are denoted as c_1 and c_2, respectively. Define $T = W_1 - W_2 = V_1 - V_2$, which is a difference between two gravitational potentials, thus being harmonic outside S, and regular at infinity. Its boundary value on S is $c_1 - c_2$. Set $U = V = T$ in Green's first identity for an external domain (2.20). We obtain

$$\int_{v_e} (\nabla T \cdot \nabla T) dv = -(c_1 - c_2) \int_S \left(\frac{\partial V_1}{\partial n} - \frac{\partial V_2}{\partial n} \right) dS. \qquad (2.57)$$

As V_1 and V_2 correspond to the same total attracting mass M, we obtain, according to Gauss' law of gravity (2.43),

$$\int_{v_e} (\nabla T \cdot \nabla T) dv = -(c_1 - c_2)(-4\pi GM + 4\pi GM) = 0. \qquad (2.58)$$

In more detail, this equation can be written as

$$\int_{v_e} (\nabla T \cdot \nabla T) dv = \int_{v_e} \left[\left(\frac{\partial T}{\partial x} \right)^2 + \left(\frac{\partial T}{\partial y} \right)^2 + \left(\frac{\partial T}{\partial z} \right)^2 \right] dv = 0. \qquad (2.59)$$

In the integrand, none of the three terms can be smaller than 0. In order that the integral be 0, all the 3 terms must be 0. Hence,

$$\frac{\partial T}{\partial x} = \frac{\partial T}{\partial y} = \frac{\partial T}{\partial z} = 0. \tag{2.60}$$

From this formula, we can infer that T is constant within v_e. As T vanishes at infinity, we conclude that $T = 0$ within v_e; thus $W_1 = W_2$, meaning that the gravity potential is uniquely determined. The uniqueness of gravity follows right away.

2.3   Boundary Value Problems and Uniqueness of Solution

2.3.1   Three Types of Boundary Value Problems

Our purpose is to determine the gravity field of the Earth, which is completely represented by the gravity potential. By definition, the gravity potential of the Earth at a point P of coordinate (x, y, z) is

$$W = G \int_v \frac{\rho}{l} dv + \frac{1}{2}\Omega^2(x^2 + y^2), \tag{2.61}$$

where v is the body of the Earth, ρ the Earth's density as function of position, l the distance between P and the volume element dv, Ω the Earth's rotation rate (the rotation rate of the Earth-fixed frame $Oxyz$), and the z-axis is chosen to be the rotation axis of the Earth.

In (2.61), the first term is the gravitational potential, and the second term is the centrifugal potential. The computation of the centrifugal potential is easy once the Earth's rotation rate and axis are determined accurately, which is indeed the case at present; the Earth rotation is monitored with extreme accuracy using the very long baseline interferometry (VLBI) technique. In order to compute the gravitational potential based on the definition in (2.61), the Earth's shape and internal density distribution must be known with sufficient accuracy, which is not the case at present. Recently, with the advance of satellite and airborne technologies, the Earth's surface is being mapped more and more accurately with finer and finer resolution, but the advance in the determination of the Earth's density is not as rapid. Global models of density distribution within the Earth are mainly determined using data of seismic waves and the Earth's free oscillations collected after Earthquakes. However, due to the non-uniqueness of solution and the insufficiency of data, assumptions of material property and structure must be introduced. Therefore, there is no hope in the foreseeable future to have a global model of the Earth's density to the detail and accuracy for computing the gravitational potential with competing accuracy as compared to that determined using gravity measurements. By contrary, gravity data are frequently used to study the density distribution in the Earth's interior in geophysics and geology. However, as the gravity is an integral of the density, the same gravity data set may correspond to more than one density distribution. Hence, gravity data are insufficient to determine the density distribution uniquely, and other data, such as data of geological structure, must be used.

In reality, the Earth's external gravity field is nowadays determined using satellite orbit data, satellite gravimetry data, and surface data, where surface data may be deduced from measurements made on the Earth's surface, or from airborne or satellite altimetry data. In this book, we focus on the use of surface data for explaining the fundamental concepts and the theory of the gravity field.

For explanatory purpose, we assume that the gravity is measured all over the Earth's surface. After removing the centrifugal acceleration, we obtain the gravitational attraction all over the Earth's surface. Conceptually, according to the properties of the gravitational potential, if we can determine a unique function, which is harmonic outside the Earth, regular at infinity, and its gradient is equal to the gravitational attraction on the Earth's surface, this function must be the gravitational potential. The determination of the gravitational potential in this way is a kind of boundary value problem. In reality, different boundary value problems are developed for different kinds of data. Here we focus on the general concepts of the boundary value problems concerned.

A boundary value problem in potential theory is the determination of a solution of Laplace's equation within a space domain, which satisfies some conditions specified at the boundary of the domain. When the space domain is enclosed within its boundary, the problem is an internal boundary value problem. When the space domain is outside its boundary and extends to infinity, the problem is an external boundary value problem. In the theory of the Earth's external gravity field, external boundary value problems are relevant. The following are three kinds of most common boundary value problems relevant to the theory of the Earth's gravity field.

First-type boundary value problem, also known as Dirichlet's problem: Determination of a function V, which is harmonic outside the boundary, regular at infinity, and satisfies the condition $V = f$ on the boundary, where f is a known function.

Second-type boundary value problem, also known as Neumann's problem: Determination of a function V, which is harmonic outside the boundary, regular at infinity, and satisfies the condition $\partial V / \partial n = f$ on the boundary, where $\hat{e}_n$ is the outward normal to the boundary, and f is a known function.

Third-type boundary value problem, also known as Robin's problem: Determination of a function V, which is harmonic outside the boundary, regular at infinity, and satisfies the condition $\alpha V + \beta(\partial V / \partial n) = f$ on the boundary, where $\hat{e}_n$ is the outward normal to the boundary, and f is a known function.

The above are all boundary value problems of Laplace's equation for an external domain, which are relevant to the studies of the Earth's external gravity field. We note that in a second-type boundary value problem of Laplace's equation for an internal domain, the boundary value must satisfy (2.42); otherwise, there would be no solution.

2.3.2 Uniqueness of Solution

In order to determine the gravitational potential by solving a boundary value problem, it must be shown that the solution is unique. Let v_e be a space domain that extends to infinity with a closed surface S as its inner boundary. Let T be a harmonic function in the domain v_e and regular at infinity. Set $U = V = T$ in Green's first identity for an external domain (2.20), and we obtain

$$\int_{v_e} (\nabla T \cdot \nabla T) dv = -\int_S T \frac{\partial T}{\partial n} dS. \tag{2.62}$$

This is the equation we will use to study the uniqueness of the solutions of the boundary value problems.

For the first-type boundary value problem, we assume that there exist two solutions regular at infinity and satisfying the same boundary condition and use T to represent their difference. The boundary condition for T is then $T = 0$ on S. Hence, according to (2.62),

$$\int_{v_e} (\nabla T \cdot \nabla T) dv = \int_{v_e} \left[\left(\frac{\partial T}{\partial x} \right)^2 + \left(\frac{\partial T}{\partial y} \right)^2 + \left(\frac{\partial T}{\partial z} \right)^2 \right] dv = 0. \tag{2.63}$$

For the same logic as in the proof of Stokes' theorem, we conclude that $T = 0$ in v_e. Hence, the two solutions must be identical, meaning that the solution is unique.

The proof for the second-type boundary value problem is almost identical. The only difference is that the boundary condition for T is replaced by the boundary condition for $\partial T/\partial n$, still leading to (2.63).

For the third-type boundary value problem, the boundary condition for T becomes $\alpha T + \beta(\partial T/\partial n) = 0$. We thus have $\partial T/\partial n = -(\alpha/\beta)T$. The substitution of it into (2.62) leads to

$$\int_{v_e} (\nabla T \cdot \nabla T) dv = \int_S \frac{\alpha}{\beta} T^2 dS. \tag{2.64}$$

When α and β have opposite signs, the left-hand side cannot be smaller than 0, and the right-hand side cannot be larger than 0, thus both sides must be 0 for the identity to hold. This proves the solution to be unique only for the case when α and β have opposite signs.

In physical geodesy, a third-type boundary value problem with α and β having the same sign is to be solved, where the solution is indeed not unique and can be made unique with further constraint. Normally, the process of finding an analytic solution serves as a proof for the existence and uniqueness of the solution. This will be made clear in the context of the problem we are going to solve.

2.4   Solution of the First-Type Boundary Value Problem

2.4.1   Method of Green's Function

The solution is to be built based on (2.50), which we write here in the form of

$$
V = \frac{1}{4\pi} \int_S \left[-\frac{1}{l}\frac{\partial V}{\partial n} + V\frac{\partial}{\partial n}\left(\frac{1}{l}\right) \right] dS. \tag{2.65}
$$

As this formula is obtained from Green's third identity, it requires that V has continuous derivatives up to order two. Although V was assumed to be the gravitational potential in (2.50), this equation holds for any harmonic function V outside and on S and regular at infinity. Let U be another such function. We can obtain, from Green's second identity for an external domain (2.21),

$$
0 = \int_S \left(U\frac{\partial V}{\partial n} - V\frac{\partial U}{\partial n} \right) dS. \tag{2.66}
$$

By adding (2.65) and (2.66), we obtain

$$
V = \frac{1}{4\pi} \int_S \left[\frac{\partial V}{\partial n}\left(U - \frac{1}{l}\right) - V\frac{\partial}{\partial n}\left(U - \frac{1}{l}\right) \right] dS. \tag{2.67}
$$

This is the starting equation to solve for the harmonic function V outside S, and the function

$$
G = U - \frac{1}{l} \tag{2.68}
$$

is referred to as Green's function for this particular problem.

If we can find a harmonic function U such that $G = 0$ on S and regular at infinity, we have

$$
V = -\frac{1}{4\pi} \int_S V\frac{\partial G}{\partial n} dS, \tag{2.69}
$$

which is the solution of the first-type boundary value problem of Laplace's equation.

Conceptually, imposing different requirements to G may lead to the solutions of the other two types of boundary value problem of Laplace's equation.

2.4.2   Poisson Integral

We now seek the solution of the first-type external boundary value problem of Laplace's equation for a spherical boundary. As shown in Fig. 2.3, we denote the

Fig. 2.3 First-type external
boundary value problem for a
spherical boundary

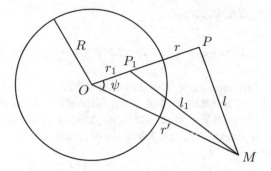

sphere as S, its center as O, and its radius as R. We assume the harmonic function
V is known on S and seek its solution at an external point P.

Select a point P_1 on OP, such that $rr_1 = R^2$, where r and r_1 are the distance of
P and P_1 to O, respectively. Such a point P_1 is called conjugate of the point P. We
express G and U as functions of M, which is another point outside S, and denote
the distances of M to P and P_1 as l and l_1, respectively. The Green's function we
are seeking is then

$$G = \frac{R}{r}\frac{1}{l_1} - \frac{1}{l},\tag{2.70}$$

which corresponds to

$$U = \frac{R}{r}\frac{1}{l_1}.\tag{2.71}$$

The proof of (2.70) and (2.71) includes two parts. The first part is to prove that U
is harmonic outside S and regular at infinity. The second part is to prove that $G = 0$
on S.

Since the expression of U is the same as the gravitational potential of a point
mass located at the point P_1, with the exception that the constant coefficient Gm
is replaced by another constant coefficient R/r, the proof of the first part follows
directly from the properties of the gravitational potential.

For the second part, we use the relations

$$l^2 = r^2 + r'^2 - 2rr'\cos\psi,\tag{2.72}$$

$$l_1^2 = r_1^2 + r'^2 - 2r_1r'\cos\psi,\tag{2.73}$$

where r' is the distance of M to O. As P_1 is the conjugate point of P, we have
$r_1 = R^2/r$. We also have $r' = R$ on S. Hence, when M is on S, we obtain from the

two relations above

$$l_1^2 = \frac{R^4}{r^2} + R^2 - 2\frac{R^3}{r}\cos\psi = \frac{R^2}{r^2}(R^2 + r^2 - 2Rr\cos\psi) = \frac{R^2}{r^2}l^2. \quad (2.74)$$

The conclusion $G = 0$ follows right away by substituting this formula into (2.70).

Now we know that G is the Green's function of the problem. However, in order to compute V at P according to (2.69), we must derive a concrete formula for $\partial G/\partial n$. In this case, $\hat{e}_n$ is the same as $\hat{e}_{r'}$. Hence,

$$\frac{\partial G}{\partial n} = \left(\frac{\partial G}{\partial r'}\right)_{r'=R} = -\frac{R}{r}\left(\frac{1}{l_1^2}\frac{\partial l_1}{\partial r'}\right)_{r'=R} + \left(\frac{1}{l^2}\frac{\partial l}{\partial r'}\right)_{r'=R}. \quad (2.75)$$

According to (2.73),

$$\left(\frac{1}{l_1^2}\frac{\partial l_1}{\partial r'}\right)_{r'=R} = \left(\frac{r' - r_1\cos\psi}{l_1^3}\right)_{r'=R} = \left(\frac{R - r_1\cos\psi}{l_1^3}\right)_{r'=R}. \quad (2.76)$$

As $r_1 = R^2/r$, and l_1 is provided by (2.74) when $r' = R$, we further have

$$\left(\frac{1}{l_1^2}\frac{\partial l_1}{\partial r'}\right)_{r'=R} = \left[\frac{R - (R^2/r)\cos\psi}{(R^3/r^3)l^3}\right]_{r'=R} = \left[\frac{r(r^2 - Rr\cos\psi)}{R^2 l^3}\right]_{r'=R}. \quad (2.77)$$

According to (2.72),

$$\left(\frac{1}{l^2}\frac{\partial l}{\partial r'}\right)_{r'=R} = \left(\frac{r' - r\cos\psi}{l^3}\right)_{r'=R} = \left(\frac{R - r\cos\psi}{l^3}\right)_{r'=R}. \quad (2.78)$$

We finally obtain, by substituting the last two formulae into (2.75),

$$\frac{\partial G}{\partial n} = -\left(\frac{r^2 - Rr\cos\psi}{Rl^3}\right)_{r'=R} + \left(\frac{R - r\cos\psi}{l^3}\right)_{r'=R} = -\left(\frac{r^2 - R^2}{Rl^3}\right)_{r'=R}. \quad (2.79)$$

By substituting (2.79) into (2.69), we obtain the solution of V, which we write as

$$V = \frac{r^2 - R^2}{4\pi R}\int_S \frac{V}{l^3}dS = \frac{R(r^2 - R^2)}{4\pi}\int_\omega \frac{V}{l^3}d\omega, \quad (2.80)$$

where ω represents a unit sphere, i.e., a sphere of radius 1, and $dS = R^2 d\omega$, $d\omega = \sin\theta d\theta d\lambda$, where θ and λ are the spherical coordinates, i.e., the polar angle and longitude, respectively. We emphasize that l is the distance between P and dS. This formula is referred to as Poisson integral.

In (2.80), the integrand is the product between a quantity V assumed to be known and a function $1/l^3$. The integral is referred to as a convolution between V and $1/l^3$. The function $1/l^3$ is referred to as kernel of the integral or convolution. In this book, many such kinds of integral or convolution will be derived, and the term *kernel* will be repeatedly used. Anyway, the function that the term *kernel* refers to will always be explicitly indicated.

2.5 Gradient of the Gravitational Attraction in Spherical Coordinates

2.5.1 Gradients of Scalars and Vectors and Their Coordinate Transformations

We have already defined the gradient of a scalar, such that the gravitational attraction is the gradient of the gravitational potential. In this subsection, we introduce the general definition of gradients of both scalars and vectors and derive the formulae of their coordinate transformations between two Cartesian coordinate systems.

Let $Oxyz$ and $P\xi\eta\zeta$ be two Cartesian coordinate systems as shown in Fig. 2.4. As usual, we use $(\hat{e}_\xi, \hat{e}_x)$ to denote the angle between the ξ- and x-axes, and so forth. Two of these angles are shown in Fig. 2.4 with the aid of a coordinate system $Px'y'x'$ with axes parallel to those of $Oxyz$. According to the definition of the angles among the coordinate axes, we have

$$\begin{cases} \hat{e}_\xi = \cos(\hat{e}_\xi, \hat{e}_x)\hat{e}_x + \cos(\hat{e}_\xi, \hat{e}_y)\hat{e}_y + \cos(\hat{e}_\xi, \hat{e}_z)\hat{e}_z\,, \\ \hat{e}_\eta = \cos(\hat{e}_\eta, \hat{e}_x)\hat{e}_x + \cos(\hat{e}_\eta, \hat{e}_y)\hat{e}_y + \cos(\hat{e}_\eta, \hat{e}_z)\hat{e}_z\,, \\ \hat{e}_\zeta = \cos(\hat{e}_\zeta, \hat{e}_x)\hat{e}_x + \cos(\hat{e}_\zeta, \hat{e}_y)\hat{e}_y + \cos(\hat{e}_\zeta, \hat{e}_z)\hat{e}_z\,. \end{cases} \qquad (2.81)$$

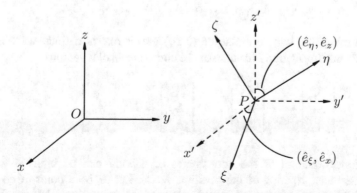

Fig. 2.4 Coordinate transformation between two Cartesian coordinate systems

Similarly,

$$\begin{cases} \hat{e}_x = \cos(\hat{e}_\xi, \hat{e}_x)\hat{e}_\xi + \cos(\hat{e}_\eta, \hat{e}_x)\hat{e}_\eta + \cos(\hat{e}_\zeta, \hat{e}_x)\hat{e}_\zeta, \\ \hat{e}_y = \cos(\hat{e}_\xi, \hat{e}_y)\hat{e}_\xi + \cos(\hat{e}_\eta, \hat{e}_y)\hat{e}_\eta + \cos(\hat{e}_\zeta, \hat{e}_y)\hat{e}_\zeta, \\ \hat{e}_z = \cos(\hat{e}_\xi, \hat{e}_z)\hat{e}_\xi + \cos(\hat{e}_\eta, \hat{e}_z)\hat{e}_\eta + \cos(\hat{e}_\zeta, \hat{e}_z)\hat{e}_\zeta. \end{cases} \tag{2.82}$$

Define

$$N = \begin{bmatrix} \cos(\hat{e}_\xi, \hat{e}_x) & \cos(\hat{e}_\xi, \hat{e}_y) & \cos(\hat{e}_\xi, \hat{e}_z) \\ \cos(\hat{e}_\eta, \hat{e}_x) & \cos(\hat{e}_\eta, \hat{e}_y) & \cos(\hat{e}_\eta, \hat{e}_z) \\ \cos(\hat{e}_\zeta, \hat{e}_x) & \cos(\hat{e}_\zeta, \hat{e}_y) & \cos(\hat{e}_\zeta, \hat{e}_z) \end{bmatrix}. \tag{2.83}$$

We can write (2.81) and (2.82) shortly as

$$\begin{bmatrix} \hat{e}_\xi \\ \hat{e}_\eta \\ \hat{e}_\zeta \end{bmatrix} = N \begin{bmatrix} \hat{e}_x \\ \hat{e}_y \\ \hat{e}_z \end{bmatrix}, \qquad \begin{bmatrix} \hat{e}_x \\ \hat{e}_y \\ \hat{e}_z \end{bmatrix} = N^T \begin{bmatrix} \hat{e}_\xi \\ \hat{e}_\eta \\ \hat{e}_\zeta \end{bmatrix}. \tag{2.84}$$

As these are inverse transformations to each other, we have

$$N^{-1} = N^T, \tag{2.85}$$

and hence,

$$NN^T = I, \tag{2.86}$$

i.e., N is an orthogonal matrix.

A vector independent of coordinate system can be expressed in either $Oxyz$ or $P\xi\eta\zeta$,

$$\vec{A} = A_x\hat{e}_x + A_y\hat{e}_y + A_z\hat{e}_z = A_\xi\hat{e}_\xi + A_\eta\hat{e}_\eta + A_\zeta\hat{e}_\zeta. \tag{2.87}$$

By substituting the two formulae of (2.84) into it one at a time, we obtain the coordinate transformation formulae of the components of a vector,

$$\begin{bmatrix} A_\xi \\ A_\eta \\ A_\zeta \end{bmatrix} = N \begin{bmatrix} A_x \\ A_y \\ A_z \end{bmatrix}, \qquad \begin{bmatrix} A_x \\ A_y \\ A_z \end{bmatrix} = N^T \begin{bmatrix} A_\xi \\ A_\eta \\ A_\zeta \end{bmatrix}. \tag{2.88}$$

The transformation of the coordinates of a point can be inferred from the coordinate transformation of the position vector. Let M be a point of coordinate (x, y, z) and (ξ, η, ζ) in the two coordinate systems, respectively. We use $\vec{r}_O^M$ and $\vec{r}_P^M$ to denote the position vectors of M with respect to O and P, respectively, and

$\overrightarrow{r}_O^P$ the position vector of P with respect to O. We then have

$$\overrightarrow{r}_O^M - \overrightarrow{r}_O^P = \overrightarrow{r}_P^M .\tag{2.89}$$

Considering that

$$\begin{cases} \overrightarrow{r}_O^M = x\hat{e}_x + y\hat{e}_y + z\hat{e}_z , \\ \overrightarrow{r}_O^P = x_O^P\hat{e}_x + y_O^P\hat{e}_y + z_O^P\hat{e}_z , \\ \overrightarrow{r}_P^M = \xi\hat{e}_\xi + \eta\hat{e}_\eta + \zeta\hat{e}_\zeta , \end{cases}\tag{2.90}$$

we obtain, according to (2.88),

$$\begin{bmatrix} \xi \\ \eta \\ \zeta \end{bmatrix} = N \begin{bmatrix} x - x_O^P \\ y - y_O^P \\ z - z_O^P \end{bmatrix} , \quad \begin{bmatrix} x - x_O^P \\ y - y_O^P \\ z - z_O^P \end{bmatrix} = N^T \begin{bmatrix} \xi \\ \eta \\ \zeta \end{bmatrix} .\tag{2.91}$$

We can further obtain from these equations

$$\frac{\partial(\xi, \eta, \zeta)}{\partial(x, y, z)} = \begin{bmatrix} \dfrac{\partial\xi}{\partial x} & \dfrac{\partial\eta}{\partial x} & \dfrac{\partial\zeta}{\partial x} \\ \dfrac{\partial\xi}{\partial y} & \dfrac{\partial\eta}{\partial y} & \dfrac{\partial\zeta}{\partial y} \\ \dfrac{\partial\xi}{\partial z} & \dfrac{\partial\eta}{\partial z} & \dfrac{\partial\zeta}{\partial z} \end{bmatrix} = N^T , \quad \frac{\partial(x, y, z)}{\partial(\xi, \eta, \zeta)} = \begin{bmatrix} \dfrac{\partial x}{\partial\xi} & \dfrac{\partial y}{\partial\xi} & \dfrac{\partial z}{\partial\xi} \\ \dfrac{\partial x}{\partial\eta} & \dfrac{\partial y}{\partial\eta} & \dfrac{\partial z}{\partial\eta} \\ \dfrac{\partial x}{\partial\zeta} & \dfrac{\partial y}{\partial\zeta} & \dfrac{\partial z}{\partial\zeta} \end{bmatrix} = N ,\tag{2.92}$$

which are essential for forthcoming formulations.

The three components of the gradient of a scalar defined in $Oxyz$ and $P\xi\eta\zeta$ are

$$\nabla\phi = \begin{bmatrix} (\nabla\phi)_x \\ (\nabla\phi)_y \\ (\nabla\phi)_z \end{bmatrix} = \begin{bmatrix} \dfrac{\partial\phi}{\partial x} \\ \dfrac{\partial\phi}{\partial y} \\ \dfrac{\partial\phi}{\partial z} \end{bmatrix} , \quad \nabla\phi = \begin{bmatrix} (\nabla\phi)_\xi \\ (\nabla\phi)_\eta \\ (\nabla\phi)_\zeta \end{bmatrix} = \begin{bmatrix} \dfrac{\partial\phi}{\partial\xi} \\ \dfrac{\partial\phi}{\partial\eta} \\ \dfrac{\partial\phi}{\partial\zeta} \end{bmatrix} .\tag{2.93}$$

We write them shortly as

$$\frac{\partial\phi}{\partial(x, y, z)} = \begin{bmatrix} \dfrac{\partial\phi}{\partial x} \\ \dfrac{\partial\phi}{\partial y} \\ \dfrac{\partial\phi}{\partial z} \end{bmatrix} , \quad \frac{\partial\phi}{\partial(\xi, \eta, \zeta)} = \begin{bmatrix} \dfrac{\partial\phi}{\partial\xi} \\ \dfrac{\partial\phi}{\partial\eta} \\ \dfrac{\partial\phi}{\partial\zeta} \end{bmatrix} .\tag{2.94}$$

The coordinate transformation between them can be derived as follows:

$$
\begin{bmatrix} \dfrac{\partial \phi}{\partial \xi} \\[2mm] \dfrac{\partial \phi}{\partial \eta} \\[2mm] \dfrac{\partial \phi}{\partial \zeta} \end{bmatrix} = \begin{bmatrix} \dfrac{\partial \phi}{\partial x}\dfrac{\partial x}{\partial \xi} + \dfrac{\partial \phi}{\partial y}\dfrac{\partial y}{\partial \xi} + \dfrac{\partial \phi}{\partial z}\dfrac{\partial z}{\partial \xi} \\[2mm] \dfrac{\partial \phi}{\partial x}\dfrac{\partial x}{\partial \eta} + \dfrac{\partial \phi}{\partial y}\dfrac{\partial y}{\partial \eta} + \dfrac{\partial \phi}{\partial z}\dfrac{\partial z}{\partial \eta} \\[2mm] \dfrac{\partial \phi}{\partial x}\dfrac{\partial x}{\partial \zeta} + \dfrac{\partial \phi}{\partial y}\dfrac{\partial y}{\partial \zeta} + \dfrac{\partial \phi}{\partial z}\dfrac{\partial z}{\partial \zeta} \end{bmatrix}
$$

$$
= \begin{bmatrix} \dfrac{\partial x}{\partial \xi} & \dfrac{\partial y}{\partial \xi} & \dfrac{\partial z}{\partial \xi} \\[2mm] \dfrac{\partial x}{\partial \eta} & \dfrac{\partial y}{\partial \eta} & \dfrac{\partial z}{\partial \eta} \\[2mm] \dfrac{\partial x}{\partial \zeta} & \dfrac{\partial y}{\partial \zeta} & \dfrac{\partial z}{\partial \zeta} \end{bmatrix} \begin{bmatrix} \dfrac{\partial \phi}{\partial x} \\[2mm] \dfrac{\partial \phi}{\partial y} \\[2mm] \dfrac{\partial \phi}{\partial z} \end{bmatrix} = N \begin{bmatrix} \dfrac{\partial \phi}{\partial x} \\[2mm] \dfrac{\partial \phi}{\partial y} \\[2mm] \dfrac{\partial \phi}{\partial z} \end{bmatrix}, \tag{2.95}
$$

which can be written briefly as

$$
\frac{\partial \phi}{\partial (\xi, \eta, \zeta)} = \frac{\partial (x, y, z)}{\partial (\xi, \eta, \zeta)} \frac{\partial \phi}{\partial (x, y, z)} = N \frac{\partial \phi}{\partial (x, y, z)}. \tag{2.96}
$$

The gradient of a vector $\vec{A}$ has nine components. In $Oxyz$, the components are

$$
\begin{cases} (\nabla \vec{A})_{xx} = \dfrac{\partial A_x}{\partial x}, \ (\nabla \vec{A})_{xy} = \dfrac{\partial A_y}{\partial x}, \ (\nabla \vec{A})_{xz} = \dfrac{\partial A_z}{\partial x}; \\[2mm] (\nabla \vec{A})_{yx} = \dfrac{\partial A_x}{\partial y}, \ (\nabla \vec{A})_{yy} = \dfrac{\partial A_y}{\partial y}, \ (\nabla \vec{A})_{yz} = \dfrac{\partial A_z}{\partial y}; \\[2mm] (\nabla \vec{A})_{zx} = \dfrac{\partial A_x}{\partial z}, \ (\nabla \vec{A})_{zy} = \dfrac{\partial A_y}{\partial z}, \ (\nabla \vec{A})_{zz} = \dfrac{\partial A_z}{\partial z}. \end{cases} \tag{2.97}
$$

In $P\xi\eta\zeta$, the components are

$$
\begin{cases} (\nabla \vec{A})_{\xi\xi} = \dfrac{\partial A_\xi}{\partial \xi}, \ (\nabla \vec{A})_{\xi\eta} = \dfrac{\partial A_\eta}{\partial \xi}, \ (\nabla \vec{A})_{\xi\zeta} = \dfrac{\partial A_\zeta}{\partial \xi}; \\[2mm] (\nabla \vec{A})_{\eta\xi} = \dfrac{\partial A_\xi}{\partial \eta}, \ (\nabla \vec{A})_{\eta\eta} = \dfrac{\partial A_\eta}{\partial \eta}, \ (\nabla \vec{A})_{\eta\zeta} = \dfrac{\partial A_\zeta}{\partial \eta}; \\[2mm] (\nabla \vec{A})_{\zeta\xi} = \dfrac{\partial A_\xi}{\partial \zeta}, \ (\nabla \vec{A})_{\zeta\eta} = \dfrac{\partial A_\eta}{\partial \zeta}, \ (\nabla \vec{A})_{\zeta\zeta} = \dfrac{\partial A_\zeta}{\partial \zeta}. \end{cases} \tag{2.98}
$$

We arrange the components briefly in the form of a matrix as

$$
\begin{cases}
\dfrac{\partial(A_x, A_y, A_z)}{\partial(x, y, z)} = \begin{bmatrix}
\dfrac{\partial A_x}{\partial x} & \dfrac{\partial A_y}{\partial x} & \dfrac{\partial A_z}{\partial x} \\[2mm]
\dfrac{\partial A_x}{\partial y} & \dfrac{\partial A_y}{\partial y} & \dfrac{\partial A_z}{\partial y} \\[2mm]
\dfrac{\partial A_x}{\partial z} & \dfrac{\partial A_y}{\partial z} & \dfrac{\partial A_z}{\partial z}
\end{bmatrix} , \\[12mm]
\dfrac{\partial(A_\xi, A_\eta, A_\zeta)}{\partial(\xi, \eta, \zeta)} = \begin{bmatrix}
\dfrac{\partial A_\xi}{\partial \xi} & \dfrac{\partial A_\eta}{\partial \xi} & \dfrac{\partial A_\zeta}{\partial \xi} \\[2mm]
\dfrac{\partial A_\xi}{\partial \eta} & \dfrac{\partial A_\eta}{\partial \eta} & \dfrac{\partial A_\zeta}{\partial \eta} \\[2mm]
\dfrac{\partial A_\xi}{\partial \zeta} & \dfrac{\partial A_\eta}{\partial \zeta} & \dfrac{\partial A_\zeta}{\partial \zeta}
\end{bmatrix} .
\end{cases}
\tag{2.99}
$$

A first step to derive the coordinate transformation between them is (expand to matrix form to see clearly the equation for each component)

$$
\frac{\partial(A_\xi, A_\eta, A_\zeta)}{\partial(\xi, \eta, \zeta)} = \frac{\partial(x, y, z)}{\partial(\xi, \eta, \zeta)} \frac{\partial(A_\xi, A_\eta, A_\zeta)}{\partial(x, y, z)} = N \frac{\partial(A_\xi, A_\eta, A_\zeta)}{\partial(x, y, z)} .
\tag{2.100}
$$

For the next step, we take only one column as an example

$$
\frac{\partial A_\xi}{\partial(x, y, z)} = \begin{bmatrix}
\dfrac{\partial A_\xi}{\partial x} \\[2mm]
\dfrac{\partial A_\xi}{\partial y} \\[2mm]
\dfrac{\partial A_\xi}{\partial z}
\end{bmatrix}
$$

$$
= \begin{bmatrix}
\dfrac{\partial A_x}{\partial x} \cos(\hat{e}_\xi, \hat{e}_x) + \dfrac{\partial A_y}{\partial x} \cos(\hat{e}_\xi, \hat{e}_y) + \dfrac{\partial A_z}{\partial x} \cos(\hat{e}_\xi, \hat{e}_z) \\[2mm]
\dfrac{\partial A_x}{\partial y} \cos(\hat{e}_\xi, \hat{e}_x) + \dfrac{\partial A_y}{\partial y} \cos(\hat{e}_\xi, \hat{e}_y) + \dfrac{\partial A_z}{\partial y} \cos(\hat{e}_\xi, \hat{e}_z) \\[2mm]
\dfrac{\partial A_x}{\partial z} \cos(\hat{e}_\xi, \hat{e}_x) + \dfrac{\partial A_y}{\partial z} \cos(\hat{e}_\xi, \hat{e}_y) + \dfrac{\partial A_z}{\partial z} \cos(\hat{e}_\xi, \hat{e}_z)
\end{bmatrix} ,
\tag{2.101}
$$

which can be written briefly as

$$
\frac{\partial A_\xi}{\partial(x, y, z)} = \begin{bmatrix}
\dfrac{\partial A_x}{\partial x} & \dfrac{\partial A_y}{\partial x} & \dfrac{\partial A_z}{\partial x} \\[2mm]
\dfrac{\partial A_x}{\partial y} & \dfrac{\partial A_y}{\partial y} & \dfrac{\partial A_z}{\partial y} \\[2mm]
\dfrac{\partial A_x}{\partial z} & \dfrac{\partial A_y}{\partial z} & \dfrac{\partial A_z}{\partial z}
\end{bmatrix} \begin{bmatrix}
\cos(\hat{e}_\xi, \hat{e}_x) \\[1mm]
\cos(\hat{e}_\xi, \hat{e}_y) \\[1mm]
\cos(\hat{e}_\xi, \hat{e}_z)
\end{bmatrix} .
\tag{2.102}
$$

Repeating the process for the other two columns, and then putting together, we obtain

$$
\frac{\partial(A_\xi, A_\eta, A_\zeta)}{\partial(x, y, z)} =
\begin{bmatrix}
\dfrac{\partial A_x}{\partial x} & \dfrac{\partial A_y}{\partial x} & \dfrac{\partial A_z}{\partial x} \\[6pt]
\dfrac{\partial A_x}{\partial y} & \dfrac{\partial A_y}{\partial y} & \dfrac{\partial A_z}{\partial y} \\[6pt]
\dfrac{\partial A_x}{\partial z} & \dfrac{\partial A_y}{\partial z} & \dfrac{\partial A_z}{\partial z}
\end{bmatrix}
\begin{bmatrix}
\cos(\hat{e}_\xi, \hat{e}_x) & \cos(\hat{e}_\eta, \hat{e}_x) & \cos(\hat{e}_\zeta, \hat{e}_x) \\
\cos(\hat{e}_\xi, \hat{e}_y) & \cos(\hat{e}_\eta, \hat{e}_y) & \cos(\hat{e}_\zeta, \hat{e}_y) \\
\cos(\hat{e}_\xi, \hat{e}_z) & \cos(\hat{e}_\eta, \hat{e}_y) & \cos(\hat{e}_\zeta, \hat{e}_y)
\end{bmatrix}
$$

$$
= \frac{\partial(A_x, A_y, A_z)}{\partial(x, y, z)} N^T . \tag{2.103}
$$

Substitute this equation into (2.102). We obtain the result desired

$$
\frac{\partial(A_\xi, A_\eta, A_\zeta)}{\partial(\xi, \eta, \zeta)} = N \frac{\partial(A_\xi, A_\eta, A_\zeta)}{\partial(x, y, z)} = N \frac{\partial(A_x, A_y, A_z)}{\partial(x, y, z)} N^T . \tag{2.104}
$$

2.5.2   Spherical Coordinates

As shown in Fig. 2.5, we introduce a spherical coordinate system $Or\theta\lambda$ associated with $Oxyz$ as usual. The transformations of coordinates of a point between them are

$$
x = r \sin\theta \cos\lambda , \qquad y = r \sin\theta \sin\lambda , \qquad z = r \cos\theta , \tag{2.105}
$$

Fig. 2.5 Spherical coordinate system

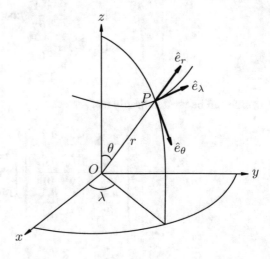

and

$$r = (x^2 + y^2 + z^2)^{1/2}, \qquad \theta = \arccos \frac{z}{(x^2 + y^2 + z^2)^{1/2}}, \qquad \lambda = \arctan \frac{y}{x}.$$

(2.106)

The transformations between base vectors of the two coordinate systems are

$$\begin{cases} \hat{e}_r = \sin\theta \cos\lambda \, \hat{e}_x + \sin\theta \sin\lambda \, \hat{e}_y + \cos\theta \, \hat{e}_z, \\ \hat{e}_\theta = \cos\theta \cos\lambda \, \hat{e}_x + \cos\theta \sin\lambda \, \hat{e}_y - \sin\theta \, \hat{e}_z, \\ \hat{e}_\lambda = -\sin\lambda \, \hat{e}_x + \cos\lambda \, \hat{e}_y; \end{cases}$$

(2.107)

and

$$\begin{cases} \hat{e}_x = \sin\theta \cos\lambda \, \hat{e}_r + \cos\theta \cos\lambda \, \hat{e}_\theta - \sin\lambda \, \hat{e}_\lambda, \\ \hat{e}_y = \sin\theta \sin\lambda \, \hat{e}_r + \cos\theta \sin\lambda \, \hat{e}_\theta + \cos\lambda \, \hat{e}_\lambda, \\ \hat{e}_z = \cos\theta \, \hat{e}_r - \sin\theta \, \hat{e}_\theta. \end{cases}$$

(2.108)

Define

$$N_S = \begin{bmatrix} \sin\theta \cos\lambda & \sin\theta \sin\lambda & \cos\theta \\ \cos\theta \cos\lambda & \cos\theta \sin\lambda & -\sin\theta \\ -\sin\lambda & \cos\lambda & 0 \end{bmatrix}.$$

(2.109)

We can write (2.107) and (2.108) briefly as

$$\begin{bmatrix} \hat{e}_r \\ \hat{e}_\theta \\ \hat{e}_\lambda \end{bmatrix} = N_S \begin{bmatrix} \hat{e}_x \\ \hat{e}_y \\ \hat{e}_z \end{bmatrix}, \qquad \begin{bmatrix} \hat{e}_x \\ \hat{e}_y \\ \hat{e}_z \end{bmatrix} = N_S^T \begin{bmatrix} \hat{e}_r \\ \hat{e}_\theta \\ \hat{e}_\lambda \end{bmatrix}.$$

(2.110)

As these are inverse transformations to each other, we have

$$N_S^T = N_S^{-1},$$

(2.111)

and hence,

$$N_S N_S^T = I,$$

(2.112)

i.e., N_S is an orthogonal matrix.

Since $Or\theta\lambda$ is a curvilinear coordinate system, the derivatives of r, θ, and λ with respect to x, y, and z are no longer constants. For example, according to (2.106), we have

$$\frac{\partial r}{\partial x} = \frac{x}{(x^2 + y^2 + z^2)^{1/2}} = \sin\theta \cos\lambda.$$

(2.113)

Repeat all instances and then put together. We obtain

$$\frac{\partial(r,\theta,\lambda)}{\partial(x,y,z)} = \begin{bmatrix} \frac{\partial r}{\partial x} & \frac{\partial \theta}{\partial x} & \frac{\partial \lambda}{\partial x} \\ \frac{\partial r}{\partial y} & \frac{\partial \theta}{\partial y} & \frac{\partial \lambda}{\partial y} \\ \frac{\partial r}{\partial z} & \frac{\partial \theta}{\partial z} & \frac{\partial \lambda}{\partial z} \end{bmatrix} = \begin{bmatrix} \sin\theta\cos\lambda & \dfrac{\cos\theta\cos\lambda}{r} & -\dfrac{\sin\lambda}{r\sin\theta} \\ \sin\theta\cos\lambda & \dfrac{\cos\theta\sin\lambda}{r} & \dfrac{\cos\lambda}{r\sin\theta} \\ \cos\theta & -\dfrac{\sin\theta}{r} & 0 \end{bmatrix}. \quad (2.114)$$

Another property of a curvilinear coordinate system is that its base vectors are functions of coordinates, i.e., $\hat{e}_r$, $\hat{e}_\theta$, and $\hat{e}_\lambda$ are functions of r, θ, and λ. For example, the rate of change of $\hat{e}_\lambda$ as a function of λ is, according to (2.107),

$$\frac{\partial \hat{e}_\lambda}{\partial \lambda} = -\cos\lambda\,\hat{e}_x - \sin\lambda\,\hat{e}_y$$

$$= -\cos\lambda(\sin\theta\cos\lambda\,\hat{e}_r + \cos\theta\cos\lambda\,\hat{e}_\theta - \sin\lambda\,\hat{e}_\lambda)$$

$$\quad - \sin\lambda(\sin\theta\sin\lambda\,\hat{e}_r + \cos\theta\sin\lambda\,\hat{e}_\theta + \cos\lambda\,\hat{e}_\lambda)$$

$$= -\sin\theta\,\hat{e}_r - \cos\theta\,\hat{e}_\theta. \quad (2.115)$$

Repeat all instances and then put together. We obtain

$$\frac{\partial(\hat{e}_r,\hat{e}_\theta,\hat{e}_\lambda)}{\partial(r,\theta,\lambda)} = \begin{bmatrix} \frac{\partial \hat{e}_r}{\partial r} & \frac{\partial \hat{e}_\theta}{\partial r} & \frac{\partial \hat{e}_\lambda}{\partial r} \\ \frac{\partial \hat{e}_r}{\partial \theta} & \frac{\partial \hat{e}_\theta}{\partial \theta} & \frac{\partial \hat{e}_\lambda}{\partial \theta} \\ \frac{\partial \hat{e}_r}{\partial \lambda} & \frac{\partial \hat{e}_\theta}{\partial \lambda} & \frac{\partial \hat{e}_\lambda}{\partial \lambda} \end{bmatrix} = \begin{bmatrix} 0 & 0 & 0 \\ \hat{e}_\theta & -\hat{e}_r & 0 \\ \sin\theta\,\hat{e}_\lambda & \cos\theta\,\hat{e}_\lambda & -\sin\theta\,\hat{e}_r - \cos\theta\,\hat{e}_\theta \end{bmatrix}.$$

$$(2.116)$$

A vector $\vec{A}$ independent of coordinate system can be expressed in either $Oxyz$ or $Or\theta\lambda$,

$$\vec{A} = A_x\hat{e}_x + A_y\hat{e}_y + A_z\hat{e}_z = A_r\hat{e}_r + A_\theta\hat{e}_\theta + A_\lambda\hat{e}_\lambda. \quad (2.117)$$

By substituting (2.107) and (2.108) into it one at a time, we obtain the coordinate transformations of the components

$$\begin{cases} A_r = \sin\theta\cos\lambda\,A_x + \sin\theta\sin\lambda\,A_y + \cos\theta\,A_z, \\ A_\theta = \cos\theta\cos\lambda\,A_x + \cos\theta\sin\lambda\,A_y - \sin\theta\,,\ A_z \\ A_\lambda = -\sin\lambda\,A_x + \cos\lambda\,A_y, \end{cases} \quad (2.118)$$

and

$$\begin{cases} A_x = \sin\theta\cos\lambda\, A_r + \cos\theta\cos\lambda\, A_\theta - \sin\lambda\, A_\lambda\,, \\ A_y = \sin\theta\sin\lambda\, A_r + \cos\theta\sin\lambda\, A_\theta + \cos\lambda\, A_\lambda\,, \\ A_z = \cos\theta\, A_r - \sin\theta\, A_\theta\,. \end{cases} \tag{2.119}$$

These can be written briefly in matrix form as

$$\begin{bmatrix} A_r \\ A_\theta \\ A_\lambda \end{bmatrix} = N_S \begin{bmatrix} A_x \\ A_y \\ A_z \end{bmatrix}\,, \qquad \begin{bmatrix} A_x \\ A_y \\ A_z \end{bmatrix} = N_S^T \begin{bmatrix} A_r \\ A_\theta \\ A_\lambda \end{bmatrix}\,. \tag{2.120}$$

Finally, we express

$$\frac{\partial(A_x, A_y, A_z)}{\partial(r, \theta, \lambda)} = \begin{bmatrix} \dfrac{\partial A_x}{\partial r} & \dfrac{\partial A_y}{\partial r} & \dfrac{\partial A_z}{\partial r} \\[2mm] \dfrac{\partial A_x}{\partial \theta} & \dfrac{\partial A_y}{\partial \theta} & \dfrac{\partial A_z}{\partial \theta} \\[2mm] \dfrac{\partial A_x}{\partial \lambda} & \dfrac{\partial A_y}{\partial \lambda} & \dfrac{\partial A_z}{\partial \lambda} \end{bmatrix} \tag{2.121}$$

in terms of

$$\frac{\partial(A_r, A_\theta, A_\lambda)}{\partial(r, \theta, \lambda)} = \begin{bmatrix} \dfrac{\partial A_r}{\partial r} & \dfrac{\partial A_\theta}{\partial r} & \dfrac{\partial A_\lambda}{\partial r} \\[2mm] \dfrac{\partial A_r}{\partial \theta} & \dfrac{\partial A_\theta}{\partial \theta} & \dfrac{\partial A_\lambda}{\partial \theta} \\[2mm] \dfrac{\partial A_r}{\partial \lambda} & \dfrac{\partial A_\theta}{\partial \lambda} & \dfrac{\partial A_\lambda}{\partial \lambda} \end{bmatrix} \quad \text{and} \quad \begin{bmatrix} A_r \\ A_\theta \\ A_\lambda \end{bmatrix} \tag{2.122}$$

for later reference. We still take one column as an example

$$\begin{cases} \dfrac{\partial A_x}{\partial r} = \sin\theta\cos\lambda\dfrac{\partial A_r}{\partial r} + \cos\theta\cos\lambda\dfrac{\partial A_\theta}{\partial r} - \sin\lambda\dfrac{\partial A_\lambda}{\partial r}\,, \\[3mm] \dfrac{\partial A_x}{\partial \theta} = \sin\theta\cos\lambda\dfrac{\partial A_r}{\partial \theta} + \cos\theta\cos\lambda\dfrac{\partial A_\theta}{\partial \theta} - \sin\lambda\dfrac{\partial A_\lambda}{\partial \theta} \\[2mm] \qquad\quad + \cos\theta\cos\lambda A_r - \sin\theta\cos\lambda A_\theta\,, \\[3mm] \dfrac{\partial A_x}{\partial \lambda} = \sin\theta\cos\lambda\dfrac{\partial A_r}{\partial \lambda} + \cos\theta\cos\lambda\dfrac{\partial A_\theta}{\partial \lambda} - \sin\lambda\dfrac{\partial A_\lambda}{\partial \lambda} \\[2mm] \qquad\quad - \sin\theta\sin\lambda A_r - \cos\theta\sin\lambda A_\theta - \cos\lambda A_\lambda\,. \end{cases} \tag{2.123}$$

This can be written in matrix form as

$$
\begin{bmatrix} \dfrac{\partial A_x}{\partial r} \\[2mm] \dfrac{\partial A_x}{\partial \theta} \\[2mm] \dfrac{\partial A_x}{\partial \lambda} \end{bmatrix} = \begin{bmatrix} \dfrac{\partial A_r}{\partial r} & \dfrac{\partial A_\theta}{\partial r} & \dfrac{\partial A_\lambda}{\partial r} \\[2mm] \dfrac{\partial A_r}{\partial \theta} & \dfrac{\partial A_\theta}{\partial \theta} & \dfrac{\partial A_\lambda}{\partial \theta} \\[2mm] \dfrac{\partial A_r}{\partial \lambda} & \dfrac{\partial A_\theta}{\partial \lambda} & \dfrac{\partial A_\lambda}{\partial \lambda} \end{bmatrix} \begin{bmatrix} \sin\theta\cos\lambda \\[2mm] \cos\theta\cos\lambda \\[2mm] -\sin\lambda \end{bmatrix}
$$

$$
+ \begin{bmatrix} 0 \\ \cos\theta\cos\lambda \\ -\sin\theta\sin\lambda \end{bmatrix} A_r + \begin{bmatrix} 0 \\ -\sin\theta\cos\lambda \\ -\cos\theta\sin\lambda \end{bmatrix} A_\theta + \begin{bmatrix} 0 \\ 0 \\ -\cos\lambda \end{bmatrix} A_\lambda .
$$

$$(2.124)$$

Repeat for the other two columns, and then put together. We obtain

$$
\frac{\partial(A_x, A_y, A_z)}{\partial(r, \theta, \lambda)} = \frac{\partial(A_r, A_\theta, A_\lambda)}{\partial(r, \theta, \lambda)} N_S + N_S^r A^r + N_S^\theta A_\theta + N_S^\lambda A_\lambda ,
\tag{2.125}
$$

where

$$
N_S^r = \begin{bmatrix} 0 & 0 & 0 \\ \cos\theta\cos\lambda & \cos\theta\sin\lambda & -\sin\theta \\ -\sin\theta\sin\lambda & \sin\theta\cos\lambda & 0 \end{bmatrix} ,
\tag{2.126}
$$

$$
N_S^\theta = \begin{bmatrix} 0 & 0 & 0 \\ -\sin\theta\cos\lambda & -\sin\theta\sin\lambda & -\cos\theta \\ -\cos\theta\sin\lambda & \cos\theta\cos\lambda & 0 \end{bmatrix} ,
\tag{2.127}
$$

and

$$
N_S^\lambda = \begin{bmatrix} 0 & 0 & 0 \\ 0 & 0 & 0 \\ -\cos\lambda & -\sin\lambda & 0 \end{bmatrix} .
\tag{2.128}
$$

2.5.3   Gradients of Scalars and Vectors in Spherical Coordinates

In a spherical coordinate system, the gradients of scalars and vectors at a point P are defined based on a local Cartesian coordinate system $P\xi\eta\zeta$, where $\hat{e}_\xi$, $\hat{e}_\eta$, and $\hat{e}_\zeta$ coincide with $\hat{e}_r$, $\hat{e}_\theta$, and $\hat{e}_\lambda$ at the point P. See Fig. 2.6. When transformations between $Oxyz$ and $P\xi\eta\zeta$ are considered, we have

$$
N = N_S .
\tag{2.129}
$$

Fig. 2.6 Definition of
gradient in a spherical
coordinate system

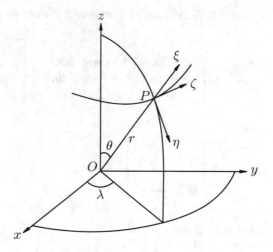

In the definition of gradients, $\hat{e}_\xi$, $\hat{e}_\eta$, and $\hat{e}_\zeta$ are considered fixed, just as if P is considered a fixed point. The three components of the gradient of a scalar in the spherical coordinate system are defined as

$$\nabla\phi = \begin{bmatrix} (\nabla\phi)_r \\ (\nabla\phi)_\theta \\ (\nabla\phi)_\lambda \end{bmatrix} = \begin{bmatrix} (\nabla\phi)_\xi \\ (\nabla\phi)_\eta \\ (\nabla\phi)_\zeta \end{bmatrix} = \frac{\partial\phi}{\partial(\xi, \eta, \zeta)}. \tag{2.130}$$

This implies that, even though $Or\theta\lambda$ is a curvilinear coordinate system, the three components of the gradient of a scalar are still the rates of change along straight lines. They can be obtained as follows:

$$\nabla\phi = N_S \frac{\partial\phi}{\partial(x, y, z)} = N_S \frac{\partial(r, \theta, \lambda)}{\partial(x, y, z)} \frac{\partial\phi}{\partial(r, \theta, \lambda)} = M_S \frac{\partial\phi}{\partial(r, \theta, \lambda)}, \tag{2.131}$$

where

$$M_S = N_S \frac{\partial(r, \theta, \lambda)}{\partial(x, y, z)}. \tag{2.132}$$

Substitute (2.109) and (2.114) into it. We obtain

$$
M_S = \begin{bmatrix} \sin\theta\cos\lambda & \sin\theta\sin\lambda & \cos\theta \\ \cos\theta\cos\lambda & \cos\theta\sin\lambda & -\sin\theta \\ -\sin\lambda & \cos\lambda & 0 \end{bmatrix} \begin{bmatrix} \sin\theta\cos\lambda & \dfrac{\cos\theta\cos\lambda}{r} & -\dfrac{\sin\lambda}{r\sin\theta} \\ \sin\theta\cos\lambda & \dfrac{\cos\theta\sin\lambda}{r} & \dfrac{\cos\lambda}{r\sin\theta} \\ \cos\theta & -\dfrac{\sin\theta}{r} & 0 \end{bmatrix}
$$

$$
= \begin{bmatrix} 1 & 0 & 0 \\ 0 & \dfrac{1}{r} & 0 \\ 0 & 0 & \dfrac{1}{r\sin\theta} \end{bmatrix}. \tag{2.133}
$$

Hence, finally,

$$
\nabla\phi = M_S \frac{\partial\phi}{\partial(r,\theta,\lambda)} = \begin{bmatrix} 1 & 0 & 0 \\ 0 & \dfrac{1}{r} & 0 \\ 0 & 0 & \dfrac{1}{r\sin\theta} \end{bmatrix} \begin{bmatrix} \dfrac{\partial\phi}{\partial r} \\ \dfrac{\partial\phi}{\partial\theta} \\ \dfrac{\partial\phi}{\partial\lambda} \end{bmatrix} = \begin{bmatrix} \dfrac{\partial\phi}{\partial r} \\ \dfrac{1}{r}\dfrac{\partial\phi}{\partial\theta} \\ \dfrac{1}{r\sin\theta}\dfrac{\partial\phi}{\partial\lambda} \end{bmatrix}. \tag{2.134}
$$

We note that this result can be inferred easily by observing that the lengths of infinitesimal line segments along $\hat{e}_r$, $\hat{e}_\theta$, and $\hat{e}_\lambda$ are dr, $rd\theta$, and $r\sin\theta d\lambda$, respectively.

Similar to the case of a scalar, the nine components of the gradient of a vector are defined as

$$
\nabla\vec{A} = \begin{bmatrix} (\nabla\vec{A})_{rr} & (\nabla\vec{A})_{r\theta} & (\nabla\vec{A})_{r\lambda} \\ (\nabla\vec{A})_{\theta r} & (\nabla\vec{A})_{\theta\eta} & (\nabla\vec{A})_{\theta\lambda} \\ (\nabla\vec{A})_{\lambda r} & (\nabla\vec{A})_{\lambda\eta} & (\nabla\vec{A})_{\lambda\lambda} \end{bmatrix} = \begin{bmatrix} (\nabla\vec{A})_{\xi\xi} & (\nabla\vec{A})_{\xi\eta} & (\nabla\vec{A})_{\xi\zeta} \\ (\nabla\vec{A})_{\eta\xi} & (\nabla\vec{A})_{\eta\eta} & (\nabla\vec{A})_{\eta\zeta} \\ (\nabla\vec{A})_{\zeta\xi} & (\nabla\vec{A})_{\zeta\eta} & (\nabla\vec{A})_{\zeta\zeta} \end{bmatrix}
$$

$$
= \frac{\partial(A_\xi, A_\eta, A_\zeta)}{\partial(\xi,\eta,\zeta)}. \tag{2.135}
$$

Their expressions in terms of spherical coordinates can be obtained as

$$
\nabla\vec{A} = \frac{\partial(A_\xi, A_\eta, A_\zeta)}{\partial(\xi,\eta,\zeta)} = N_S \frac{\partial(A_x, A_y, A_z)}{\partial(x,y,z)} N_S^T = N_S \frac{\partial(r,\theta,\lambda)}{\partial(x,y,z)} \frac{\partial(A_x, A_y, A_z)}{\partial(r,\theta,\lambda)} N_S^T
$$

$$
= M_S \left[\frac{\partial(A_r, A_\theta, A_\lambda)}{\partial(r,\theta,\lambda)} N_S + N_S^r A_r + N_S^\theta A_\theta + N_S^\lambda A_\lambda \right] N_S^T
$$

$$
= M_S \frac{\partial(A_r, A_\theta, A_\lambda)}{\partial(r,\theta,\lambda)} + M_S N_S^r N_S^T A_r + M_S N_S^\theta N_S^T A_\theta + M_S N_S^\lambda N_S^T A_\lambda. \tag{2.136}
$$

Substitute (2.133), (2.109), and (2.126) to (2.128) into (2.136). We finally obtain

$$
\nabla \vec{A} =
\begin{bmatrix}
\dfrac{\partial A_r}{\partial r} & \dfrac{\partial A_\theta}{\partial r} \\[2mm]
\dfrac{1}{r}\dfrac{\partial A_r}{\partial \theta} - \dfrac{A_\theta}{r} & \dfrac{1}{r}\dfrac{\partial A_\theta}{\partial \theta} + \dfrac{A_r}{r} \\[2mm]
\dfrac{1}{r\sin\theta}\dfrac{\partial A_r}{\partial \lambda} - \dfrac{A_\lambda}{r} & \dfrac{1}{r\sin\theta}\dfrac{\partial A_\theta}{\partial \lambda} - \dfrac{\cot\theta\,A_\lambda}{r}
\end{bmatrix}
$$

$$
\times
\begin{bmatrix}
\dfrac{\partial A_\lambda}{\partial r} \\[2mm]
\dfrac{1}{r}\dfrac{\partial A_\lambda}{\partial \theta} \\[2mm]
\dfrac{1}{r\sin\theta}\dfrac{\partial A_\lambda}{\partial \lambda} + \dfrac{A_r}{r} + \dfrac{\cot\theta\,A_\theta}{r}
\end{bmatrix}. \tag{2.137}
$$

2.5.4 Divergence and Curl of a Vector

The divergence defined in $Oxyz$ and $P\xi\eta\zeta$ are

$$
\nabla \cdot \vec{A} = \frac{\partial A_x}{\partial x} + \frac{\partial A_y}{\partial y} + \frac{\partial A_z}{\partial z}
\quad \text{and} \quad
\nabla \cdot \vec{A} = \frac{\partial A_\xi}{\partial \xi} + \frac{\partial A_\eta}{\partial \eta} + \frac{\partial A_\zeta}{\partial \zeta}, \tag{2.138}
$$

respectively. Clearly, $\nabla \cdot \vec{A}$ is the sum of the three diagonal terms of $\nabla \vec{A}$. Hence,

$$
\nabla \cdot \vec{A} = \frac{\partial A_r}{\partial r} + \frac{1}{r}\frac{\partial A_\theta}{\partial \theta} + \frac{1}{r\sin\theta}\frac{\partial A_\lambda}{\partial \lambda} + 2\frac{A_r}{r} + \frac{\cot\theta\,A_\theta}{r}. \tag{2.139}
$$

The curl defined in $Oxyz$ and $P\xi\eta\zeta$ are, in terms of components,

$$
\nabla \times \vec{A} =
\begin{bmatrix}
\dfrac{\partial A_z}{\partial y} - \dfrac{\partial A_y}{\partial z} \\[2mm]
\dfrac{\partial A_x}{\partial z} - \dfrac{\partial A_z}{\partial x} \\[2mm]
\dfrac{\partial A_y}{\partial x} - \dfrac{\partial A_x}{\partial y}
\end{bmatrix}
\quad \text{and} \quad
\nabla \times \vec{A} =
\begin{bmatrix}
\dfrac{\partial A_\zeta}{\partial \eta} - \dfrac{\partial A_\eta}{\partial \zeta} \\[2mm]
\dfrac{\partial A_\xi}{\partial \zeta} - \dfrac{\partial A_\zeta}{\partial \xi} \\[2mm]
\dfrac{\partial A_\eta}{\partial \xi} - \dfrac{\partial A_\xi}{\partial \eta}
\end{bmatrix}, \tag{2.140}
$$

respectively. Each term corresponds to an element of $\nabla \vec{A}$. We obtain

$$
\nabla \times \vec{A} =
\begin{bmatrix}
\dfrac{1}{r}\dfrac{\partial A_\lambda}{\partial \theta} - \dfrac{1}{r\sin\theta}\dfrac{\partial A_\theta}{\partial \lambda} + \dfrac{\cot\theta\,A_\lambda}{r} \\[2mm]
\dfrac{1}{r\sin\theta}\dfrac{\partial A_r}{\partial \lambda} - \dfrac{\partial A_\lambda}{\partial r} - \dfrac{A_\lambda}{r} \\[2mm]
\dfrac{\partial A_\theta}{\partial r} - \dfrac{1}{r}\dfrac{\partial A_r}{\partial \theta} + \dfrac{A_\theta}{r}
\end{bmatrix}. \tag{2.141}
$$

2.5.5 Application to the Gravitational Field

For the gravitational field, (2.134) implies that, in spherical coordinates, the gravitational attraction is the gradient of the gravitational potential in the form of

$$
\overrightarrow{F} =
\begin{bmatrix}
\dfrac{\partial V}{\partial r} \\[2mm]
\dfrac{1}{r}\dfrac{\partial V}{\partial \theta} \\[2mm]
\dfrac{1}{r\sin\theta}\dfrac{\partial V}{\partial \lambda}
\end{bmatrix}.
\tag{2.142}
$$

The substitution of this formula into (2.137) yields the gradient of the gravitational attraction expressed in terms of the gravitational potential

$$
\nabla\overrightarrow{F} =
\begin{bmatrix}
\dfrac{\partial^2 V}{\partial r^2} & \dfrac{1}{r}\dfrac{\partial^2 V}{\partial r\partial\theta}-\dfrac{1}{r^2}\dfrac{\partial V}{\partial\theta} \\[3mm]
\dfrac{1}{r}\dfrac{\partial^2 V}{\partial r\partial\theta}-\dfrac{1}{r^2}\dfrac{\partial V}{\partial\theta} & \dfrac{1}{r^2}\dfrac{\partial^2 V}{\partial\theta^2}+\dfrac{1}{r}\dfrac{\partial V}{\partial r} \\[3mm]
\dfrac{1}{r\sin\theta}\dfrac{\partial^2 V}{\partial r\partial\lambda} & \dfrac{1}{r^2\sin\theta}\dfrac{\partial V}{\partial\lambda} \quad \dfrac{1}{r^2\sin\theta}\dfrac{\partial^2 V}{\partial\theta\partial\lambda}-\dfrac{\cos\theta}{r^2\sin^2\theta}\dfrac{\partial V}{\partial\lambda}
\end{bmatrix}
$$

$$
\times
\begin{bmatrix}
\dfrac{1}{r\sin\theta}\dfrac{\partial^2 V}{\partial r\partial\lambda}-\dfrac{1}{r^2\sin\theta}\dfrac{\partial V}{\partial\lambda} \\[3mm]
\dfrac{1}{r^2\sin\theta}\dfrac{\partial^2 V}{\partial\theta\partial\lambda}-\dfrac{\cos\theta}{r^2\sin^2\theta}\dfrac{\partial V}{\partial\lambda} \\[3mm]
\dfrac{1}{r^2\sin^2\theta}\dfrac{\partial^2 V}{\partial\lambda^2}+\dfrac{1}{r}\dfrac{\partial V}{\partial r}+\dfrac{\cot\theta}{r^2}\dfrac{\partial V}{\partial\theta}
\end{bmatrix}.
\tag{2.143}
$$

This is in fact the second order gradient of the gravitational potential in spherical coordinates.

With $\overrightarrow{F} = \nabla V$, it can be verified that $\nabla\cdot\overrightarrow{F} = \Delta V$ and $\nabla\times\overrightarrow{F} = 0$ in both $Oxyz$ and $O\xi\eta\zeta$ according to (2.93), (2.138), and (2.140). The gradient, divergence, and curl in $Or\theta\lambda$ are defined in $O\xi\eta\zeta$, but expressed in terms of r, θ, and λ. Hence, the righteousness of these relations in $O\xi\eta\zeta$ implies that they also hold in $Or\theta\lambda$. Actually, they hold in any coordinate system.

The Laplacian applied to a scalar in spherical coordinates can be obtained by replacing $\vec{A}$ in (2.139) by $\vec{F}$ provided in (2.142)

$$
\begin{aligned}
\Delta V &= \frac{\partial}{\partial r}\left(\frac{\partial V}{\partial r}\right) + \frac{1}{r}\frac{\partial}{\partial \theta}\left(\frac{1}{r}\frac{\partial V}{\partial \theta}\right) + \frac{1}{r\sin\theta}\frac{\partial}{\partial \lambda}\left(\frac{1}{r\sin\theta}\frac{\partial V}{\partial \lambda}\right) \\
&\quad + 2\frac{1}{r}\left(\frac{\partial V}{\partial r}\right) + \frac{\cot\theta}{r}\left(\frac{1}{r}\frac{\partial V}{\partial \theta}\right) \\
&= \frac{\partial^2 V}{\partial r^2} + \frac{2}{r}\frac{\partial V}{\partial r} + \frac{1}{r^2}\left(\frac{\partial^2 V}{\partial \theta^2} + \cot\theta\frac{\partial V}{\partial \theta} + \frac{1}{\sin^2\theta}\frac{\partial^2 V}{\partial \lambda^2}\right),
\end{aligned} \tag{2.144}
$$

which is the sum of the diagonal terms of (2.143). A more straightforward derivation of this formula will be provided in the next chapter.

Reference

Guan, Z. L., & Ning, J. S. (1981). *Earth's shape and external gravity field* (Vol. 2) (In Chinese). Press of Surveying and Mapping, Beijing, China.

Further Reading

Blakely, R. J. (1996). *Potential theory in gravity and magnetic applications* (Chapters 2 and 3). Cambridge University Press.

Otero, J. (1998). A uniqueness theorem for a Robin boundary value problem of physical geodesy. *Quarterly of Applied Mathematics, 56*(2), 245–257.

Spherical Harmonics

3

Abstract

This chapter systematically presents the Legendre and associated Legendre functions and spherical harmonics, which should be considered as essential knowledge of geodesy. The approach is to derive a unique and finite solution of Laplace's equation in the spherical coordinate system based on Fourier's method of variable separation. The expression of the Earth's external gravitational potential, along with the gravitational attraction and its gradient, is formulated in terms of a spherical harmonic series. In particular, the azimuthal average of spherical harmonics and the convergence of the spherical harmonic series of a function defined on a sphere, i.e., the Laplace series, are formulated based on fundamental calculus. As an illustrative example of application, the Poisson integral is derived by solving Dirichlet's boundary value problem outside a sphere using spherical harmonic series.

3.1 Separation of Variables of Laplace's Equation in Spherical Coordinates

3.1.1 Laplace's Equation in Spherical Coordinates

The Earth's shape is very close to a sphere, and the adoption of a spherical coordinate system is natural. Spherical coordinates are related to Cartesian coordinates in the form of

$$x = r \sin\theta \cos\lambda, \quad y = r \sin\theta \sin\lambda, \quad z = r \cos\theta. \tag{3.1}$$

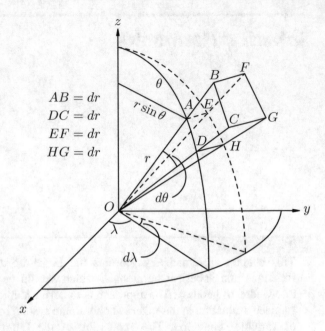

Fig. 3.1 Derivation of Laplace's equation in spherical coordinates

$$AB = dr$$
$$DC = dr$$
$$EF = dr$$
$$HG = dr$$

The most straightforward way to derive Laplace's equation in spherical coordinates is to use the chain rule of derivative based on (3.1), e.g.,

$$\frac{\partial V}{\partial x} = \frac{\partial V}{\partial r}\frac{\partial r}{\partial x} + \frac{\partial V}{\partial \theta}\frac{\partial \theta}{\partial x} + \frac{\partial V}{\partial \lambda}\frac{\partial \lambda}{\partial x}. \tag{3.2}$$

This derivation is lengthy and tedious.

In Sect. 2.5, we already provided a derivation while formulating the gradient of the gravitational attraction. There is a need to read the whole section to understand the derivation. However, that section is relatively independent of the other parts of the book. The only relevant content is the expression of the gradient of the gravitational attraction in spherical harmonics in Sect. 3.5.5. Here we provide a more concise derivation for the readers who decide to jump the gradient of the gravitational attraction.

We use Green's second identity for an internal domain (2.14). Assuming $U = 1$, we have

$$\int_{v_i} \Delta V dv = \int_S \frac{\partial V}{\partial n} dS, \tag{3.3}$$

where v_i is the interior of the boundary S and $\hat{e}_n$ the unit outward normal vector of S. As shown in Fig. 3.1, we apply this equation to the infinitesimal hexahedron $ABCDEFGH$, whose surfaces are all equi-value surfaces of coordinates. $AEHD$ and $BFGC$ are equi-r surfaces, where the coordinate r is denoted as r and $r + dr$, respectively; $AEFB$ and $DHGC$ are equi-θ surfaces, where the coordinate θ is denoted as θ and $\theta + d\theta$, respectively; $ABCD$ and $EFGH$ are equi-λ surfaces, where the coordinate λ is denoted as λ and $\lambda + d\lambda$, respectively. By applying (3.3) to this hexahedron, we obtain

$$\Delta V = \frac{1}{v} \sum_{i=1}^{6} \left(\frac{\partial V}{\partial n} \right)_i S_i , \tag{3.4}$$

where v is the volume of the hexahedron, $i = 1, 2, \ldots, 6$ represent the 6 surfaces, $(\partial V / \partial n)_i$ is the value of $\partial V / \partial n$ on the ith surface, and S_i is the area of the ith surface.

On $AEHD$, $\hat{e}_n$ is in the opposite direction of $\hat{e}_r$. Hence,

$$\left(\frac{\partial V}{\partial n} \right)_{AEHD} = -\frac{\partial V}{\partial r} . \tag{3.5}$$

The area of this surface is $r^2 \sin\theta d\theta d\lambda$. Hence,

$$\left(\frac{\partial V}{\partial n} \right)_{AEHD} S_{AEHD} = -r^2 \sin\theta \frac{\partial V}{\partial r} d\theta d\lambda . \tag{3.6}$$

The surface $BFGC$ has the same θ, λ and $d\theta$, $d\lambda$ as $AEHD$. The difference is that the coordinate r has an increment dr. Considering that, in contrary to $AEHD$, $\hat{e}_n$ is in the same direction of $\hat{e}_r$, we have

$$\left(\frac{\partial V}{\partial n} \right)_{BFGC} S_{BFGC} = r^2 \sin\theta \frac{\partial V}{\partial r} d\theta d\lambda + \frac{\partial}{\partial r} \left(r^2 \sin\theta \frac{\partial V}{\partial r} \right) dr d\theta d\lambda . \tag{3.7}$$

On $AEFB$, $\hat{e}_n$ is in the opposite direction of $\hat{e}_\theta$. As the length of an infinitesimal line segment corresponding to $d\theta$ along $\hat{e}_\theta$ is $rd\theta$, we have

$$\left(\frac{\partial V}{\partial n} \right)_{AEFB} = -\frac{1}{r} \frac{\partial V}{\partial \theta} . \tag{3.8}$$

The area of this surface is $r \sin\theta dr d\lambda$. Hence,

$$\left(\frac{\partial V}{\partial n} \right)_{AEFB} S_{AEFB} = -\sin\theta \frac{\partial V}{\partial \theta} dr d\lambda . \tag{3.9}$$

The surface $DHGC$ has the same r, λ and dr, $d\lambda$ as $AEFB$. The difference is that the coordinate θ has an increment $d\theta$. Considering that, in contrary to $AEFB$, $\hat{e}_n$ is in the same direction of $\hat{e}_\theta$, we have

$$\left(\frac{\partial V}{\partial n}\right)_{BFGC} S_{BFGC} = \sin\theta \frac{\partial V}{\partial \theta} dr d\lambda + \frac{\partial}{\partial \theta}\left(\sin\theta \frac{\partial V}{\partial \theta}\right) dr d\theta d\lambda. \qquad (3.10)$$

On $ABCD$, $\hat{e}_n$ is in the opposite direction of $\hat{e}_\lambda$. As the length of an infinitesimal line segment corresponding to $d\lambda$ along $\hat{e}_\lambda$ is $r\sin\theta d\lambda$, we have

$$\left(\frac{\partial V}{\partial n}\right)_{ABCD} = -\frac{1}{r\sin\theta}\frac{\partial V}{\partial \lambda}. \qquad (3.11)$$

The area of this surface is $r dr d\theta$. Hence,

$$\left(\frac{\partial V}{\partial n}\right)_{ABCD} S_{ABCD} = -\frac{1}{\sin\theta}\frac{\partial V}{\partial \lambda} dr d\theta. \qquad (3.12)$$

The surface $EFGH$ has the same r, θ and dr, $d\theta$ as $ABCD$. The difference is that the coordinate λ has an increment $d\lambda$. Considering that, in contrary to $ABCD$, $\hat{e}_n$ is in the same direction of $\hat{e}_\lambda$, we have

$$\left(\frac{\partial V}{\partial n}\right)_{EFGH} S_{EFGH} = \frac{1}{\sin\theta}\frac{\partial V}{\partial \lambda} dr d\theta + \frac{\partial}{\partial \lambda}\left(\frac{1}{\sin\theta}\frac{\partial V}{\partial \lambda}\right) dr d\theta d\lambda. \qquad (3.13)$$

With the volume of the hexahedron to be $v = r^2 \sin\theta dr d\theta d\lambda$, by substituting (3.6), (3.7), (3.9), (3.10), (3.12), and (3.13) into (3.4), we obtain

$$\Delta V = \frac{1}{r^2 \sin\theta}\left[\frac{\partial}{\partial r}\left(r^2 \sin\theta \frac{\partial V}{\partial r}\right) + \frac{\partial}{\partial \theta}\left(\sin\theta \frac{\partial V}{\partial \theta}\right) + \frac{\partial}{\partial \lambda}\left(\frac{1}{\sin\theta}\frac{\partial V}{\partial \lambda}\right)\right]$$

$$= \frac{1}{r^2}\frac{\partial}{\partial r}\left(r^2 \frac{\partial V}{\partial r}\right) + \frac{1}{r^2}\left[\frac{1}{\sin\theta}\frac{\partial}{\partial \theta}\left(\sin\theta \frac{\partial V}{\partial \theta}\right) + \frac{1}{\sin^2\theta}\frac{\partial^2 V}{\partial \lambda^2}\right]. \qquad (3.14)$$

Laplace's equation in spherical coordinates is then

$$\frac{1}{r^2}\frac{\partial}{\partial r}\left(r^2 \frac{\partial V}{\partial r}\right) + \frac{1}{r^2}\left[\frac{1}{\sin\theta}\frac{\partial}{\partial \theta}\left(\sin\theta \frac{\partial V}{\partial \theta}\right) + \frac{1}{\sin^2\theta}\frac{\partial^2 V}{\partial \lambda^2}\right] = 0. \qquad (3.15)$$

3.1.2 Method of Solution by Separating Variables

We express the solution of Laplace's partial differential equation (3.15) as a product of functions depending on different variables to reduce the partial differential equation to a number of ordinary differential equations depending on different variables. This approach of separating variables is called Fourier's method.

First, we separate the dependence on r from those on θ and λ by assuming

$$V(r, \theta, \lambda) = f(r)Y(\theta, \lambda). \tag{3.16}$$

The substitution of it into (3.15) yields

$$\frac{Y}{r^2}\frac{d}{dr}\left(r^2\frac{df}{dr}\right) + \frac{f}{r^2}\left[\frac{1}{\sin\theta}\frac{\partial}{\partial\theta}\left(\sin\theta\frac{\partial Y}{\partial\theta}\right) + \frac{1}{\sin^2\theta}\frac{\partial^2 Y}{\partial\lambda^2}\right] = 0. \tag{3.17}$$

By multiplying both sides by $r^2/(fY)$ and then moving the second term to the right-hand side, we obtain

$$\frac{1}{f}\frac{d}{dr}\left(r^2\frac{df}{dr}\right) = -\frac{1}{Y}\left[\frac{1}{\sin\theta}\frac{\partial}{\partial\theta}\left(\sin\theta\frac{\partial Y}{\partial\theta}\right) + \frac{1}{\sin^2\theta}\frac{\partial^2 Y}{\partial\lambda^2}\right]. \tag{3.18}$$

In this equation, the left-hand side is a function of r, while the right-hand side is a function of θ and λ. The only possibility for the equality to hold is that the left and right-hand sides are equal to the same constant, say α. We thus obtain

$$\frac{d}{dr}\left(r^2\frac{df}{dr}\right) - \alpha f = 0, \tag{3.19}$$

$$\frac{1}{Y}\left[\frac{1}{\sin\theta}\frac{\partial}{\partial\theta}\left(\sin\theta\frac{\partial Y}{\partial\theta}\right) + \frac{1}{\sin^2\theta}\frac{\partial^2 Y}{\partial\lambda^2}\right] + \alpha = 0. \tag{3.20}$$

In order to separate the dependence on θ and λ of (3.20), we assume

$$Y(\theta, \lambda) = B(\theta)L(\lambda). \tag{3.21}$$

This leads to

$$\frac{1}{B}\frac{1}{\sin\theta}\frac{d}{d\theta}\left(\sin\theta\frac{dB}{d\theta}\right) + \frac{1}{L}\frac{1}{\sin^2\theta}\frac{d^2 L}{d\lambda^2} + \alpha = 0. \tag{3.22}$$

By multiplying both sides by $\sin^2\theta$ and then moving the second term to the right-hand side, we obtain

$$\frac{1}{B}\sin\theta\frac{d}{d\theta}\left(\sin\theta\frac{dB}{d\theta}\right) + \alpha\sin^2\theta = -\frac{1}{L}\frac{d^2 L}{d\lambda^2}. \tag{3.23}$$

This is another equation with left and right-hand sides depending on different variables. In order for the equality to hold, the left and right-hand sides must equal the same constant, say β. We obtain

$$\sin\theta \frac{d}{d\theta}\left(\sin\theta \frac{dB}{d\theta}\right) + (\alpha\sin^2\theta - \beta)B = 0, \qquad (3.24)$$

$$\frac{d^2L}{d\lambda^2} + \beta L = 0. \qquad (3.25)$$

We have thus separated Laplace's partial differential equation into three ordinary differential equations (3.19), (3.24), and (3.25), which we are going to solve.

For convenience, we substitute the variable θ in (3.24) by $x = \cos\theta$. With $\sin\theta = (1 - x^2)^{1/2}$ and $d/d\theta = (dx/d\theta)(d/dx) = -\sin\theta(d/dx) = -(1 - x^2)^{1/2}(d/dx)$, (3.24) becomes

$$(1 - x^2)\frac{d}{dx}\left[(1 - x^2)\frac{dB}{dx}\right] + [\alpha(1 - x^2) - \beta]B = 0. \qquad (3.26)$$

By dividing both sides by $1 - x^2$, we finally obtain

$$\frac{d}{dx}\left[(1 - x^2)\frac{dB}{dx}\right] + \left(\alpha - \frac{\beta}{1 - x^2}\right)B = 0, \qquad (3.27)$$

which is referred to as associated Legendre equation. When $\beta = 0$, the associated Legendre equation degenerates to the Legendre equation

$$\frac{d}{dx}\left[(1 - x^2)\frac{dB^0}{dx}\right] + \alpha B^0 = 0. \qquad (3.28)$$

3.1.3 The Concept of Eigenvalue Problem

While separating variables, we have introduced two constant parameters α and β. Conceptually, no constraint has been imposed on these parameters, and thus they are considered arbitrary. However, when the background physics of the problem is considered, these parameters can no longer be taken arbitrarily.

We first solve (3.25) and meanwhile determine β. We know that the gravitational potential is a single-valued function. For any fixed values of r and θ, $2k\pi + \lambda$ correspond to the same location in space when $k = 0, 1, 2, \ldots$, and hence the gravitational potential must have the same value. This requires the function L depending on the coordinate λ to have the property

$$L(\lambda) = L(2k\pi + \lambda). \qquad (3.29)$$

This property implies that L must be a periodical function with a period of 2π, thus being referred to as periodical condition. While solving (3.25), the value of β

must be chosen so that the solution satisfies the periodical condition. It can be easily verified that (3.25) has solutions satisfying the periodical condition only when

$$\beta = k^2,$$ (3.30)

and there are two linearly independent solutions,

$$L_1 = \cos k\lambda, \quad L_2 = \sin k\lambda.$$ (3.31)

When the sign of k in these solutions is reversed, the results are still solutions, but these are linearly dependent on the original ones. Hence, we assume $k \geq 0$ hereafter.

The problem of solving a differential equation with a parameter for non-trivial solutions when the parameter takes certain values is referred to as an eigenvalue problem. The solutions are called eigenfunctions, and the values of the parameter are called eigenvalues. For the above problem, $\cos k\lambda$ and $\sin k\lambda$ are eigenfunctions, and k^2 are eigenvalues. In this problem, one eigenvalue corresponds to two eigenfunctions when $k \neq 0$.

Substitute (3.30) into (3.27) and (3.28). We obtain

$$\frac{d}{dx}\left[(1-x^2)\frac{dB}{dx}\right] + \left(\alpha - \frac{k^2}{1-x^2}\right)B = 0,$$ (3.32)

$$\frac{d}{dx}\left[(1-x^2)\frac{dB^0}{dx}\right] + \alpha B^0 = 0.$$ (3.33)

The later is copied to here for convenience of further reference. By definition, the domain of x is $[-1, 1]$. As the gravitational potential must be finite in the domain, the function B must be finite in $[-1, 1]$, which is a kind of restriction called finite condition. In subsequent sections, we will solve the Legendre and associated Legendre equations under the finite condition and meanwhile determine the values that the parameter α may take. These are also eigenvalue problems.

3.2 Legendre Function

3.2.1 Power Series Solutions of the Legendre Equation

We first solve the simpler Legendre equation, which we rewrite here in the form of

$$(1-x^2)\frac{d^2B^0}{dx^2} - 2x\frac{dB^0}{dx} + \alpha B^0 = 0.$$ (3.34)

Let the power series solution to be

$$B^0 = \sum_{i=0}^{\infty} C_i x^i.$$ (3.35)

We have

$$\frac{dB^0}{dx} = \sum_{i=0}^{\infty} i C_i x^{i-1}, \tag{3.36}$$

$$\frac{d^2 B^0}{dx^2} = \sum_{i=0}^{\infty} i(i-1) C_i x^{i-2}. \tag{3.37}$$

By substituting the above three formulae into (3.34), we obtain

$$\sum_{i=0}^{\infty} i(i-1) C_i x^{i-2} + \sum_{i=0}^{\infty} [-i(i-1)C_i - 2i C_i + \alpha C_i] x^i = 0. \tag{3.38}$$

In the first term, we use a "new" i to represent the "old" $i-2$ to obtain

$$\sum_{i=0}^{\infty} i(i-1) C_i x^{i-2} = \sum_{i=0}^{\infty} (i+2)(i+1) C_{i+2} x^i. \tag{3.39}$$

Substitute it back to (3.38). We obtain

$$\sum_{i=0}^{\infty} \{(i+2)(i+1)C_{i+2} - [i(i+1) - \alpha]C_i\} x^i = 0. \tag{3.40}$$

Evidently, in order for the power series to vanish for any x, all the coefficients must vanish

$$(i+2)(i+1)C_{i+2} - [i(i+1) - \alpha]C_i = 0. \tag{3.41}$$

A recurrence formula for the coefficients can then be derived

$$C_{i+2} = \frac{i(i+1) - \alpha}{(i+2)(i+1)} C_i. \tag{3.42}$$

The Legendre equation is a second order homogeneous linear ordinary differential equation, which has two linearly independent solutions. We see that when $C_0 \neq 0$ and $C_1 = 0$, $C_{2i+1} = 0$ and when $C_0 = 0$ and $C_1 \neq 0$, $C_{2i} = 0$. Hence, the two linearly independent solutions can be written as

$$B_1^0 = \sum_{i=0}^{\infty} C_{2i} x^{2i}, \quad B_2^0 = \sum_{i=0}^{\infty} C_{2i+1} x^{2i+1}. \tag{3.43}$$

The coefficients C_{2i} and C_{2i+1} are computed from C_0 and C_1, respectively, using (3.42). It is evident that B_1^0 includes only even power terms, and B_2^0 includes only odd power terms.

3.2.2 Convergence of the Power Series Solutions

We now study the convergence properties of the two power series in (3.43). If a series converges to a finite value, it is said to be finite. Otherwise, the series is infinite or does not converge.

It is shown in most calculus textbooks that the convergence radius of a power series like (3.35) is

$$\delta = \lim_{i \to \infty} \left| \frac{C_i}{C_{i+1}} \right|, \tag{3.44}$$

if the limit exists.

In (3.43), we can take B_1^0 as a power series of x^2 and B_2^0 the product between x and a power series of x^2. Hence, for these two power series, the ratios of coefficients corresponding to that in (3.44) are the ratios between successive coefficients. As the coefficients of these two series have the same recurrence formula (3.42), they have the same convergence radius

$$\delta = \lim_{i \to \infty} \left| \frac{C_i}{C_{i+2}} \right| = \lim_{i \to \infty} \left| \frac{(i+2)(i+1)}{i(i+1) + \alpha} \right| = 1. \tag{3.45}$$

This means that in $(-1, 1)$, both the power series in (3.43) converges to a finite value; hence, both solutions are finite.

The problem is that when $x = \pm 1$, both series in (3.43) are infinite, meaning that the solutions are infinite. In order to prove this, we first estimate the magnitudes of the coefficients C_i. According to (3.42),

$$C_i = \frac{(i-2)(i-1) - \alpha}{i(i-1)} C_{i-2} = \frac{1}{i} \left[1 - \frac{\alpha}{(i-1)(i-2)} \right] (i-2)C_{i-2}. \tag{3.46}$$

Using this formula repeatedly, we get

$$C_{i-2} = \frac{1}{i-2} \left[1 - \frac{\alpha}{(i-3)(i-4)} \right] (i-4)C_{i-4}, \tag{3.47}$$

$$\cdots \cdots$$

$$C_{N+2} = \frac{1}{N+2} \left[1 - \frac{\alpha}{(N+1)N} \right] NC_N. \tag{3.48}$$

By substituting (3.47) to (3.48) into (3.46), we obtain

$$C_i = \frac{1}{i}\left[1 - \frac{\alpha}{(i-1)(i-2)}\right]\left[1 - \frac{\alpha}{(i-3)(i-4)}\right]\cdots\left[1 - \frac{\alpha}{(N+1)N}\right]NC_N,$$

$$(3.49)$$

where i must be large enough so that $i - 2 \geq N$. Evidently, the values within all the brackets are positive and smaller than 1 when $\alpha > 0$ and larger than 1 when $\alpha < 0$. Therefore, if we can show that both series B_1^0 and B_2^0 are infinite when $\alpha > 0$, we can also conclude that they are infinite when $\alpha < 0$. Hence, we assume $\alpha > 0$ below.

We first analyze the orders of magnitude of the coefficients. Define

$$\Pi = \left[1 - \frac{\alpha}{(i-1)(i-2)}\right]\left[1 - \frac{\alpha}{(i-3)(i-4)}\right]\cdots\left[1 - \frac{\alpha}{(N+1)N}\right].$$

$$(3.50)$$

We chose a value of N sufficiently large, so that $1 - \alpha/[(N+1)N] > 0$. Then, for any $i + 2 > N$, we have

$$\Pi > \left[1 - \frac{\alpha}{(N+1)N}\right]\cdots\left[1 - \frac{\alpha}{(i-3)(i-4)}\right]\left[1 - \frac{\alpha}{(i-1)(i-2)}\right]\cdots.$$

$$(3.51)$$

Take the logarithm on both sides

$$\ln\Pi > \ln\left[1 - \frac{\alpha}{(N+1)N}\right] + \cdots + \ln\left[1 - \frac{\alpha}{(i-3)(i-4)}\right]$$

$$+ \ln\left[1 - \frac{\alpha}{(i-1)(i-2)}\right] + \cdots.$$

$$(3.52)$$

As

$$\lim_{i\to\infty}\frac{\alpha}{(i-1)(i-2)} = 0,$$

$$(3.53)$$

we have

$$\lim_{i\to\infty}\left\{\ln\left[1 - \frac{\alpha}{(i-1)(i-2)}\right]\bigg/\left[-\frac{\alpha}{(i-1)(i-2)}\right]\right\} = 1,$$

$$(3.54)$$

which can be considered as an alternative form of the formula $\lim_{x\to0}[\ln(1+x)]/x = 1$. Hence, the infinite series at the right-hand side of (3.52) converges or diverges in the same way as the series

$$\frac{\alpha}{(N+1)N} + \cdots + \frac{\alpha}{(i-3)(i-4)} + \frac{\alpha}{(i-1)(i-2)} + \cdots, \qquad (3.55)$$

which does converge, and the proof can be found in most text books of calculus. Therefore, the infinite series at the right-hand side of (3.52) does converge. We denote its sum as S. We then have $\ln \Pi > S$, and thus $\Pi > e^S$. Denote $M = e^S N C_N$. As N is fixed, M is also fixed. It can then be concluded from (3.49) and (3.50) that when i is sufficiently large,

$$C_i > \frac{M}{i}. \qquad (3.56)$$

For the two infinite series in (3.43), according to (3.56), we have

$$C_{2i} > \frac{M}{2i}, \qquad C_{2i+1} > \frac{M}{2i+1}. \qquad (3.57)$$

The convergence of an infinite series is independent of its first terms of finite numbers. Furthermore, the two series in (3.43) can be regarded as power series of x^2 and thus could be considered as positive series. Hence, if the series

$$y_1^0(x) = \sum_{i=1}^{\infty} \frac{M}{2i} x^{2i} \quad \text{and} \quad y_2^0(x) = \sum_{i=1}^{\infty} \frac{M}{2i+1} x^{2i+1} \qquad (3.58)$$

are infinite, the two series in (3.43) are infinite.

Explicit formulae in closed-form of the sums of the two series in (3.58) can be derived using the following two well-known Taylor series:

$$\begin{cases} \ln(1+x) = x - \dfrac{x^2}{2} + \dfrac{x^3}{3} - \dfrac{x^4}{4} + \cdots, \\ \ln(1-x) = -x - \dfrac{x^2}{2} - \dfrac{x^3}{3} - \dfrac{x^4}{4} - \cdots. \end{cases} \qquad (3.59)$$

By adding them together or subtracting one from another, we obtain

$$\begin{cases} \dfrac{1}{2}[\ln(1+x) + \ln(1-x)] = -\sum_{i \to \infty} \dfrac{x^{2i}}{2i}, \\ \dfrac{1}{2}[\ln(1+x) - \ln(1-x)] = \sum_{i \to \infty} \dfrac{x^{2i+1}}{2i+1}. \end{cases} \qquad (3.60)$$

By comparing these with (3.58), we finally obtain

$$
\begin{cases}
y_1^0(x) = -\dfrac{1}{2}[\ln(1+x) + \ln(1-x)], \\[2mm]
y_2^0(x) = \dfrac{1}{2}[\ln(1+x) - \ln(1-x)].
\end{cases}
\tag{3.61}
$$

Evidently, when $x \to \pm 1$, $y_1^0(x) \to +\infty$ and $y_2^0(x) \to \pm\infty$, meaning that both $y_1^0(x)$ and $y_2^0(x)$ are infinite when $x = \pm 1$. Based on the discussions throughout this subsection, we conclude that $B_1^0(x)$ and $B_2^0(x)$ are infinite when $x = \pm 1$

3.2.3 Legendre Function

What we are solving for are finite solutions of the Legendre equation in the domain $[-1, 1]$. However, we have concluded that although the two linearly independent solutions of the equation, $B_1^0(x)$ and $B_2^0(x)$, are finite in $(-1, 1)$, they are infinite when $x = \pm 1$. Nevertheless, there exists an occasion when this conclusion breaks. This happens when $\alpha = n(n + 1)$, where n is an integer not less than 0. In this instance, the recurrence relation (3.42) for the coefficients of the power series solutions becomes

$$
C_{i+2} = \frac{i(i+1) - n(n+1)}{(i+2)(i+1)} C_i .
\tag{3.62}
$$

Evidently, it can be seen from this relation that $C_{n+2} = C_{n+4} = \cdots = 0$; thus one of the two infinite power series in (3.43) becomes a polynomial. When n is even, B_1^0 becomes a polynomial, but B_2^0 remains an infinite power series. When n is odd, B_2^0 becomes a polynomial, but B_1^0 remains an infinite power series. Both polynomials are finite within $[-1, 1]$. Therefore, the eigenvalues of the Legendre equation under the finite condition are $n(n + 1)$, $n = 0, 1, 2, \ldots$. The eigenfunctions are the corresponding polynomials denoted as $P_n(x)$, which is referred to as Legendre polynomial or Legendre function of degree n, where the coefficient of its highest power term is defined as (the reason will be explained later in Sect. 3.2.5)

$$
C_n = \frac{(2n)!}{2^n (n!)^2} .
\tag{3.63}
$$

We now derive an explicit formula for $P_n(x)$. Set $i = n - 2$ in (3.62), and then substitute (3.63) into it. We obtain

$$
C_{n-2} = \frac{n(n-1)}{(n-2)(n-1) - n(n+1)} C_n = -\frac{(2n-2)!}{2^n (n-1)!(n-2)!} .
\tag{3.64}
$$

We can also obtain by repeatedly applying the above formula

$$C_{n-4} = \frac{(n-2)(n-3)}{(n-4)(n-3) - n(n+1)} C_{n-2} = \frac{(2n-4)!}{2^n \times 2 \times (n-2)!(n-4)!}.$$

(3.65)

$$C_{n-6} = \frac{(n-4)(n-5)}{(n-6)(n-5) - n(n+1)} C_{n-4} = -\frac{(2n-6)!}{2^n \times 3! \times (n-3)!(n-6)!}.$$

(3.66)

$$\cdots\cdots$$

A general expression can be obtained by induction

$$C_{n-2s} = (-1)^s \frac{(2n-2s)!}{2^n s!(n-s)!(n-2s)!}, \quad s = 0, 1, 2, \ldots, \left[\frac{n}{2}\right],$$

(3.67)

where $[n/2]$ represents the maximum integer not larger than $n/2$. Finally, we obtain an expression of the Legendre function $P_n(x)$ as the solution of $B^0(x)$,

$$B^0(x) = P_n(x),$$

(3.68)

$$P_n(x) = \sum_{s=0}^{[n/2]} (-1)^s \frac{(2n-2s)!}{2^n s!(n-s)!(n-2s)!} x^{n-2s}.$$

(3.69)

Some lower degree Legendre functions are tabulated in Table 3.1. Their graphs are shown in Fig. 3.2. As a polynomial of degree n, it will be shown that $P_n(x)$ has n zeros of multiplicity 1 in $(-1, 1)$. Hence $x = -1$ and $x = 1$ are not its zeros. Actually, we will show that $P_n(\pm 1) = (\pm 1)^n$. It can also be seen from the figure that $(d/d\theta)P_n(\cos\theta) = 0$ when $\theta = 0$ or π, whose proof will be mentioned at a later stage.

Table 3.1 Some lower degree Legendre functions

$P_0(x) = 1$	$P_0(\cos\theta) = 1$
$P_1(x) = x$	$P_1(\cos\theta) = \cos\theta$
$P_2(x) = \frac{3}{2}x^2 - \frac{1}{2}$	$P_2(\cos\theta) = \frac{3}{2}\cos^2\theta - \frac{1}{2}$
$P_3(x) = \frac{5}{2}x^3 - \frac{3}{2}x$	$P_3(\cos\theta) = \frac{5}{2}\cos^3\theta - \frac{3}{2}\cos\theta$
$P_4(x) = \frac{35}{8}x^4 - \frac{15}{4}x^2 + \frac{3}{8}$	$P_4(\cos\theta) = \frac{35}{8}\cos^4\theta - \frac{15}{4}\cos^2\theta + \frac{3}{8}$

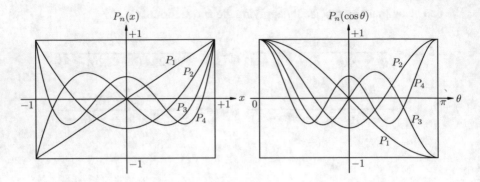

Fig. 3.2 Some Legendre functions

Some properties of the Legendre function can be readily inferred from (3.69), such as

$$P_n(-x) = (-1)^n P_n(x) \tag{3.70}$$

and

$$P_{2n}(0) = (-1)^n \frac{(2n)!}{2^{2n}(n!)^2}, \qquad P_{2n+1}(0) = 0, \tag{3.71}$$

where the value of $P_n(0)$ is equal to the constant term in (3.69).

3.2.4   Rodrigues' Formula

Another expression of the Legendre function is

$$P_n(x) = \frac{1}{2^n n!} \frac{d^n}{dx^n} (x^2 - 1)^n, \tag{3.72}$$

called Rodrigues' formula. It can be proven using the well-known binomial theorem

$$(x^2 - 1)^n = \sum_{s=0}^{n} \frac{(-1)^s n!}{s!(n-s)!} x^{2n-2s}, \tag{3.73}$$

whence

$$\frac{1}{2^n n!} \frac{d^n}{dx^n} (x^2 - 1)^n = \sum_{s=0}^{n} \frac{(-1)^s}{2^n s!(n-s)!} \frac{d^n}{dx^n} x^{2n-2s}. \tag{3.74}$$

Evidently, $(d^n/dx^n)x^{2n-2s}$ vanishes when $2n - 2s < n$. Hence, in the above formula, only the terms $s < [n/2]$ do not vanish. We have

$$\frac{1}{2^n n!}\frac{d^n}{dx^n}(x^2 - 1)^n = \sum_{s=0}^{[n/2]}(-1)^s \frac{(2n - 2s)!}{2^n s!(n - s)!(n - 2s)!}x^{n-2s}. \tag{3.75}$$

We see that (3.73) follows by comparing this formula with (3.72).

3.2.5 A Generating Function

Let $(1+t^2-2xt)^{-1/2}$ be a function of t, and consider x as a parameter. The Legendre function appears in its power series expansion as follows:

$$(1 + t^2 - 2xt)^{-1/2} = \sum_{n=0}^{\infty} P_n(x)t^n, \qquad |t| < 1. \tag{3.76}$$

Hence, the function $(1 + t^2 - 2xt)^{-1/2}$ is referred to as the generating function of $P_n(x)$. The Legendre function was actually defined through this power series expansion, called generating equation, and (3.63) is a consequence of this definition.

To prove (3.76), we use the power series expansion

$$(1 - z)^{-1/2} = \sum_{i=0}^{\infty} \frac{(2i)!}{2^{2i}(i!)^2}z^i. \tag{3.77}$$

Let $z = t(2x - t)$. We obtain

$$(1 + t^2 - 2xt)^{-1/2} = \sum_{i=0}^{\infty} \frac{(2i)!}{2^{2i}(i!)^2}t^i(2x - t)^i. \tag{3.78}$$

In the binomial theorem

$$(a + b)^i = \sum_{s=0}^{i} \frac{i!}{s!(i - s)!}a^{i-s}b^s, \tag{3.79}$$

let $a = 2x$ and $b = -t$. We obtain

$$(2x - t)^i = \sum_{s=0}^{i}(-1)^s \frac{i!}{s!(i - s)!}(2x)^{i-s}t^s. \tag{3.80}$$

Substitute it into (3.78). We obtain

$$(1+t^2-2xt)^{-1/2} = \sum_{i=0}^{\infty} \frac{(2i)!}{2^{2i}(i!)^2} t^i \sum_{s=0}^{i} (-1)^s \frac{i!}{s!(i-s)!} (2x)^{i-s} t^s$$

$$= \sum_{i=0}^{\infty} \sum_{s=0}^{i} (-1)^s \frac{(2i)!}{2^{i+s} s! i! (i-s)!} x^{i-s} t^{i+s} . \tag{3.81}$$

Replacing the summation index i by $n = i + s$, we then have $0 \le s \le [n/2]$, and

$$(1+t^2-2xt)^{-1/2} = \sum_{n=0}^{\infty} \sum_{s=0}^{[n/2]} (-1)^s \frac{(2n-2s)!}{2^n s!(n-s)!(n-2s)!} x^{n-2s} t^n . \tag{3.82}$$

We see that (3.76) can be obtained by comparing this formula with the expression of $P_n(x)$ given by (3.69).

Another expression of $P_n(\cos\theta)$ can be derived based on (3.76) after replacing x by $\cos\theta$

$$(1+t^2-2t\cos\theta)^{-1/2} = \sum_{n=0}^{\infty} P_n(\cos\theta) t^n . \tag{3.83}$$

We express the left-hand side as a power series using (3.77)

$$(1+t^2-2t\cos\theta)^{-1/2} = (1-te^{i\theta})^{-1/2}(1-te^{-i\theta})^{-1/2}$$

$$= \sum_{k=0}^{\infty} \frac{(2k)!}{2^{2k}(k!)^2} e^{ik\theta} t^k \sum_{s=0}^{\infty} \frac{(2s)!}{2^{2s}(s!)^2} e^{-is\theta} t^s$$

$$= \sum_{k=0}^{\infty} \sum_{s=0}^{\infty} \frac{(2k)!(2s)!}{2^{2(k+s)}(k!s!)^2} e^{i(k-s)\theta} t^{k+s} . \tag{3.84}$$

Replace the summation index k by $n = k + s$. It follows from $k \ge 0$ that $s \le n$. The above equation becomes

$$(1+t^2-2t\cos\theta)^{-1/2} = \sum_{n=0}^{\infty} \sum_{s=0}^{n} \frac{(2n-2s)!(2s)!}{2^{2n}[(n-s)!s!]^2} e^{i(n-2s)\theta} t^n . \tag{3.85}$$

We obtain, by comparing this formula with (3.83),

$$P_n(\cos\theta) = \sum_{s=0}^{n} \frac{(2n-2s)!(2s)!}{2^{2n}[(n-s)!s!]^2} e^{i(n-2s)\theta} . \tag{3.86}$$

Since $P_n(\cos\theta)$ is a real function, the sum of all imaginary terms at the right-hand side of this equation must vanish. So we finally obtain

$$P_n(\cos\theta) = \sum_{s=0}^{n} \frac{(2n-2s)!(2s)!}{2^{2n}[(n-s)!s!]^2} \cos[(n-2s)\theta]. \tag{3.87}$$

We now study the range of $P_n(\cos\theta)$. By setting $\cos\theta = \pm 1$ in (3.83), we obtain

$$(1 + t^2 \mp 2t)^{-1/2} = (1 \mp t)^{-1} = \sum_{n=0}^{\infty}(\pm 1)^n t^n. \tag{3.88}$$

This leads to

$$P_n(\pm 1) = (-1)^n. \tag{3.89}$$

According to this formula, when $\theta = 0$, (3.87) becomes

$$P_n(1) = \sum_{s=0}^{n} \frac{(2n-2s)!(2s)!}{2^{2n}[(n-s)!s!]^2} = 1. \tag{3.90}$$

Note that all terms under the summation sign in this formula are positive. As $|\cos[(n-2s)\theta]| \le 1$, we conclude from this formula and (3.87) that

$$-1 \le P_n(x) = P_n(\cos\theta) \le 1. \tag{3.91}$$

Finally, as the domain of convergence of the series (3.88) is $|t| < 1$, it follows from (3.91) that the domain of convergence of the series (3.76) is $|t| < 1$ for any $-1 \le x \le 1$.

A very important application of (3.76) is the development of the reciprocal of a distance into a series of Legendre functions. As shown in Fig. 3.3, the reciprocal of

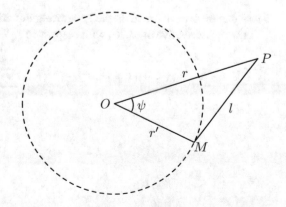

Fig. 3.3 Expression of the inverse of a distance into a series of Legendre functions

the distance between M and P is

$$\frac{1}{l} = (r^2 + r'^2 - 2rr' \cos \psi)^{-1/2} . \tag{3.92}$$

According to (3.76), we obtain

$$\frac{1}{l} = \begin{cases} \dfrac{1}{r}\left[1 + \left(\dfrac{r'}{r}\right)^2 - 2\dfrac{r'}{r}\cos\psi\right]^{-1/2} = \displaystyle\sum_{n=0}^{\infty} \dfrac{r'^n}{r^{n+1}} P_n(\cos\psi), \ r > r' ; \\[4mm] \dfrac{1}{r'}\left[1 + \left(\dfrac{r}{r'}\right)^2 - 2\dfrac{r}{r'}\cos\psi\right]^{-1/2} = \displaystyle\sum_{n=0}^{\infty} \dfrac{r^n}{r'^{n+1}} P_n(\cos\psi), \ r < r' . \end{cases}$$
$$\tag{3.93}$$

3.2.6 Some Recurrence Formulae

In practice, very high degree Legendre functions are computed. This is most efficiently done using recurrence formulae. Some of the most useful recurrence formulae of $P_n(x)$ are

$$(n + 1)P_{n+1}(x) - (2n + 1)x P_n(x) + n P_{n-1}(x) = 0 , \tag{3.94}$$

$$\frac{d}{dx}P_{n+1}(x) = x\frac{d}{dx}P_n(x) + (n + 1)P_n(x) , \tag{3.95}$$

$$\frac{d}{dx}P_{n-1}(x) = x\frac{d}{dx}P_n(x) - n P_n(x) , \tag{3.96}$$

$$\frac{d}{dx}P_{n+1}(x) - \frac{d}{dx}P_{n-1}(x) = (2n + 1)P_n(x) , \tag{3.97}$$

$$(1 - x^2)\frac{d}{dx}P_n(x) = n P_{n-1}(x) - nx P_n(x) . \tag{3.98}$$

These can be proven based on the generating equation (3.76).

Take the derivative of (3.76) with respect to t. We obtain

$$(x - t)(1 + t^2 - 2tx)^{-3/2} = \sum_{n=0}^{\infty} n P_n(x)t^{n-1} . \tag{3.99}$$

Multiply both sides by $(1 + t^2 - 2tx)$, and then apply (3.76) to the left-hand side. We obtain

$$(x - t) \sum_{n=0}^{\infty} P_n(x) t^n = (1 + t^2 - 2tx) \sum_{n=0}^{\infty} n P_n(x) t^{n-1} . \qquad (3.100)$$

Gather terms of the same degree of t. We obtain

$$\sum_{n=0}^{\infty} n P_n(x) t^{n-1} - \sum_{n=0}^{\infty} (2n + 1) x P_n(x) t^n + \sum_{n=0}^{\infty} (n + 1) P_n(x) t^{n+1} = 0 . \qquad (3.101)$$

Some rearrangement of summation indices leads it to

$$\sum_{n=0}^{\infty} (n + 1) P_{n+1}(x) t^n - \sum_{n=0}^{\infty} (2n + 1) x P_n(x) t^n + \sum_{n=1}^{\infty} n P_{n-1}(x) t^n = 0 . \qquad (3.102)$$

As t is arbitrary, the sum of the coefficients of t^n must vanish in order for the identity to hold, which leads to (3.94).

Take the derivative of (3.76) with respect to x. We obtain

$$t(1 + t^2 - 2tx)^{-3/2} = \sum_{n=0}^{\infty} \frac{d}{dx} P_n(x) t^n . \qquad (3.103)$$

Multiply both sides by $(1 + t^2 - 2tx)$, and then apply (3.76) at the left-hand side. We obtain

$$t \sum_{n=0}^{\infty} P_n(x) t^n = (1 + t^2 - 2tx) \sum_{n=0}^{\infty} \frac{d}{dx} P_n(x) t^n . \qquad (3.104)$$

Gather all terms of the same degree of t. We obtain

$$\sum_{n=0}^{\infty} P_n(x) t^{n+1} = \sum_{n=0}^{\infty} \frac{d}{dx} P_n(x) t^n + \sum_{n=0}^{\infty} \frac{d}{dx} P_n(x) t^{n+2} - 2x \sum_{n=0}^{\infty} \frac{d}{dx} P_n(x) t^{n+1} . $$
$$\qquad (3.105)$$

This formula can be further written as

$$\sum_{n=0}^{\infty} P_n(x) t^{n+1} = \sum_{n=0}^{\infty} \frac{d}{dx} P_{n+1}(x) t^{n+1} + \sum_{n=0}^{\infty} \frac{d}{dx} P_{n-1}(x) t^{n+1}$$

$$- 2x \sum_{n=0}^{\infty} \frac{d}{dx} P_n(x) t^{n+1} . \qquad (3.106)$$

As t is arbitrary, in order for the identity to hold, the sum of the coefficients of t^{n+1} must vanish. We obtain

$$P_n(x) = \frac{d}{dx} P_{n+1}(x) + \frac{d}{dx} P_{n-1}(x) - 2x \frac{d}{dx} P_n(x) . \tag{3.107}$$

Take the derivative of (3.94) with respect to x, and then eliminate $d P_{n-1}(x)/dx$ with the above formula. We obtain (3.95).

The rest can be proven as follows: (3.96) can be obtained by eliminating $d P_{n+1}(x)/dx$ from (3.95) and (3.107), (3.97) can be obtained by eliminating $d P_n(x)/dx$ from (3.95) and (3.96), and (3.98) can be obtained by replacing n by $n-1$ in (3.95) and then eliminating $d P_{n-1}(x)/dx$ with (3.96).

Finally, for easy reference, we rewrite the recurrence formulae (3.94)–(3.98) in terms of θ. With $x = \cos\theta$, $(1-x^2)^{1/2} = \sin\theta$ and $d/dx = -(1/\sin\theta)d/d\theta$, we have

$$(n+1) P_{n+1}(\cos\theta) - (2n+1)\cos\theta P_n(\cos\theta) + n P_{n-1}(\cos\theta) = 0 , \tag{3.108}$$

$$\frac{d}{d\theta} P_{n+1}(\cos\theta) = \cos\theta \frac{d}{d\theta} P_n(\cos\theta) - (n+1)\sin\theta P_n(\cos\theta) , \tag{3.109}$$

$$\frac{d}{d\theta} P_{n-1}(\cos\theta) = \cos\theta \frac{d}{d\theta} P_n(\cos\theta) + n\sin\theta P_n(\cos\theta) , \tag{3.110}$$

$$\frac{d}{d\theta} P_{n+1}(\cos\theta) - \frac{d}{d\theta} P_{n-1}(\cos\theta) = -(2n+1)\sin\theta P_n(\cos\theta) , \tag{3.111}$$

$$-\sin\theta \frac{d}{d\theta} P_n(\cos\theta) = n P_{n-1}(\cos\theta) - n\cos\theta P_n(\cos\theta) . \tag{3.112}$$

As a supplement, we reproduce the ordinary differential equation satisfied by $P_n(\cos\theta)$

$$\frac{d^2}{d\theta^2} P_n(\cos\theta) + \cot\theta \frac{d}{d\theta} P_n(\cos\theta) + n(n+1) P_n(\cos\theta) = 0 , \tag{3.113}$$

which looks similar to a recurrence formula. It is used often in formulations.

The property $d P_n(\cos\theta)/d\theta = 0$ when $\theta = 0$ or π as seen in Fig. 3.2 can be proven using (3.109) or (3.111) with $\sin\theta = 0$, $\cos\theta = 1$ or -1, and $d P_0(\cos\theta)/d\theta = d P_1(\cos\theta)/d\theta = 0$.

The formulae (3.108) and (3.109) or (3.111) are the appropriate ones to compute $P_n(\cos\theta)$ and $d P_n(\cos\theta)/d\theta$ recurrently, since no division by $\sin\theta$ or $\cos\theta$, which may be infinitesimally small, is present in them. Here we also provide a formula for computing $d^2 P_n(\cos\theta)/d\theta^2$,

$$
\frac{d^2}{d\theta^2} P_{n+1}(\cos\theta) - \frac{d^2}{d\theta^2} P_{n-1}(\cos\theta)
$$

$$
= -(2n+1)\left[\cos\theta\, P_n(\cos\theta) + \sin\theta \frac{d}{d\theta} P_n(\cos\theta)\right], \tag{3.114}
$$

which is derived by taking derivative of both sides of (3.111) with respect to θ. Initial values of $P_0(\cos\theta)$, $P_1(\cos\theta)$, $dP_0(\cos\theta)/d\theta$, $dP_1(\cos\theta)/d\theta$, $d^2 P_0(\cos\theta)/d\theta^2$, and $d^2 P_1(\cos\theta)/d\theta^2$ can be assigned based on Table 3.1. Based on the recurrence formulae, we can judge that the magnitudes of $P_n(\cos\theta)$, $dP_n(\cos\theta)/d\theta$, and $d^2 P_n(\cos\theta)/d\theta^2$ are on the order of ~ 1, $\sim n$, and $\sim n^2$.

3.3 Associated Legendre Function

3.3.1 A Relation Between the Legendre and Associated Legendre Equations

We first derive a relation between the associated Legendre equation and the Legendre equation, so that the solution of the former could be inferred from that of the later.

We write the associated Legendre equation (3.32) in the form of

$$
(1-x^2)\frac{d^2 B}{dx^2} - 2x\frac{dB}{dx} + \left(\alpha - \frac{k^2}{1-x^2}\right) B = 0. \tag{3.115}
$$

Define a function $\zeta(x)$ by

$$
B(x) = (1-x^2)^{k/2}\zeta(x). \tag{3.116}
$$

We have

$$
\frac{dB}{dx} = (1-x^2)^{k/2}\frac{d\zeta}{dx} - kx(1-x^2)^{k/2-1}\zeta, \tag{3.117}
$$

$$
\frac{d^2 B}{dx^2} = (1-x^2)^{k/2}\frac{d^2\zeta}{dx^2} - 2kx(1-x^2)^{k/2-1}\frac{d\zeta}{dx}
$$

$$
- k\left[(1-x^2)^{k/2-1} - 2\left(\frac{k}{2}-1\right)x^2(1-x^2)^{k/2-2}\right]\zeta. \tag{3.118}
$$

Substitute the above three formulae into (3.115). After some rearrangement, we divide both sides by $(1 - x^2)^{k/2}$ to obtain

$$(1 - x^2)\frac{d^2\zeta}{dx^2} - 2(k + 1)x\frac{d\zeta}{dx} + [\alpha - k(k + 1)]\zeta = 0. \tag{3.119}$$

Take the derivative of both sides. We obtain

$$(1 - x^2)\frac{d^3\zeta}{dx^3} - 2(k + 2)x\frac{d^2\zeta}{dx^2} + [\alpha - (k + 1)(k + 2)]\frac{d\zeta}{dx} = 0. \tag{3.120}$$

While comparing these two equations, we see that the later can be obtained by replacing k and ζ in the former by $k + 1$ and $d\zeta/dx$, respectively. As the former is identical to the Legendre equation when $k = 0$, we can then conclude that it can be obtained by taking the derivative of the Legendre equation k times. Hence, we have $\zeta = d^k B_0/dx^k$, and according to (3.116), we further have

$$B = (1 - x^2)^{k/2}\frac{d^k B^0}{dx^k}. \tag{3.121}$$

This relation tells that the two linearly independent solutions of the associated Legendre equation can be determined based on the two linearly independent solutions of the Legendre equation.

3.3.2 Associated Legendre Function

Substitute the two linearly independent power series solutions of the Legendre equation given by (3.43) into (3.121). We obtain the two linearly independent power series solutions of the associated Legendre equation

$$\begin{cases} B_1 = (1 - x^2)^{k/2} \displaystyle\sum_{2i \geq k} \frac{(2i)!}{(2i - k)!} C_{2i} x^{2i-k}, \\[4mm] B_2 = (1 - x^2)^{k/2} \displaystyle\sum_{2i+1 \geq k} \frac{(2i + 1)!}{(2i - k + 1)!} C_{2i+1} x^{2i-k+1}, \end{cases} \tag{3.122}$$

where C_{2i} and C_{2i+1} are still determined using (3.42). The limitations $2i \geq k$ and $2i + 1 \geq k$ are due to the fact that after taking the derivative of (3.43) k times, the terms $2i < k$ in B_1^0 and the terms $2i + 1 < k$ in B_2^0 vanish.

Similar to the cases of B_1^0 and B_2^0, it can be easily verified that the convergence radius of both B_1 and B_2 is 1, meaning that both B_1 and B_2 are finite in $(-1, 1)$. However, as will be shown below, both of them are infinite when $x = \pm 1$.

Here we take only the case when k is even. The case when k is odd is similar. Assume $k \geq 2$. Substitute the summation index i in (3.122) by $i + k/2$. We obtain

$$\begin{cases} B_1 = (1 - x^2)^{k/2} \displaystyle\sum_{i=0}^{\infty} \frac{(2i + k)!}{(2i)!} C_{2i+k} x^{2i}\,, \\[3mm] B_2 = (1 - x^2)^{k/2} \displaystyle\sum_{i-0}^{\infty} \frac{(2i + k + 1)!}{(2i + 1)!} C_{2i+k+1} x^{2i+1}\,. \end{cases} \tag{3.123}$$

When i is large enough, we have, according to (3.56),

$$\begin{cases} \dfrac{(2i + k)!}{(2i)!} C_{2i+k} > \dfrac{(2i + k - 1)!}{(2i)!} M > \dfrac{M}{2i}\,, \\[3mm] \dfrac{(2i + k + 1)!}{(2i + 1)!} C_{2i+k+1} > \dfrac{(2i + k)!}{(2i + 1)!} M > \dfrac{M}{2i + 1}\,. \end{cases} \tag{3.124}$$

Taking reference to (3.57) and the derivation below it, we can readily conclude that the two series in (3.123) are infinite when $x = \pm 1$.

The same as for the Legendre equation, we are solving for finite solutions of the associated Legendre equation in the domain $[-1, 1]$, but its two linearly independent power series solutions are infinite when $x = \pm 1$, although they are finite in $(-1, 1)$. The same special instance happens for this conclusion to break, namely, when $\alpha = n(n + 1)$. In this case, $B^0(x) = P_n(x)$, and we denote $B(x)$ as $P_n^k(x)$, i.e.,

$$B(x) = P_n^k(x)\,. \tag{3.125}$$

According to (3.121), we obtain[1]

$$P_n^k(x) = (1 - x^2)^{k/2} \frac{d^k}{dx^k} P_n(x)\,, \tag{3.126}$$

which is referred to as associated Legendre function of degree n and order k. As $P_n(x)$ is a polynomial of degree n, $P_n^k(x)$ vanishes when $n < k$. Hence, we have $n \geq k$. Actually, this can also be inferred from the limitations of summations, $2i \geq k$ and $2i + 1 \geq k$, in the power series in (3.122), where the lower limits of $2i$ and $2i + 1$ are n.

By substituting the expressions of the Legendre function (3.69) and (3.72) into (3.126), we obtain the corresponding expressions for the associated Legendre

---

[1] An alternative definition in the literature includes a factor $(-1)^k$ at the right-hand side. Hence, there is a need to clarify or verify the definition while reading.

function:

$$P_n^k(x) = (1 - x^2)^{k/2} \sum_{s=0}^{[(n-k)/2]} \frac{(2n - 2s)!}{2^n s!(n - s)!(n - 2s - k)!} x^{n-k-2s} ,$$ (3.127)

$$P_n^k(x) = \frac{1}{2^n n!} (1 - x^2)^{k/2} \frac{d^{n+k}}{dx^{n+k}} (x^2 - 1)^n .$$ (3.128)

When $k = 0$, the associated Legendre function $P_n^k(x)$ degenerates to the Legendre function $P_n(x)$. In the literature, the term *Legendre function* is also used to represent both the associated Legendre and Legendre functions. By substituting $x = \cos\theta$ into (3.127), we obtain

$$P_n^k(\cos\theta) = \sin^k\theta \sum_{s=0}^{[(n-k)/2]} \frac{(2n - 2s)!}{2^n s!(n - s)!(n - 2s - k)!} \cos^{n-k-2s}\theta .$$ (3.129)

Some lower degree and order associated Legendre functions are listed in Table 3.2. If $k \neq 0$, the factor $\sin^k\theta$ equals 1 at the equator and decreases toward

Table 3.2 Some lower degree and order associated Legendre functions

$P_0(x) = 1$	$P_0(\cos\theta) = 1$
$P_1(x) = x$	$P_1(\cos\theta) = \cos\theta$
$P_1^1(x) = (1 - x^2)^{1/2}$	$P_1^1(\cos\theta) = \sin\theta$
$P_2(x) = \frac{3}{2}x^2 - \frac{1}{2}$	$P_2(\cos\theta) = \frac{3}{2}\cos^2\theta - \frac{1}{2}$
$P_2^1(x) = 3(1 - x^2)^{1/2}x$	$P_2^1(\cos\theta) = 3\sin\theta\cos\theta$
$P_2^2(x) = 3(1 - x^2)$	$P_2^2(\cos\theta) = 3\sin^2\theta$
$P_3(x) = \frac{5}{2}x^3 - \frac{3}{2}x$	$P_3(\cos\theta) = \frac{5}{2}\cos^3\theta - \frac{3}{2}\cos\theta$
$P_3^1(x) = (1 - x^2)^{1/2}\left(\frac{15}{2}x^2 - \frac{3}{2}\right)$	$P_3^1(\cos\theta) = \sin\theta\left(\frac{15}{2}\cos^2\theta - \frac{3}{2}\right)$
$P_3^2(x) = 15(1 - x^2)x$	$P_3^2(\cos\theta) = 15\sin^2\theta\cos\theta$
$P_3^3(x) = 15(1 - x^2)^{3/2}$	$P_3^3(\cos\theta) = 15\sin^3\theta$
$P_4(x) = \frac{35}{8}x^4 - \frac{15}{4}x^2 + \frac{3}{8}$	$P_4(\cos\theta) = \frac{35}{8}\cos^4\theta - \frac{15}{4}\cos^2\theta + \frac{3}{8}$
$P_4^1(x) = (1 - x^2)^{1/2}\left(\frac{35}{2}x^3 - \frac{15}{2}x\right)$	$P_4^1(\cos\theta) = \sin\theta\left(\frac{35}{2}\cos^3\theta - \frac{15}{2}\cos\theta\right)$
$P_4^2(x) = (1 - x^2)\left(\frac{105}{2}x^2 - \frac{15}{2}\right)$	$P_4^2(\cos\theta) = \sin^2\theta\left(\frac{105}{2}\cos^2\theta - \frac{15}{2}\right)$
$P_4^3(x) = 105(1 - x^2)^{3/2}x$	$P_4^3(\cos\theta) = 105\sin^3\theta\cos\theta$
$P_4^4(x) = 105(1 - x^2)^2$	$P_4^4(\cos\theta) = 105\sin^4\theta$

the pole, where it equals 0; the larger k is, the quicker the decease is. In Fig. 3.4, we show the graphs of $P_{20}^k(\cos\theta)$ for $k = 0, 1, \ldots, 20$, which are scaled to fit the height of the boxes for clarity. The increasing effect of the factor $\sin^k\theta$ as k increases can be seen in this figure. There is a region around the poles where $P_{20}^k(\cos\theta)$ remains extremely small, and the size of the region increases as k increases. Therefore, when k is extremely large, $P_n^k(\cos\theta)$ is significantly different from 0 only in a narrow region around the equator.

The following properties can be inferred right away from (3.127):

$$P_n^k(-x) = (-1)^{n-k} P_n^k(x);\qquad(3.130)$$

$$P_n^k(0) = \begin{cases} (-1)^{(n-k)/2}\dfrac{(n+k)!}{2^n[(n-k)/2]![(n+k)/2]!}, & \text{when } n-k \text{ is even}; \\ 0, & \text{when } n-k \text{ is odd}; \end{cases}$$
$$(3.131)$$

$$P_n^k(\pm 1) = 0,\qquad(3.132)$$

where the value of $P_n^k(0)$ is the constant term of (3.127).

The associated Legendre function $P_n^k(\cos\theta)$ can be understood as a sum in the form of $\sum_{s=0}^{n} C_s \cos s\theta$ or $\sin\theta \sum_{s=0}^{n-1} C_s \sin s\theta$ when k is even, and $\sum_{s=0}^{n} C_s \sin s\theta$ or $\sin\theta \sum_{s=0}^{n-1} C_s \cos s\theta$ when k is odd. This can be realized based on the relations

$$\cos^m\theta = \begin{cases} \dfrac{2}{2^m}\displaystyle\sum_{s=0}^{(m-1)/2}\dfrac{m!}{s!(m-s)!}\cos[(m-2s)\theta], & \text{when } m \text{ is odd}; \\ \dfrac{1}{2^m}\dfrac{m!}{[(m/2)!]^2}+\dfrac{2}{2^m}\displaystyle\sum_{s=0}^{m/2-1}\dfrac{n!}{s!(m-s)!}\cos[(m-2s)\theta], & \text{when } m \text{ is even}; \end{cases}$$
$$(3.133)$$

and

$$\sin^m\theta = \begin{cases} \dfrac{2}{2^m}\displaystyle\sum_{s=0}^{(m-1)/2}\dfrac{m!}{s!(m-s)!}\sin[(m-2s)\theta], & \text{when } m \text{ is odd}; \\ \dfrac{1}{2^m}\dfrac{m!}{[(m/2)!]^2}+\dfrac{2}{2^m}\displaystyle\sum_{s=0}^{m/2-1}\dfrac{(-1)^{m/2-s}n!}{s!(m-s)!}\cos[(m-2s)\theta], & \text{when } m \text{ is even}, \end{cases}$$
$$(3.134)$$

together with the product-to-sum formulae of sine and cosine functions.

The associated Legendre equation remains unchanged when k is replaced by $-k$. Hence, it can be conjectured that if we replace k by $-k$ in (3.127) or (3.128), the function $P_n^{-k}(x)$ obtained remains a solution of the associated Legendre equation. Actually, $P_n^{-k}(x)$ is different from $P_n^k(x)$ only by a constant factor. A proof is provided below.

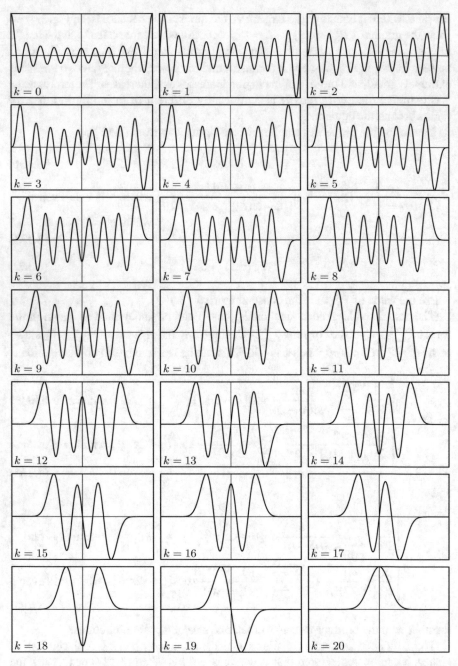

Fig. 3.4 The associated Legendre function $P_{20}^k(\cos\theta)$ scaled to fit the height of the boxes for clarity

Assume $k > 0$. We obtain, from (3.128),

$$P_n^{-k}(x) = \frac{1}{2^n n!}(1 - x^2)^{-k/2}\frac{d^{n-k}}{dx^{n-k}}(x^2 - 1)^n .$$ (3.135)

According to the derivative formula

$$\frac{d^m}{dx^m}(uv) = \sum_{i=0}^{m}\frac{m!}{i!(m-i)!}\frac{d^i u}{dx^i}\frac{d^{m-i}v}{dx^{m-i}} ,$$ (3.136)

we obtain

$$\frac{d^{n+k}}{dx^{n+k}}(x^2 - 1)^n = \frac{d^{n+k}}{dx^{n+k}}[(x-1)^n(x+1)^n]$$

$$= \sum_{i=0}^{n+k}\frac{(n+k)!}{i!(n+k-i)!}\frac{d^i}{dx^i}(x-1)^n\frac{d^{n+k-i}}{dx^{n+k-i}}(x+1)^n .$$ (3.137)

When $i > n$, we have $d^i(x-1)^n/dx^i = 0$. When $i < k$, we have $n+k-i > n$, and thus $d^{n+k-i}(x+1)^n/dx^{n+k-i} = 0$. We obtain

$$\frac{d^{n+k}}{dx^{n+k}}(x^2 - 1)^n = \sum_{i=k}^{n}\frac{(n+k)!}{i!(n+k-i)!}\frac{n!}{(n-i)!}(x-1)^{n-i}\frac{n!}{(n-k)!}(x+1)^{i-k}$$

$$= \sum_{j=0}^{n-k}\frac{(n+k)!}{(j+k)!(n-j)!}\frac{n!}{(n-k-j)!}(x-1)^{n-k-j}\frac{n!}{j!}(x+1)^j$$

$$= \frac{(n+k)!}{(n-k)!}(x^2 - 1)^{-k}\sum_{j=0}^{n-k}\frac{(n-k)!}{j!(n-k-j)!}\frac{n!}{(n-j)!}(x-1)^{n-j}$$

$$\times \frac{n!}{(j+k)!}(x+1)^{j+k}$$

$$= \frac{(n+k)!}{(n-k)!}(x^2 - 1)^{-k}\sum_{j=0}^{n-k}\frac{(n-k)!}{j!(n-k-j)!}\frac{d^j}{dx^j}(x-1)$$

$$\times \frac{d^{n-k-j}}{dx^{n-k-j}}(x+1)$$

$$= \frac{(n+k)!}{(n-k)!}(x^2 - 1)^{-k}\frac{d^{n-k}}{dx^{n-k}}\left[(x-1)^n(x+1)^n\right]$$

$$= (-1)^k\frac{(n+k)!}{(n-k)!}(1 - x^2)^{-k}\frac{d^{n-k}}{dx^{n-k}}(x^2 - 1)^n .$$ (3.138)

By comparing (3.128) and (3.135) taking into account of the above formula, we obtain

$$P_n^{-k}(x) = (-1)^k \frac{(n-k)!}{(n+k)!} P_n^k(x) \,. \tag{3.139}$$

Thus, we proved the conclusion that $P_n^{-k}(x)$ is different from $P_n^k(x)$ only by a constant factor.

3.3.3 Zeros of an Associated Legendre Function

We first show that $P_n^k(x)$ has $n - k$ zeros of multiplicity 1 in $(-1, 1)$. According to the relation between $P_n^k(x)$ and $P_n^{-k}(x)$ given by (3.139) and the expression of $P_n^{-k}(x)$ given by (3.135), we have

$$P_n^k(x) = (-1)^k \frac{(n+k)!}{2^n n!(n-k)!} (1-x^2)^{-k/2} \frac{d^{n-k}}{dx^{n-k}} (x^2-1)^n \,. \tag{3.140}$$

Evidently, $P_n^k(x)$ and the polynomial $(d^{n-k}/dx^{n-k})(x^2-1)^n$ have the same zeros in $(-1, 1)$.

As $(x^2-1)^n$ is a polynomial of degree $2n$, it has at most $2n$ zeros. However, both -1 and 1 are its zeros of multiplicity n, it cannot have other zeros. It can be easily verified that both -1 and 1 are zeros of multiplicity $n - 1$ of $(d/dx)(x^2-1)^n$, zeros of multiplicity $n - 2$ of $(d^2/dx^2)(x^2-1)^n$, ..., and zeros of multiplicity k of $(d^{n-k}/dx^{n-k})(x^2-1)^n$. This process of inference requires that the order of derivative to be smaller than n, which is the reason to chose the expression (3.140) over the standard definition (3.128).

According to Rolle's theorem, as $(x^2-1)^n$ is zero at the boundaries of the closed domain $[-1, 1]$, there must be a zero point of $(d/dx)(x^2-1)^n$ in $(-1, 1)$, which we denote as x'. As $(d/dx)(x^2-1)^n$ is a polynomial of degree $2n-1$, it has $2n-1$ zeros at most. However, as -1 and 1 are its zeros of multiplicity $n - 1$, it can have at most $(2n-1) - 2(n-1) = 1$ zero in $(-1, 1)$. This means that x' is its only zero different from -1 and 1, and it is of multiplicity 1. Again, according to Rolle's theorem, as $-1, x'$ and 1 are zeros of $(d/dx)(x^2-1)^n$, $(d^2/dx^2)(x^2-1)^n$ must have a zero x'' in $(-1, x')$ and a zero x''' in $(x', 1)$. As $(d^2/dx^2)(x^2-1)^n$ is a polynomial of degree $2n - 2$, it has $2n - 2$ zeros at most. However, as -1 and 1 are its zeros of multiplicity $n - 2$, it can have at most $(2n-2) - 2(n-2) = 2$ zeros in $(-1, 1)$. This means that x'' and x''' are its only zeros different from -1 and 1, and they are of multiplicity 1. The continuation of this chain of inference leads to the conclusion that $(d^{n-k}/dx^{n-k})(x^2-1)^n$ has, and has only, $n - k$ zeros in $(-1, 1)$, which are all of multiplicity 1, which is equivalent to the conclusion that the associated Legendre function $P_n^k(x)$ has $n - k$ zeros of multiplicity 1 in $(-1, 1)$. This can be seen in Fig. 3.4 for the case $n = 20$.

A degenerate case is the Legendre function $P_n(x)$, which has n zeros of multiplicity 1 in $(-1, 1)$, meaning that ± 1 are not its zero. In fact, we already know that $P_n(\pm 1) = (-1)^n$.

We now look back at (3.128), which states that $P_n^k(x)$ is the product of $(1-x^2)^{k/2}$ and a polynomial of degree $n - k$. Therefore, the $n - k$ zeros of $P_n^k(x)$ in $(-1, 1)$ are, and are the only, zeros of the polynomial, and all of them are of multiplicity 1. Hence, if we denote the zeros as x_i, $i = 1, 2, \ldots, n - k$, we can write $P_n^k(x)$ as

$$P_n^k(x) = \frac{(2n)!}{2^n n!(n-k)!}(1-x^2)^{k/2}\prod_{i=1}^{n-k}(x - x_i)\,, \tag{3.141}$$

where $(2n)!/[2^n n!(n-k)!]$ is the coefficient of x^{n-k} in the polynomial.

According to (3.130), if x_i is a zero of $P_n^k(x)$, $-x_i$ is also. When $n - k$ is even, $x = 0$ is not a zero of $P_n^k(x)$. When $n - k$ is odd, $x = 0$ is a zero of $P_n^k(x)$. Hence, (3.141) can be written, using only the positive zeros, as

$$P_n^k(x) = \frac{(2n)!}{2^n n!(n-k)!}(1-x^2)^{k/2}x^\nu\prod_{i=1}^{(n-k-\nu)/2}(x^2 - x_i^2)\,, \tag{3.142}$$

where ν is defined as

$$\nu = \begin{cases} 0, & n - k \text{ even;} \\ 1, & n - k \text{ odd.} \end{cases} \tag{3.143}$$

3.3.4 Recurrence Formulae

Similar to $P_n(x)$, it is much more practical to use recurrence formulae to compute $P_n^k(x)$ when n and k are large. Some fundamental recurrence formulae for $P_n^k(x)$ are

$$(2n + 1)x P_n^k(x) = (n + k)P_{n-1}^k(x) + (n - k + 1)P_{n+1}^k(x)\,, \tag{3.144}$$

$$(2n + 1)(1 - x^2)^{1/2}P_n^k(x) = P_{n+1}^{k+1}(x) - P_{n-1}^{k+1}(x)\,, \tag{3.145}$$

$$(2n + 1)(1 - x^2)^{1/2}P_n^k(x) = (n + k)(n + k - 1)P_{n-1}^{k-1}(x)$$
$$-(n - k + 2)(n - k + 1)P_{n+1}^{k-1}(x)\,, \tag{3.146}$$

$$(1 - x^2)\frac{d}{dx}P_n^k(x) = (1 - x^2)^{1/2}P_n^{k+1}(x) - kx P_n^k(x)\,, \tag{3.147}$$

$$(2n + 1)(1 - x^2)\frac{d}{dx}P_n^k(x) = (n + 1)(n + k)P_{n-1}^k(x) - n(n - k + 1)P_{n+1}^k(x),$$
$$(3.148)$$

$$(1 - x^2)\frac{d}{dx}P_n^k(x) = (n + k)P_{n-1}^k(x) - nxP_n^k(x).$$
$$(3.149)$$

These can be proven using the recurrence formulae of $P_n(x)$.

Take k times the derivative of (3.94) with respect to x. We obtain

$$(n + 1)\frac{d^k}{dx^k}P_{n-1}(x) - (2n + 1)x\frac{d^k}{dx^k}P_n(x) - (2n + 1)k\frac{d^{k-1}}{dx^{k-1}}P_n(x)$$

$$+ n\frac{d^k}{dx^k}P_{n-1}(x) = 0.$$
$$(3.150)$$

Take $k - 1$ times the derivative of (3.97) with respect to x. We obtain

$$\frac{d^k}{dx^k}P_{n+1}(x) - \frac{d^k}{dx^k}P_{n-1}(x) = (2n + 1)\frac{d^{k-1}}{dx^{k-1}}P_n(x).$$
$$(3.151)$$

Eliminate $d^{k-1}P_n(x)/dx^{k-1}$ from the above two formulae. We obtain

$$(2n+1)x\frac{d^k}{dx^k}P_n(x) = (n+k)\frac{d^k}{dx^k}P_{n-1}(x) + (n-k+1)\frac{d^k}{dx^k}P_{n+1}(x).$$
$$(3.152)$$

By multiplying both sides by $(1 - x^2)^{k/2}$, we obtain (3.144) according to the definition of $P_n^k(x)$ in (3.128).

Take the derivative of (3.151) with respect to x, and then multiply both sides by $(1 - x^2)^{(k+1)/2}$. We obtain (3.145) according to the definition of $P_n^k(x)$ in (3.128).

Eliminate $xP_n(x)$ from (3.94) and (3.98). We obtain

$$(2n + 1)(1 - x^2)\frac{d}{dx}P_n(x) = n(n + 1)P_{n-1}(x) - n(n + 1)P_{n+1}(x).$$
$$(3.153)$$

Take $k - 1$ times the derivative with respect to x. We obtain

$$(2n + 1)(1 - x^2)\frac{d^k}{dx^k}P_n(x) - 2(k - 1)(2n + 1)x\frac{d^{k-1}}{dx^{k-1}}P_n(x)$$

$$- (k - 2)(k - 1)(2n + 1)\frac{d^{k-2}}{dx^{k-2}}P_n(x)$$

$$= n(n + 1)\frac{d^{k-1}}{dx^{k-1}}P_{n-1}(x) - n(n + 1)\frac{d^{k-1}}{dx^{k-1}}P_{n+1}(x).$$
$$(3.154)$$

Replace k by $k-1$ in (3.151) and (3.152), and then substitute into the above formula. We obtain

$$(2n+1)(1-x^2)\frac{d^k}{dx^k}P_n(x) = (n+k)(n+k-1)\frac{d^{k-1}}{dx^{k-1}}P_{n-1}(x)$$

$$-(n-k+2)(n-k+1)\frac{d^{k-1}}{dx^{k-1}}P_{n+1}(x)\,.$$

$$(3.155)$$

After multiplying both sides by $(1-x^2)^{(k-1)/2}$, we obtain (3.146) according to the definition of $P_n^k(x)$ in (3.128).

Take the derivative of (3.128) with respect to x. We obtain

$$\frac{d}{dx}P_n^k(x) = (1-x^2)^{k/2}\frac{d^{k+1}}{dx^{k+1}}P_n(x) - k(1-x^2)^{k/2-1}\frac{d^k}{dx^k}P_n(x)\,. \qquad (3.156)$$

Multiply both sides by $1-x^2$. We obtain (3.147) according to (3.128).

The last two formulae can be derived from the first two. Here we leave the derivation to the readers.

One thing to note is that in the recurrence relations, $P_n^k(x) = 0$ when $k > n$.

For easy reference, we rewrite the recurrence formulae (3.94) to (3.98) in terms of θ. With $x = \cos\theta$, $(1-x^2)^{1/2} = \sin\theta$ and $d/dx = -(1/\sin\theta)d/d\theta$, we have

$$(2n+1)\cos\theta\, P_n^k(\cos\theta) = (n+k)P_{n-1}^k(\cos\theta) + (n-k+1)P_{n+1}^k(\cos\theta)\,, \qquad (3.157)$$

$$(2n+1)\sin\theta\, P_n^k(\cos\theta) = P_{n+1}^{k+1}(\cos\theta) - P_{n-1}^{k+1}(\cos\theta)\,, \qquad (3.158)$$

$$(2n+1)\sin\theta\, P_n^k(\cos\theta) = (n+k)(n+k-1)P_{n-1}^{k-1}(\cos\theta)$$

$$-(n-k+2)(n-k+1)P_{n+1}^{k-1}(\cos\theta)\,, \qquad (3.159)$$

$$-\sin\theta\frac{d}{d\theta}P_n^k(\cos\theta) = \sin\theta\, P_n^{k+1}(\cos\theta) - k\cos\theta\, P_n^k(\cos\theta)\,, \qquad (3.160)$$

$$-(2n+1)\sin\theta\frac{d}{d\theta}P_n^k(\cos\theta) = (n+1)(n+k)P_{n-1}^k(\cos\theta)$$

$$-n(n-k+1)P_{n+1}^k(\cos\theta)\,, \qquad (3.161)$$

$$-\sin\theta\frac{d}{d\theta}P_n^k(\cos\theta) = (n+k)P_{n-1}^k(\cos\theta) - n\cos\theta\, P_n^k(\cos\theta)\,. \qquad (3.162)$$

As a supplement, we reproduce the ordinary differential equation satisfied by $P_n^k(\cos\theta)$,

$$\frac{d^2}{d\theta^2} P_n^k(\cos\theta) + \cot\theta \frac{d}{d\theta} P_n^k(\cos\theta) - \frac{k^2}{\sin^2\theta} P_n^k(\cos\theta) = -n(n+1)P_n^k(\cos\theta),$$

(3.163)

which looks similar to a recurrence formula.

In the computation of $P_n^k(\cos\theta)$, the best choice is to use (3.157), which does not involve any division by $\sin\theta$ or $\cos\theta$, with $P_k^k(\cos\theta)$ and $P_{k+1}^k(\cos\theta)$ obtained from (3.127) as

$$P_k^k(\cos\theta) = \frac{(2k)!}{2^k k!} \sin^k\theta, \quad P_{k+1}^k(\cos\theta) = (2k+1)\cos\theta\, P_k^k(\cos\theta).$$

(3.164)

The second one can be obtained by setting $n = k$ in (3.157) and using $P_{k-1}^k(x) = 0$.

None of the formulae provided so far is optimal for computing $(d/d\theta)P_n^k(\cos\theta)$. We now derive a formula for it that does not involve any division by $\sin\theta$ or $\cos\theta$. By adding (3.160) and (3.162), we obtain

$$2\sin\theta \frac{d}{d\theta} P_n^k(\cos\theta) = (n+k)\left[\cos\theta\, P_n^k(\cos\theta) - P_n^{k-1}(\cos\theta)\right] - \sin\theta\, P_n^{k+1}(\cos\theta),$$

(3.165)

Solve for $\cos\theta\, P_n^k(\cos\theta)$ from (3.157), and then substitute the result into this equation. We obtain

$$2\sin\theta \frac{d}{d\theta} P_n^k(\cos\theta) = \frac{(n+k)(n-k+1)}{2n+1}\left[P_{n+1}^k(\cos\theta) - P_n^{k-1}(\cos\theta)\right]$$
$$- \sin\theta\, P_n^{k+1}(\cos\theta),$$

(3.166)

Substitute k by $k-1$ in (3.158), and then substitute the result into this equation. We obtain

$$2\frac{d}{d\theta} P_n^k(\cos\theta) = (n+k)(n-k+1)P_n^{k-1}(\cos\theta) - P_n^{k+1}(\cos\theta),$$

(3.167)

which is the formula we are seeking. For the special case $k = 0$, a formula can be inferred from (3.160)

$$\frac{d}{d\theta} P_n(\cos\theta) = -P_n^1(\cos\theta).$$

(3.168)

When $n = k$, the second term in (3.167) vanishes, and we have

$$\frac{d}{d\theta} P_k^k(\cos\theta) = k P_k^{k-1}(\cos\theta). \tag{3.169}$$

The forms of these formulae do not change after taking derivatives of any order with respect to θ. We have

$$2\frac{d^{i+1}}{d\theta^{i+1}} P_n^k(\cos\theta) = (n+k)(n-k+1)\frac{d^i}{d\theta^i} P_n^{k-1}(\cos\theta) - \frac{d^i}{d\theta^i} P_n^{k+1}(\cos\theta), \tag{3.170}$$

$$\frac{d^{i+1}}{d\theta^{i+1}} P_n(\cos\theta) = -\frac{d^i}{d\theta^i} P_n^1(\cos\theta), \tag{3.171}$$

$$\frac{d^{i+1}}{d\theta^{i+1}} P_k^k(\cos\theta) = k\frac{d^i}{d\theta^i} P_k^{k-1}(\cos\theta). \tag{3.172}$$

These formulae can be used to compute $(d^{i+1}/d\theta^{i+1}) P_n^k(\cos\theta)$ from the values of $(d^i/d\theta^i) P_n^k(\cos\theta)$.

In physical geodesy, the fully normalized associated Legendre function is usually used, which is the topic of Sect. 3.5.6. The use of the above-mentioned recurrence formulae will be further discussed there.

3.3.5 Orthogonality

The Concept of Orthogonality
First, we define a symbol

$$\delta_{mn} = \begin{cases} 0, & m \neq n; \\ 1, & m = n, \end{cases} \tag{3.173}$$

called Kronecker δ. Another Kronecker delta to be used later is defined by

$$\delta_k = \begin{cases} 0, & k \neq 0; \\ 1, & k = 0, \end{cases} \tag{3.174}$$

which is a short form of δ_{k0}.

Let $\phi_1(x), \phi_2(x), \cdots$ to be a sequence of functions defined in $a \leq x \leq b$. If

$$\int_a^b \phi_n(x)\phi_m(x)W(x)dx = \delta_{mn}\lambda_n = \begin{cases} 0, & m \neq n; \\ \lambda_n, & m = n, \end{cases} \tag{3.175}$$

we say that they are orthogonal in $a \leq x \leq b$, where $W(x)$ is called weight.

Basic Orthogonality 1

The most useful orthogonality formula of the associated Legendre function is

$$\int_{-1}^{1} P_n^k(x) P_m^k(x) dx = \frac{2\delta_{mn}}{2n+1} \frac{(n+k)!}{(n-k)!} = \begin{cases} 0, & m \neq n; \\ \dfrac{2}{2n+1} \dfrac{(n+k)!}{(n-k)!}, & m = n. \end{cases}$$

(3.176)

When $k = 0$, this formula degenerates to

$$\int_{-1}^{1} P_n(x) P_m(x) dx = \frac{2\delta_{mn}}{2n+1} = \begin{cases} 0, & m \neq n; \\ \dfrac{2}{2n+1}, & m = n. \end{cases}$$

(3.177)

Substitute $x = \cos\theta$ into (3.176) and (3.177), we obtain their expressions in θ,

$$\int_0^\pi P_n^k(\cos\theta) P_m^k(\cos\theta) \sin\theta d\theta = \frac{2\delta_{mn}}{2n+1} \frac{(n+k)!}{(n-k)!},$$

(3.178)

$$\int_0^\pi P_n(\cos\theta) P_m(\cos\theta) \sin\theta d\theta = \frac{2\delta_{mn}}{2n+1}.$$

(3.179)

The proof of (3.176) is provided in the following paragraphs.

When $m \neq n$, we obtain the differential equations satisfied by $P_n^k(x)$ and $P_m^k(x)$ to be, by setting $\alpha = n(n+1)$ and $\alpha = m(m+1)$ in (3.32),

$$\frac{d}{dx}\left[(1-x^2)\frac{dP_n^k(x)}{dx}\right] + \left[n(n+1) - \frac{k^2}{1-x^2}\right] P_n^k(x) = 0,$$

(3.180)

$$\frac{d}{dx}\left[(1-x^2)\frac{dP_m^k(x)}{dx}\right] + \left[m(m+1) - \frac{k^2}{1-x^2}\right] P_m^k(x) = 0.$$

(3.181)

Multiply the first equation by $P_m^k(x)$ and the second by $P_n^k(x)$, and then subtract the second from the first. We obtain

$$[n(n+1) - m(m+1)] P_n^k(x) P_m^k(x)$$

$$= P_n^k(x) \frac{d}{dx}\left[(1-x^2)\frac{dP_m^k(x)}{dx}\right] - P_m^k(x) \frac{d}{dx}\left[(1-x^2)\frac{dP_n^k(x)}{dx}\right]$$

$$= \frac{d}{dx}\left\{(1-x^2)\left[P_n^k(x)\frac{dP_m^k(x)}{dx} - P_m^k(x)\frac{dP_n^k(x)}{dx}\right]\right\}.$$

(3.182)

Integrate both sides over $-1 \leq x \leq 1$. We obtain

$$[n(n+1) - m(m+1)] \int_{-1}^{1} P_n^k(x) P_m^k(x) dx = 0. \qquad (3.183)$$

When $m \neq n$, $n(n+1) - m(m+1) \neq 0$. Hence,

$$\int_{-1}^{1} P_n^k(x) P_m^k(x) dx = 0, \qquad m \neq n. \qquad (3.184)$$

This is the $m \neq n$ portion of (3.176).

When $m = n$, we obtain, according to (3.128) and (3.140),

$$\int_{-1}^{1} [P_n^k(x)]^2 dx = \frac{(-1)^k}{2^{2n} (n!)^2} \frac{(n+k)!}{(n-k)!} J, \qquad (3.185)$$

where J is an integral defined as

$$J = \int_{-1}^{1} \frac{d^{n+k}}{dx^{n+k}} (x^2 - 1)^n \frac{d^{n-k}}{dx^{n-k}} (x^2 - 1)^n dx. \qquad (3.186)$$

By integrating by parts, J can be reduced to

$$J = \left[\frac{d^{n+k}}{dx^{n+k}} (x^2 - 1)^n \frac{d^{n-k-1}}{dx^{n-k-1}} (x^2 - 1)^n \right]_{-1}^{1}$$
$$- \int_{-1}^{1} \frac{d^{n+k+1}}{dx^{n+k+1}} (x^2 - 1)^n \frac{d^{n-k-1}}{dx^{n-k-1}} (x^2 - 1)^n dx. \qquad (3.187)$$

The first term vanishes, thus,

$$J = - \int_{-1}^{1} \frac{d^{n+k+1}}{dx^{n+k+1}} (x^2 - 1)^n \frac{d^{n-k-1}}{dx^{n-k-1}} (x^2 - 1)^n dx. \qquad (3.188)$$

Integrate by parts again. We obtain

$$J = \int_{-1}^{1} \frac{d^{n+k+2}}{dx^{n+k+2}} (x^2 - 1)^n \frac{d^{n-k-2}}{dx^{n-k-2}} (x^2 - 1)^n dx, \qquad (3.189)$$

$$\cdots$$

$$J = (-1)^{n-k} \int_{-1}^{1} \left[\frac{d^{2n}}{dx^{2n}} (x^2 - 1)^n \right] (x^2 - 1)^n dx. \qquad (3.190)$$

As $(x^2 - 1)^n$ is a polynomial of degree $2n$, $d^{2n}(x^2 - 1)^n/dx^{2n} = (2n)!$. Substitute it into the above formula. We obtain

$$J = (-1)^{n-k}(2n)! \int_{-1}^{1} (x^2 - 1)^n dx = (-1)^k (2n)! \int_{-1}^{1} (1 - x^2)^n dx \,. \qquad (3.191)$$

Finally, substitute the above formula into (3.185). We obtain

$$\int_{-1}^{1} [P_n^k(x)]^2 dx = \frac{(2n)!}{2^{2n}(n!)^2} \frac{(n+k)!}{(n-k)!} \int_{-1}^{1} (1 - x^2)^n dx \,. \qquad (3.192)$$

Now we only have to compute the integral in this formula. We have

$$\int_{-1}^{1} (1 - x^2)^n dx = \int_{\pi}^{0} \sin^{2n} \theta \, d \cos \theta$$

$$= \left(\sin^{2n} \theta \cos \theta \right)_{\pi}^{0} - \int_{\pi}^{0} \cos \theta \, d \sin^{2n} \theta$$

$$= -2n \int_{\pi}^{0} \cos^2 \theta \sin^{2n-1} \theta \, d\theta$$

$$= 2n \int_{\pi}^{0} (1 - \sin^2 \theta) \sin^{2n-2} \theta \, d \cos \theta$$

$$= 2n \left(\int_{\pi}^{0} \sin^{2n-2} \theta \, d \cos \theta - \int_{\pi}^{0} \sin^{2n} \theta \, d \cos \theta \right)$$

$$= 2n \left[\int_{-1}^{1} (1 - x^2)^{n-1} dx - \int_{-1}^{1} (1 - x^2)^n dx \right] \,. \qquad (3.193)$$

We obtain from this formula

$$\int_{-1}^{1} (1 - x^2)^n dx = \frac{2n}{2n+1} \int_{-1}^{1} (1 - x^2)^{n-1} dx \,. \qquad (3.194)$$

We can further obtain by repeatedly using this formula

$$\int_{-1}^{1} (1 - x^2)^{n-1} dx = \frac{2(n-1)}{2(n-1)+1} \int_{-1}^{1} (1 - x^2)^{n-2} dx \,, \qquad (3.195)$$

$$\cdots \cdots$$

$$\int_{-1}^{1} (1 - x^2) dx = \frac{2}{3} \int_{-1}^{1} dx = \frac{2}{3} \times 2 \,. \qquad (3.196)$$

By substituting (3.195) to (3.196) into (3.194), we obtain

$$\int_{-1}^{1} (1 - x^2)^n dx = \frac{2n[2(n-1)][2(n-2)] \cdots \cdots 2}{(2n+1)(2n-1)(2n-3) \cdots \cdots 3} \times 2 = \frac{2^{2n}(n!)^2}{(2n+1)!} \times 2.$$

(3.197)

Finally, by substituting this formula into (3.192), we obtain the desired result

$$\int_{-1}^{1} [P_n^k(x)]^2 dx = \frac{2}{2n+1} \frac{(n+k)!}{(n-k)!}.$$

(3.198)

This is the $m = n$ portion of (3.176).

Basic Orthogonality 2
Another fundamental orthogonality formula of the associated Legendre function is

$$\int_{-1}^{1} P_n^k(x) P_n^l(x) \frac{dx}{1 - x^2} = \frac{\delta_{kl}}{k} \frac{(n+k)!}{(n-k)!} = \begin{cases} \frac{1}{k} \frac{(n+k)!}{(n-k)!}, & l = k; \\ 0, & l \neq k. \end{cases}$$

(3.199)

With $x = \cos\theta$, it can be written as

$$\int_0^\pi P_n^k(\cos\theta) P_n^l(\cos\theta) \frac{1}{\sin\theta} d\theta = \frac{\delta_{kl}}{k} \frac{(n+k)!}{(n-k)!}.$$

(3.200)

It is evident that the special case $k = l = 0$ is not permitted. The proof of (3.199) is provided in the following paragraphs.

When $l \neq k$, $P_n^k(x)$ and $P_n^l(x)$ satisfy

$$\frac{d}{dx}\left[(1 - x^2)\frac{dP_n^k(x)}{dx}\right] + \left[n(n+1) - \frac{k^2}{1-x^2}\right] P_n^k(x) = 0,$$

(3.201)

$$\frac{d}{dx}\left[(1 - x^2)\frac{dP_n^l(x)}{dx}\right] + \left[n(n+1) - \frac{l^2}{1-x^2}\right] P_n^l(x) = 0,$$

(3.202)

respectively. Multiply the first one by $P_n^l(x)$ and the second one by $P_n^k(x)$, and then subtract one from another. We obtain

$$(k^2 - l^2) P_n^k(x) P_n^l(x) \frac{1}{1-x^2}$$

$$= P_n^l(x)\frac{d}{dx}\left[(1 - x^2)\frac{dP_n^k(x)}{dx}\right] - P_n^k(x)\frac{d}{dx}\left[(1 - x^2)\frac{dP_n^l(x)}{dx}\right]$$

$$= \frac{d}{dx}\left\{(1 - x^2)\left[P_n^l(x)\frac{dP_n^k(x)}{dx} - P_n^k(x)\frac{dP_n^l(x)}{dx}\right]\right\}.$$

(3.203)

Integrate both sides over x in the domain $-1 \leq x \leq 1$. We obtain

$$(k^2 - l^2) \int_{-1}^{1} P_n^k(x) P_n^l(x) \frac{dx}{1 - x^2} = 0 . \tag{3.204}$$

As $l \neq k$, thus $(k^2 - l^2) \neq 0$, this formula is the $l \neq k$ portion of (3.199).
  According to (3.128), we have

$$\int_{-1}^{1} [P_n^k(x)]^2 \frac{dx}{1 - x^2} = \frac{1}{2^{2n}(n!)^2} J , \tag{3.205}$$

where

$$J = \int_{-1}^{1} (1 - x^2)^{k-1} \left[\frac{d^{n+k}}{dx^{n+k}} (x^2 - 1)^n \right]^2 dx . \tag{3.206}$$

Let

$$G = (1 - x^2)^{k-1} \frac{d^{n+k}}{dx^{n+k}} (x^2 - 1)^n . \tag{3.207}$$

We can write (3.206) as

$$J = \int_{-1}^{1} G \frac{d^{n+k}}{dx^{n+k}} (x^2 - 1)^n dx . \tag{3.208}$$

As $G(\pm 1) = 0$, by integrating by parts, we obtain

$$J = \left[G \frac{d^{n+k-1}}{dx^{n+k-1}} (x^2 - 1)^n \right]_{-1}^{1} - \int_{-1}^{1} \frac{dG}{dx} \frac{d^{n+k-1}}{dx^{n+k-1}} (x^2 - 1)^n dx$$

$$= - \int_{-1}^{1} \frac{dG}{dx} \frac{d^{n+k-1}}{dx^{n+k-1}} (x^2 - 1)^n dx . \tag{3.209}$$

As G includes the factor $(1 - x^2)^{k-1}$, we have

$$\left(\frac{d^m G}{dx^m} \right)_{x=\pm 1} = 0 , \qquad m = 0, 1, 2 \ldots \ldots, k - 2 . \tag{3.210}$$

Hence, we can repeatedly integrate (3.209) by parts to obtain

$$J = (-1)^{k-1} \int_{-1}^{1} \frac{d^{k-1}G}{dx^{k-1}} \frac{d^{n+1}}{dx^{n+1}} (x^2 - 1)^n dx . \tag{3.211}$$

By integrating by parts once again, we obtain

$$J = (-1)^{k-1} \left\{ \left[\frac{d^{k-1}G}{dx^{k-1}} \frac{d^n}{dx^n} (x^2 - 1)^n \right]_{-1}^{1} - \int_{-1}^{1} \frac{d^k G}{dx^k} \frac{d^n}{dx^n} (x^2 - 1)^n dx \right\} . \tag{3.212}$$

By substituting this formula into (3.205) and then using (3.72), we obtain

$$\int_{-1}^{1} [P_n^k(x)]^2 \frac{dx}{1-x^2} = \frac{(-1)^{k-1}}{2^n n!} \left[\frac{d^{k-1}G}{dx^{k-1}} P_n(x) \right]_{-1}^{1} - \frac{(-1)^{k-1}}{2^n n!} \int_{-1}^{1} \frac{d^k G}{dx^k} P_n(x) dx . \tag{3.213}$$

As G is a polynomial of degree $2(k-1) + 2n - (n+k) = n+k-2$, $d^k G/dx^k$ should be a polynomial of degree $n-2$. As the Legendre function $P_m(x)$ is a polynomial of degree m, we can express $d^k G/dx^k$ as a linear combination of Legendre polynomial up to degree $n-2$, i.e.,

$$\frac{d^k G}{dx^k} = \sum_{m=0}^{n-2} a_m P_m(x) , \tag{3.214}$$

where the coefficients a_m are constants. By substituting (3.214) into the second term at the right-hand side of (3.213) and then using (3.177), we conclude that this term vanishes. Hence, (3.213) becomes

$$\int_{-1}^{1} [P_n^k(x)]^2 \frac{dx}{1 - x^2} = \frac{(-1)^{k-1}}{2^n n!} \left[\frac{d^{k-1}G}{dx^{k-1}} P_n(x) \right]_{-1}^{1} . \tag{3.215}$$

The substitution of $P_n(\pm 1) = (\pm 1)^n$ into this formula yields

$$\int_{-1}^{1} [P_n^k(x)]^2 \frac{dx}{1 - x^2} = \frac{(-1)^{k-1}}{2^n n!} \left[\left(\frac{d^{k-1}G}{dx^{k-1}} \right)_{x=1} - (-1)^n \left(\frac{d^{k-1}G}{dx^{k-1}} \right)_{x=-1} \right] . \tag{3.216}$$

The rest is to formulate $\left(d^{k-1}G/dx^{k-1}\right)_{x=\pm 1}$. First, we have

$$
\left(\frac{d^{k-1}G}{dx^{k-1}}\right)_{x=\pm 1}
$$

$$
= \frac{d^{k-1}}{dx^{k-1}}\left[(1-x^2)^{k-1}\frac{d^{n+k}}{dx^{n+k}}(x^2-1)^n\right]_{x=\pm 1}
$$

$$
= \sum_{l=0}^{k-1}\frac{(k-1)!}{l!(k-1-l)!}\left[\frac{d^l}{dx^l}(1-x^2)^{k-1}\frac{d^{n+k+k-1-l}}{dx^{n+k+k-1-l}}(x^2-1)^n\right]_{x=\pm 1}.
$$

$$
\tag{3.217}
$$

It can be seen that only when $l = k-1$, the term $d^l(1-x^2)^{k-1}/dx^l$ does not vanish when $x \pm 1$. Hence, the above formula can be written as

$$
\left(\frac{d^{k-1}G}{dx^{k-1}}\right)_{x=\pm 1} = \left[\frac{d^{k-1}}{dx^{k-1}}(1-x^2)^{k-1}\frac{d^{n+k}}{dx^{n+k}}(x^2-1)^n\right]_{x=\pm 1}. \tag{3.218}
$$

According to (3.72) and (3.89), we have

$$
\left[\frac{d^{k-1}}{dx^{k-1}}(1-x^2)^{k-1}\right]_{x=\pm 1} = (-1)^{k-1}2^{k-1}(k-1)!P_{k-1}(\pm 1)
$$

$$
= (\mp 1)^{k-1}2^{k-1}(k-1)!. \tag{3.219}
$$

We also have

$$
\left[\frac{d^{n+k}}{dx^{n+k}}(x^2-1)^n\right]_{x=\pm 1}
$$

$$
= \sum_{l=0}^{n+k}\frac{(n+k)!}{l!(n+k-l)!}\left[\frac{d^l}{dx^l}(x-1)^n\frac{d^{n+k-l}}{dx^{n+k-l}}(x+1)^n\right]_{x=\pm 1}. \tag{3.220}
$$

When $x = 1$, $d^l(x-1)^n/dx^l$ is not vanishing only when $l = n$. Hence,

$$
\left[\frac{d^{n+k}}{dx^{n+k}}(x^2-1)^n\right]_{x=1} = \frac{(n+k)!}{n!k!}\left[\frac{d^n}{dx^n}(x-1)^n\frac{d^k}{dx^k}(x+1)^n\right]_{x=1}
$$

$$
= \frac{(n+k)!}{n!k!}n!\frac{n!}{(n-k)!}2^{n-k}
$$

$$
= \frac{n!}{k!}2^{n-k}\frac{(n+k)!}{(n-k)!}. \tag{3.221}
$$

Furthermore, when $x = -1$, $d^{n+k-1}(x+1)^n/dx^{n+k-1}$ is not vanishing only when $n + k - l = n$. Hence,

$$\left[\frac{d^{n+k}}{dx^{n+k}}(x^2 - 1)^n\right]_{x=-1} = \frac{(n+k)!}{k!n!}\left[\frac{d^k}{dx^k}(x-1)^n\frac{d^n}{dx^n}(x+1)^n\right]_{x=-1}$$

$$= \frac{(n+k)!}{k!n!}\frac{n!}{(n-k)!}(-2)^{n-k}n!$$

$$= (-1)^{n-k}\frac{n!}{k!}2^{n-k}\frac{(n+k)!}{(n-k)!}. \tag{3.222}$$

By substituting (3.219), (3.221), and (3.222) into (3.218), we obtain

$$\left(\frac{d^{k-1}G}{dx^{k-1}}\right)_{x=\pm 1} = (\mp 1)^{k-1}2^{k-1}(k-1)!(\pm 1)^{n-k}\frac{n!}{k!}2^{n-k}\frac{(n+k)!}{(n-k)!}$$

$$= (\mp 1)^{k-1}(\pm 1)^{n-k}2^{n-1}\frac{n!}{k}\frac{(n+k)!}{(n-k)!}. \tag{3.223}$$

Finally, the substitution of this formula into (3.216) yields

$$\int_{-1}^{1}[P_n^k(x)]^2\frac{dx}{1-x^2} = \frac{(-1)^{k-1}}{2^n n!}\left[(-1)^{k-1}2^{n-1}\frac{n!}{k}\frac{(n+k)!}{(n-k)!}\right.$$

$$\left. - (-1)^n(-1)^{n-k}2^{n-1}\frac{n!}{k}\frac{(n+k)!}{(n-k)!}\right]$$

$$= \frac{1}{k}\frac{(n+k)!}{(n-k)!}. \tag{3.224}$$

This is the portion $l = k$ of (3.199).

Some Other Formulae

Two extra orthogonality relations of the (associated) Legendre function can be derived from (3.178). The differential equation satisfied by $P_n^k(\cos\theta)$ (3.163) can be written as

$$\frac{1}{\sin\theta}\frac{d}{d\theta}\left[\sin\theta\frac{d}{d\theta}P_n^k(\cos\theta)\right] + \left[n(n+1) - \frac{k^2}{\sin^2\theta}\right]P_n^k(\cos\theta) = 0, \tag{3.225}$$

from which we can obtain

$$\frac{1}{\sin\theta}\frac{d}{d\theta}\left[\sin\theta P_m^k(\cos\theta)\frac{d}{d\theta}P_n^k(\cos\theta)\right] + n(n+1)P_n^k(\cos\theta)P_m^k(\cos\theta)$$

$$= \frac{k^2}{\sin^2\theta}P_n^k(\cos\theta)P_m^k(\cos\theta) + \frac{d}{d\theta}P_n^k(\cos\theta)\frac{d}{d\theta}P_m^k(\cos\theta). \tag{3.226}$$

Multiply both sides by $\sin \theta$, and then integrate over $0 \leq \theta \leq \pi$. We obtain

$$\int_0^\pi \left[\frac{k^2}{\sin^2 \theta} P_n^k(\cos \theta) P_m^k(\cos \theta) + \frac{d}{d\theta} P_n^k(\cos \theta) \frac{d}{d\theta} P_m^k(\cos \theta) \right] \sin \theta \, d\theta$$

$$= n(n+1) \int_0^\pi P_n^k(\cos \theta) P_m^k(\cos \theta) \sin \theta \, d\theta$$

$$= \frac{2n(n+1)\delta_{mn}}{2n+1} \frac{(n+k)!}{(n-k)!}. \tag{3.227}$$

This is the first formula we are proving. The second one is obtained by integrating the equation

$$\frac{d}{d\theta} \left[\cos \theta P_n^k(\cos \theta) P_m^k(\cos \theta) \right] = -P_n^k(\cos \theta) P_m^k(\cos \theta) \sin \theta$$

$$+ \cos \theta \frac{d}{d\theta} \left[P_n^k(\cos \theta) P_m^k(\cos \theta) \right] \tag{3.228}$$

over $0 \leq \theta \leq \pi$, yielding

$$\int_0^\pi \cos \theta \frac{d}{d\theta} \left[P_n^k(\cos \theta) P_m^k(\cos \theta) \right] d\theta$$

$$= \int_0^\pi P_n^k(\cos \theta) P_m^k(\cos \theta) \sin \theta \, d\theta + \left[\cos \theta P_n^k(\cos \theta) P_m^k(\cos \theta) \right]_0^\pi$$

$$= \frac{2\delta_{mn}}{2n+1} \frac{(n+k)!}{(n-k)!} - \delta_k[(-1)^{m+n} + 1], \tag{3.229}$$

where δ_k is a Kronecker δ defined in (3.174). A useful integral formula can be obtained from (3.227) and (3.200)

$$\int_0^\pi \left[\frac{d}{d\theta} P_n^k(\cos \theta) \right]^2 \sin \theta \, d\theta = \frac{2n(n+1)}{2n+1} \frac{(n+k)!}{(n-k)!} - \int_0^\pi \frac{k^2}{\sin \theta} \left[P_n^k(\cos \theta) \right]^2 d\theta$$

$$= \frac{2n(n+1)}{2n+1} \frac{(n+k)!}{(n-k)!} - k \frac{(n+k)!}{(n-k)!}. \tag{3.230}$$

3.4   Spherical Harmonics

3.4.1   Expansion of a Harmonic Function as Spherical Harmonics Series

While we were solving Laplace's equation, we have separated the harmonic function V as the product of three functions $f(r)$, $B(\theta)$, and $L(\lambda)$. We have already obtained $B(\theta)$ and $L(\lambda)$. Now we solve for $f(r)$.

Substitute the eigenvalue $\alpha = n(n+1)$ into the equation of $f(r)$ (3.19), we obtain

$$\frac{d}{dr}\left(r^2\frac{df}{dr}\right) - n(n+1)f = 0.\tag{3.231}$$

It can be easily verified that its two linearly independent solutions are

$$f_1 = r^n, \qquad f_2 = \frac{1}{r^{n+1}}.\tag{3.232}$$

As $f_1 \to 0$ and $f_2 \to \infty$ when $r \to 0$ and $f_1 \to \infty$ and $f_2 \to 0$ when $r \to \infty$, f_1 is suitable in studying the solution of Laplace's equation in the interior of a sphere with center on the origin of the coordinate system $Or\theta\lambda$ and f_2 in the exterior of a sphere.

Evidently, all possible combinations $f(r)B(\theta)L(\lambda)$ are solutions of Laplace's equation. Within a sphere with center on the origin of the coordinate system $Or\theta\lambda$, the most general finite solution based on separation of variables is

$$V = \sum_{n=0}^{\infty}\sum_{k=0}^{n} r^n (A_n^k \cos k\lambda + B_n^k \sin k\lambda) P_n^k(\cos\theta).\tag{3.233}$$

In the exterior of a sphere, the most general finite solution based on separation of variables is

$$V = \sum_{n=0}^{\infty}\sum_{k=0}^{n} \frac{1}{r^{n+1}}(A_n^k \cos k\lambda + B_n^k \sin k\lambda) P_n^k(\cos\theta).\tag{3.234}$$

Within a spherical shell with center on the origin of the coordinate system $Or\theta\lambda$, the most general finite solution based on separation of variables is

$$V = \sum_{n=0}^{\infty}\sum_{k=0}^{n} r^n (A_n^k \cos k\lambda + B_n^k \sin k\lambda) P_n^k(\cos\theta)$$

$$+ \sum_{n=0}^{\infty}\sum_{k=0}^{n} \frac{1}{r^{n+1}}(C_n^k \cos k\lambda + D_n^k \sin k\lambda) P_n^k(\cos\theta).\tag{3.235}$$

In these formulae, $P_n^k(\cos\theta)\cos k\lambda$ and $P_n^k(\cos\theta)\sin k\lambda$ are referred to as surface spherical harmonics of degree n and order k, and their products with r or $1/r$ are referred to as solid spherical harmonics. The term *spherical harmonics* is also used in more general senses. When $k = 0$, the coefficients A_n^k and C_n^k are also denoted as A_n and C_n, respectively. The coefficients B_n^0 and D_n^0 are meaningless, as they multiply with 0.

The Earth's external gravitational potential is expressed in the form of (3.234).

The spherical harmonic series (3.233)–(3.235) can also be expressed in the complex form. The main component in these expansions is the surface spherical harmonics of degree n,

$$Y_n(\theta, \lambda) = \sum_{k=0}^{n}(A_n^k\cos k\lambda + B_n^k\sin k\lambda)P_n^k(\cos\theta)\,. \tag{3.236}$$

With Euler's formula

$$e^{ik\lambda} = \cos k\lambda + i\sin k\lambda\,, \tag{3.237}$$

and taking into account of (3.139), we realize that (3.236) can be expressed in the complex form

$$Y_n(\theta, \lambda) = \sum_{k=-n}^{n} C_n^k P_n^k(\cos\theta)e^{ik\lambda}\,. \tag{3.238}$$

Now we express C_n^k in terms of A_n^k and B_n^k. To do so, we reduce (3.238) to the form of (3.236):

$$Y_n(\theta, \lambda) = \sum_{k=-n}^{n} C_n^k(\cos k\lambda + i\sin k\lambda)P_n^k(\cos\theta)$$

$$= \sum_{k=0}^{n}(C_n^k\cos k\lambda + iC_n^k\sin k\lambda)P_n^k(\cos\theta)$$

$$+ \sum_{k=-n}^{-1}(C_n^k\cos k\lambda + iC_n^k\sin k\lambda)P_n^k(\cos\theta)$$

$$= \sum_{k=0}^{n}(C_n^k\cos k\lambda + iC_n^k\sin k\lambda)P_n^k(\cos\theta)$$

$$+ \sum_{k=1}^{n}(C_n^{-k}\cos k\lambda - iC_n^{-k}\sin k\lambda)P_n^{-k}(\cos\theta)$$

$$= \sum_{k=0}^{n} (C_n^k \cos k\lambda + i C_n^k \sin k\lambda) P_n^k (\cos \theta)$$

$$+ \sum_{k=1}^{n} (C_n^{-k} \cos k\lambda - i C_n^{-k} \sin k\lambda)(-1)^k \frac{(n-k)!}{(n+k)!} P_n^k (\cos \theta)$$

$$= C_n P_n (\cos \theta) + \sum_{k=1}^{n} \left\{ \left[C_n^k + (-1)^k \frac{(n-k)!}{(n+k)!} C_n^{-k} \right] \cos k\lambda \right.$$

$$\left. + i \left[C_n^k - (-1)^k \frac{(n-k)!}{(n+k)!} C_n^{-k} \right] \sin k\lambda \right\} P_n^k (\cos \theta) . \qquad (3.239)$$

We obtain, by comparing (3.239) with (3.236),

$$\begin{cases} A_n = C_n \, ; \\ A_n^k = C_n^k + (-1)^k \dfrac{(n-k)!}{(n+k)!} C_n^{-k} \, , \qquad k \geq 1 \, ; \\ B_n^k = i \left[C_n^k - (-1)^k \dfrac{(n-k)!}{(n+k)!} C_n^{-k} \right] , \; k \geq 1 \, . \end{cases} \qquad (3.240)$$

For a real function, A_n^k and B_n^k are real. Hence,

$$\begin{cases} \mathrm{Im}(C_n) = 0 \, , \\ \mathrm{Im}(C_n^k) + (-1)^k \dfrac{(n-k)!}{(n+k)!} \mathrm{Im}(C_n^{-k}) = 0 \, , \\ \mathrm{Re}(C_n^k) - (-1)^k \dfrac{(n-k)!}{(n+k)!} \mathrm{Re}(C_n^{-k}) = 0 \, , \end{cases} \qquad (3.241)$$

where Re and Im are used to indicate the real and imaginary components, respectively. In this case, (3.240) can be simplified to

$$\begin{cases} A_n = \mathrm{Re}(C_n) \, ; \\ A_n^k = 2\mathrm{Re}(C_n^k) \, , \quad k \geq 1 \, ; \\ B_n^k = -2\mathrm{Im}(C_n^k) \, , \; k \geq 1 \, . \end{cases} \qquad (3.242)$$

3.4.2 Geometrical Properties

When $k = 0$, a surface spherical harmonics degenerates to $P_n(\cos \theta)$, which has n zeros in $0° < \theta < 180°$, and changes sign at all zeros. Hence, $P_n(\cos \theta)$ divides the sphere into $n + 1$ zones of alternate signs, thus being referred to as zonal spherical harmonics. In Fig. 3.5a, we show $P_3(\cos \theta)$ with zeros $\theta = 50.8°, 90°$, and $129.2°$.

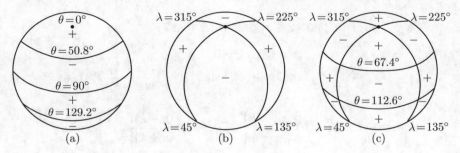

Fig. 3.5 Graphs of some spherical harmonics

When $k = n$, $P_n^n(\cos\theta)$ has no zero in $0° < \theta < 180°$. It equals zero only at $\theta = 0°$ and $180°$. However, $\cos n\lambda$ or $\sin n\lambda$ has $2n$ zeros in $0° \leq \lambda < 360°$ and changes sign at all zeros. Hence, $P_n^n(\cos\theta)\cos n\lambda$ or $P_n^n(\cos\theta)\sin n\lambda$ divides the sphere into $2n$ sectors of alternate signs, thus being referred to as sectorial spherical harmonics. In Fig. 3.5b, we show $P_2^2(\cos\theta)\cos 2\lambda$ with zeros $\lambda = 45°$, $135°$, $225°$, and $315°$.

When $0 < k < n$, $P_n^k(\cos\theta)$ has $n - k$ zeros in $0° < \theta < 180°$, and $\cos k\lambda$ or $\sin k\lambda$ has $2k$ zeros in $0° \leq \lambda < 360°$. Both $P_n^k(\cos\theta)$ and $\cos k\lambda$ or $\sin k\lambda$ change sign at their zeros. Hence, $P_n^k(\cos\theta)\cos k\lambda$ or $P_n^k(\cos\theta)\sin k\lambda$ divides the sphere into $2k(n - k + 1)$ quadrangles (triangles at the poles), thus being referred to as tesseral spherical harmonics. In Fig. 3.5c, we show $P_4^2(\cos\theta)\cos 2\lambda$ with zeros $\theta = 0°$, $67.4°$, $112.6°$, and $180°$, and $\lambda = 45°$, $135°$, $225°$, and $315°$.

3.4.3 Orthogonality

Orthogonality of spherical harmonics can be derived from the orthogonality of associated Legendre function and those of cosine and sine functions. The orthogonality of associated Legendre function is given in (3.178). The cosine and sine have the following well-known orthogonality relations:

$$\int_0^{2\pi} \cos k\lambda \cos l\lambda d\lambda = \delta_{lk}(1 + \delta_k)\pi = \begin{cases} 0, & l \neq k; \\ \pi, & l = k \neq 0; \\ 2\pi, & l = k = 0; \end{cases} \tag{3.243}$$

$$\int_0^{2\pi} \cos k\lambda \sin l\lambda d\lambda = 0; \tag{3.244}$$

$$\int_0^{2\pi} \sin k\lambda \sin l\lambda d\lambda = \delta_{lk}(1 - \delta_k)\pi = \begin{cases} 0, & l \neq k; \\ \pi, & l = k \neq 0; \\ 0, & l = k = 0, \end{cases} \tag{3.245}$$

where δ_{mn} and δ_k are the Kronecker symbols defined in (3.173) and (3.174).

We express the orthogonality relations of spherical harmonics as integrals on a unit sphere ω, where the area element is $d\omega = \sin\theta d\theta d\lambda$. For brevity, we introduce the symbols

$$C_n^k(\theta, \lambda) = P_n^k(\cos\theta)\cos k\lambda, \qquad S_n^k(\theta, \lambda) = P_n^k(\cos\theta)\sin k\lambda. \qquad (3.246)$$

The orthogonality relations of spherical harmonics can then be obtained using (3.178) and (3.243)–(3.245) as follows:

$$\int_\omega C_n^k(\theta, \lambda) C_m^l(\theta, \lambda) d\omega = \delta_{mn}\delta_{lk}(1 + \delta_k)\frac{2\pi}{2n+1}\frac{(n+k)!}{(n-k)!}$$

$$= \begin{cases} 0, & m \neq n \text{ or } l \neq k; \\ \dfrac{2\pi}{2n+1}\dfrac{(n+k)!}{(n-k)!}, & m = n \text{ and } l = k \neq 0; \\ \dfrac{4\pi}{2n+1}\dfrac{(n+k)!}{(n-k)!}, & m = n \text{ and } l = k = 0; \end{cases} \qquad (3.247)$$

$$\int_\omega C_n^k(\theta, \lambda) S_m^l(\theta, \lambda) d\omega = 0, \qquad (3.248)$$

$$\int_\omega S_n^k(\theta, \lambda) S_m^l(\theta, \lambda) d\omega = \delta_{mn}\delta_{lk}(1 - \delta_k)\frac{2\pi}{2n+1}\frac{(n+k)!}{(n-k)!}$$

$$= \begin{cases} 0, & m \neq n \text{ or } l \neq k; \\ \dfrac{2\pi}{2n+1}\dfrac{(n+k)!}{(n-k)!}, & m = n \text{ and } l = k \neq 0; \\ 0, & m = n \text{ and } l = k = 0. \end{cases} \qquad (3.249)$$

In order to derive the orthogonality relations in the complex form, we use (3.139) to obtain

$$P_m^k(\cos\theta) = (-1)^k \frac{(m+k)!}{(m-k)!} P_m^{-k}(\cos\theta). \qquad (3.250)$$

By substituting it into (3.178), we obtain

$$\int_0^\pi P_n^k(\cos\theta) P_m^{-k}(\cos\theta) \sin\theta d\theta = (-1)^k \frac{2\delta_{mn}}{2n+1} = \begin{cases} 0, & m \neq n; \\ (-1)^k \dfrac{2}{2n+1}, & m = n. \end{cases} \qquad (3.251)$$

With

$$\int_0^{2\pi} e^{ik\lambda} e^{-il\lambda} d\lambda = 2\pi \delta_{lk} = \begin{cases} 0, & l \neq k; \\ 2\pi, & l = k, \end{cases} \tag{3.252}$$

we obtain

$$\int_\omega \left[P_n^k(\cos\theta) e^{ik\lambda} \right] \left[P_m^{-k}(\cos\theta) e^{-il\lambda} \right] d\omega = (-1)^k \delta_{mn} \delta_{lk} \frac{4\pi}{2n+1}$$

$$= \begin{cases} 0, & m \neq n \ \text{ or } \ l \neq k; \\ (-1)^k \dfrac{4\pi}{2n+1}, & m = n \ \text{ and } \ l = k. \end{cases} \tag{3.253}$$

In summary, the orthogonality relations are (3.247)–(3.249) and (3.253).

3.4.4   Addition Theorem and Azimuthal Average of Spherical Harmonics

We will come up with the Legendre function $P_n(\cos\psi)$, where ψ is the angle made by two points on a sphere, P and M, with respect to the center of the sphere. We denote the coordinates of P and M as (θ, λ) and (θ', λ'), respectively, as shown in Fig. 3.6. The formula of $\cos\psi$ is derived in Appendix A. According to (A.9), we have

$$\cos\psi = \cos\theta \cos\theta' + \sin\theta \sin\theta' \cos(\lambda - \lambda'). \tag{3.254}$$

We attempt to express $P_n(\cos\psi)$ in terms of spherical harmonics of θ, λ and θ', λ'. It would be quite complicated to do the derivation by substituting (3.254) into the expression of $P_n(\cos\psi)$. Here we start from the variable separation approach of Laplace's equation.

Assume M is a fixed point. Let V be a harmonic function within the sphere as function of r, θ, and λ. It can be expressed as a spherical harmonic series

$$V = \sum_{n=0}^{\infty} r^n Y_n(\theta, \lambda), \tag{3.255}$$

where $Y_n(\theta, \lambda)$ is a surface spherical harmonics defined as

$$Y_n(\theta, \lambda) = \sum_{k=0}^{n} (A_n^k \cos k\lambda + B_n^k \sin k\lambda) P_n^k(\cos\theta). \tag{3.256}$$

Fig. 3.6 An angle made by
two points on the sphere with
respect to the center of the
sphere

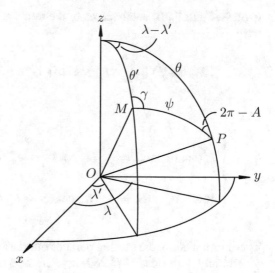

Actually, $Y_n(\theta, \lambda)$ is a function of the position of P on the surface. We also define
a spherical coordinate system with M as the pole, $Or\psi\gamma$,[2] where (ψ, γ) is the
coordinate of P. In this coordinate system, V can also be expressed as a spherical
harmonic series

$$V = \sum_{n=0}^{\infty} r^n Y_n(\psi, \gamma), \qquad (3.257)$$

where $Y_n(\psi, \gamma)$ is a surface spherical harmonics defined as

$$Y_n(\psi, \gamma) = \sum_{k=0}^{n} (D_n^k \cos k\gamma + E_n^k \sin k\gamma) P_n^k(\cos \psi). \qquad (3.258)$$

In order that (3.255) and (3.257) express the same harmonic function, $Y_n(\theta, \lambda)$ and
$Y_n(\psi, \gamma)$ must be equal. We have

$$\sum_{k=0}^{n} (D_n^k \cos k\gamma + E_n^k \sin k\gamma) P_n^k(\cos \psi) = \sum_{k=0}^{n} (A_n^k \cos k\lambda + B_n^k \sin k\lambda) P_n^k(\cos \theta).$$

$$(3.259)$$

Therefore, every term at the left-hand side can be expressed as a linear combination
of the terms at the right-hand side, and every term at the right-hand side can be

---

[2] As compared to $Or\theta\lambda$, it is more appropriate to use $2\pi - \gamma$ instead of γ. However, using γ also
serves our purpose; using $2\pi - \gamma$ is equivalent to change the sign of the coefficient of $\sin k\gamma$.

expressed as a linear combination of the terms at the left-hand side. Particularly we have

$$P_n(\cos\psi) = \sum_{k=0}^{n}(F_n^k \cos k\lambda + G_n^k \sin k\lambda)P_n^k(\cos\theta)\,, \tag{3.260}$$

and

$$\begin{cases} P_n^l(\cos\theta)\cos l\lambda = \displaystyle\sum_{k=0}^{n}(I_n^k \cos k\gamma + J_n^k \sin k\gamma)P_n^k(\cos\psi)\,, \\[4mm] P_n^l(\cos\theta)\sin l\lambda = \displaystyle\sum_{k=0}^{n}(K_n^k \cos k\gamma + L_n^k \sin k\gamma)P_n^k(\cos\psi)\,. \end{cases} \tag{3.261}$$

These formulae are the starting point of our derivation.

Multiply both sides of (3.260) by $P_n^l(\cos\theta)\cos l\lambda$ or $P_n^l(\cos\theta)\sin l\lambda$, and then integrate over a unit sphere. According to the orthogonality relations of spherical harmonics, we obtain

$$\begin{cases} F_n^l = \dfrac{1}{1+\delta_l}\dfrac{2n+1}{2\pi}\dfrac{(n-l)!}{(n+l)!}\displaystyle\int_\omega P_n(\cos\psi)P_n^l(\cos\theta)\cos l\lambda d\omega\,, \\[4mm] G_n^l = \dfrac{2n+1}{2\pi}\dfrac{(n-l)!}{(n+l)!}\displaystyle\int_\omega P_n(\cos\psi)P_n^l(\cos\theta)\sin l\lambda d\omega\,. \end{cases} \tag{3.262}$$

Substitute (3.261) into it. We obtain

$$\begin{cases} F_n^l = \dfrac{2}{1+\delta_l}\dfrac{(n-l)!}{(n+l)!}I_n\,, \\[4mm] G_n^l = 2\dfrac{(n-l)!}{(n+l)!}K_n\,. \end{cases} \tag{3.263}$$

Thus the problem of computing F_n^l and G_n^l becomes a problem of computing I_n and K_n, which can be inferred directly from (3.261).

In (3.261), we consider the particular case $\psi = 0$ and consequently $\theta = \theta'$ and $\lambda = \lambda'$. We then have

$$P_n^k(\cos\psi) = P_n^k(1) = \begin{cases} 0\,, k \neq 0\,; \\ 1\,, k = 0\,. \end{cases} \tag{3.264}$$

Hence, only the $k = 0$ terms remain, and we have

$$\begin{cases} I_n = P_n^l(\cos\theta')\cos l\lambda'\,, \\ K_n = P_n^l(\cos\theta')\sin l\lambda'\,. \end{cases} \tag{3.265}$$

The results of F_n^l and G_n^l are obtained by substituting this formula into (3.263)

$$
\begin{cases}
F_n^l = \dfrac{2}{1 + \delta_l} \dfrac{(n-l)!}{(n+l)!} P_n^l(\cos\theta')\cos l\lambda', \\[2mm]
G_n^l = 2\dfrac{(n-l)!}{(n+l)!} P_n^l(\cos\theta')\sin l\lambda'.
\end{cases}
\tag{3.266}
$$

Finally, replace the index l by k in (3.266), and then substitute it into (3.260). By treating the term $k = 0$ separately, we obtain the addition theorem

$$
P_n(\cos\psi) = P_n(\cos\theta)P_n(\cos\theta')
$$

$$
+ 2\sum_{k=1}^{n} \frac{(n-k)!}{(n+k)!} P_n^k(\cos\theta) P_n^k(\cos\theta')\cos k(\lambda - \lambda'),
\tag{3.267}
$$

which can be written concisely as

$$
P_n(\cos\psi) = \sum_{k=0}^{n} \frac{2}{1 + \delta_k} \frac{(n-k)!}{(n+k)!} P_n^k(\cos\theta) P_n^k(\cos\theta')\cos k(\lambda - \lambda'),
\tag{3.268}
$$

where δ_k is a Kronecker symbol defined in (3.174). A complex form of the addition theorem can be obtained using (3.250)

$$
P_n(\cos\psi) = P_n(\cos\theta)P_n(\cos\theta')
$$

$$
+ \sum_{k=1}^{n}(-1)^k P_n^k(\cos\theta) P_n^{-k}(\cos\theta')\left[e^{ik(\lambda-\lambda')} + e^{-ik(\lambda-\lambda')}\right]
$$

$$
= \sum_{k=-n}^{n}(-1)^k P_n^k(\cos\theta) P_n^{-k}(\cos\theta')e^{ik(\lambda-\lambda')}
$$

$$
= \sum_{k=-n}^{n}(-1)^k \left[P_n^k(\cos\theta)e^{ik\lambda}\right]\left[P_n^{-k}(\cos\theta')e^{-ik\lambda'}\right].
\tag{3.269}
$$

By setting $\theta' = \theta$ and $\lambda' = \lambda$ in (3.268), we obtain an identity of the associated Legendre function,

$$
\sum_{k=1}^{n} \frac{2}{1 + \delta_k} \frac{(n-k)!}{(n+k)!} [P_n^k(\cos\theta)]^2 = 1.
\tag{3.270}
$$

This identity can be used to check the accuracy of the associated Legendre function computed numerically.

The formulae of the azimuthal averages of the surface spherical harmonics, which will be used later in this book, can be derived incidentally based on the derivation of the addition theorem. By integrating the two formulae of (3.261) with respect to γ over 0 to 2π, we see that only the $k = 0$ term is non-vanishing. By further taking reference to (3.265), we obtain

$$
\begin{cases}
\dfrac{1}{2\pi} \displaystyle\int_0^{2\pi} P_n^l(\cos\theta)\cos l\lambda\, d\gamma = I_n P_n(\cos\psi) = P_n(\cos\psi) P_n^l(\cos\theta')\cos l\lambda'\,, \\[2ex]
\dfrac{1}{2\pi} \displaystyle\int_0^{2\pi} P_n^l(\cos\theta)\sin l\lambda\, d\gamma = K_n P_n(\cos\psi) = P_n(\cos\psi) P_n^l(\cos\theta')\sin l\lambda'\,.
\end{cases}
$$

$$(3.271)$$

This is an integral over the circle that P draws around M while keeping ψ fixed. As shown in Fig. 3.6, γ is the azimuth of P with respect to M, i.e., the angle from the north to P as seen at M.

In the derivation, we can swap M and P, θ' and θ, λ' and λ, as well as γ and the azimuth of M with respect to P denoted as A. This yields

$$
\begin{cases}
\dfrac{1}{2\pi} \displaystyle\int_0^{2\pi} P_n^l(\cos\theta')\cos l\lambda'\, dA = P_n(\cos\psi) P_n^l(\cos\theta)\cos l\lambda\,, \\[2ex]
\dfrac{1}{2\pi} \displaystyle\int_0^{2\pi} P_n^l(\cos\theta')\sin l\lambda'\, dA = P_n(\cos\psi) P_n^l(\cos\theta)\sin l\lambda\,.
\end{cases}
$$

$$(3.272)$$

We have shown in Fig. 3.6 the angle $2\pi - A$, which is more convenient than showing A.

The results (3.271) and (3.272) are the azimuthal averages of the surface spherical harmonics.

3.5   Spherical Harmonic Series of the Gravitational Potential

3.5.1   Spherical Harmonic Series of the Gravitational Potential of a Solid Body

The most important application of spherical harmonics in physical geodesy is the representation of the external gravitational field. As shown in Fig. 3.7, we consider the gravitational potential V of a solid body v at a point P. By definition, we have

$$
V = G \int_v \frac{\rho}{l}\, dv\,.
$$

$$(3.273)$$

We first express (3.273) as a spherical harmonic series. We draw a sphere centered on the origin O with the smallest possible radius that encloses the entire body v within it, which is called Brillouin sphere, and denote its radius as R_B.

Fig. 3.7 Gravitational potential of a solid body

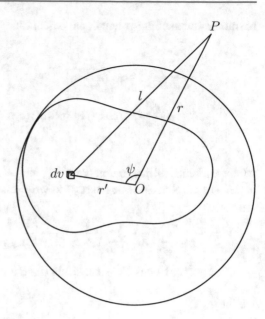

Outside the Brillouin sphere, we can express $1/l$ as the following series according to (3.93):

$$\frac{1}{l} = \frac{1}{r}\left[1 + \left(\frac{r'}{r}\right)^2 + 2\frac{r'}{r}\cos\psi\right]^{-1/2} = \sum_{n=0}^{\infty}\frac{r'^n}{r^{n+1}}P_n(\cos\psi). \tag{3.274}$$

Write the addition theorem as

$$P_n(\cos\psi) = \sum_{k=0}^{n}\frac{2}{1+\delta_k}\frac{(n-k)!}{(n+k)!}$$

$$\times P_n^k(\cos\theta)P_n^k(\cos\theta')(\cos k\lambda\cos k\lambda' + \sin k\lambda\sin k\lambda'), \tag{3.275}$$

and then substitute it into (3.274). We obtain

$$\frac{1}{l} = \sum_{n=0}^{\infty}\sum_{k=0}^{n}\frac{2}{1+\delta_k}\frac{(n-k)!}{(n+k)!}\frac{r'^n}{r^{n+1}}P_n^k(\cos\theta)$$

$$\times P_n^k(\cos\theta')(\cos k\lambda\cos k\lambda' + \sin k\lambda\sin k\lambda'), \tag{3.276}$$

where (r, θ, λ) is the coordinate of the point P and (r', θ', λ') that of the volume element dv. Substitute (3.276) into (3.273). We obtain a primitive spherical harmonic series of the gravitational potential:

$$
V = G \sum_{n=0}^{\infty} \sum_{k=0}^{n} \frac{2}{1 + \delta_k} \frac{(n-k)!}{(n+k)!}
$$

$$
\times \int_v \rho \frac{r'^n}{r^{n+1}} P_n^k(\cos \theta) P_n^k(\cos \theta')(\cos k\lambda \cos k\lambda' + \sin k\lambda \sin k\lambda') dv \,,
$$

$$(3.277)$$

where variables of integration are r', θ', and λ'. Denote the total mass of the body as M. We introduce a scale length a to write this formula as

$$
V = \frac{GM}{r} \sum_{n=0}^{\infty} \left(\frac{a}{r}\right)^n \sum_{k=0}^{n} \frac{2}{1 + \delta_k} \frac{(n-k)!}{(n+k)!} \int_v \frac{\rho}{M} \left(\frac{r'}{a}\right)^n P_n^k(\cos \theta) P_n^k(\cos \theta')
$$

$$
\times (\cos k\lambda \cos k\lambda' + \sin k\lambda \sin k\lambda') dv \,.
$$

$$(3.278)$$

Define

$$
\begin{cases}
C_n^k = \dfrac{2}{1 + \delta_k} \dfrac{(n-k)!}{(n+k)!} \displaystyle\int_v \dfrac{\rho}{M} \left(\dfrac{r'}{a}\right)^n P_n^k(\cos \theta') \cos k\lambda' dv \,, \\[4mm]
S_n^k = \dfrac{2}{1 + \delta_k} \dfrac{(n-k)!}{(n+k)!} \displaystyle\int_v \dfrac{\rho}{M} \left(\dfrac{r'}{a}\right)^n P_n^k(\cos \theta') \sin k\lambda' dv \,.
\end{cases}
$$

$$(3.279)$$

We can write the gravitational potential as

$$
V = \frac{GM}{r} \sum_{n=0}^{\infty} \left(\frac{a}{r}\right)^n \sum_{k=0}^{n} (C_n^k \cos k\lambda + S_n^k \sin k\lambda) P_n^k(\cos \theta) \,,
$$

$$(3.280)$$

which is often referred to as Stokes series. The coefficients C_n^k and S_n^k are referred to as potential coefficients or Stokes coefficients.

Finally, we note that (3.280) is guaranteed to converge, thus being meaningful, only outside the Brillouin sphere, where (3.274) holds for all possible values of r'.

3.5.2 Properties of Some Lower Degree and Order Potential Coefficients

According to (3.279), the potential coefficients depend on the shape, size, and density distribution of the body. Of particular importance are the lower degree and order coefficients up to 2, which are related to the mass, the center of mass, and the moment of inertia tensor. According to Table 3.2, substitute the explicit expressions of $P_n^k(\cos \theta')$ into (3.279), and then convert the spherical coordinates r', θ', and λ'

Table 3.3 Expressions of some lower degree and order potential coefficients

$$C_0 = \int_v \frac{\rho}{M} dv \qquad\qquad = \frac{1}{M} \int_v \rho dv$$

$$C_1 = \int_v \frac{\rho}{M} \frac{r'}{a} \cos\theta' dv \qquad = \frac{1}{Ma} \int_v \rho z' dv$$

$$C_1^1 = \int_v \frac{\rho}{M} \frac{r'}{a} \sin\theta' \cos\lambda' dV \qquad = \frac{1}{Ma} \int_v \rho x' dv$$

$$S_1^1 = \int_v \frac{\rho}{M} \frac{r'}{a} \sin\theta' \sin\lambda' dv \qquad = \frac{1}{Ma} \int_v \rho y' dv$$

$$C_2 = \int_v \frac{\rho}{M} \left(\frac{r'}{a}\right)^2 \left(\frac{3}{2}\cos^2\theta' - 1\right) dv \qquad = \frac{1}{2Ma^2} \int_v \rho(2z'^2 - x'^2 - y'^2) dv$$

$$C_2^1 = \frac{1}{3} \int_v \frac{\rho}{M} \left(\frac{r'}{a}\right)^2 (3\sin\theta'\cos\theta')\cos\lambda' dv \qquad = \frac{1}{Ma^2} \int_v x'z' dV$$

$$S_2^1 = \frac{1}{3} \int_v \frac{\rho}{M} \left(\frac{r'}{a}\right)^2 (3\sin\theta'\cos\theta')\sin\lambda' dv \qquad = \frac{1}{Ma^2} \int_v \rho y'z' dv$$

$$C_2^2 = \frac{1}{12} \int_v \frac{\rho}{M} \left(\frac{r'}{a}\right)^2 (3\sin^2\theta')\cos2\lambda' dv \qquad = \frac{1}{4Ma^2} \int_v \rho(x'^2 - y'^2) dv$$

$$S_2^2 = \frac{1}{12} \int_v \frac{\rho}{M} \left(\frac{r'}{a}\right)^2 (3\sin^2\theta')\sin2\lambda' dv \qquad = \frac{1}{2Ma^2} \int_v \rho x'y' dv$$

to Cartesian coordinates x', y', and z'. We obtain the explicit integral expressions of some lower degree and order potential coefficients as listed in Table 3.3.

In Table 3.3, every integral has a specific physical meaning. The mass of the body is

$$M = \int_v \rho dv . \qquad (3.281)$$

The coordinates of the center of mass are

$$X_0 = \frac{1}{M} \int_v \rho x' dv , \qquad Y_0 = \frac{1}{M} \int_v \rho y' dv , \qquad Z_0 = \frac{1}{M} \int_v \rho z' dv . \qquad (3.282)$$

The moments of inertia are

$$A = \int_v \rho(y'^2 + z'^2) dv , \qquad B = \int_v \rho(z'^2 + x'^2) dv , \qquad C = \int_v \rho(x'^2 + y'^2) dv . \qquad (3.283)$$

The products of inertia are

$$D = \int_v \rho x'y' dv , \qquad E = \int_v \rho y'z' dv , \qquad F = \int_v \rho z'x' dv . \qquad (3.284)$$

According to (3.281)–(3.284), the potential coefficients listed in Table 3.3 can be written as

$$
\begin{cases}
C_0 = 1 ; \\
C_1^0 = \dfrac{Z_0}{a} , & C_1^1 = \dfrac{X_0}{a} , \quad S_1^1 = \dfrac{Y_0}{a} ; \\
C_2 = -\dfrac{C - (A+B)/2}{Ma^2} , & C_2^1 = \dfrac{F}{Ma^2} , \quad S_2^1 = \dfrac{E}{Ma^2} , \quad C_2^2 = \dfrac{B-A}{4Ma^2} , \quad S_2^2 = \dfrac{D}{2Ma^2} .
\end{cases}
$$

$$(3.285)$$

These explain the physical meaning of the lower degree and order potential coefficients.

3.5.3   The Center of Mass and Principal Moment of Inertia Coordinate System

If the origin of the coordinate system O is set on the center of mass of the solid body, we have $X_0 = Y_0 = Z_0 = 0$, and hence, $C_1 = C_1^1 = S_1^1 = 0$. This means that in the center of mass coordinate system, the first degree potential coefficients vanish.

The directions of coordinate axes can also be particularly chosen, such that $D = E = F = 0$, resulting in $C_2^1 = S_2^1 = S_2^2 = 0$, which is to be proven in subsequent paragraphs.

As shown in Fig. 3.8, we start the proof by studying the moment of inertia J_u around the u-axis. We have

$$
J_u = \int_v \rho d^2 dv ,
$$

$$(3.286)$$

Fig. 3.8 Moment of inertia around an arbitrary axis

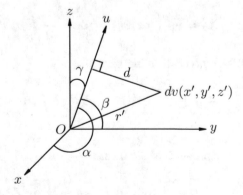

where ρ is the density and d the distance from the volume element dv to the u-axis. We will first express J_u as a function of A, B, C, D, E, and F and the angles between the u-axis and the three coordinate axes, α, β, and γ.

The position vector of the volume element dv is

$$\vec{r} = x'\hat{e}_x + y'\hat{e}_y + z'\hat{e}_z , \tag{3.287}$$

where (x', y', z') is the coordinate of the volume element dv. The unit vector along the u-axis is

$$\hat{e}_u = \cos\alpha\,\hat{e}_x + \cos\beta\,\hat{e}_y + \cos\gamma\,\hat{e}_z . \tag{3.288}$$

The length of the projection of $\vec{r}$ along the u-axis is then

$$|\vec{r} \cdot \hat{e}_u| = |x'\cos\alpha + y'\cos\beta + z'\cos\gamma| . \tag{3.289}$$

According to this formula, we can express d^2 as a function of x', y', z' and α, β, γ

$$\begin{aligned}
d^2 =& r^2 - |\vec{r} \cdot \hat{e}_u|^2 \\
=& x'^2 + y'^2 + z'^2 - (x'\cos\alpha + y'\cos\beta + z'\cos\gamma)^2 \\
=& (1 - \cos^2\alpha)x'^2 + (1 - \cos^2\beta)y'^2 + (1 - \cos^2\gamma)z'^2 \\
& - 2x'y'\cos\alpha\cos\beta - 2y'z'\cos\beta\cos\gamma - 2z'x'\cos\gamma\cos\alpha .
\end{aligned} \tag{3.290}$$

As

$$\cos^2\alpha + \cos^2\beta + \cos^2\gamma = 1 , \tag{3.291}$$

we can further reduce the expression of d^2 to

$$\begin{aligned}
d^2 =& (\cos^2\beta + \cos^2\gamma)x'^2 + (\cos^2\gamma + \cos^2\alpha)y'^2 + (\cos^2\alpha + \cos^2\beta)z'^2 \\
& - 2x'y'\cos\alpha\cos\beta - 2y'z'\cos\beta\cos\gamma - 2z'x'\cos\gamma\cos\alpha \\
=& \cos^2\alpha(y'^2 + z'^2) + \cos^2\beta(z'^2 + x'^2) + \cos^2\gamma(x'^2 + y'^2) \\
& - 2x'y'\cos\alpha\cos\beta - 2y'z'\cos\beta\cos\gamma - 2z'x'\cos\gamma\cos\alpha .
\end{aligned} \tag{3.292}$$

Finally, substitute this formula into (3.286). With (3.283) and (3.284), we obtain

$$\begin{aligned}
J_u =& A\cos^2\alpha + B\cos^2\beta + C\cos^2\gamma - 2D\cos\alpha\cos\beta - 2E\cos\beta\cos\gamma \\
& - 2F\cos\gamma\cos\alpha .
\end{aligned} \tag{3.293}$$

This is the expression of J_u as a function of A, B, C, D, E, F and α, β, γ, which we are looking for.

In order to study how J_u varies as a function of the direction of the u-axis described by the angles α, β, and γ, we choose a point K on the u-axis, such that $OK = 1/J_u^{1/2}$. The coordinates of the point K are then

$$x = \frac{\cos \alpha}{J_u^{1/2}}, \qquad y = \frac{\cos \beta}{J_u^{1/2}}, \qquad z = \frac{\cos \gamma}{J_u^{1/2}}. \tag{3.294}$$

As the direction of the u-axis varies, the position of K also varies, and the trace of K completely describes the dependence of J_u on the direction of the u-axis. We obtain, by dividing both sides of (3.293) by J_u and then using (3.294),

$$Ax^2 + By^2 + Cz^2 - 2Dxy - 2Eyz - 2Fzx = 1, \tag{3.295}$$

which is the equation of a quadratic surface representing the trace of the point K. As J_u does not vanish, nor does it tend to infinitely large, this surface is closed. Hence, it is an ellipsoid.

An ellipsoid has three axes perpendicular to one another, called principal axes, such that if they are chosen as the axes of the coordinate system, the products between different coordinates vanish in (3.295). Hence, in this particular coordinate system, the Eq. (3.295) simplifies to

$$Ax^2 + By^2 + Cz^2 = 1, \tag{3.296}$$

which implies $D = E = F = 0$, i.e., the products of inertia vanish, and $C_2^1 = S_2^1 = S_2^2 = 0$. The moments of inertia around these axes are referred to as principle moments of inertia. Hence the coordinate system is referred to as principal moment of inertia coordinate system.

In summary, in the center of mass coordinate system, the gravitational potential can be written as

$$V = \frac{GM}{r} \left[1 + \sum_{n=2}^{\infty} \left(\frac{a}{r} \right)^n \sum_{k=0}^{n} (C_n^k \cos k\lambda + S_n^k \sin k\lambda) P_n^k (\cos \theta) \right], \tag{3.297}$$

where the degree 1 term is excludes as it vanishes, and in the center of mass and principal moment of inertia coordinate system, the following degree 2 coefficients vanish:

$$C_2^1 = 0, \qquad S_2^1 = 0, \qquad S_2^2 = 0. \tag{3.298}$$

It is noted that (3.297) is a simplification of (3.280) due to the choice of a particular coordinate system. Except the fact that the center of the Brillouin sphere follows the origin of the coordinate system, this choice of coordinate system does not further influence the property of convergence.

In the derivation, a scale length a is included so that the potential coefficients are dimensionless. Hence, its choice is somewhat arbitrary, and it could be longer or shorter than the radius of the Brillouin sphere. Different choices of a lead to different potential coefficients, but the sum and convergence property of the series remain unchanged. The Earth can be approximated by an ellipsoid of revolution, called the reference Earth ellipsoid. Its minor axis is the symmetry axis and is along the rotation axis of the Earth, and its major axis is in the equator. In the representation of the Earth's gravitational field, the scale length a is customarily chosen to be the semi-major axis.

In the special case when the Earth is considered as a spherically symmetrical body, (3.297) degenerates to $V = GM/r$, which equals the gravitational potential of a point mass M located at the center of the Earth.

3.5.4 MacCullagh's Formula

We assume $Oxyz$ to be a center of mass coordinate system. The gravitational potential (3.280) can be written as, according to (3.285) and Table 3.3,

$$
V = \frac{GM}{r}\left\{1 + \frac{1}{2Mr^2}\left[(A+B-2C)\left(\frac{3}{2}\cos^2\theta - \frac{1}{2}\right) + 2F(3\sin\theta\cos\theta)\cos\lambda \right.\right.
$$
$$
\left. + 2E(3\sin\theta\cos\theta)\sin\lambda + \frac{B-A}{2}(3\sin^2\theta)\cos 2\lambda + D(3\sin^2\theta)\sin 2\lambda\right]
$$
$$
\left. + O\left(\frac{a}{r}\right)^3\right\},
\tag{3.299}
$$

where $O(a/r)^3$ represents quantities on the order of $(a/r)^3$ or smaller. We further assume that the point (r, θ, λ) is on the u-axis as shown in Fig. 3.8, thus $\gamma = \theta$, $\cos\alpha = \sin\theta\cos\lambda$, and $\cos\beta = \sin\theta\sin\lambda$. We obtain, from (3.293),

$$
J_u = A\sin^2\theta\cos^2\lambda + B\sin^2\theta\sin^2\lambda + C\cos^2\theta
$$
$$
- 2D\sin^2\theta\cos\lambda\sin\lambda - 2E\cos\theta\sin\theta\sin\lambda - 2F\cos\theta\sin\theta\cos\lambda.
\tag{3.300}
$$

By substituting (3.300) into (3.299), we obtain

$$V = \frac{GM}{r} \left\{ 1 + \frac{1}{2Mr^2} \left[(A + B - 2C) \left(\frac{3}{2} \cos^2 \theta - \frac{1}{2} \right) + \frac{B - A}{2} (3 \sin^2 \theta) \cos 2\lambda \right. \right.$$

$$\left. \left. + 3A \sin^2 \theta \cos^2 \lambda + 3B \sin^2 \theta \sin^2 \lambda + 3C \cos^2 \theta - 3J_u \right] + O \left(\frac{a}{r} \right)^3 \right\}$$

$$= \frac{GM}{r} \left[1 + \frac{1}{2Mr^2} (A + B + C - 3J_u) + O \left(\frac{a}{r} \right)^3 \right], \tag{3.301}$$

which is referred to as MacCullagh's formula.

3.5.5 Gravitational Attraction and Its Gradient in Spherical Harmonic Series

The formula of the gravitational attraction to be computed from gravitational potential has been provided in (2.142). Here we infer it directly, which is quite straightforward.

According to Fig. 3.9, the lengths of infinitesimal line segments along the directions of the base vectors $\hat{e}_r$, $\hat{e}_\theta$, and $\hat{e}_\lambda$ are dr, $rd\theta$, and $r \sin \theta d\lambda$, respectively. The projections of gravitational attraction along these directions are the directional derivatives of gravitational potential along them. Hence, it follows immediately that

$$F_r = \frac{\partial V}{\partial r}, \qquad F_\theta = \frac{1}{r} \frac{\partial V}{\partial \theta}, \qquad F_\lambda = \frac{1}{r \sin \theta} \frac{\partial V}{\partial \lambda}. \tag{3.302}$$

Fig. 3.9 Directions of the base vectors of a spherical coordinate system

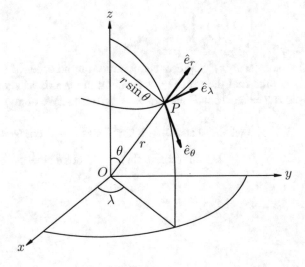

We use the spherical harmonic series of the gravitational potential given in (3.280), which includes also the degree 1 term. By substituting (3.280) into (3.302), it follows that

$$F_r = -\frac{GM}{r^2} \sum_{n=0}^{\infty} (n+1) \left(\frac{a}{r}\right)^n \sum_{k=0}^{n} (C_n^k \cos k\lambda + S_n^k \sin k\lambda) P_n^k(\cos\theta), \qquad (3.303)$$

$$F_\theta = \frac{GM}{r^2} \sum_{n=1}^{\infty} \left(\frac{a}{r}\right)^n \sum_{k=0}^{n} (C_n^k \cos k\lambda + S_n^k \sin k\lambda) \frac{d}{d\theta} P_n^k(\cos\theta), \qquad (3.304)$$

$$F_\lambda = \frac{GM}{r^2} \sum_{n=1}^{\infty} \left(\frac{a}{r}\right)^n \sum_{k=0}^{n} (S_n^k \cos k\lambda - C_n^k \sin k\lambda) \frac{k}{\sin\theta} P_n^k(\cos\theta), \qquad (3.305)$$

where the degree 0 term does not contribute to F_θ and F_λ, as it is independent of θ and λ.

The gradient of the gravitational attraction, or the second order gradient of the gravitational potential, to be computed from the gravitational potential is provided in (2.143). By substituting (3.280) into it, we obtain its spherical harmonic series. Here we write the result component by component (a superscript G is used to indicate gradient).

$$F_{rr}^G = \frac{GM}{r^3} \sum_{n=0}^{\infty} (n+1)(n+2) \left(\frac{a}{r}\right)^n \sum_{k=0}^{n} (C_n^k \cos k\lambda + S_n^k \sin k\lambda) P_n^k(\cos\theta),$$
$$(3.306)$$

$$F_{r\theta}^G = F_{\theta r}^G = -\frac{GM}{r^3} \sum_{n=0}^{\infty} (n+2) \left(\frac{a}{r}\right)^n \sum_{k=0}^{n} (C_n^k \cos k\lambda + S_n^k \sin k\lambda) \frac{d}{d\theta} P_n^k(\cos\theta),$$
$$(3.307)$$

$$F_{r\lambda}^G = F_{\lambda r}^G = -\frac{GM}{r^3} \sum_{n=0}^{\infty} (n+2) \left(\frac{a}{r}\right)^n \sum_{k=0}^{n} (S_n^k \cos k\lambda - C_n^k \sin k\lambda) \frac{k}{\sin\theta} P_n^k(\cos\theta),$$
$$(3.308)$$

$$F_{\theta\theta}^G = \frac{GM}{r^3} \sum_{n=0}^{\infty} \left(\frac{a}{r}\right)^n \sum_{k=0}^{n} (C_n^k \cos k\lambda + S_n^k \sin k\lambda) \left[\frac{d^2}{d\theta^2} P_n^k(\cos\theta) - (n+1) P_n^k(\cos\theta)\right],$$
$$(3.309)$$

$$F_{\theta\lambda}^G = F_{\lambda\theta}^G = \frac{GM}{r^3} \sum_{n=0}^{\infty} \left(\frac{a}{r}\right)^n \sum_{k=0}^{n} (S_n^k \cos k\lambda - C_n^k \sin k\lambda)$$

$$\times \frac{k}{\sin\theta} \left[\frac{d}{d\theta} P_n^k(\cos\theta) - \cot\theta P_n^k(\cos\theta)\right], \qquad (3.310)$$

$$F_{\lambda\lambda}^G = \frac{GM}{r^3} \sum_{n=0}^{\infty} \left(\frac{a}{r}\right)^n \sum_{k=0}^{n} (C_n^k \cos k\lambda + S_n^k \sin k\lambda)$$

$$\times \left\{ \cot\theta \frac{d}{d\theta} P_n^k(\cos\theta) - \left[(n+1) + \frac{k^2}{\sin^2\theta} \right] P_n^k(\cos\theta) \right\} . \qquad (3.311)$$

The formulation of the components of the centrifugal acceleration as well as its gradient in the spherical coordinate system using the centrifugal potential given in (1.43) is similar and is left to the readers.

3.5.6   Fully Normalized Associated Legendre Function and Its Computation

A normalization of the associated Legendre function is a scaling by a degree and order dependent factor, say, $\bar{P}_n^k(\cos\theta) = \alpha_n^k P_n^k(\cos\theta)$. The term *normalized spherical harmonics* is also correspondingly used.

In the representation of the Earth's gravitational potential, the normalization is defined by

$$\begin{cases} \dfrac{1}{4\pi} \displaystyle\int_\omega \left[\bar{P}_n^k(\cos\theta) \cos k\lambda \right]^2 d\omega = 1 , \ k \geq 0 ; \\ \dfrac{1}{4\pi} \displaystyle\int_\omega \left[\bar{P}_n^k(\cos\theta) \sin k\lambda \right]^2 d\omega = 1 , \ k > 0 , \end{cases} \qquad (3.312)$$

where the left-hand sides are the averages of $\left[\bar{P}_n^k(\cos\theta) \cos k\lambda \right]^2$ and $\left[\bar{P}_n^k(\cos\theta) \sin k\lambda \right]^2$ over a unit sphere. It can be easily obtained that

$$\bar{P}_n^k(\cos\theta) = \left[\frac{2(2n+1)}{1+\delta_k} \frac{(n-k)!}{(n+k)!} \right]^{1/2} P_n^k(\cos\theta) , \qquad (3.313)$$

which is referred to as fully normalized associated Legendre function or shortly fully normalized Legendre function. The gravitational potential is expressed as a series of fully normalized spherical harmonics as

$$V = \frac{GM}{r} \left[1 + \sum_{n=2}^{\infty} \left(\frac{a}{r}\right)^n \sum_{k=0}^{n} (\bar{C}_n^k \cos k\lambda + \bar{S}_n^k \sin k\lambda) \bar{P}_n^k(\cos\theta) \right] , \qquad (3.314)$$

where the fully normalized spherical harmonic coefficients $\bar{C}_n^k$ and $\bar{S}_n^k$ are related to the un-normalized ones in the following form:

$$\begin{cases} C_n^k = \left[\dfrac{2(2n+1)}{1+\delta_k} \dfrac{(n-k)!}{(n+k)!} \right]^{1/2} \bar{C}_n^k, \ k \geq 0; \\[3mm] S_n^k = \left[\dfrac{2(2n+1)}{1+\delta_k} \dfrac{(n-k)!}{(n+k)!} \right]^{1/2} \bar{S}_n^k, \ k > 0. \end{cases} \tag{3.315}$$

The logic behind the normalization is that the left-hand sides of (3.312) are considered as the overall magnitudes of $\bar{P}_n^k(\cos\theta)\cos k\lambda$ and $\bar{P}_n^k(\cos\theta)\sin k\lambda$, such that the normalized coefficients $\bar{C}_n^k$ and $\bar{S}_n^k$ in (3.314) represent the overall magnitudes of signal of the respective terms.[3]

In practical computations, $\bar{P}_n^k(\cos\theta)$ is normally used. As the order of magnitude of $P_n^k(\cos\theta)$ varies much more strongly with n and k, its values exceed the range of double precision variables of a programming language much more quickly. Therefore, the recurrence formulae used, (3.164) and (3.157), should be converted to those for $\bar{P}_n^k(\cos\theta)$. Here we provide the formulae necessary for the computation and leave the derivation to the readers, which is straightforward. The recurrence formula (3.157) can be developed to

$$\bar{P}_{n+1}^k(\cos\theta) = \left(1/\alpha_{n+1}^k \right) \left[\cos\theta\, \bar{P}_n^k(\cos\theta) - \alpha_n^k \bar{P}_{n-1}^k(\cos\theta) \right] \tag{3.316}$$

with

$$\alpha_n^k = \left[\frac{(n-k)(n+k)}{(2n-1)(2n+1)} \right]^{1/2}. \tag{3.317}$$

These formulae can be used to compute $\bar{P}_{k+2}^k(\cos\theta)$, $\bar{P}_{k+3}^k(\cos\theta)$, $\cdots$ assuming $\bar{P}_k^k(\cos\theta)$ and $\bar{P}_{k+1}^k(\cos\theta)$ are precomputed. A recurrence formula for $\bar{P}_k^k(\cos\theta)$

---

[3] In the representation of geomagnetic potential, the normalization is defined by

$$\begin{cases} \dfrac{1}{4\pi} \displaystyle\int_\omega \left[\bar{P}_n^k(\cos\theta)\cos k\lambda \right]^2 d\omega = \dfrac{1}{2n+1}, \ k \geq 0; \\[3mm] \dfrac{1}{4\pi} \displaystyle\int_\omega \left[\bar{P}_n^k(\cos\theta)\sin k\lambda \right]^2 d\omega = \dfrac{1}{2n+1}; \ k > 0. \end{cases}$$

As the right-hand sides still depend on n, $\bar{P}_n^k(\cos\theta)$ is called semi-normalized associated Legendre function or shortly semi-normalized Legendre function. It is

$$\bar{P}_n^k(\cos\theta) = \left[\frac{2}{1+\delta_k} \frac{(n-k)!}{(n+k)!} \right]^{1/2} P_n^k(\cos\theta).$$

can be obtained based on the formula of $P_k^k(\cos\theta)$ given in (3.164),

$$\bar{P}_k^k(\cos\theta) = \beta_k \sin\theta\, \bar{P}_{k-1}^{k-1}(\cos\theta) \tag{3.318}$$

with

$$\beta = \left[1 + \frac{0.5}{k}\right]^{1/2}. \tag{3.319}$$

These formulae can be used to compute $\bar{P}_2^2(\cos\theta)$, $\bar{P}_3^3(\cos\theta)$, $\cdots$ with the initial value $\bar{P}_1^1(\cos\theta) = \sqrt{3}\sin\theta$. Due to the presence of δ_k in the normalization factor in (3.313), $\bar{P}_0^0(\cos\theta) = 1$ is a special case, and it cannot be used to compute $\bar{P}_1^1(\cos\theta)$ based on the above formulae. The recurrence formula for $\bar{P}_{k+1}^k(\cos\theta)$ can be obtained by setting $n = k$ in (3.316)

$$\bar{P}_{k+1}^k(\cos\theta) = \left(1/\alpha_{k+1}^k\right)\cos\theta\, \bar{P}_k^k(\cos\theta). \tag{3.320}$$

When computation is to be done for multiple values of θ, the factors α_n^k, $1/\alpha_n^k$, and β_k should be precomputed to save computation time. The formula for checking the accuracy of computation (3.270) becomes

$$\sum_{k=1}^{n}[\bar{P}_n^k(\cos\theta)]^2 = 2n+1. \tag{3.321}$$

The relative error of computation is

$$\left\{\sum_{k=1}^{n}[\bar{P}_n^k(\cos\theta)]^2 - (2n+1)\right\}(2n+1)^{-1}. \tag{3.322}$$

The recurrence formula for the derivative of $\bar{P}_n^k(\cos\theta)$ with respect to θ can be derived from (3.170) to (3.172) by setting $i = 0$,

$$\frac{d}{d\theta}\bar{P}_n^k(\cos\theta) = (1+\delta_{k-1})^{1/2}d_n^{k-1}\bar{P}_n^{k-1}(\cos\theta) - d_n^k\bar{P}_n^{k+1}(\cos\theta), \tag{3.323}$$

$$\frac{d}{d\theta}\bar{P}_n(\cos\theta) = -\sqrt{2}d_n^0\bar{P}_n^1(\cos\theta), \tag{3.324}$$

$$\frac{d}{d\theta}\bar{P}_k^k(\cos\theta) = (1+\delta_{k-1})^{1/2}d_k^{k-1}\bar{P}_k^{k-1}(\cos\theta), \tag{3.325}$$

with

$$d_n^k = \frac{1}{2}[(n + k + 1)(n - k)]^{1/2}. \tag{3.326}$$

The procedure is to first compute $d\bar{P}_n(\cos\theta)/d\theta$ and $d\bar{P}_k^k(\cos\theta)/d\theta$ and then compute $d\bar{P}_n^k(\cos\theta)/d\theta$ for $k = 1, 2, \ldots$ and $n = k + 1, k + 2, \ldots$, where $(1 + \delta_{k-1})^{1/2}$ equals $\sqrt{2}$ when $k = 1$ and 1 when $k \neq 1$. The precision of computation is controlled by that of $\bar{P}_n^k(\cos\theta)$ as known data. If we replace $\bar{P}_n^k(\cos\theta)$ by $d\bar{P}_n^k(\cos\theta)/d\theta$ as known data, we obtain $d^2\bar{P}_n^k(\cos\theta)/d\theta^2$. More generally, if we replace $\bar{P}_n^k(\cos\theta)$ by $d^i\bar{P}_n^k(\cos\theta)/d\theta^i$ as known data, we obtain $d^{i+1}\bar{P}_n^k(\cos\theta)/d\theta^{i+1}$. A formula for checking the correctness of program can be derived from (3.321),

$$\sum_{k=1}^n \bar{P}_n^k(\cos\theta) \frac{d}{d\theta} \bar{P}_k^k(\cos\theta) = 0. \tag{3.327}$$

Actually, even $\bar{P}_n^k(\cos\theta)$ cannot be computed using the recurrence formulae as is when n and k are large, and some more tricks are required to not exceed the range of double precision variables of a programming language.

3.6 Spherical Harmonic Series of a Functions on a Sphere

3.6.1 Definition of the Series

We understand the spherical harmonics series as an extension of the Fourier series. Let $f(x)$ be a function with a period of 2π and is piecewise smooth. Its Fourier series is defined as

$$S(x) = \sum_{k=0}^{\infty}(a_k \cos kx + b_k \sin kx), \tag{3.328}$$

where the coefficients are

$$\begin{cases} a_k = \dfrac{1}{1 + \delta_k} \dfrac{1}{\pi} \displaystyle\int_0^{2\pi} f(x) \cos kx\, dx, \\[2mm] a_k = \dfrac{1}{1 + \delta_k} \dfrac{1}{\pi} \displaystyle\int_0^{2\pi} f(x) \sin kx\, dx. \end{cases} \tag{3.329}$$

The case $k = 0$ in the second formula is meaningless. Here we include it for similarity to the first one. The Fourier series (3.328) is convergent everywhere, and

$$S(x) = \frac{1}{2}[f_+(x) + f_-(x)], \tag{3.330}$$

where $f_+(x)$ and $f_-(x)$ are the right- and left-handed limits, respectively.

If $f(x)$ is a function with a period of 2π, piecewise smooth and continuous, we have

$$f(x) = \sum_{k=0}^{\infty} (a_k \cos kx + b_k \sin kx), \tag{3.331}$$

where the coefficients are defined in (3.329), which can be formally obtained from (3.331) by multiplying both sides by $\cos kx$ or $\sin kx$ and then integrating in $[0, 2\pi]$ making use of the orthogonality relations of $\cos kx$ and $\sin kx$.

Although we assume that the readers are familiar with Fourier series, it will be derived as a degenerate case of spherical harmonic series.

The surface spherical harmonics $P_n^k(\cos\theta)\cos k\lambda$ and $P_n^k(\cos\theta)\sin k\lambda$ have orthogonality relations on a sphere in similar forms to those of $\cos kx$ and $\sin kx$ in the internal $[0, 2\pi]$. Hence, the spherical harmonic series of a function $f(\theta, \lambda)$ on a sphere is defined as

$$S(\theta, \lambda) = \sum_{n=0}^{\infty} \sum_{k=0}^{n} (A_n^k \cos k\lambda + B_n^k \sin k\lambda) P_n^k(\cos\theta), \tag{3.332}$$

where the coefficients are

$$\begin{cases} A_n^k = \dfrac{1}{1+\delta_k} \dfrac{2n+1}{2\pi} \dfrac{(n-k)!}{(n+k)!} \displaystyle\int_\omega f(\theta', \lambda') P_n^k(\cos\theta') \cos k\lambda' d\omega, \\[2mm] B_n^k = \dfrac{1}{1+\delta_k} \dfrac{2n+1}{2\pi} \dfrac{(n-k)!}{(n+k)!} \displaystyle\int_\omega f(\theta', \lambda') P_n^k(\cos\theta') \sin k\lambda' d\omega. \end{cases} \tag{3.333}$$

The factor $1/(1 + \delta_k)$ in the expression of B_n^k does not contribute to the result, as B_n itself multiplies with 0. Inclusion of it keeps the expressions of A_n^k and B_n^k in similar forms. At present, we do not know yet when $S(\theta, \lambda)$ converges and what function it converges to.

Substitute (3.333) into (3.332). We obtain

$$S(\theta, \lambda) = \sum_{n=0}^{\infty} \sum_{k=0}^{n} \frac{1}{1+\delta_k} \frac{2n+1}{2\pi} \frac{(n-k)!}{(n+k)!} \left\{ \left[\int_\omega f(\theta', \lambda') P_n^k(\cos\theta') \cos k\lambda' d\omega \right] \cos k\lambda \right.$$

$$\left. + \left[\int_\omega f(\theta', \lambda') P_n^k(\cos\theta') \sin k\lambda' d\omega \right] \sin k\lambda \right\} P_n^k(\cos\theta). \tag{3.334}$$

According to the addition theorem, this formula can be simplified to

$$S(\theta, \lambda) = \sum_{n=0}^{\infty} \frac{2n+1}{2\pi} \int_{\omega} f(\theta', \lambda') P_n(\cos \psi) d\omega, \tag{3.335}$$

where ψ is a function of θ, λ and θ', λ', and the integration is performed for the variables θ' and λ'.

The spherical harmonic series of a function on a sphere is referred to as Laplace series, which can be written in the forms of (3.332), (3.334), or (3.335). In the rest of this section, we will provide a proof of its convergence following Guo (2000), which relies on very fundamental mathematics.

3.6.2 Partial Sum of the Series

We establish a new spherical coordinate system on the sphere, $O\psi\gamma$, where the point of coordinate (θ, λ) is the pole, ψ the polar angle, and γ the longitude. We can then transform the variables of integration θ' and λ' in (3.335) to ψ and γ. We obtain

$$S(\theta, \lambda) = \sum_{n=0}^{\infty} \frac{2n+1}{4\pi} \int_0^{2\pi} d\gamma \int_0^{\pi} f(\psi, \gamma) P_n(\cos \psi) \sin \psi d\psi. \tag{3.336}$$

We substitute ψ by x according to

$$x = \cos \psi \tag{3.337}$$

and define $\phi(x)$ as

$$\phi(x) = \phi(\cos \psi) = \frac{1}{2\pi} \int_0^{2\pi} f(\psi, \gamma) d\gamma. \tag{3.338}$$

We can then write (3.336) as

$$S(\theta, \lambda) = \sum_{n=0}^{\infty} \frac{2n+1}{2} \int_{-1}^{1} \phi(x) P_n(x) dx. \tag{3.339}$$

We truncate (3.339) up to degree N. As the summation and integration signs can be switched, we obtain

$$S_N(\theta, \lambda) = \frac{1}{2} \int_{-1}^{1} \phi(x) \sum_{n=0}^{N} (2n+1) P_n(x) dx. \tag{3.340}$$

Add $(2n + 1)P_n(x)$ to both sides of the recurrence formula (3.94). We can write it in the form of

$$(2n + 1)(1 - x)P_n(x) = -(n + 1)[P_{n+1}(x) - P_n(x)] + n[P_n(x) - P_{n-1}(x)].$$
(3.341)

Set $n = 0, 1, 2, \ldots, N$ to obtain $N + 1$ formulae, and then add them together. We obtain

$$\sum_{n=0}^{N}(2n + 1)P_n(x) = -(N + 1)\frac{P_{N+1}(x) - P_N(x)}{1 - x}.$$
(3.342)

Substituting this formula into (3.340), we obtain an alternative expression of the partial sum

$$S_N(\theta, \lambda) = -\frac{1}{2}\int_{-1}^{1}\phi(x)(N + 1)\frac{P_{N+1}(x) - P_N(x)}{1 - x}dx.$$
(3.343)

Another form of the integrand can be found using the recurrence formulae (3.95) and (3.96). Set $n = N$ in (3.95) and $n = N + 1$ in (3.96), and then add together to obtain an equation, which can be solved to obtain

$$(N + 1)\frac{P_{N+1}(x) - P_N(x)}{1 - x} = -\left[\frac{d}{dx}P_{N+1}(x) + \frac{d}{dx}P_N(x)\right].$$
(3.344)

3.6.3 An Auxiliary Formula

In order to study the limit of $S_N(\theta, \lambda)$ as $N \to \infty$, we first prove a limit formula.

Let $g(x)$ be an arbitrary function defined in $[-1, 1]$. We define a number G as follows:

$$G = \int_{-1}^{1}\left[g(x') - \sum_{n=0}^{N}\frac{2n + 1}{2}P_n(x')\int_{-1}^{1}g(x)P_n(x)dx\right]^2 dx'.$$
(3.345)

It can be further derived using the orthogonality relation (3.177) as

$$G = \int_{-1}^{1}[g(x')]^2 dx' - 2\sum_{n=0}^{N}\frac{2n + 1}{2}\left[\int_{-1}^{1}g(x)P_n(x)dx\right]^2$$

$$+ \int_{-1}^{1}\left[\sum_{n=0}^{N}\frac{2n + 1}{2}P_n(x')\int_{-1}^{1}g(x)P_n(x)dx\right]^2 dx'$$

$$= \int_{-1}^{1} [g(x')]^2 dx' - 2 \sum_{n=0}^{N} \frac{2n+1}{2} \left[\int_{-1}^{1} g(x) P_n(x) dx \right]^2$$

$$+ \sum_{n=0}^{N} \left(\frac{2n+1}{2} \right)^2 \int_{-1}^{1} [P_n(x')]^2 dx' \left[\int_{-1}^{1} g(x) P_n(x) dx \right]^2$$

$$= \int_{-1}^{1} [g(x)]^2 dx - \sum_{n=0}^{N} \frac{2n+1}{2} \left[\int_{-1}^{1} g(x) P_n(x) dx \right]^2 . \tag{3.346}$$

From now on, we assume that $\int_{-1}^{1} [g(x)]^2 dx$ is integrable. As $G \geq 0$, we obtain

$$\sum_{n=0}^{N} \frac{2n+1}{2} \left[\int_{-1}^{1} g(x) P_n(x) dx \right]^2 \leq \int_{-1}^{1} [g(x)]^2 dx . \tag{3.347}$$

Therefore, the positive term series

$$\sum_{n=0}^{N} \frac{2n+1}{2} \left[\int_{-1}^{1} g(x) P_n(x) dx \right]^2 \tag{3.348}$$

converges.

Number series have the following property: let $\sum_{n=0}^{\infty} u_n$ and $\sum_{n=0}^{\infty} v_n$ be positive term series, and there exists a number $k > 0$, such that $\lim_{n \to \infty} u_n/v_n = k$. The two series are then both convergent or divergent. Now let $\sum_{n=0}^{\infty} u_n$ to be convergent, and $\sum_{n=0}^{\infty} v_n = \sum_{n=0}^{\infty} 1/n$, which is known to be divergent. We then have $\lim_{n \to \infty} u_n/v_n = \lim_{n \to \infty} nu_n = 0$. If this identity does not hold, $\sum_{n=0}^{\infty} u_n$ would be divergent. Apply this property to (3.348). We obtain

$$\lim_{n \to \infty} \frac{n(2n+1)}{2} \left[\int_{-1}^{1} g(x) P_n(x) dx \right]^2 = 0 . \tag{3.349}$$

Write $[n(2n+1)]/2$ as $n^2[1 + 1/(2n)]$. We have

$$\lim_{n \to \infty} n^2 \left[\int_{-1}^{1} g(x) P_n(x) dx \right]^2 = 0 . \tag{3.350}$$

Finally, we obtain the required formula

$$\lim_{n \to \infty} n \int_{-1}^{1} g(x) P_n(x) dx = 0 . \tag{3.351}$$

We mention again that this formula holds under the condition that $\int_{-1}^{1} [g(x)]^2 dx$ is integrable.

3.6.4 Convergence of the Series

Now we study the limit $\lim_{N \to \infty} S_N(\theta, \lambda)$. A first thought might be to apply (3.351) to (3.343). However, as the function $g(x) = \phi(x)/(1 - x) \to \infty$ as $x \to 1$, $\int_{-1}^{1} [g(x)]^2 dx$ is not necessarily integrable, and some modification of (3.343) is required.

Define a new function

$$g(x) = \begin{cases} 0, & 1 - \epsilon < x \leq 1 ; \\ \dfrac{\phi(x)}{1 - x}, & -1 \leq x \leq 1 - \epsilon , \end{cases} \tag{3.352}$$

where ϵ is a small positive number. We assume $\int_{-1}^{1} [\phi(x)]^2 dx$ is integrable from now on. Then $\int_{-1}^{1} [g(x)]^2 dx$ is also integrable. In this case, we can use the above defined $g(x)$ to rewrite (3.343) as

$$S_N(\theta, \lambda) = -\frac{N+1}{2} \int_{-1}^{1} g(x) [P_{N+1}(x) - P_N(x)] dx$$

$$- \frac{1}{2} \int_{1-\epsilon}^{1} \phi(x)(N+1) \frac{P_{N+1}(x) - P_N(x)}{1 - x} dx . \tag{3.353}$$

Take the limit $N \to \infty$ on both sides. As the first term on the right-hand side tends to 0 according to (3.351), we obtain

$$S(\theta, \lambda) = \lim_{N \to \infty} S_N(\theta, \lambda) = - \lim_{N \to \infty} \frac{1}{2} \int_{1-\epsilon}^{1} \phi(x)(N+1) \frac{P_{N+1}(x) - P_N(x)}{1 - x} dx . \tag{3.354}$$

We further assume that $\phi(x)$ is piecewise smooth, such that it can be treated as constant within $[1 - \epsilon, 1]$. According to (3.338), we define

$$\phi(1) = \lim_{x \to 1} \phi(x) = \lim_{\psi \to 0} \frac{1}{2\pi} \int_{0}^{2\pi} f(\psi, \gamma) d\gamma . \tag{3.355}$$

We can then write (3.354) as

$$S(\theta, \lambda) = -\phi(1) \lim_{N \to \infty} \frac{1}{2} \int_{1-\epsilon}^{1} (N+1) \frac{P_{N+1}(x) - P_N(x)}{1-x} dx. \qquad (3.356)$$

In such a way, we have taken $\phi(x)$ out of the integral sign.

In order to evaluate the integral in (3.356), we extend it into the domain $[-1, 1]$. Define

$$h(x) = \begin{cases} 0, & 1-\epsilon < x \le 1; \\ \dfrac{1}{1-x}, & -1 \le x \le 1-\epsilon. \end{cases} \qquad (3.357)$$

As $\int_{-1}^{1} [h(x)]^2 dx$ is integrable, we have, according to (3.351),

$$\lim_{N \to \infty} \int_{-1}^{1} h(x)(N+1)[P_{N+1}(x) - P_N(x)] dx = 0. \qquad (3.358)$$

By substituting (3.357) into this formula, we obtain

$$\lim_{N \to \infty} \int_{-1}^{1-\epsilon} (N+1) \frac{P_{N+1}(x) - P_N(x)}{1-x} dx = 0. \qquad (3.359)$$

This formula can be used to rewrite (3.356) as

$$S(\theta, \lambda) = -\phi(1) \lim_{N \to \infty} \frac{1}{2} \int_{-1}^{1} (N+1) \frac{P_{N+1}(x) - P_N(x)}{1-x} dx. \qquad (3.360)$$

Finally, with (3.344), we obtain

$$\begin{aligned} S(\theta, \lambda) &= \phi(1) \lim_{N \to \infty} \frac{1}{2} \int_{-1}^{1} \left[\frac{d}{dx} P_{N+1}(x) + \frac{d}{dx} P_N(x) \right] dx \\ &= \frac{1}{2} \phi(1) \lim_{N \to \infty} [P_{N+1}(x) + P_N(x)]_{-1}^{1} \\ &= \phi(1), \end{aligned} \qquad (3.361)$$

where $P_N(\pm 1) = (\pm 1)^N$ is used. This formula provides the sum of the Laplace series, which holds under the conditions that $\phi(x)$ is piecewise smooth and $\int_{-1}^{1} [\phi(x)]^2 dx$ is integrable.

Our last task is to express $\phi(1)$ using $f(\theta, \lambda)$. Assume that $f(\theta, \lambda)$ is finite and piecewise smooth. Then, $\phi(x)$ is piecewise smooth and $\int_{-1}^{1} [\phi(x)]^2 dx$ is integrable. According to (3.338), $\phi(1)$ can be understood as the average of $f(\theta', \lambda')$ on an infinitesimal circle around the point (θ, λ). Hence, at places where $f(\theta, \lambda)$ is continuous, we have

$$S(\theta, \lambda) = \phi(1) = f(\theta, \lambda).$$ (3.362)

At places where $f(\theta, \lambda)$ has a jump across a smooth line but continuous on both sides of the line, we have

$$S(\theta, \lambda) = \phi(1) = \frac{1}{2}[f_1(\theta, \lambda) + f_2(\theta, \lambda)],$$ (3.363)

where $f_1(\theta, \lambda)$ and $f_2(\theta, \lambda)$ are the limits of $f(\theta, \lambda)$ on the two sides of the smooth line. This formula would be different if the discontinuity is more complicated. For example, the readers may consider a vertex of a broken line across which $f(\theta, \lambda)$ is discontinuous.

3.6.5 Expression of a Function as a Spherical Harmonic Series

Based on results of the previous subsection, a finite, piecewise smooth, and continuous function on a sphere can be expressed as the following spherical harmonic series:

$$f(\theta, \lambda) = \sum_{n=0}^{\infty} \sum_{k=0}^{n} (A_n^k \cos k\lambda + B_n^k \sin k\lambda) P_n^k(\cos \theta);$$ (3.364)

$$\begin{cases} A_n^k = \dfrac{1}{1+\delta_k} \dfrac{2n+1}{2\pi} \dfrac{(n-k)!}{(n+k)!} \displaystyle\int_{\omega} f(\theta, \lambda) P_n^k(\cos \theta) \cos k\lambda d\omega, \\ B_n^k = \dfrac{1}{1+\delta_k} \dfrac{2n+1}{2\pi} \dfrac{(n-k)!}{(n+k)!} \displaystyle\int_{\omega} f(\theta, \lambda) P_n^k(\cos \theta) \sin k\lambda d\omega. \end{cases}$$ (3.365)

The complex form of it is

$$f(\theta, \lambda) = \sum_{n=0}^{\infty} \sum_{k=-n}^{n} C_n^k P_n^k(\cos \theta) e^{ik\lambda},$$ (3.366)

$$C_n^k = (-1)^k \frac{2n+1}{4\pi} \int_{\omega} f(\theta, \lambda) P_n^{-k}(\cos \theta) e^{-ik\lambda} d\omega.$$ (3.367)

In the degenerate case when f is a function of θ alone and independent of λ, the series expression is simplified to, after setting $x = \cos\theta$,

$$f(x) = \sum_{n=0}^{\infty} A_n P_n(x), \tag{3.368}$$

$$A_n = \frac{2n+1}{2} \int_{-1}^{1} f(x) P_n(x) dx, \tag{3.369}$$

which is called Legendre series of $f(x)$.

The spherical harmonic series (3.364) can be written as

$$f(\theta, \lambda) = \sum_{k=0}^{\infty} \left[\left(\sum_{n=k}^{\infty} A_n^k P_n^k(\cos\theta) \right) \cos k\lambda + \left(\sum_{n=k}^{\infty} B_n^k P_n^k(\cos\theta) \right) \sin k\lambda \right], \tag{3.370}$$

which is the form used for computation using fast Fourier transform (FFT). Here we use it to examine the degenerate case when f is a function of λ alone. In this case, the coefficients of $\cos k\lambda$ and $\sin k\lambda$ must be independent of θ. Hence, (3.370) can be written as

$$f(\lambda) = \sum_{k=0}^{\infty} (a_k \cos k\lambda + b_k \sin k\lambda). \tag{3.371}$$

As this relation is guaranteed to hold, we can use the orthogonality relations of the sine and cosine functions to obtain the coefficients

$$\begin{cases} a_k = \dfrac{1}{1+\delta_k} \dfrac{1}{\pi} \displaystyle\int_{0}^{2\pi} f(\lambda) \cos k\lambda d\lambda, \\[2mm] b_k = \dfrac{1}{1+\delta_k} \dfrac{1}{\pi} \displaystyle\int_{0}^{2\pi} f(\lambda) \sin k\lambda d\lambda. \end{cases} \tag{3.372}$$

In this way, we have proven the Fourier series expansion of a function as a special case of the spherical harmonic series.

A three dimensional function can also be expressed as a spherical harmonic series. As an example, we consider a finite and smooth function $f(x, \theta, \lambda)$ defined in $-1 \le x \le 1$, $0 \le \theta \le \pi$, and $0 \le \lambda \le 2\pi$. We first take $f(x, \theta, \lambda)$ as a function of θ and λ and consider x as a parameter. We can express $f(x, \theta, \lambda)$ as a spherical harmonic series with coefficients depending on x.

$$f(x, \theta, \lambda) = \sum_{n=0}^{\infty} \sum_{k=0}^{n} \left[A_n^k(x) \cos k\lambda + B_n^k(x) \sin k\lambda \right] P_n^k(\cos \theta) , \qquad (3.373)$$

$$\begin{cases} A_n^k(x) = \dfrac{1}{1 + \delta_k} \dfrac{2n + 1}{2\pi} \dfrac{(n - k)!}{(n + k)!} \displaystyle\int_\omega f(x, \theta, \lambda) P_n^k(\cos \theta) \cos k\lambda d\omega , \\[4mm] B_n^k(x) = \dfrac{1}{1 + \delta_k} \dfrac{2n + 1}{2\pi} \dfrac{(n - k)!}{(n + k)!} \displaystyle\int_\omega f(x, \theta, \lambda) P_n^k(\cos \theta) \sin k\lambda d\omega . \end{cases}$$
$$(3.374)$$

Next, we express $A_n^k(x)$ and $B_n^k(x)$ as Legendre series:

$$\begin{cases} A_n^k(x) = \displaystyle\sum_{m=0}^{\infty} {}_m A_n^k P_m(x) , \\[4mm] B_n^k(x) = \displaystyle\sum_{m=0}^{\infty} {}_m B_n^k P_m(x) ; \end{cases} \qquad (3.375)$$

$$\begin{cases} {}_m A_n^k = \dfrac{2n + 1}{2} \displaystyle\int_{-1}^{1} A_n^k(x) P_m(x) dx , \\[4mm] {}_m B_n^k = \dfrac{2n + 1}{2} \displaystyle\int_{-1}^{1} B_n^k(x) P_m(x) dx . \end{cases} \qquad (3.376)$$

Substitute (3.375) and (3.376) into (3.373) and (3.374), and we obtain

$$f(x, \theta, \lambda) = \sum_{m=0}^{\infty} \sum_{n=0}^{\infty} \sum_{k=0}^{n} \left({}_m A_n^k \cos k\lambda + {}_m B_n^k \sin k\lambda \right) P_m(x) P_n^k(\cos \theta) ;$$
$$(3.377)$$

$$\begin{cases} {}_m A_n^k = \dfrac{2m+1}{1+\delta_k} \dfrac{2n+1}{4\pi} \dfrac{(n-k)!}{(n+k)!} \displaystyle\int_{-1}^{1} \left[P_m(x) \int_\omega f(x, \theta, \lambda) P_n^k(\cos \theta) \cos k\lambda d\omega \right] dx , \\[4mm] {}_m B_n^k = \dfrac{2m+1}{1+\delta_k} \dfrac{2n+1}{4\pi} \dfrac{(n-k)!}{(n+k)!} \displaystyle\int_{-1}^{1} \left[P_m(x) \int_\omega f(x, \theta, \lambda) P_n^k(\cos \theta) \sin k\lambda d\omega \right] dx . \end{cases}$$
$$(3.378)$$

It is left to the readers to infer the convergence of (3.368), (3.371), and (3.377) based on the convergence of the Laplace series.

3.6.6 Application: An Alternative Derivation of the Poisson Integral

As an application, we use the spherical harmonic series expansion of a function defined on a sphere to solve the external first-type boundary value problem, which we have solved using Green's function technique.

We use V to denote the solution of Laplace's equation and R the radius of the sphere. The value of V on the sphere can be expressed as the following spherical harmonic series:

$$V_R = \sum_{n=0}^{\infty} \sum_{k=0}^{n} (A_n^k \cos k\lambda + B_n^k \sin k\lambda) P_n^k(\cos \theta), \tag{3.379}$$

$$\begin{cases} A_n^k = \dfrac{1}{1+\delta_k} \dfrac{2n+1}{2\pi} \dfrac{(n-k)!}{(n+k)!} \displaystyle\int_{\omega} V_R(\theta', \lambda') P_n^k(\cos \theta') \cos k\lambda' d\omega, \\[3mm] B_n^k = \dfrac{1}{1+\delta_k} \dfrac{2n+1}{2\pi} \dfrac{(n-k)!}{(n+k)!} \displaystyle\int_{\omega} V_R(\theta', \lambda') P_n^k(\cos \theta') \sin k\lambda' d\omega. \end{cases} \tag{3.380}$$

The expression of V outside the sphere is

$$V = \sum_{n=0}^{\infty} \sum_{k=0}^{n} \left(\frac{R}{r}\right)^{n+1} (A_n^k \cos k\lambda + B_n^k \sin k\lambda) P_n^k(\cos \theta). \tag{3.381}$$

We see that (3.381) reduces to (3.379) when $r = R$, thus being the solution.

The solution in terms of spherical harmonic series is thus (3.380) and (3.381). By substituting (3.380) into (3.381), we obtain a single expression of the solution in terms of spherical harmonic series

$$V = \sum_{n=0}^{\infty} \frac{2n+1}{4\pi} \left(\frac{R}{r}\right)^{n+1} \left\{ \int_{\omega} \left[V_R(\theta', \lambda') \sum_{k=0}^{n} \frac{2}{1+\delta_k} \frac{(n-k)!}{(n+k)!} \right.\right.$$

$$\times P_n^k(\cos \theta) P_n^k(\cos \theta')(\cos k\lambda \cos k\lambda' + \sin k\lambda \sin k\lambda') \Big] d\omega \Big\}$$

$$= \sum_{n=0}^{\infty} \frac{2n+1}{4\pi} \left(\frac{R}{r}\right)^{n+1} \int_{\omega} V_R(\theta', \lambda') P_n(\cos \psi) d\omega, \tag{3.382}$$

where the addition theorem is used to derive the second identity, and $\psi(\theta, \lambda, \theta', \lambda')$ is the angle between the two points (θ, λ) and (θ', λ') with respect to the center of the sphere. The variables of integration are θ' and λ'. We further write (3.382) in the form of

$$V = \frac{1}{4\pi} \int_\omega V_R G(r, \psi) d\omega = \frac{1}{4\pi R^2} \int_S V_R G(r, \psi) dS, \qquad (3.383)$$

where

$$G(r, \psi) = \sum_{n=0}^{\infty} (2n+1) \left(\frac{R}{r}\right)^{n+1} P_n(\cos \psi). \qquad (3.384)$$

We now derive an expression in closed-form for $G(r, \psi)$. For brevity, let

$$x = \frac{R}{r}. \qquad (3.385)$$

Set $t = \cos \psi$ in the generating equation of Legendre function, (3.76), multiply both sides of it by $2x^{1/2}$, and then take the derivative with respect to x. We obtain

$$\frac{1 - x^2}{x^{1/2}(1 + x^2 - 2x \cos \psi)^{3/2}} = \sum_{n=0}^{\infty} (2n+1) x^{n-1/2} P_n(\cos \psi). \qquad (3.386)$$

It can be readily obtained using this formula that

$$G(r, \psi) = \sum_{n=0}^{\infty} (2n+1) x^{n+1} P_n(\cos \psi) = \frac{x(1 - x^2)}{(1 + x^2 - 2x \cos \psi)^{3/2}}$$

$$= \frac{R(r^2 - R^2)}{(r^2 + R^2 - 2rR \cos \psi)^{3/2}}. \qquad (3.387)$$

Substitute this formula into (3.383). We obtain the Poisson integral

$$V = \frac{R(r^2 - R^2)}{4\pi} \int_\omega \frac{V_R}{(r^2 + R^2 - 2rR \cos \psi)^{3/2}} d\omega = \frac{R(r^2 - R^2)}{4\pi} \int_\omega \frac{V}{l^3} d\omega, \qquad (3.388)$$

where, as defined in (2.72),

$$l = (r^2 + R^2 - 2rR \cos \psi)^{1/2}. \qquad (3.389)$$

The result (3.388) is identical to (2.80). In the second integral, we have removed the subscript R of the function V under the integral and implicitly understand V as its value on the sphere of radius R.

This example shows that spherical harmonics can be used as a tool to solve boundary value problems. We will use this approach extensively in the solution of boundary value problems in gravity field determination.

Reference

Guo, J. Y. (2000). A simple proof for the convergence of the laplace series (in Chinese). *Chinese Journal of Mathematics for Technology, 16*(4), 108–110.

Further Reading

Bosch, W. (2000). On the computation of derivatives of Legendre functions, *Physics and Chemistry of the Earth (A), 25*, 655–659. https://doi.org/10.1016/S1464-1895(00)00101-0

Fukushima, T. (2012a). Numerical computation of spherical harmonics of arbitrary degree and order by extending exponent of floating-point numbers. *Journal of Geodesy, 86*, 271–285. https://doi.org/10.1007/s00190-011-0519-2

Fukushima, T. (2012b). Numerical computation of spherical harmonics of arbitrary degree and order by extending exponent of floating point numbers: II first-, second-, and third-order derivatives. *Journal of Geodesy, 86*, 1019–1028. http://dx.doi.org/10.1007/s00190-012-0561-8

Hobson, E. W. (1931). *The theory of spherical and ellipsoidal Harmonics* (Chapters 1–4). Cambridge University Press.

The Normal Gravity Field and Reference Earth Ellipsoid

4

Abstract

This chapter formulates the normal gravity field of the Earth. It is the gravity field of a rotating oblate ellipsoid of revolution as an approximation of the Earth, called a reference Earth ellipsoid, whose surface is an equipotential surface of gravity among other properties. A formulation different from other books in the English literature is adopted. Based on Stokes' theorem, the normal gravity field is treated as that of a Maclaurin ellipsoid with a homoeoid superimposed on the surface. The normal gravity potential and gravity usually found in the literature are derived, which are in closed-form expressions on the surface and are in spherical harmonic series outside the surface. In particular, the second order approximate formulae are derived systematically. Finally, the determination of the parameters of the reference Earth ellipsoid, called Geodetic Reference System (GRS), is addressed.

4.1 Basic Concepts

4.1.1 The Normal Gravity Field and Reference Earth Ellipsoid

By definition, the Earth's gravitational potential and attraction are irregular functions of position, and the shape of geoid is also irregular. Conceptually, the Earth's normal gravity field is an approximation of the Earth's gravity field and is described analytically using simple functions. Currently, in physical geodesy, the Earth's normal gravity field is defined as the gravity field of a fictitious Earth of regular shape, called the reference Earth ellipsoid.

The reference Earth ellipsoid is a rotating solid body as an approximation of the real Earth; a more specific consideration is that the surface of the reference Earth ellipsoid is an approximation of the geoid. The gravity potential and gravity of the reference Earth ellipsoid are referred to as normal gravity potential and normal gravity, respectively. Based on observations, the shape of the geoid is very close to an ellipsoid of revolution with its minor axis along the Earth's rotation axis. Hence the reference Earth ellipsoid is defined to have the following properties: (1) its surface is an ellipsoid of revolution with its minor axis along the Earth's rotational axis, (2) its rotation rate is equal to that of the Earth, (3) its mass is equal to that of the Earth, (4) its surface is an equipotential surface of normal gravity, (5) the normal gravity potential on the surface of the reference Earth ellipsoid is equal to the gravity potential of the Earth on the geoid, and (6) the center of the reference Earth ellipsoid is on the center of mass of the Earth.

Traditionally, the reference Earth ellipsoid thus defined is referred to as the mean Earth ellipsoid, as it is practically a global mean, but the term *reference Earth ellipsoid* is used in a less restricted sense, which could be an ellipsoid that best fits the geoid of one or several countries. At present, as global measurements from satellites are always used, any reference Earth ellipsoid could be considered as a realization of the mean Earth ellipsoid. Hence, we choose to use the term *reference Earth ellipsoid* alone for all realizations of ellipsoidal bodies as approximations of the Earth.

As already mentioned, the external gravity field of the reference Earth ellipsoid is referred to as the Earth's normal gravity field. Other nomenclatures include normal gravitational potential and normal gravitational attraction. The normal gravity potential is also briefly referred to as normal potential. The difference between the real and normal gravity fields is referred to as disturbing gravity field.

In order to formulate the Earth's normal gravity field, we have to solve the following mathematical problem: given a reference Earth ellipsoid with semi-major axis a, semi-minor axis b, rotation rate Ω, and normal potential at its surface, W_0, derive the formula for the external normal potential U. As the centrifugal potential is known and can be subtracted from the normal potential to obtain the normal gravitational potential, the problem is equivalent to derive the formula of the normal gravitational potential outside the reference Earth ellipsoid when it is known on the surface of the reference Earth ellipsoid. This is a first-type external boundary value problem of an ellipsoid of revolution, where the solution is unique. Many textbooks followed this approach and introduce ellipsoidal harmonics to solve this boundary value problem.

In this book, we follow an alternative approach to solve for the normal potential when a, b, Ω, and the Earth's mass M are known, where, according to Stokes' theorem, the solution is also uniquely determined. By comprising this approach with the one summarized in the previous paragraph, we can already realize that the parameters a, b, Ω, W_0, and M are not independent, as an unique solution can be obtained either with a, b, Ω, and W_0 or with a, b, Ω, and M; therefore a relation among them will be derived. In the approach we chose, a hypothetical mass distribution within the reference Earth ellipsoid with a total mass of M is devised so that the surface be an equipotential surface of normal gravity. Instead of ellipsoidal harmonics, this approach involves the formulation of the Maclaurin ellipsoid, which is considered to pertain to the knowledge of theoretical geodesy. As the hypothetical mass distribution constitutes a homogeneous ellipsoid of revolution and a material surface of ellipsoidal homoeoid of revolution, we start by formulating the gravitational fields of these configurations in the subsequent sections. This approach was adopted by He (1957) and Levallois (1970).

From now on, we will briefly use the term *reference Earth ellipsoid* to indicate both its surface and body. Normally, this does not lead to any confusion. For example, in *the gravitational potential of the reference Earth ellipsoid*, the term refers to the body, and in *on the reference Earth ellipsoid*, the term refers to the surface. When a confusion is likely, we will state the *surface* or *body* explicitly.

Some parameters of the reference Earth ellipsoid are to be used in the formulation. The semi-axes a and b describe both the size and the shape of the reference Earth ellipsoid. The dimensionless parameters, the flattening α, first eccentricity e, and second eccentricity e', defined by

$$\alpha = \frac{a-b}{a}, \quad e^2 = \frac{a^2 - b^2}{a^2} \quad \text{and} \quad e'^2 = \frac{a^2 - b^2}{b^2}, \tag{4.1}$$

respectively, are used to describe only the shape. The distance between the center and a focus, $E = (a^2 - b^2)^{1/2}$, is referred to as linear eccentricity. Another dimensionless parameter of the reference Earth ellipsoid to be used is

$$m = \frac{\Omega^2 a^2 b}{GM} = \frac{\Omega^2 a^3}{GM(1 + e'^2)^{1/2}}, \tag{4.2}$$

which is approximately the ratio between the centrifugal acceleration and gravitational attraction at the equator and is on the order of $1/300$ in magnitude. The flattening α is on the same order of magnitude as m and both e^2 and e'^2 on the order of 2α.

4.1.2 Geocentric and Geodetic Coordinates

The reference Earth ellipsoid is used as a reference surface for representing position. Two coordinate systems, geocentric and geodetic, are most used.

Before defining coordinate systems, we first introduce the definitions of poles, equator, and meridian, as shown in Fig. 4.1. The minor axis intersects the surface of the reference Earth ellipsoid at two points; the one at north is called north pole, and the one at south is called south pole. The plane perpendicular to the minor axis and passing through the center of the reference Earth ellipsoid is called equatorial plane, and its intersection with the surface of the reference Earth ellipsoid is called equator, which is a circle of radius a. A half plane extending to infinity with the minor axis as its side is called a meridional plane, and its intersection with the surface of the reference Earth ellipsoid is called meridian, which is a half ellipse of semi-major and semi-minor axes a and b.

The geocentric Cartesian coordinate system $Oxyz$ is defined such that O is at the center of the reference Earth ellipsoid, z points to the north pole, x points to

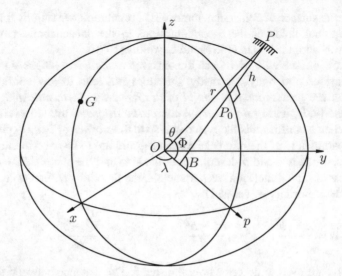

Fig. 4.1 Geocentric and geodetic coordinates

the intersection point between the equator and a specific meridian called prime meridian, which is customarily chosen as the Greenwich meridian (a meridian passing through the Greenwich Observatory denoted as G in Fig. 4.1), and y is defined by the right-hand rule. A spherical coordinate system $Or\theta\lambda$ associated with $Oxyz$ is defined as usual. For any point P, the Cartesian coordinates are evident. The spherical coordinates are shown in Fig. 4.1. We call $\Phi = 90° - \theta$ geocentric latitude and λ longitude. We specifically mention that the longitude λ is the angle from the prime meridional plane to the meridional plane passing through the point P, positive eastward.

The geodetic coordinates of a point P are defined based on a straight line passing through it and perpendicular to the reference Earth ellipsoid, which intersects the reference Earth ellipsoid at P_0 and makes an angle B with the equatorial plane. This line is within the meridional plane passing through P, just as the line OP. Geodetic coordinates are then the geodetic latitude B, longitude λ, and geodetic height $h = P_0P$. Geodetic latitude is positive if P is north of the equatorial plane and negative if P is south. The geodetic height is positive if P is above the reference Earth ellipsoid and negative if P is below.

Evidently, a meridian is a curved line connecting all points on the reference Earth ellipsoid with the same longitude. A curved line connecting all points on the reference Earth ellipsoid with the same (geocentric or geodetic) latitude is called parallel, which is within a plane parallel to the equatorial plane.

Another line of importance is the prime vertical. For a point P, the prime vertical is defined as the intersection line between the reference Earth ellipsoid and a plane passing through P and perpendicular to the meridian; it can be inferred that this plane contains the perpendicular line from P to the reference Earth ellipsoid. If P

is on the reference Earth ellipsoid, the parallel and prime vertical passing through it are tangential to each other.

Ellipsoidal geometry is an integrated part of the science of geodesy and is often taught in a separate course named like *Ellipsoidal Geodesy* or *Geometric Geodesy*. Besides the points, planes, lines, and coordinate systems defined in this subsection, some elementary ellipsoidal geometry necessary to this book is presented in Appendix A, so that this book be self-sufficient.

4.2   Internal Gravitational Field of a Homogeneous Ellipsoid of Revolution

We use the Cartesian coordinate system $Oxyz$ as shown in Fig. 4.1. We use ρ to denote the density of the ellipsoid and (ξ, η, ζ) the Cartesian coordinate of any point on its surface. The equation of its surface is

$$\frac{\xi^2 + \eta^2}{a^2} + \frac{\zeta^2}{b^2} = 1. \tag{4.3}$$

We first compute the gravitational attraction and then use the gravitational attraction to compute the gravitational potential.

By definition, the components of the gravitational attraction at a point P of coordinate (x, y, z) are

$$F_x = -G\rho \int_v \frac{x - \xi}{l^3} dv, \quad F_y = -G\rho \int_v \frac{y - \eta}{l^3} dv, \quad F_z = -G\rho \int_v \frac{z - \zeta}{l^3} dv, \tag{4.4}$$

where the domain v is the inside of the ellipsoid, (ξ, η, ζ) the coordinate of the volume element dv, and l the distance between P and dv. The variables of integration are ξ, η, and ζ.

4.2.1   The Gravitational Attraction Components F_x and F_y

Based on the symmetry of the problem, the expressions of the gravitational attraction components F_x and F_y are similar. We provide only a detailed formulation for F_x. The expression of F_y is inferred according to symmetry.

Define a spherical coordinate system with origin at the point P, polar axis along the x-axis, and denote the radial distance, polar angle, and longitude using r, ψ, and γ, respectively. The variables of integration ξ, η, and ζ in (4.4) are related to r, ψ, and γ as

$$\xi = x + r\cos\psi, \quad \eta = y + r\sin\psi\cos\gamma, \quad \zeta = z + r\sin\psi\sin\gamma. \tag{4.5}$$

The volume element expressed using r, ψ, and γ is

$$dv = r^2 \sin \psi \, dr d\psi d\gamma \,. \tag{4.6}$$

Substitute the above two formulae into the first formula of (4.4). We obtain

$$F_x = G\rho \int_0^\pi d\psi \int_0^{2\pi} d\gamma \int_0^{r_1} \sin \psi \cos \psi \, dr \,, \tag{4.7}$$

where r_1 is the distance between the point P and the surface of the ellipsoid, which is a function of ψ and γ, and is determined using the equation of the surface of the ellipsoid. After integration over r, we obtain

$$F_x = G\rho \int_0^\pi d\psi \int_0^{2\pi} r_1 \sin \psi \cos \psi \, d\gamma \,. \tag{4.8}$$

In order to further develop (4.8), we have to evaluate r_1 first. We realize that in the spherical coordinate system $Pr\psi\gamma$, the coordinate of any point on the surface of the ellipsoid is (r_1, ψ, γ). Hence, r_1 is the solution of r of the equation obtained by substituting (4.5) into the equation of the surface of the ellipsoid (4.3). This implies that r_1 satisfies the equation

$$\frac{(x + r_1 \cos \psi)^2 + (y + r_1 \sin \psi \cos \gamma)^2}{a^2} + \frac{(z + r_1 \sin \psi \sin \gamma)^2}{b^2} = 1 \,. \tag{4.9}$$

This is a second order algebraic equation of r_1, which can be developed to

$$Lr_1^2 + 2Mr_1 + N = 0 \,, \tag{4.10}$$

where (noting that here M is used to represent an expression, but not a mass)

$$\begin{cases} L = \dfrac{\cos^2 \psi}{a^2} + \dfrac{\sin^2 \psi \cos^2 \gamma}{a^2} + \dfrac{\sin^2 \psi \sin^2 \gamma}{b^2} \,, \\ M = \dfrac{x \cos \psi}{a^2} + \dfrac{y \sin \psi \cos \gamma}{a^2} + \dfrac{z \sin \psi \sin \gamma}{b^2} \,, \\ N = \dfrac{x^2}{a^2} + \dfrac{y^2}{a^2} + \dfrac{z^2}{b^2} - 1 \,. \end{cases} \tag{4.11}$$

Evidently, L is always positive, N is negative when P is within the ellipsoid, and 0 when P is on the surface of the ellipsoid. Hence, the discriminant $M^2 - LN$ is always positive and not smaller than M^2, and the equation has two real solutions

$$r_1 = \frac{-M \pm (M^2 - LN)^{1/2}}{L} \,, \tag{4.12}$$

one positive and one negative. The one we need is positive

$$r_1 = \frac{-M + (M^2 - LN)^{1/2}}{L}.$$

(4.13)

Substitute it into (4.8). We obtain

$$F_x = -G\rho \int_0^\pi d\psi \int_0^{2\pi} \frac{M}{L} \sin\psi \cos\psi \, d\gamma$$

$$+ G\rho \int_0^\pi d\psi \int_0^{2\pi} \frac{(M^2 - LN)^{1/2}}{L} \sin\psi \cos\psi \, d\gamma \, .$$

(4.14)

It can be shown that the second integral in (4.14) vanishes. For a pair of points (ψ_1, γ_1) and $(\pi - \psi_1, \pi + \gamma_1)$, L and N have the same values, and M has the same absolute value with opposite signs. Hence, $(M^2 - LN)^{1/2}/L$ has the same value at these two points, and $\sin\psi \cos\psi$ has the same absolute value with opposite signs. Consequently, the integrand of the second integral in (4.14) has the same absolute value with opposite signs at these two points. Based on this analysis, we can express the integral as the sum of four as follows (omitting integrand for brevity):

$$\int_0^{\pi/2} d\psi \int_0^\pi d\gamma + \int_0^{\pi/2} d\psi \int_\pi^{2\pi} d\gamma + \int_{\pi/2}^\pi d\psi \int_0^\pi d\gamma + \int_{\pi/2}^\pi d\psi \int_\pi^{2\pi} d\gamma \, .$$

(4.15)

We see that the first and fourth ones cancel each other, the second and third ones cancel each other, and hence, the integral vanishes. Therefore, (4.14) is simplified to

$$F_x = -G\rho \int_0^\pi d\psi \int_0^{2\pi} \frac{M}{L} \sin\psi \cos\psi \, d\gamma \, .$$

(4.16)

After substituting the expression of M given by (4.11) into it, we obtain

$$F_x = - G\rho \frac{x}{a^2} \int_0^\pi d\psi \int_0^{2\pi} \frac{1}{L} \sin\psi \cos^2\psi \, d\gamma$$

$$- G\rho \frac{y}{a^2} \int_0^\pi d\psi \int_0^{2\pi} \frac{1}{L} \sin^2\psi \cos\psi \cos\gamma \, d\gamma$$

$$- G\rho \frac{z}{b^2} \int_0^\pi d\psi \int_0^{2\pi} \frac{1}{L} \sin^2\psi \cos\psi \sin\gamma \, d\gamma \, .$$

(4.17)

We can also show that the last two integrals in (4.17) vanish. To do so, we just need to consider a pair of points (ψ_1, γ_1) and $(\psi_1, \pi + \gamma_1)$ and perform a similar analysis as in the previous paragraph. Hence, the expression of F_x is simplified to

$$F_x = -G\rho \frac{x}{a^2} \int_0^\pi d\psi \int_0^{2\pi} \frac{1}{L} \sin\psi \cos^2\psi d\gamma \,. \tag{4.18}$$

The expression of L can be further developed as follows:

$$\begin{aligned} L &= \frac{\cos^2\psi + \sin^2\psi \cos^2\gamma}{a^2} + \frac{\sin^2\psi \sin^2\gamma}{b^2} \\ &= \frac{1}{a^2} + \left(\frac{1}{b^2} - \frac{1}{a^2}\right) \sin^2\psi \sin^2\gamma \\ &= \frac{1}{a^2}\left(1 + e'^2 \sin^2\psi \sin^2\gamma\right), \end{aligned} \tag{4.19}$$

where e' is the second eccentricity defined in (4.1). Substitute (4.19) into (4.18). We obtain

$$F_x = -G\rho x \int_0^\pi d\psi \int_0^{2\pi} \frac{\sin\psi \cos^2\psi}{1 + e'^2 \sin^2\psi \sin^2\gamma} d\gamma \,. \tag{4.20}$$

In the integrand, if we replace ψ by $\pi - \psi$ and/or replace γ by $\pi - \gamma$, $\pi + \gamma$, or $2\pi - \gamma$, the value of the integrand remains unchanged. We can thus express the integral as the sum of eight equal integrals, which we write in the form of

$$F_x = -8G\rho x \int_0^{\pi/2} \left[\sin\psi \cos^2\psi \int_0^{\pi/2} \frac{d\gamma}{1 + e'^2 \sin^2\psi \sin^2\gamma} \right] d\psi \,. \tag{4.21}$$

The integral over γ in (4.21) can be evaluated as follows. Substitute the variable γ by t according to $t = \cot\gamma$. We have

$$\sin^2\gamma = \frac{1}{1 + t^2} \,. \tag{4.22}$$

As $dt/d\gamma = -1/\sin^2\gamma$, we also have

$$d\gamma = -\sin^2\gamma dt = -\frac{dt}{1 + t^2} \,. \tag{4.23}$$

Based on the above two formulae, we have

$$
\int_0^{\pi/2} \frac{d\gamma}{1 + e'^2 \sin^2 \psi \sin^2 \gamma} = -\int_\infty^0 \frac{dt}{1 + e'^2 \sin^2 \psi + t^2}
$$

$$
= -\left\{ \arctan\left[\frac{t}{(1 + e'^2 \sin^2 \psi)^{1/2}} \right] \middle/ (1 + e'^2 \sin^2 \psi)^{1/2} \right\}_\infty^0
$$

$$
- \frac{\pi}{2(1 + e'^2 \sin^2 \psi)^{1/2}} \, . \tag{4.24}
$$

Substitute this formula into (4.21). We obtain

$$
F_x = -4\pi G\rho x \int_0^{\pi/2} \frac{\sin \psi \cos^2 \psi}{(1 + e'^2 \sin^2 \psi)^{1/2}} d\psi \, . \tag{4.25}
$$

To evaluate the integral over ψ in (4.25), we perform a variable substitution according to

$$
\cos \psi = \frac{(1 + e'^2)^{1/2}}{e'} \sin \phi \, . \tag{4.26}
$$

Taking the differential on both sides leads to

$$
\sin \psi d\psi = -\frac{(1 + e'^2)^{1/2}}{e'} \cos \phi d\phi \, . \tag{4.27}
$$

We can further obtain from (4.26)

$$
\sin^2 \psi = 1 - \cos^2 \psi = 1 - \frac{1 + e'^2}{e'} \sin^2 \phi \, . \tag{4.28}
$$

Substitute the above three formulae into (4.25). We obtain

$$
F_x = 4\pi G\rho x \frac{1 + e'^2}{e'^3} \int_{\arcsin\left[e'/(1+e'^2)^{1/2}\right]}^0 \sin^2 \phi d\phi
$$

$$
= 4\pi G\rho x \frac{1 + e'^2}{e'^3} \left[\frac{1}{2}(\phi - \sin \phi \cos \phi) \right]_{\arcsin\left[e'/(1+e'^2)^{1/2}\right]}^0
$$

$$
= -2\pi G\rho x \frac{1 + e'^2}{e'^3} \left[\arcsin \frac{e'}{(1 + e'^2)^{1/2}} - \frac{e'}{(1 + e'^2)^{1/2}} \right.
$$

$$
\left. \times \cos\left(\arcsin \frac{e'}{(1 + e'^2)^{1/2}} \right) \right] \, . \tag{4.29}
$$

The result (4.29) can be further simplified. Let

$$\arcsin \frac{e'}{(1 + e'^2)^{1/2}} = \alpha .\tag{4.30}$$

We then have

$$\sin \alpha = \frac{e'}{(1 + e'^2)^{1/2}} , \qquad \cos \alpha = \frac{1}{(1 + e'^2)^{1/2}} ;\tag{4.31}$$

$$\tan \alpha = e' , \qquad \alpha = \arctan e' .\tag{4.32}$$

We can finally simplify (4.29) to

$$F_x = -2\pi G \rho x \frac{1 + e'^2}{e'^3} \left(\arctan e' - \frac{e'}{1 + e'^2} \right) .\tag{4.33}$$

According to the symmetry of the ellipsoid of revolution, we similarly have

$$F_y = -2\pi G \rho y \frac{1 + e'^2}{e'^3} \left(\arctan e' - \frac{e'}{1 + e'^2} \right) .\tag{4.34}$$

4.2.2   The Gravitational Attraction Component F_z

A large portion of the formulation of the gravitational attraction component F_z is similar to the case of F_x in the previous subsection. Here we introduce a spherical coordinate system with origin on P and polar axis parallel to the z-axis and also denote the polar angle and longitude as ψ and γ, respectively. The variables of integration ξ, η, and ζ in (4.4) are then related to r, ψ, and γ as

$$\xi = x + r \sin \psi \cos \gamma , \qquad \eta = y + r \sin \psi \sin \gamma , \qquad \zeta = z + r \cos \psi .\tag{4.35}$$

Substitute these formulae into the third formula of (4.4). We obtain, considering that the volume element is the same as (4.6),

$$F_z = G\rho \int_0^\pi d\psi \int_0^{2\pi} d\gamma \int_0^{r_1} \sin \psi \cos \psi dr = G\rho \int_0^\pi d\psi \int_0^{2\pi} r_1 \sin \psi \cos \psi .\tag{4.36}$$

Here r_1 is also the distance between the point P and the surface of the ellipsoid, which satisfies the equation.

$$\frac{(x + r_1 \sin \psi \cos \gamma)^2 + (y + r_1 \sin \psi \sin \gamma)^2}{a^2} + \frac{(z + r_1 \cos \psi)^2}{b^2} = 1 .\tag{4.37}$$

Some algebraic operations lead to the second order equation

$$Lr_1^2 + 2Mr_1 + N = 0,$$ (4.38)

where (again, noting that here M is used to represent an expression, but not a mass)

$$\begin{cases} L = \dfrac{\sin^2 \psi}{a^2} + \dfrac{\cos^2 \psi}{b^2}, \\ M = \dfrac{x \sin \psi \cos \gamma}{a^2} + \dfrac{y \sin \psi \sin \gamma}{a^2} + \dfrac{z \cos \psi}{b^2}, \\ N = \dfrac{x^2}{a^2} + \dfrac{y^2}{a^2} + \dfrac{z^2}{b^2} - 1. \end{cases}$$ (4.39)

The positive root of (4.38) we need is

$$r_1 = \frac{-M + (M^2 - LN)^{1/2}}{L}.$$ (4.40)

Substitute it into (4.36). We obtain

$$F_z = -G\rho \int_0^\pi d\psi \int_0^{2\pi} \frac{M}{L} \sin \psi \cos \psi d\gamma$$

$$+ G\rho \int_0^\pi d\psi \int_0^{2\pi} \frac{(M^2 - LN)^{1/2}}{L} \sin \psi \cos \psi d\gamma.$$ (4.41)

Similar to the case of F_x, the second integral can be shown to vanish. Substitute the formula of M given by (4.39) into this formula. We obtain

$$F_z = -G\rho \frac{x}{a^2} \int_0^\pi d\psi \int_0^{2\pi} \frac{1}{L} \sin^2 \psi \cos \psi \cos \gamma d\gamma$$

$$- G\rho \frac{y}{a^2} \int_0^\pi d\psi \int_0^{2\pi} \frac{1}{L} \sin^2 \psi \cos \psi \sin \gamma d\gamma$$

$$- G\rho \frac{z}{b^2} \int_0^\pi d\psi \int_0^{2\pi} \frac{1}{L} \sin \psi \cos^2 \psi d\gamma.$$ (4.42)

It can be shown that the first two integrals in this formula vanish. Hence,

$$F_z = -G\rho \frac{z}{b^2} \int_0^\pi d\psi \int_0^{2\pi} \frac{1}{L} \sin \psi \cos^2 \psi d\gamma.$$ (4.43)

Write L in the form

$$L = \frac{\sin^2 \psi}{b^2(1 + e'^2)} + \frac{\cos^2 \psi}{b^2} = \frac{1 + e'^2 \cos^2 \psi}{b^2(1 + e'^2)}, \tag{4.44}$$

and then substitute it into (4.43). We obtain

$$F_z = -G\rho z(1 + e'^2) \int_0^{\pi} d\psi \int_0^{2\pi} \frac{\sin \psi \cos^2 \psi}{1 + e'^2 \cos^2 \psi} d\gamma. \tag{4.45}$$

As the integrand is independent of γ, the integration over γ can be evaluated first. We obtain

$$F_z = -2\pi G\rho z(1 + e'^2) \int_0^{\pi} \frac{\sin \psi \cos^2 \psi}{1 + e'^2 \cos^2 \psi} d\psi. \tag{4.46}$$

Considering that the integrand has the same value at the pair of points $\psi = \psi_1$ and $\psi = \pi - \psi_1$, we can write this formula as

$$F_z = -4\pi G\rho z(1 + e'^2) \int_0^{\pi/2} \frac{\sin \psi \cos^2 \psi}{1 + e'^2 \cos^2 \psi} d\psi. \tag{4.47}$$

Let

$$\cos \psi = \frac{1}{e'} \tan \phi. \tag{4.48}$$

Taking the differential on both sides yields

$$\sin \psi d\psi = \frac{d\phi}{e' \cos^2 \phi}. \tag{4.49}$$

Substitute the above two formulae into (4.47). We finally obtain

$$F_z = 4\pi G\rho z \frac{1 + e'^2}{e'^3} \int_{\arctan e'}^{0} \left(\frac{1}{\cos^2 \phi} - 1 \right) d\phi$$

$$= 4\pi G\rho z \frac{1 + e'^2}{e'^3} (\tan \phi - \phi)_{\arctan e'}^{0}$$

$$= -4\pi G\rho z \frac{1 + e'^2}{e'^3} (e' - \arctan e'). \tag{4.50}$$

4.2.3 Gravitational Potential in the Interior

According to (4.33), (4.34), and (4.50), the gravitational attraction of a homogeneous ellipsoid of revolution in the interior is

$$\vec{F} = -Px\hat{e}_x - Py\hat{e}_y - Qz\hat{e}_z , \qquad (4.51)$$

where P and Q are constant parameters defined by

$$\begin{cases} P = 2\pi G\rho \dfrac{1+e'^2}{e'^3} \left(\arctan e' - \dfrac{e'}{1+e'^2} \right) , \\ Q = 4\pi G\rho \dfrac{1+e'^2}{e'^3} (e' - \arctan e') . \end{cases} \qquad (4.52)$$

Evidently, the gravitational potential in the interior should be

$$V_I = -\frac{1}{2}Px^2 - \frac{1}{2}Py^2 - \frac{1}{2}Qz^2 + K , \qquad (4.53)$$

where K is the gravitational potential at the center,

$$K = G\rho \int_v \frac{1}{r} dv . \qquad (4.54)$$

The rest is to derive the formula of K. Define a spherical coordinate system with origin at the center of the ellipsoid and polar axis along the symmetry axis, and denote the radial distance, polar angle, and longitude as r, ψ, and γ, respectively. We have for the coordinates of the volume element dv

$$\xi = r \sin \psi \cos \gamma , \qquad \eta = r \sin \psi \sin \gamma , \qquad \zeta = r \cos \psi . \qquad (4.55)$$

The expression of the volume element is the same as (4.6). Hence,

$$K = G\rho \int_0^\pi d\psi \int_0^{2\pi} d\gamma \int_0^{r_1} r \sin \psi dr = \frac{1}{2}G\rho \int_0^\pi d\psi \int_0^{2\pi} r_1^2 \sin \psi d\gamma , \qquad (4.56)$$

where r_1 is the distance from the center to the surface. Naturally, r_1 satisfies the equation

$$\frac{(r_1 \sin \psi \cos \gamma)^2 + (r_1 \sin \psi \sin \gamma)^2}{a^2} + \frac{(r_1 \cos \psi)^2}{b^2} = 1 , \qquad (4.57)$$

which has the solution

$$r_1^2 = \frac{a^2}{1 + e'^2 \cos^2 \psi}. \tag{4.58}$$

Substitute this formula into (4.56). We obtain

$$K = \frac{1}{2} G \rho a^2 \int_0^\pi d\psi \int_0^{2\pi} \frac{\sin \psi d\gamma}{1 + e'^2 \cos^2 \psi} = \pi G \rho a^2 \int_0^\pi \frac{\sin \psi d\psi}{1 + e'^2 \cos^2 \psi}. \tag{4.59}$$

Substitute the variable according to $t = e' \cos \psi$. We obtain

$$K = -\pi G \rho \frac{a^2}{e'} \int_{e'}^{-e'} \frac{dt}{1 + t^2} = 2\pi G \rho \frac{a^2}{e'} \arctan e'. \tag{4.60}$$

In summary, the gravitational potential in the interior is given by (4.53), (4.52), and (4.60).

Later on, we will need to substitute the density ρ by the mass M. Their relation is

$$M = \frac{4}{3} \pi \rho a^2 b = \frac{4}{3} \pi \rho \frac{a^3}{(1 + e'^2)^{1/2}}. \tag{4.61}$$

4.3   External Gravitational Field of a Homogeneous Ellipsoid of Revolution

Following MacMillan (1958), we first compute the gravitational attraction on the symmetry axis, then we compute the gravitational potential on the symmetry axis by integrating the gravitational attraction, and finally, we infer the gravitational potential outside the ellipsoid from that on the symmetry axis.

As in the case for an internal domain, we use the Cartesian coordinate system $Oxyz$ as shown in Fig. 4.1. We use ρ to denote the density of the ellipsoid and (ξ, η, ζ) the Cartesian coordinate of any point on its surface. The equation of its surface is again

$$\frac{\xi^2 + \eta^2}{a^2} + \frac{\zeta^2}{b^2} = 1. \tag{4.62}$$

As shown in Fig. 4.2, we consider a point P on the z-axis with coordinate $(0, 0, z)$, where the coordinate z can also be understood as the distance between the point P and the center of the ellipsoid O. We make use of (1.84) by dividing the ellipsoid into a series of thin layers perpendicular to the z-axis and treat each layer as a disk of surface density $\mu = \rho d\zeta$, where ζ is the z-coordinate of the layer and $d\zeta$ the

Fig. 4.2 Gravity of a
homogeneous ellipsoid of
revolution

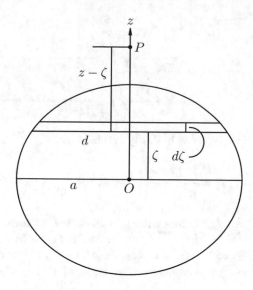

thickness of the layer. Considering that in (1.84), z is the distance of P to the
disk and a the radius of the disk, here, with the symbols defined in Fig. 4.2, the
gravitational attraction at the point P along the z-axis is

$$F = 2\pi G\rho \int_{-b}^{b} \left\{ \frac{z-\zeta}{[(z-\zeta)^2 + d^2]^{1/2}} - 1 \right\} d\zeta, \qquad (4.63)$$

where $z - \zeta$ is the distance from P to the disk and d the radius of the disk

$$d = \left[a^2 - \frac{a^2}{b^2}\zeta^2 \right]^{1/2}. \qquad (4.64)$$

Substitute (4.64) into (4.63). We obtain

$$F = 2\pi G\rho \left\{ \int_{-b}^{b} \frac{z-\zeta}{[(z-\zeta)^2 + a^2 - (a^2/b^2)\zeta^2]^{1/2}} d\zeta - 2b \right\}. \qquad (4.65)$$

The trick to evaluate the integral is to express $z - \zeta$ as

$$z - \zeta = \frac{[-z + \zeta - (a^2/b^2)\zeta] + (a^2/b^2)z}{a^2/b^2 - 1}. \qquad (4.66)$$

According to (4.66), we separate the integral in (4.65) into two parts,

$$F = 2\pi G\rho (I_1 + I_2 - 2b), \qquad (4.67)$$

where

$$I_1 = \frac{b^2}{a^2 - b^2} \int_{-b}^{b} \frac{-z + \zeta - (a^2/b^2)\zeta}{[(z - \zeta)^2 + a^2 - (a^2/b^2)\zeta^2]^{1/2}} d\zeta \,, \tag{4.68}$$

$$I_2 = \frac{a^2 z}{a^2 - b^2} \int_{-b}^{b} \frac{1}{[(z - \zeta)^2 + a^2 - (a^2/b^2)\zeta^2]^{1/2}} d\zeta \,. \tag{4.69}$$

For I_1, it can be easily verified that

$$I_1 = \frac{b^2}{a^2 - b^2} \left\{ \left[(z - \zeta)^2 + a^2 - (a^2/b^2)\zeta^2 \right]^{1/2} \right\}_{-b}^{b} = -\frac{2b^3}{a^2 - b^2} \,. \tag{4.70}$$

For I_2, we first write it in an alternative form to use the indefinite integral formula $\int (a^2 - x^2)^{-1/2} dx = \arcsin(x/a) + C$ and then further use the trigonometric formulae $\arcsin x = \arctan[x/(1-x^2)^{1/2}]$, $\arctan x - \arctan y = \arctan[(x-y)/(1+xy)]$, and $\arctan x = 2\arctan\{x/[1 + (1 + x^2)^{1/2}]\}$. We have

$$I_2 = \frac{a^2 b z}{(a^2 - b^2)^{3/2}} \int_{-b}^{b} \left[\frac{a^2 b^2 (a^2 - b^2 + z^2)}{(a^2 - b^2)^2} - \left(\frac{b^2 z}{a^2 - b^2} + \zeta \right)^2 \right]^{-1/2} d\zeta$$

$$= \left[\arcsin \frac{(a^2 - b^2)\zeta + b^2 z}{ab(a^2 - b^2 + z^2)^{1/2}} \right]_{-b}^{b}$$

$$= \arcsin \frac{a^2 - b^2 + bz}{a(a^2 - b^2 + z^2)^{1/2}} - \arcsin \frac{-a^2 + b^2 + bz}{a(a^2 - b^2 + z^2)^{1/2}}$$

$$= \arctan \frac{a^2 - b^2 + bz}{(a^2 - b^2)^{1/2}(z - b)} - \arctan \frac{-a^2 + b^2 + bz}{(a^2 - b^2)^{1/2}(z + b)}$$

$$= \arctan \frac{2z(a^2 - b^2)^{1/2}}{z^2 + b^2 - a^2}$$

$$= 2 \arctan \frac{(a^2 - b^2)^{1/2}}{z} \,. \tag{4.71}$$

Finally, substitute (4.70) and (4.71) into (4.67). We obtain

$$F = 4\pi G\rho \left[-b - \frac{b^3}{a^2 - b^2} + \frac{a^2 b z}{(a^2 - b^2)^{3/2}} \arctan \frac{(a^2 - b^2)^{1/2}}{z} \right]$$

$$= 4\pi G\rho \frac{a^2 b}{a^2 - b^2} \left[-1 + \frac{z}{(a^2 - b^2)^{1/2}} \arctan \frac{(a^2 - b^2)^{1/2}}{z} \right]$$

$$= \frac{3GM}{a^2 - b^2} \left[-1 + \frac{z}{(a^2 - b^2)^{1/2}} \arctan \frac{(a^2 - b^2)^{1/2}}{z} \right] \,, \tag{4.72}$$

where M is the mass of the ellipsoid provided by (4.61). By using the Taylor series

$$\arctan x = \sum_{n=0}^{\infty} (-1)^n \frac{x^{2n+1}}{2n+1},$$ (4.73)

we obtain

$$
\begin{aligned}
F &= \frac{3GM}{a^2 - b^2} \left\{ -1 + \frac{z}{(a^2 - b^2)^{1/2}} \sum_{n=0}^{\infty} (-1)^n \frac{1}{2n+1} \left[\frac{(a^2-b^2)^{1/2}}{z} \right]^{2n+1} \right\} \\
&= \frac{3GMz}{(a^2-b^2)^{3/2}} \sum_{n=1}^{\infty} (-1)^n \frac{1}{2n+1} \left[\frac{(a^2-b^2)^{1/2}}{z} \right]^{2n+1} \\
&= \frac{3GMz}{(a^2-b^2)^{3/2}} \sum_{n=1}^{\infty} (-1)^n \frac{1}{2n+1} \frac{(a^2-b^2)^{n+1/2}}{z^{2n+1}}] \\
&= GM \sum_{n=1}^{\infty} (-1)^n \frac{3}{2n+1} \frac{(a^2-b^2)^{n-1}}{z^{2n}} \\
&= GM \sum_{n=0}^{\infty} (-1)^{n+1} \frac{3}{2n+3} \frac{(a^2-b^2)^n}{z^{2n+2}} \\
&= GM \sum_{n=0}^{\infty} \frac{3(-1)^{n+1} e^{2n}}{2n+3} \frac{a^{2n}}{z^{2n+2}},
\end{aligned}
$$ (4.74)

where e is the first eccentricity defined in (4.1)

The gravitational potential on the symmetry axis can be obtained by integrating F. As $dV/dz = F$ and $V_{z \to \infty} = 0$, we have

$$
\begin{aligned}
V = \int_{\infty}^{z} F(z') dz' &= -GM \sum_{n=0}^{\infty} \frac{3(-1)^{n+1} e^{2n}}{(2n+1)(2n+3)} \frac{a^{2n}}{z^{2n+1}} \\
&= \frac{GM}{z} \left[1 - \sum_{n=1}^{\infty} \frac{3(-1)^{n+1} e^{2n}}{(2n+1)(2n+3)} \left(\frac{a}{z} \right)^{2n} \right].
\end{aligned}
$$ (4.75)

Now we consider the general expression of the gravitational potential of the ellipsoid in the exterior, which is denoted as V_E. In the spherical coordinate system $Or\theta\lambda$, where O is at the center of the ellipsoid, the general spherical harmonic series of V_E is in the form of (3.297). Due to the symmetry with respect to the minor axis, V_E is independent of λ, and hence only the order 0 terms of the spherical harmonic series are non-vanishing. Furthermore, the ellipsoid is symmetrical about the equatorial plane, and thus only the even degree terms are non-vanishing. On the symmetry axis where $r = z$, the degenerate expression of V_E is exactly (4.75). Hence, we can infer

$$V_E = \frac{GM}{r} \left[1 - \sum_{n=1}^{\infty} \frac{3(-1)^{n+1} e^{2n}}{(2n+1)(2n+3)} \left(\frac{a}{r}\right)^{2n} P_{2n}(\cos\theta) \right], \tag{4.76}$$

which degenerates to (4.75) when $\theta = 0$.

4.4   Gravitational Potential of a Homogeneous Ellipsoidal Homoeoid of Revolution

In this section, we consider a particular kind of shell: (1) its outer and inner surfaces are all ellipsoid of revolution, (2) the major and minor axes of the outer and inner surfaces overlap, and (3) the outer and inner surfaces are similar, i.e., if the semi-major and semi-minor axes of the outer surface are a and b, respectively, the semi-major and semi-minor axes of the inner surface can be expressed as αa and αb, respectively. Evidently, the first and second eccentricities of the two surfaces are identical. Such a shell is referred to as an ellipsoidal homoeoid of revolution.

4.4.1   Gravitational Potential in the Interior

We denote the semi-major axes of the outer and inner surfaces as a and αa, respectively. The gravitational potential of the homoeoid is the difference between those of the ellipsoids with semi-major axes a and αa. According to (4.53), (4.52), and (4.60), we obtain

$$V_I = 2\pi G\rho a^2 \frac{1-\alpha^2}{e'^2} \arctan e', \tag{4.77}$$

which is a constant, and thus the gravitational attraction in the interior vanishes. The mass of the shell is

$$M = \frac{4}{3}\pi\rho(a^2 b - \alpha^2 a^2 b) = \frac{4}{3}\pi\rho a^2 b(1-\alpha^2) = \frac{4}{3}\pi\rho a^3 \frac{1-\alpha^2}{(1+e'^2)^{1/2}}. \tag{4.78}$$

Substitute the density ρ by the mass M in (4.77). We obtain

$$V_I = \frac{3}{2a} GM \frac{(1+e'^2)^{1/2}}{e'} \frac{1-\alpha^2}{1-\alpha^3} \arctan e'$$

$$= \frac{3}{2a} GM \frac{(1+e'^2)^{1/2}}{e'} \frac{1+\alpha}{1+\alpha+\alpha^2} \arctan e', \tag{4.79}$$

where the factorization formula $1 - \alpha^3 = (1 - \alpha)(1 + \alpha + \alpha^2)$ is used.

Let $\alpha \to 1$. We assume that $\rho \to \infty$ such that M remains unchanged. The shell becomes a material surface, and we have

$$V_I = \frac{GM}{a} \frac{(1 + e'^2)^{1/2}}{e'} \arctan e', \tag{4.80}$$

which is a constant. As the gravitational potential of a material surface is continuous across the surface, we conclude that the gravitational potential of this material surface is constant on the surface, which will be used later to devise a solid body that generates the normal gravity field.

4.4.2 Gravitational Potential in the Exterior

Substitute the mass M by density ρ in (4.76). We obtain the gravitational potential of a homogeneous ellipsoid in the exterior expressed using density

$$V_E = \frac{4\pi G \rho a^3}{3r(1 + e'^2)^{1/2}} \left[1 - \sum_{n=1}^{\infty} \frac{3(-1)^{n+1}}{(2n+1)(2n+3)} e'^{2n} \left(\frac{a}{r}\right)^{2n} P_{2n}(\cos\theta) \right]. \tag{4.81}$$

Again, the gravitational potential of the homoeoid is the difference between those of the ellipsoids with semi-major axes a and αa. We have

$$\begin{aligned}
V_E &= \frac{4\pi G \rho a^3}{3r(1 + e'^2)^{1/2}} \left[1 - \sum_{n=1}^{\infty} \frac{3(-1)^{n+1}}{(2n+1)(2n+3)} e'^{2n} \left(\frac{a}{r}\right)^{2n} P_{2n}(\cos\theta) \right] \\
&\quad - \frac{4\pi G \rho \alpha^3 a^3}{3r(1 + e'^2)^{1/2}} \left[1 - \sum_{n=1}^{\infty} \frac{3(-1)^{n+1}}{(2n+1)(2n+3)} \alpha^{2n} e'^{2n} \left(\frac{a}{r}\right)^{2n} P_{2n}(\cos\theta) \right] \\
&= \frac{4\pi G \rho a^3}{3r(1 + e'^2)^{1/2}} \left[1 - \alpha^3 - \sum_{n=1}^{\infty} \frac{3(-1)^{n+1}}{(2n+1)(2n+3)} (1 - \alpha^{2n+3}) \right. \\
&\quad \left. \times e'^{2n} \left(\frac{a}{r}\right)^{2n} P_{2n}(\cos\theta) \right] \\
&= \frac{4\pi G \rho a^3 (1 - \alpha^3)}{3r(1 + e'^2)^{1/2}} \left[1 - \sum_{n=1}^{\infty} \frac{3(-1)^{n+1}}{(2n+1)(2n+3)} \frac{1 - \alpha^{2n+3}}{1 - \alpha^3} \right. \\
&\quad \left. \times e'^{2n} \left(\frac{a}{r}\right)^{2n} P_{2n}(\cos\theta) \right].
\end{aligned} \tag{4.82}$$

Substitute the density ρ by mass M according to (4.78). We obtain

$$V_E = \frac{GM}{r} \left[1 - \sum_{n=1}^{\infty} \frac{3(-1)^{n+1}}{(2n+1)(2n+3)} \frac{1-\alpha^{2n+3}}{1-\alpha^3} e^{2n} \left(\frac{a}{r}\right)^{2n} P_{2n}(\cos\theta) \right]. \tag{4.83}$$

Let $\alpha \to 1$. We assume that $\rho \to \infty$ such that M remains unchanged. The shell becomes a material surface. According to L'Hôpital's rule, we have

$$\lim_{\alpha \to 1} \frac{1-\alpha^{2n+3}}{1-\alpha^3} = \lim_{\alpha \to 1} \frac{-(2n+3)\alpha^{2n+2}}{-3\alpha^2} = \frac{2n+3}{3}. \tag{4.84}$$

We obtain finally

$$V_E = \frac{GM}{r} \left[1 - \sum_{n=1}^{\infty} \frac{(-1)^{n+1}}{2n+1} e^{2n} \left(\frac{a}{r}\right)^{2n} P_{2n}(\cos\theta) \right]. \tag{4.85}$$

Based on the conclusion for the interior, we already know that the value of this series is constant on the surface and is given in (4.80).

4.4.3   Surface Density

In the previous two sections, we defined a material surface by assuming $\alpha \to 1$ to a homogeneous ellipsoidal homoeoid of revolution while forcing the total mass to remain unchanged. Its gravitational potentials in the interior and exterior have already been formulated. A particular property is that its gravitational potentials in the interior are constant; therefore, it is also constant on the surface, as the gravitational potential of a material surface is continuous across it.

Here we derive the formula of its surface density. We start by analyzing an extreme thin shell. As shown in Fig. 4.3, let Φ and B be the geocentric and geodetic latitudes. The density of the shell can be expressed as

$$\mu = \rho \, PP_1 \cos(B - \Phi), \tag{4.86}$$

where PP_1 is the distance between P and P_1. The formulae of r and $\cos(B - \Phi)$ are derived in Appendix A. According to (A.27), we have

$$PP_1 = r_P - r_{P_1} = (1-\alpha)r_P = (1-\alpha) \left(\frac{a^4 \cos^2 B + b^4 \sin^2 B}{a^2 \cos^2 B + b^2 \sin^2 B} \right)^{1/2}. \tag{4.87}$$

Fig. 4.3 Density of a
material surface of ellipsoidal
homoeoid of revolution

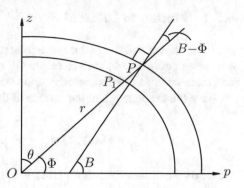

By substituting (A.28) and (4.87) into (4.86), we obtain the surface density

$$\mu = \rho(1 - \alpha)(a^2 \cos^2 B + b^2 \sin^2 B)^{1/2}. \tag{4.88}$$

Substitute the density of the shell, ρ, by its mass M according to (4.78). We obtain

$$\mu = \frac{3M(1 - \alpha)}{4\pi a^2 b(1 - \alpha^3)}(a^2 \cos^2 B + b^2 \sin^2 B)^{1/2}$$

$$= \frac{3M}{4\pi a^2 b(1 + \alpha + \alpha^2)}(a^2 \cos^2 B + b^2 \sin^2 B)^{1/2}. \tag{4.89}$$

Let $\alpha \to 1$ while assuming M remains unchanged. We obtain the final result

$$\mu = \frac{M}{4\pi a^2 b}(a^2 \cos^2 B + b^2 \sin^2 B)^{1/2}. \tag{4.90}$$

4.5   The Earth's Normal Gravity Field

In this section, we devise a mass distribution within an ellipsoid of revolution
rotating with the constant rate Ω around its minor axis, such that its total mass
is equal to that of the Earth M, and its surface is an equipotential surface of
gravity of itself. According to Stokes' theorem, the external gravitational potential
is independent of the detail of mass distribution and is uniquely determined by Ω,
M, and the semi-major and semi-minor axes a and b.

4.5.1 Maclaurin Ellipsoid

It can be shown that the surface of a rotating homogeneous ellipsoid of revolution
can be an equipotential surface of gravity of itself, provided that the rotation rate
and mass satisfy a relation, which we are going to derive in this subsection.

We write the equation of the surface of the ellipsoid of revolution as

$$F(x, y, z) = \frac{x^2}{a^2} + \frac{y^2}{a^2} + \frac{z^2}{b^2} - 1 = 0. \tag{4.91}$$

Based on the discussion below (1.18), we know that the vector

$$\vec{n} = \frac{\partial F}{\partial x}\hat{e}_x + \frac{\partial F}{\partial y}\hat{e}_y + \frac{\partial F}{\partial z}\hat{e}_z = 2\left(\frac{x}{a^2}\hat{e}_x + \frac{y}{a^2}\hat{e}_y + \frac{z}{b^2}\hat{e}_z\right) \tag{4.92}$$

is normal to the surface. According to (4.51) and (1.14), the gravity on the surface
is

$$\vec{g} = (\Omega^2 - P)x\hat{e}_x + (\Omega^2 - P)y\hat{e}_y - Qz\hat{e}. \tag{4.93}$$

If the surface is an equipotential surface of gravity, the gravity $\vec{g}$ should be
perpendicular to it, meaning that $\vec{n}$ and $\vec{g}$ are parallel. Hence, we should have

$$\frac{x/a^2}{(\Omega^2 - P)x} = \frac{y/a^2}{(\Omega^2 - P)y} = \frac{z/b^2}{-Qz}. \tag{4.94}$$

This chain of equations does have a solution

$$(\Omega^2 - P)a^2 = -Qb^2, \tag{4.95}$$

which is independent of the variables x, y, and z. This fact implies that the surface
is indeed an equipotential surface of gravity provided the relation (4.95) holds. Such
an ellipsoid is called the Maclaurin ellipsoid.

Write (4.95) in the form of

$$\Omega^2 = P - \frac{Q}{1 + e'^2}, \tag{4.96}$$

and then substitute (4.52) into it. We obtain

$$\frac{\Omega^2}{2\pi G\rho_{\text{Mac}}} = \frac{3 + e'^2}{e'^3}\arctan e' - \frac{3}{e'^2}, \tag{4.97}$$

where we used ρ_{Mac} to denote the density. Substitute the density by the mass M_{Mac} according to (4.61). We obtain

$$M_{\text{Mac}} = \frac{2\Omega^2 a^3 e'^3}{3G \left[(3 + e'^2)\arctan e' - 3e'\right](1 + e'^2)^{1/2}}.$$ (4.98)

For an ellipsoid with a rotation rate Ω, the mass must be M_{Mac} in order for its surface to be an equipotential surface of gravity of itself. In the more general sense, (4.98) is a relation among the mass, rotation rate, size, and shape of a Maclaurin ellipsoid.

4.5.2 Normal Gravitational Potential

By definition, the normal potential is the gravity potential of such an ellipsoid of revolution: (1) its mass is M, (2) it rotates around its minor axis with a constant rotation rate Ω, and (3) its surface is an equipotential surface of gravity of itself. The surface of the Maclaurin ellipsoid is an equipotential surface of gravity of itself, but its mass M_{Mac} is related to Ω according to (4.98). We cannot expect that M_{Mac} happens to be M. However, such an ellipsoid can be devised as a Maclaurin ellipsoid superimposed with an ellipsoidal homoeoid on its surface. Thus we write the normal potential as

$$U = V_{\text{Mac}} + Q + V_{\text{Homo}},$$ (4.99)

where V_{Mac} and Q are the gravitational and centrifugal potentials of the Maclaurin ellipsoid with semi-major and semi-minor axes a and b and rotation rate Ω, and V_{Homo} is the gravitational potential of an ellipsoidal homoeoid with mass $M_{\text{Homo}} = M - M_{\text{Mac}}$ superimposed on the surface of the Maclaurin ellipsoid. As both $V_{\text{Mac}} + Q$ and V_{Homo} are constant on the surface of the ellipsoid, U is also constant on the surface, and the total mass is M. Hence, according to Stokes' theorem, U must be the normal potential.

According to (4.76), V_{Mac} is

$$V_{\text{Mac}} = \frac{GM_{\text{Mac}}}{r}\left[1 - \sum_{n=1}^{\infty} \frac{3(-1)^{n+1}e^{2n}}{(2n+1)(2n+3)}\left(\frac{a}{r}\right)^{2n} P_{2n}(\cos\theta)\right].$$ (4.100)

According to (4.85), V_{Homo} is

$$V_{\text{Homo}} = \frac{G(M - M_{\text{Mac}})}{r}\left[1 - \sum_{n=1}^{\infty} \frac{(-1)^{n+1}e^{2n}}{2n+1}\left(\frac{a}{r}\right)^{2n} P_{2n}(\cos\theta)\right].$$ (4.101)

The normal gravitational potential is then the sum $V_{\text{Mac}} + V_{\text{Homo}}$, which is customarily expressed as

$$V = \frac{GM}{r}\left[1 - \sum_{n=1}^{\infty} J_{2n}\left(\frac{a}{r}\right)^{2n} P_{2n}(\cos\theta)\right], \tag{4.102}$$

where

$$J_{2n} = \frac{3(-1)^{n+1}e^{2n}}{(2n+1)(2n+3)}\frac{V_{\text{Mac}}}{M} + \frac{(-1)^{n+1}e^{2n}}{2n+1}\frac{M - M_{\text{Mac}}}{M}$$

$$= \frac{(-1)^{n+1}e^{2n}[2n+3-2n(M_{\text{Mac}}/M)]}{(2n+1)(2n+3)}. \tag{4.103}$$

According to (4.98), we can write J_{2n} as

$$J_{2n} = \frac{(-1)^{n+1}e^{2n}}{(2n+1)(2n+3)}$$

$$\times \left\{2n+3 - \frac{4n\Omega^2 a^3 e'^3}{3GM\left[(3+e'^2)\arctan e' - 3e'\right](1+e'^2)^{1/2}}\right\}. \tag{4.104}$$

With the parameter m defined in (4.2), we can further write J_{2n} as

$$J_{2n} = \frac{(-1)^{n+1}e^{2n}}{(2n+1)(2n+3)}\left\{2n+3 - \frac{4ne'^3 m}{3\left[(3+e'^2)\arctan e' - 3e'\right]}\right\}. \tag{4.105}$$

4.5.3   The Reference Earth Ellipsoid

In the previous subsection, the formula of normal gravitational potential is derived using four parameters: GM, Ω, a and b (or e', or e). For the normal potential to be as close as possible to the real gravity potential of the Earth, we chose GM and Ω to equal those of the Earth. However, a and b are replaced by another two parameters: one is the normal potential on the reference Earth ellipsoid, U_0, which is assumed to equal the gravity potential of the Earth on the geoid, W_0, and the other is the dynamical form factor of the Earth, J_2, which is assumed to be the negative of the potential coefficient C_2 of the Earth,

$$J_2 = -C_2 = \frac{C - (A+B)/2}{Ma^2}. \tag{4.106}$$

In addition, the center of the reference Earth ellipsoid is assumed to be on the center of mass of the Earth, and the symmetry axis of the reference Earth ellipsoid is assumed to coincide with the Earth's rotation axis.

Evidently, to complete the theoretical formulation, we have to derive the formula to compute a and b from GM, Ω, U_0, and J_2. We first use the closed-form expressions of internal gravitational potentials of the Maclaurin ellipsoid and ellipsoidal homoeoid to derive a closed-form expression of U_0 in terms of GM, Ω, a, and b. We express the normal potential on the surface of the reference Earth ellipsoid as

$$U_0 = U_{\text{Mac}} + V_{\text{Homo}}, \tag{4.107}$$

where the symbols are self-explanatory. The gravity potential of the Maclaurin ellipsoid is, according to (4.35) and (4.95),

$$
\begin{aligned}
U_{\text{Mac}} &= -\frac{1}{2}Px^2 - \frac{1}{2}Py^2 - \frac{1}{2}Qz^2 + K + \frac{1}{2}\Omega^2(x^2 + y^2) \\
&= \frac{1}{2}(\Omega^2 - P)x^2 - \frac{1}{2}(\Omega^2 - P)y^2 - \frac{1}{2}Qz^2 + K \\
&= -\frac{1}{2}b^2 Q\left(\frac{x^2}{a^2} + \frac{y^2}{a^2} + \frac{z^2}{b^2}\right) + K \\
&= -\frac{1}{2}\frac{a^2}{1 + e'^2}Q + K .
\end{aligned}
\tag{4.108}
$$

Replace ρ by ρ_{Mac} in (4.52) and (4.60), and then substitute them into the above formula. We obtain

$$U_{\text{Mac}} = 2\pi G\rho_{\text{Mac}}\frac{a^2}{e'^3}[(1 + e'^2)\arctan e' - e']. \tag{4.109}$$

The substitution of ρ_{Mac} by M_{Mac} according to (4.61) leads to

$$U_{\text{Mac}} = \frac{3}{2}\frac{GM_{\text{Mac}}}{a}\frac{(1 + e'^2)^{1/2}}{e'^3}[(1 + e'^2)\arctan e' - e']. \tag{4.110}$$

The gravitational potential of the ellipsoidal homoeoid of revolution is, according to (4.80),

$$V_{\text{Homo}} = \frac{G(M - M_{\text{Mac}})}{a}\frac{(1 + e'^2)^{1/2}}{e'}\arctan e'. \tag{4.111}$$

Substitute these two formulae into (4.107). We obtain

$$
\begin{aligned}
U_0 &= \frac{GM}{a}\frac{(1 + e'^2)^{1/2}}{e'}\arctan e' \\
&\quad + \frac{1}{2}\frac{GM_{\text{Mac}}}{a}\frac{(1 + e'^2)^{1/2}}{e'^3}[(3 + e'^2)\arctan e' - 3e'].
\end{aligned}
\tag{4.112}
$$

Finally, we can simplify it to, according to (4.98),

$$U_0 = \frac{GM}{a} \frac{(1 + e'^2)^{1/2}}{e'} \arctan e' + \frac{1}{3}\Omega^2 a^2 . \tag{4.113}$$

This is the equation we are seeking that links U_0 to GM, Ω, a, and e' (hence also links to b).

We also need another equation to link J_2 to GM, Ω, a, and e' (hence also link to b), which is quite straightforward to derive. We have only to set $n = 1$ in (4.105) and then substitute e by e'

$$J_2 = \frac{1}{3} \frac{e'^2}{1 + e'^2} \left\{ 1 - \frac{4e'^3 m}{15[(3 + e'^2) \arctan e' - 3e']} \right\} , \tag{4.114}$$

where m is defined in (4.2).

We mention that a and e' (hence also b) can be computed from GM, Ω, U_0, and J_2 by solving (4.113) and (4.114) after substituting m by its expression in terms of a and e' given by (4.2).

Contemporarily, GM and J_2 are measured very accurately by observing satellite orbits, and Ω is also measured very accurately by observing very distant celestial bodies. Observations on or of the Earth's surface are required only for determining U_0.

Finally, according to (4.105) and (4.114), a formula of J_{2n} depending on J_2 can be derived

$$J_{2n} = \frac{(-1)^{n+1} e^{2n}}{2n + 1} \left[1 - \frac{5n}{2n + 3} \left(1 - \frac{3J_2}{e^2} \right) \right] . \tag{4.115}$$

4.5.4 Normal Gravity on the Reference Earth Ellipsoid

A closed-form expression of the normal gravity on the reference Earth ellipsoid can be derived according to the formulae of internal gravity of the Maclaurin ellipsoid and the density of the ellipsoidal homoeoid. Based on the results of previous subsections, the normal gravity on the reference Earth ellipsoid is the sum of the gravity of a Maclaurin ellipsoid and the gravitational attraction of a homoeoid

$$\vec{\gamma}_0 = \vec{g}_{\text{Mac}} + \vec{F}_{\text{Homo}} . \tag{4.116}$$

We emphasize that $\vec{\gamma}_0$ is the limit of the terms at the right-hand side while approaching the surface of the reference Earth ellipsoid from outside. As the gravitational attraction of a solid body is continuous across its surface, we can use its limit in the interior as its value on the surface. Furthermore, as the surface of a Maclaurin ellipsoid is an equipotential surface of gravity, $\vec{g}_{\text{Mac}}$ is perpendicular to it. As the gravitational attraction of a homoeoid vanishes in the inside and the gravitational attraction of a material surface has a jump across it, we can infer that $\overrightarrow{F}_{\text{Homo}}$ is the jump, which is perpendicular to the surface of the ellipsoid. Hence, $\overrightarrow{\gamma}_0$ is perpendicular to the surface of the ellipsoid, which is a property we are expecting. The absolute value of the normal gravity can then be written as

$$\gamma_0 = -\overrightarrow{g}_{\text{Mac}} \cdot \hat{e}_n - \overrightarrow{F}_{\text{Homo}} \cdot \hat{e}_n, \tag{4.117}$$

where $\hat{e}_n$ is the unit normal vector of the ellipsoid toward outside.

In the following, we derive an explicit formula for γ_0. According to (4.92) and (4.93), we have

$$
\begin{aligned}
-\overrightarrow{g}_{\text{Mac}} \cdot \hat{e}_n &= -\left(\frac{x^2}{a^4} + \frac{y^2}{a^4} + \frac{z^2}{b^4}\right)^{-1/2} \left[\frac{(\Omega^2 - P)(x^2 + y^2)}{a^2} - \frac{Qz^2}{b^2}\right] \\
&= -r\left(\frac{\sin^2\theta}{a^4} + \frac{\cos^2\theta}{b^2}\right)^{-1/2} \left[\frac{(\Omega^2 - P)\sin^2\theta}{a^2} - \frac{Q\cos^2\theta}{b^2}\right].
\end{aligned}
\tag{4.118}
$$

As (4.95) holds for a Maclaurin ellipsoid, we have

$$
\begin{aligned}
-\overrightarrow{g}_{\text{Mac}} \cdot \hat{e}_n &= r\left(\frac{\sin^2\theta}{a^4} + \frac{\cos^2\theta}{b^2}\right)^{-1/2} \left[\frac{b^2 Q \sin^2\theta}{a^4} + \frac{Q\cos^2\theta}{b^2}\right] \\
&= rQ\frac{(b^4/a^4)\tan^2\theta + 1}{[(b^4/a^4)\tan^2\theta + 1]^{1/2}(1 + \tan^2\theta)^{1/2}} \\
&= rQ\left[\frac{(b^4/a^4)\tan^2\theta + 1}{1 + \tan^2\theta}\right]^{1/2}.
\end{aligned}
\tag{4.119}
$$

The substitution of θ by B according to (A.23) yields

$$-\overrightarrow{g}_{\text{Mac}} \cdot \hat{e}_n = rQ\left[\frac{\cot^2 B + 1}{1 + (a^4/b^4)\cot^2 B}\right]^{1/2} = \frac{b^2 rQ}{\left(a^4\cos^2 B + b^4\sin^2 B\right)^{1/2}}. \tag{4.120}$$

Substitute (4.52) and (A.27) into this formula. We obtain

$$-\overrightarrow{g}_{\text{Mac}} \cdot \hat{e}_n = \frac{4\pi G\rho_{\text{Mac}} b^2(1 + e'^2)(e' - \arctan e')}{e'^3 \left(a^2\cos^2 B + b^2\sin^2 B\right)^{1/2}}. \tag{4.121}$$

The substitution of ρ_{Mac} by M_{Mac} according to (4.61) leads to

$$-\vec{g}_{\text{Mac}} \cdot \hat{e}_n = \frac{3GM_{\text{Mac}}b(1 + e'^2)(e' - \arctan e')}{a^2 e'^3 \left(a^2 \cos^2 B + b^2 \sin^2 B\right)^{1/2}} . \tag{4.122}$$

This is for the gravity of the Maclaurin ellipsoid. For the ellipsoidal homoeoid, we have

$$-\vec{F}_{\text{Homo}} \cdot \hat{e}_n = 4\pi G\mu = \frac{G(M - M_{\text{Mac}})}{a^2 b} \left(a^2 \cos^2 B + b^2 \sin^2 B\right)^{1/2} . \tag{4.123}$$

Substitute the above two formulae into (4.117). We obtain

$$\gamma_0 = \frac{GM}{a \left(a^2 \cos^2 B + b^2 \sin^2 B\right)^{1/2}} \left\{ \frac{3M_{\text{Mac}}}{Ma} \frac{1 + e'^2}{e'^3}(e' - \arctan e') \right.$$

$$\left. + \frac{1 - M_{\text{Mac}}/M}{ab} \left(a^2 \cos^2 B + b^2 \sin^2 B\right)^{1/2} \right\}$$

$$= \frac{GM}{a \left(a^2 \cos^2 B + b^2 \sin^2 B\right)^{1/2}} \left\{ \left[3\frac{b}{a} \frac{M_{\text{Mac}}}{M} \frac{1 + e'^2}{e'^3}(e' - \arctan e') \right. \right.$$

$$\left. + \frac{a}{b}\left(1 - \frac{M_{\text{Mac}}}{M}\right) \right] \cos^2 B$$

$$+ \left[3\frac{b}{a} \frac{M_{\text{Mac}}}{M} \frac{1 + e'^2}{e'^3}(e' - \arctan e') + \frac{b}{a}\left(1 - \frac{M_{\text{Mac}}}{M}\right) \right] \sin^2 B \right\}$$

$$= \frac{GM}{a \left(a^2 \cos^2 B + b^2 \sin^2 B\right)^{1/2}} \left\{ \frac{a}{b} \left[1 + \frac{3(e' - \arctan e') - e'^3}{e'^3} \frac{M_{\text{Mac}}}{M} \right] \cos^2 B \right.$$

$$\left. + \frac{b}{a} \left[1 + \frac{3(1 + e'^2)(e' - \arctan e') - e'^3}{e'^3} \frac{M_{\text{Mac}}}{M} \right] \sin^2 B \right\} . \tag{4.124}$$

Finally, we obtain, by substituting (4.98) into this formula,

$$\gamma_0 = \frac{GM}{a \left(a^2 \cos^2 B + b^2 \sin^2 B\right)^{1/2}} \left\{ \frac{a}{b} \left\{ 1 + m\frac{6(e' - \arctan e') - 2e'^3}{3[(3 + e'^2)\arctan e' - 3e']} \right\} \cos^2 B \right.$$

$$\left. + \frac{b}{a} \left\{ 1 + m\frac{6(1 + e'^2)(e' - \arctan e') - 2e'^3}{3[(3 + e'^2)\arctan e' - 3e']} \right\} \sin^2 B \right\} . \tag{4.125}$$

This is the final result of γ_0.

The expression of γ_0 can be written in a more concise form. By setting $B = 0$ in (4.125), we obtain the normal gravity at the equator,

$$\gamma_e = \frac{GM}{ab}\left\{1 + m\frac{6(e' - \arctan e') - 2e'^3}{3[(3 + e'^2)\arctan e' - 3e']}\right\}. \tag{4.126}$$

By setting $B = \pm\pi/2$ in (4.125), we obtain the normal gravity at the poles

$$\gamma_p = \frac{GM}{a^2}\left\{1 + m\frac{6(1 + e'^2)(e' - \arctan e') - 2e'^3}{3[(3 + e'^2)\arctan e' - 3e']}\right\}. \tag{4.127}$$

Based on the above two formulae, the expression of γ_0 can be written as

$$\gamma_0 = \frac{a\gamma_e \cos^2 B + b\gamma_p \sin^2 B}{(a^2 \cos^2 B + b^2 \sin^2 B)^{1/2}}, \tag{4.128}$$

which is referred to as Somigliana formula.

Substitute (4.2) into (4.126) and (4.127), we can obtain

$$a\gamma_p - b\gamma_e = \frac{2e'^2(e' - \arctan e')\Omega^2 ab}{(3 + e'^2)\arctan e' - 3e'}, \tag{4.129}$$

which relates the geometrical parameters a and b of the reference Earth ellipsoid to its physical parameters γ_p and γ_e and is referred to as Clairaut's theorem. The relation (4.114) is similar, which relates the geometrical parameters to the physical parameter J_2, and hence is also referred to as Clairaut's theorem.

The normal plumb line, i.e., the plumb line of the normal gravity, is perpendicular to the reference Earth ellipsoid. Taking reference to Fig. 4.3, the outward unit normal vector of the reference ellipsoid is

$$\hat{n} = \cos(B - \Phi)\hat{e}_r - \sin(B - \Phi)\hat{e}_\theta, \tag{4.130}$$

where $\cos(B - \Phi)$ and $\sin(B - \Phi)$ are given in (A.28) and (A.29), which are copied to here for convenience of reference,

$$\begin{cases} \cos(B - \Phi) = \dfrac{a^2 \cos^2 B + b^2 \sin^2 B}{(a^4 \cos^2 B + b^4 \sin^2 B)^{1/2}}, \\ \sin(B - \Phi) = \dfrac{(a^2 - b^2)\sin B \cos B}{(a^4 \cos^2 B + b^4 \sin^2 B)^{1/2}}. \end{cases} \tag{4.131}$$

The absolute magnitude of the normal gravity on the reference Earth ellipsoid, γ_0, is given by the Somigliana formula (4.128). The normal gravity vector on the reference

Earth ellipsoid is thus

$$\vec{\gamma}_0 = -\gamma_0 \hat{n} = -\gamma_0 \cos(B - \Phi)\hat{e}_r + \gamma_0 \sin(B - \Phi)\hat{e}_\theta . \tag{4.132}$$

4.5.5   Normal Gravity Above the Reference Earth Ellipsoid

The normal gravitational attraction is a special case of the general spherical harmonic series of gravitational attraction, (3.303) to (3.305), with $C_0 = 1$, $C_{2n} = -J_{2n}$ and all other coefficients vanishing. By adding the centrifugal acceleration, we obtain

$$\gamma_r = -\frac{GM}{r^2}\left[1 - \sum_{n=1}^{\infty}(2n+1)J_{2n}\left(\frac{a}{r}\right)^{2n} P_{2n}(\cos\theta) - \frac{mr^3}{a^2 b}\sin^2\theta \right], \tag{4.133}$$

$$\gamma_\theta = -\frac{GM}{r^2}\left[\sum_{n=1}^{\infty}J_{2n}\left(\frac{a}{r}\right)^{2n}\frac{d}{d\theta}P_{2n}(\cos\theta) - \frac{mr^3}{a^2 b}\sin\theta\cos\theta \right], \tag{4.134}$$

$$\gamma_\lambda = 0 . \tag{4.135}$$

If the geodetic latitude B and height h are provided, we have to transform them to r and θ. The derivation is provided in Appendix A, where the transformation from r and θ to B and h is also included following the approach of Borkowski (1987). The formulae of r, $\cos\theta$, and $\sin\theta$ are given in (A.40), (A.47), and (A.48), which are copied to here for reference,

$$r = \left\{ \left[h + \frac{a^2}{(a^2\cos^2 B + b^2\sin^2 B)^{1/2}} \right]^2 \cos^2 B \right.$$

$$\left. + \left[h + \frac{b^2}{(a^2\cos^2 B + b^2\sin^2 B)^{1/2}} \right]^2 \sin^2 B \right\}^{1/2} . \tag{4.136}$$

$$\cos\theta = \frac{1}{r}\left[h + \frac{b^2}{(a^2\cos^2 B + b^2\sin^2 B)^{1/2}} \right] \sin B . \tag{4.137}$$

$$\sin\theta = \frac{1}{r}\left[h + \frac{a^2}{(a^2\cos^2 B + b^2\sin^2 B)^{1/2}} \right] \cos B . \tag{4.138}$$

We mention that (4.131) is the degenerate case of the above two formulae with $h = 0$.

On the reference Earth ellipsoid, the result computed using (4.133) to (4.135) agrees with that computed using the closed-form expressions (4.131) and (4.132).

4.5.6 Graphical Representation

We show in Fig. 4.4 the relative positions of the normal gravitational attraction $\vec{F}$, centrifugal acceleration $\vec{F_c}$, and normal gravity $\vec{\gamma}_0$ on the reference Earth ellipsoid, which can be seen most clearly thus far from (4.133) to (4.135). The direction of $\vec{F_c}$ is evident based on its definition. However, the direction of $\vec{F}$ requires some explanation. In (4.134), the degree 2 term of the gravitational attraction is dominant. With $(d/d\theta)P_2(\cos\theta) = -3\sin\theta\cos\theta$, we can readily infer that the contribution of this term to γ_θ is positive on the northern hemisphere and negative on the southern hemisphere, thus pointing to the equator. Therefore, it renders the normal gravitational attraction $\vec{F}$ to deviate from the direction to the center toward the equator, or more specifically, toward the south on the northern hemisphere, and toward the north on the southern hemisphere.

In Fig. 4.5, we show a global view of equipotential surfaces of normal gravity and normal plumb lines, where the flattening and centrifugal effect are exaggerated.

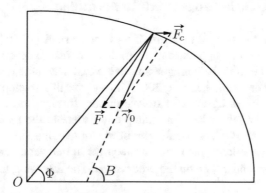

Fig. 4.4 Relative positions of the normal gravitational attraction, centrifugal acceleration, and normal gravity on the reference Earth ellipsoid

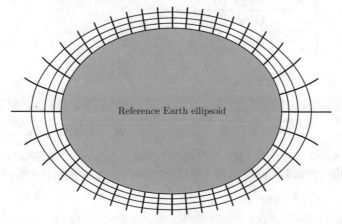

Fig. 4.5 Equipotential surfaces of normal gravity and normal plumb lines

In the rest of this paragraph, we implicitly understand the terms *equipotential surface* and *gravity* as the normal ones for brevity. First, we see that the nearer to the equator, the larger the distance between two adjacent equipotential surfaces is. This is due to the fact that the nearer to the equator, the smaller the gravity is; the distance between two adjacent equipotential surfaces is inverse proportional to the gravity. The decrease of gravity from the poles to the equator on the reference Earth ellipsoid is caused by the centrifugal acceleration, which is proportional to the distance to and points away from the rotation axis. The flattening of the reference Earth ellipsoid, which is a result of the long term action of the centrifugal force by dragging the material inside the Earth away from the rotation axis, also contributes to this decrease; the nearer to the equator, the farther from the center of mass is. We also see that the normal plumb lines convex toward the equator and concave toward the pole of the same hemisphere. This could be inferred straightforwardly according to (1.172) or (1.172) based on the fact that the gravity actually increases from the equator to the poles.

4.6   Second Order Approximate Formulae

In the preceding sections, exact formulae of the normal gravity field have been derived. In the literature, approximate formulae accurate to first, second, and even third order of the Earth's flattening α are also cited. The flattening is on the same order of magnitude as m, i.e., $\sim 1/300$. Throughout this section, when the term *order* is used to represent an order of magnitude, it refers to α, or equally m, if not explicitly stated otherwise. For example, accurate to the second order means accurate to the order of α^2, or equally αm or m^2, in magnitude.

In this section, we derive the formulae accurate to the second order. The squares of eccentricities e^2 and e'^2 are on the order of 2α. The following formulae are to be used:

$$\begin{cases} e^2 = 1 - (1-\alpha)^2 = 2\alpha - \alpha^2, \\ e'^2 = \dfrac{1 - (1-\alpha)^2}{(1-\alpha)^2} = 2\alpha + 3\alpha^2 + \cdots, \\ \dfrac{\arctan e'}{e'} = 1 - \dfrac{1}{3}e'^2 + \dfrac{1}{5}e'^4 - \dfrac{1}{7}e'^6 + \cdots. \end{cases} \tag{4.139}$$

Furthermore, the Taylor series

$$(1+x)^p = 1 + px + \frac{p(p-1)}{2}x^2 + \frac{p(p-1)(p-2)}{6}x^3 + \cdots \tag{4.140}$$

with $p = n, -1, 1/2, -1/2$, and -2 will be used. Actually, it has already been used to derive the second formula of (4.139) with $p = -2$.

4.6.1 The Normal Potential

Accurate to the second order, the spherical harmonic coefficients of the normal gravitational potential outside the reference Earth ellipsoid (4.105) can be developed as

$$J_{2n} = \frac{(-1)^{n+1}e^{2n}}{(2n+1)(2n+3)}$$

$$\times \left\{ 2n+3 - \frac{4ne'^2 m}{3\left\{(3+e'^2)\left[1-(1/3)e'^2+(1/5)e'^2-(1/7)e'^2\right]-3\right\}} \right\}$$

$$= \frac{(-1)^{n+1}e^{2n}}{(2n+1)(2n+3)} \left\{ 2n+3 - \frac{5nm}{e'^2\left[1-(6/7)e'^2\right]} \right\}$$

$$= \frac{(-1)^{n+1}(2\alpha - \alpha^2)^n}{(2n+1)(2n+3)} \left\{ 2n+3 - \frac{5nm}{(2\alpha+3\alpha^2)\left[1-(6/7)\times(2\alpha)\right]} \right\}$$

$$= \frac{2^n(-1)^{n+1}}{(2n+1)(2n+3)}\alpha^n \left(1-\frac{1}{2}\alpha\right)^n \left\{ 2n+3 - \frac{5nm}{2\alpha\left[1-(3/14)\alpha\right]} \right\}$$

$$= \frac{2^n(-1)^{n+1}}{(2n+1)(2n+3)}\alpha^{n-1} \left(1-\frac{n}{2}\alpha\right) \left\{ (2n+3)\alpha - \frac{5}{2}nm\left(1+\frac{3}{14}\alpha\right) \right\}.$$

$$(4.141)$$

Concretely, we have

$$J_2 = \frac{2}{15}\left(1-\frac{1}{2}\alpha\right)\left[5\alpha - \frac{5}{2}m\left(1+\frac{3}{14}\alpha\right)\right] = \frac{2}{3}\alpha - \frac{1}{3}m - \frac{1}{3}\alpha^2 + \frac{2}{21}\alpha m,$$

$$(4.142)$$

$$J_4 = -\frac{4}{35}\alpha(1-\alpha)\left[7\alpha - 5m\left(1+\frac{3}{14}\alpha\right)\right] = -\frac{4}{5}\alpha^2 + \frac{4}{7}\alpha m. \qquad (4.143)$$

All other coefficients are beyond the second order. Evidently, J_2 is on the first order and J_4 on the second order. Hence, the normal potential outside the reference Earth ellipsoid is

$$U = \frac{GM}{r}\left[1 - J_2\left(\frac{a}{r}\right)^2 P_2(\cos\theta) - J_4\left(\frac{a}{r}\right)^4 P_4(\cos\theta) + \frac{mr^3}{2a^2b}\sin^2\theta\right].$$

$$(4.144)$$

The normal potential on the reference Earth ellipsoid (4.113) can be developed as follows:

$$
\begin{aligned}
U_0 &= \frac{GM}{a}\left[\frac{(1+e'^2)^{1/2}}{e'}\arctan e' + \frac{\Omega^2 a^3}{3GM}\right]\\
&= \frac{GM}{a}\left[\left(1+\frac{1}{2}e'^2 - \frac{1}{8}e'^4\right)\left(1-\frac{1}{3}e'^2 + \frac{1}{5}e'^4\right) + \frac{1}{3(1-\alpha)}\frac{\Omega^2 a^2 b}{GM}\right]\\
&= \frac{GM}{a}\left[1+\frac{1}{6}e'^2 - \frac{11}{120}e'^4 + \frac{1}{3}(1+\alpha)m\right]\\
&= \frac{GM}{a}\left(1+\frac{1}{3}\alpha + \frac{1}{2}\alpha^2 - \frac{11}{30}\alpha^2 + \frac{1}{3}m + \frac{1}{3}\alpha m\right)\\
&= \frac{GM}{a}\left(1+\frac{1}{3}\alpha + \frac{1}{3}m + \frac{2}{15}\alpha^2 + \frac{1}{3}\alpha m\right).
\end{aligned}
\tag{4.145}
$$

4.6.2 Normal Gravity on the Reference Earth Ellipsoid

The normal gravity at the equator (4.126) can be developed as

$$
\begin{aligned}
\gamma_e &= \frac{GM}{ab}\left\{1+m\frac{6[(1/3)e'^2 - (1/5)e'^4 + (1/7)e'^6] - 2e'^2}{3\{(3+e'^2)[1-(1/3)e'^2 + (1/5)e'^4 - (1/7)e'^6] - 3\}}\right\}\\
&= \frac{GM}{ab}\left\{1+m\frac{-(6/5)e'^4 + (6/7)e'^6}{3[(4/15)e'^4 - (8/35)e'^6]}\right\}\\
&= \frac{GM}{ab}\left\{1-\frac{3}{2}m\frac{1-(5/7)e'^2}{1-(6/7)e'^2}\right\}\\
&= \frac{GM}{a^2}\frac{1}{1-\alpha}\left[1-\frac{3}{2}m\left(1-\frac{5}{7}e'^2\right)\left(1+\frac{6}{7}e'^2\right)\right]\\
&= \frac{GM}{a^2}(1+\alpha+\alpha^2)\left[1-\frac{3}{2}m\left(1+\frac{2}{7}\alpha\right)\right]\\
&= \frac{GM}{a^2}\left(1+\alpha - \frac{3}{2}m + \alpha^2 - \frac{27}{14}\alpha m\right).
\end{aligned}
\tag{4.146}
$$

The normal gravity at the poles (4.127) can be developed as

$$\gamma_p = \frac{GM}{a^2}\left\{1 + m\frac{6(1 + e'^2)[(1/3)e'^2 - (1/5)e'^4 + (1/7)e'^6] - 2e'^2}{3\{(3 + e'^2)[1 - (1/3)e'^2 + (1/5)e'^4 - (1/7)e'^6] - 3\}}\right\}$$

$$= \frac{GM}{a^2}\left\{1 + m\frac{(12/15)e'^4 - (12/35)e'^6}{3[(4/15)e'^4 - (8/35)e'^6]}\right\}$$

$$= \frac{GM}{a^2}\left\{1 + m\frac{1 - (3/7)e'^2}{1 - (6/7)e'^2}\right\}$$

$$= \frac{GM}{a^2}\left[1 + m\left(1 - \frac{3}{7}e'^2\right)\left(1 + \frac{6}{7}e'^2\right)\right]$$

$$= \frac{GM}{a^2}\left(1 + m + \frac{3}{7}me'^2\right)$$

$$= \frac{GM}{a^2}\left(1 + m + \frac{6}{7}\alpha m\right). \tag{4.147}$$

Define a gravimetric flattening

$$\beta = \frac{\gamma_p - \gamma_e}{\gamma_e}, \tag{4.148}$$

where $\gamma_p > \gamma_e$ is taken into account so that β be positive. It can be obtained according to (4.146) and (4.147) that

$$\beta = \frac{-\alpha + (5/2)m - \alpha^2 + (39/14)\alpha m}{1 + \alpha - (3/2)m + \alpha^2 - (27/14)\alpha m}$$

$$= \left(-\alpha + \frac{5}{2}m - \alpha^2 + \frac{39}{14}\alpha m\right)\left(1 - \alpha + \frac{3}{2}m\right)$$

$$= -\alpha + \frac{5}{2}m - \frac{17}{14}\alpha m + \frac{15}{4}m^2. \tag{4.149}$$

The general expression of the normal gravity on the reference Earth ellipsoid (4.128) can be developed as

$$\gamma_0 = \gamma_e\frac{\cos^2 B + (b/a)(\gamma_p/\gamma_e)\sin^2 B}{[\cos^2 B + (b^2/a^2)\sin^2 B]^{1/2}}$$

$$= \gamma_e\frac{\cos^2 B + (1 - \alpha)(1 + \beta)\sin^2 B}{[\cos^2 B + (1 - \alpha)^2\sin^2 B]^{1/2}}$$

$$= \gamma_e \frac{1 + (-\alpha + \beta - \alpha\beta) \sin^2 B}{[1 - (2\alpha - \alpha^2) \sin^2 B]^{1/2}}$$

$$= \gamma_e [1 + (-\alpha + \beta - \alpha\beta) \sin^2 B]$$

$$\times \left[1 + \frac{1}{2}(2\alpha - \alpha^2) \sin^2 B + \frac{3}{8}(2\alpha - \alpha^2)^2 \sin^4 B \right]$$

$$= \gamma_e \left[1 + \left(\beta - \frac{1}{2}\alpha^2 - \alpha\beta \right) \sin^2 B + \left(\frac{1}{2}\alpha^2 + \alpha\beta \right) \sin^4 B \right]$$

$$= \gamma_e \left(1 + \beta \sin^2 B - \beta_1 \sin^2 2B \right) , \tag{4.150}$$

which is referred to as Cassinis' formula, where the identity

$$\sin^4 B = \sin^2 B - \frac{1}{4} \sin^2 2B \tag{4.151}$$

has been used in the last step of formulation, and the parameter β_1 is defined as

$$\beta_1 = \frac{1}{8}\alpha^2 + \frac{1}{4}\alpha\beta . \tag{4.152}$$

We mention that β is on the first order and β_1 on the second order.

In the subsequent derivations, the identity (4.151) will be frequently used.

By substituting (4.146), (4.149), and (4.152) into (4.150), we obtain a formula of γ_0 expressed using α and m:

$$\gamma_0 = \frac{GM}{a^2} \left(1 + \alpha - \frac{3}{2}m + \alpha^2 - \frac{27}{14}\alpha m \right) \left\{ 1 + \left(-\alpha + \frac{5}{2}m - \frac{17}{14}\alpha m + \frac{15}{4}m^2 \right) \right.$$

$$\times \sin^2 B - \left[\frac{1}{8}\alpha^2 + \frac{1}{4}\alpha \left(-\alpha + \frac{5}{2}m \right) \right] \sin^2 2B \right\}$$

$$= \frac{GM}{a^2} \left[1 + \alpha - \frac{3}{2}m + \alpha^2 - \frac{27}{14}\alpha m + \left(-\alpha + \frac{5}{2}m - \alpha^2 + \frac{39}{14}\alpha m \right) \sin^2 B \right.$$

$$\left. + \left(\frac{1}{8}\alpha^2 - \frac{5}{8}\alpha m \right) \sin^2 2B \right] . \tag{4.153}$$

4.6.3   Normal Gravity Above the Reference Earth Ellipsoid

We derive a formula of the normal gravity at the vicinity of the reference Earth ellipsoid, assuming the position of P, where the normal gravity is to be computed, is provided in terms of its geodetic coordinates B, λ, and h. The normal gravity is given in (4.133) and (4.134). We write them together accurate to the second order of α:

$$\vec{\gamma} = -\frac{GM}{a^2}\left\{\left[\left(\frac{a}{r}\right)^2 - 3J_2\left(\frac{a}{r}\right)^4 P_2(\cos\theta) - 5J_4\left(\frac{a}{r}\right)^6 P_4(\cos\theta) - m\frac{r}{b}\sin^2\theta\right]\hat{e}_r\right.$$
$$\left. + \left[J_2\left(\frac{a}{r}\right)^4 \frac{d}{d\theta}P_2(\cos\theta) + J_4\left(\frac{a}{r}\right)^6 \frac{d}{d\theta}P_4(\cos\theta) - m\frac{r}{b}\sin\theta\cos\theta\right]\hat{e}_\theta\right\}.$$

$$(4.154)$$

If P is on the Earth's surface, h would not exceed ~ 10 km. Here we derive the formula assuming h/a to be on the same order of magnitudes as α, i.e., $h \sim 20$ km. Hence, the subsequent formulation will be accurate to the second order of α, m, and h/a.

First, we consider the terms with r. We have, by expressing (4.136) as a power series accurate to the second order using (4.140) with $p = 1/2$,

$$r = \left[\frac{a^4\cos^2 B + b^4\sin^2 B}{a^2\cos^2 B + b^2\sin^2 B} + 2h\frac{a^2\cos^2 B + b^2\sin^2 B}{(a^2\cos^2 B + b^2\sin^2 B)^{1/2}} + h^2\right]^{1/2}$$

$$= a\left\{\frac{\cos^2 B + (1-\alpha)^4\sin^2 B}{\cos^2 B + (1-\alpha)^2\sin^2 B} + 2\frac{h}{a}[\cos^2 B + (1-\alpha)^2\sin^2 B]^{1/2} + \left(\frac{h}{a}\right)^2\right\}^{1/2}$$

$$= a\left[\frac{1 - (4\alpha - 6\alpha^2)\sin^2 B}{1 - (2\alpha - \alpha^2)\sin^2 B} + 2\frac{h}{a}(1 - 2\alpha\sin^2 B)^{1/2} + \left(\frac{h}{a}\right)^2\right]^{1/2}$$

$$= a\left[1 - (2\alpha - 5\alpha^2)\sin^2 B - 4\alpha^2\sin^4 B + 2\frac{h}{a}(1 - \alpha\sin^2 B) + \left(\frac{h}{a}\right)^2\right]^{1/2}$$

$$= a\left[1 - \left(\alpha - \frac{5}{2}\alpha^2\right)\sin^2 B - \frac{5}{2}\alpha^2\sin^4 B + \frac{h}{a}\right]$$

$$= a\left(1 - \alpha\sin^2 B + \frac{5}{8}\alpha^2\sin^2 2B + \frac{h}{a}\right).$$

$$(4.155)$$

We remark that this formula is accurate to $(h/a)^2$; it just happens that the coefficient of this term vanishes. We then have, using (4.140) with $p = -2$,

$$\left(\frac{a}{r}\right)^2 = \left(1 - \alpha\sin^2 B + \frac{5}{8}\alpha^2\sin^2 2B + \frac{h}{a}\right)^{-2}$$

$$= 1 + 2\left(\alpha\sin^2 B - \frac{5}{8}\alpha^2\sin^2 2B - \frac{h}{a}\right) + 3\left(\alpha\sin^2 B - \frac{h}{a}\right)^2$$

$$= 1 + (2\alpha + 3\alpha^2)\sin^2 B - 2\alpha^2\sin^2 2B - 2\frac{h}{a} - 6\frac{h}{a}\alpha\sin^2 B + 3\left(\frac{h}{a}\right)^2.$$

$$(4.156)$$

As $(a/r)^4$ multiplies with the first order quantity J_2, it should also be accurate to the first order only

$$\left(\frac{a}{r}\right)^4 = 1 + 4\alpha \sin^2 B - 4\frac{h}{a}. \tag{4.157}$$

The term $(a/r)^6$ multiplies with the second order quantity J_4. Hence, even the first order terms can be neglected

$$\left(\frac{a}{r}\right)^6 = 1. \tag{4.158}$$

The term r/b multiplies with the first order quantity m. Hence, it needs to be accurate only to the first order. We have

$$\frac{r}{b} = \frac{r}{a}\frac{1}{1-\alpha} = 1 + \alpha - \alpha \sin^2 B + \frac{h}{a}. \tag{4.159}$$

Terms including $\cos \theta$ and $\sin \theta$ are multiplied at least by a first order quantity. Hence, they have to be accurate only to the first order. According to (4.137) and (4.138), we have

$$\begin{aligned}
\cos \theta &= \frac{a}{r}\left\{\frac{h}{a} + \frac{(1-\alpha)^2}{[\cos^2 B + (1-\alpha)^2 \sin^2 B]^{1/2}}\right\} \sin B \\
&= \frac{a}{r}\left[\frac{h}{a} + \frac{1-2\alpha}{(1 - 2\alpha \sin^2 B)^{1/2}}\right] \sin B \\
&= \left(1 + \alpha \sin^2 B - \frac{h}{a}\right)\left[\frac{h}{a} + (1-2\alpha)(1 + \alpha \sin^2 B)\right] \sin B \\
&= \left(1 - 2\alpha + 2\alpha \sin^2 B\right) \sin B, \tag{4.160}
\end{aligned}$$

$$\begin{aligned}
\sin \theta &= \frac{a}{r}\left[\frac{h}{a} + \frac{1}{(\cos^2 B + (1-\alpha)^2 \sin^2 B)^{1/2}}\right] \cos B \\
&= \frac{a}{r}\left[\frac{h}{a} + \frac{1}{(1 - 2\alpha \sin^2 B)^{1/2}}\right] \cos B \\
&= \left(1 + \alpha \sin^2 B - \frac{h}{a}\right)\left[\frac{h}{a} + (1 + \alpha \sin^2 B)\right] \cos B \\
&= \left(1 + 2\alpha \sin^2 B\right) \cos B. \tag{4.161}
\end{aligned}$$

The terms $P_2(\cos\theta)$ and $(d/d\theta)P_2(\cos\theta)$ are multiplied with the first order term J_2; they have to be accurate to the first order only. We have

$$P_2(\cos\theta) = \frac{3}{2}\cos^2\theta - \frac{1}{2} = \frac{3}{2}\sin^2 B - \frac{1}{2} - \frac{3}{2}\alpha\sin^2 2B, \qquad (4.162)$$

$$\frac{d}{d\theta}P_2(\cos\theta) = -3\sin\theta\cos\theta = -\left(\frac{3}{2} - 3\alpha + 6\alpha\sin^2 B\right)\sin 2B. \qquad (4.163)$$

The terms $P_4(\cos\theta)$ and $(d/d\theta)P_4(\cos\theta)$ are multiplied with the second order term J_4. Hence, even first order terms can be neglected,

$$P_4(\cos\theta) = \frac{35}{8}\cos^4\theta - \frac{15}{4}\cos^2\theta + \frac{3}{8} = \frac{5}{8}\sin^2 B - \frac{35}{32}\sin^2 2B + \frac{3}{8}, \qquad (4.164)$$

$$\frac{d}{d\theta}P_4(\cos\theta) = -\sin\theta\left(\frac{35}{2}\cos^3\theta - \frac{15}{2}\cos\theta\right) = -\left(\frac{35}{4}\sin^2 B - \frac{15}{4}\right)\sin 2B. \qquad (4.165)$$

The following two formulae are for the centrifugal acceleration:

$$\sin^2\theta = \left(1 + 4\alpha\sin^2 B\right)\cos^2 B = 1 - \sin^2 B + \alpha\sin^2 2B, \qquad (4.166)$$

$$\sin\theta\cos\theta = \left(\frac{1}{2} - \alpha + 2\alpha\sin^2 B\right)\sin 2B. \qquad (4.167)$$

Substitute (4.156) to (4.159) and (4.162) to (4.167) into (4.154). We obtain, after some tedious algebraic derivation,

$$\vec{\gamma} = -\frac{GM}{a^2}\left\{\left[1 + \frac{3}{2}J_2 - \frac{15}{8}J_4 - m - m\alpha\right.\right.$$

$$+ \left(2\alpha + 3\alpha^2 - \frac{9}{2}J_2 - 12J_2\alpha - \frac{25}{8}J_4 + m + m\alpha\right)\sin^2 B$$

$$+ \left(-2\alpha^2 + 9\alpha J_2 + \frac{175}{32}J_4 - \frac{3}{4}m\alpha\right)\sin^2 2B$$

$$+ \left.[-6J_2 - 2 - m + (-6\alpha + 18J_2 + m)\sin^2 B]\frac{h}{a} + 3\left(\frac{h}{a}\right)^2\right]\hat{e}_r$$

$$- \left[\frac{3}{2}J_2 - 3J_2\alpha - \frac{15}{4}J_4 + \frac{1}{2}m - \frac{1}{2}m\alpha\right.$$

$$+ \left.\left(\frac{3}{2}m\alpha + 12J_2\alpha + \frac{35}{4}J_4\right)\sin^2 B + \left(\frac{1}{2}m - 6J_2\right)\frac{h}{a}\right]\sin 2B\hat{e}_\theta\right\}. $$

$$(4.168)$$

Substitute (4.142) and (4.143) into it. We obtain, after some tedious algebraic derivation,

$$
\vec{\gamma} = -\frac{GM}{a^2}\left\{\left[1+\alpha-\frac{3}{2}m+\alpha^2-\frac{27}{14}\alpha m+\left(-\alpha+\frac{5}{2}m-\alpha^2+\frac{39}{14}\alpha m\right)\sin^2 B\right.\right.
$$

$$
+\left(-\frac{3}{8}\alpha^2-\frac{5}{8}\alpha m\right)\sin^2 2B
$$

$$
\left.+[-2-4\alpha+m+(6\alpha-5m)\sin^2 B]\frac{h}{a}+3\left(\frac{h}{a}\right)^2\right]\hat{e}_r
$$

$$
\left.-\left[\alpha+\frac{1}{2}\alpha^2-\frac{3}{2}m\alpha+\left(\alpha^2+\frac{5}{2}m\alpha\right)\sin^2 B+\left(-4\alpha+\frac{5}{2}m\right)\frac{h}{a}\right]\sin 2B\,\hat{e}_\theta\right\}.
$$

$$
(4.169)
$$

With (4.153), this formula can be further simplified to

$$
\vec{\gamma} = -\gamma_0\left\{\left[1-\frac{1}{2}\alpha^2\sin^2 2B-2(1+\alpha+m-2\alpha\sin^2 B)\frac{h}{a}+3\left(\frac{h}{a}\right)^2\right]\hat{e}_r\right.
$$

$$
\left.-\left[\alpha-\frac{1}{2}\alpha^2+2\alpha^2\sin^2 B+\left(-4\alpha+\frac{5}{2}m\right)\frac{h}{a}\right]\sin 2B\,\hat{e}_\theta\right\},\qquad(4.170)
$$

which is the final vector form of the normal gravity around the surface of the reference Earth ellipsoid. The absolute value of normal gravity around the reference Earth ellipsoid, $|\vec{\gamma}|$, is thus

$$
\gamma = \gamma_0\left\{\left[1-\frac{1}{2}\alpha^2\sin^2 2B-2(1+\alpha+m-2\alpha\sin^2 B)\frac{h}{a}+3\left(\frac{h}{a}\right)^2\right]^2\right.
$$

$$
\left.+\left[\alpha-\frac{1}{2}\alpha^2+2\alpha^2\sin^2 B+\left(-4\alpha+\frac{5}{2}m\right)\frac{h}{a}\right]^2\sin^2 2B\right\}^{1/2}
$$

$$
= \gamma_0\left[1-2(1+\alpha+m-2\alpha\sin^2 B)\frac{h}{a}+3\left(\frac{h}{a}\right)^2\right].\qquad(4.171)
$$

The correction for computing γ from γ_0 can be obtained from (4.171) as

$$
\gamma-\gamma_0 = -2\gamma_0(1+\alpha+m-2\alpha\sin^2 B)\frac{h}{a}+3\gamma_0\frac{h^2}{a^2}.\qquad(4.172)
$$

Substitute γ_0 by γ_e at the right-hand side according to (4.150), and then substitute β by α and m according to (4.149). We obtain

$$\gamma - \gamma_0 = -2\gamma_e \left[1 + \alpha + m + \left(-3\alpha + \frac{5}{2}m \right) \sin^2 B \right] \frac{h}{a} + 3\gamma_e \frac{h^2}{a^2} . \tag{4.173}$$

The difference between $\cos \theta$ and $\sin B$ is a first order quantity. Hence, replacing $\sin B$ by $\cos \theta$ in this formula does not influence accuracy. We obtain

$$\gamma - \gamma_0 = -2\gamma_e \left[1 + \alpha + m + \left(-3\alpha + \frac{5}{2}m \right) \cos^2 \theta \right] \frac{h}{a} + 3\gamma_e \frac{h^2}{a^2} . \tag{4.174}$$

For points on the ground, the second term could be neglected in most regions.

We mention that in (4.174), the quantities multiplied by h and $h^2/2$ are first and second order derivatives of γ with respect to h on the reference Earth ellipsoid. Based on the fact that the reference Earth ellipsoid is an equipotential surface of normal gravity, an exact closed-form expression of the first order derivative on the reference Earth ellipsoid can be obtained according to (1.164)

$$\left(\frac{\partial \gamma}{\partial h} \right)_0 = -\gamma_0 \left(\frac{1}{R_M} + \frac{1}{R_N} \right) - 2\Omega^2 , \tag{4.175}$$

where R_M and R_N are the radii of curvature of meridian and prime vertical, respectively, and are derived in Appendix A in (A.35) and (A.36).

4.7 Parameters of the Reference Earth Ellipsoid

In the formulation of the normal gravity field, we have assumed that the geometrical parameters of the reference Earth ellipsoid are known. However, in the present satellite geodesy era, the geocentric gravitational constant GM and the dynamical form factor J_2 are measured extremely accurately by observing satellite orbits. Also, Ω is measured very accurately by observing very distant celestial bodies. Hence, the four fundamental parameters of the reference Earth ellipsoid and normal gravity field are chosen to be U_0, GM, J_2, and Ω. For convenience of applications, U_0 is replaced by a. Therefore, the four initial parameters provided are a, GM, J_2, and Ω, which are referred to as Geodetic Reference System.

We mention that even GM can be measured very accurately, as G is measured in laboratories with an accuracy far below that of GM, M is determined with a much lower accuracy than GM.

The parameter m defined in (4.2) cannot be computed at first place from the initial parameters. An alternative to it is

$$q = \frac{\Omega^2 a^3}{GM} , \tag{4.176}$$

which can be computed from the initial parameters right away. The parameter m can be expressed using q and the ellipsoidal shape parameters α, e, and e' as

$$m = q(1 - \alpha) = q(1 - e^2)^{1/2} = q(1 + e'^2)^{-1/2}, \tag{4.177}$$

which can be used to substitute m by q in the formulae derived.

As an example, we derive the second order approximate formula to compute α from J_2 and q. Substitute (4.177) into (4.142), and then solve for α. We can first obtain

$$\alpha = \frac{3}{2}J_2 + \frac{1}{2}q + \frac{1}{2}\alpha^2 - \frac{9}{14}\alpha q. \tag{4.178}$$

Substitute α at the right-hand side by the whole right-hand side, and we obtain, accurate to the second order,

$$\begin{aligned}
\alpha &= \frac{3}{2}J_2 + \frac{1}{2}q + \frac{1}{2}\left(\frac{3}{2}J_2 + \frac{1}{2}q\right)^2 - \frac{9}{14}\left(\frac{3}{2}J_2 + \frac{1}{2}q\right)q \\
&= \frac{3}{2}J_2 + \frac{1}{2}q + \frac{9}{8}J_2^2 - \frac{3}{14}J_2 q - \frac{11}{56}q^2.
\end{aligned} \tag{4.179}$$

The more accurate way is to solve the exact Eq. (4.114) for e'^2 after substituting m by q and e'^2 according to (4.177), as will be done later.

The most used standard values of the initial parameters in geodetic applications are the Geodetic Reference System GRS80, with

$$a = 6378137 \, \text{m}, \tag{4.180}$$

$$GM = 3.986005 \times 10^{14} \, \text{m}^3 \, \text{s}^{-2}, \tag{4.181}$$

$$J_2 = 1.08263 \times 10^{-3}, \tag{4.182}$$

$$\Omega = 7.292115 \times 10^{-5} \, \text{rad} \, \text{s}^{-1}. \tag{4.183}$$

These are traditional values and are not the most accurate determinations. However, once these values are used as standard, they are considered to be error free. The readers are referred to the IERS (International Earth Rotational and Reference System Service) convention for a more thorough and accurate description of the data.

Other parameters can be computed. Here we use the second order approximate formulae. The result is listed below, which serves as an idea on the orders of magnitudes of the parameters:

$$q = 3.4613914 \times 10^{-3}, \tag{4.184}$$

$$1/\alpha = 298.25792, \tag{4.185}$$

$$b = 6356752.4\,\text{m}\,, \tag{4.186}$$

$$m = 3.4497860 \times 10^{-3}\,, \tag{4.187}$$

$$U_0 = 62636860\,\text{m}^2\,\text{s}^{-2}\,, \tag{4.188}$$

$$J_4 = -2.384 \times 10^{-6}\,, \tag{4.189}$$

$$\gamma_e = 9.7803272\,\text{m}\,\text{s}^{-2}\,, \tag{4.190}$$

$$\gamma_p = 9.8321860\,\text{m}\,\text{s}^{-2}\,, \tag{4.191}$$

$$\beta = 5.3022461 \times 10^{-3}\,, \tag{4.192}$$

$$\beta_1 = 5.850 \times 10^{-6}\,. \tag{4.193}$$

We particularly mention that the equi-volumetric radius is (radius of the sphere with the same volume)

$$R = (a^2 b)^{1/3} = a(1 - \alpha)^{1/3} = 6371000.8\,\text{m}\,. \tag{4.194}$$

It is noted that as terms on the order of magnitude of α^3 are neglected in the second order approximate formulae, the relative accuracy of the values listed above is different for parameters of different order of magnitudes. For example, assuming the data given in (4.180) to (4.183) to be error free, the relative accuracies of U_0 and α are on the order of $\alpha^3 \sim 4 \times 10^{-8}$ and $\alpha^2 \sim 1 \times 10^{-5}$, respectively. Especially, as J_4 and β_1 are on the order of magnitude of α^2, their relative accuracies are on the order of $\alpha \sim 3 \times 10^{-3}$, and hence, their values are given with less digits.

In practical applications, there is a need of much higher accuracy for internal consistency. Therefore, the parameters should be computed using the exact formulae. An equation for computing e'^2 can be inferred from (4.114)

$$e'^2 = 3J_2 \left(1 + e'^2\right) + \frac{4q}{15 \left(1 + e'^2\right)^{1/2}} \left[\frac{(3 + e'^2)\arctan e' - 3e'}{e'^5}\right]^{-1}, \tag{4.195}$$

which is to be solved iteratively. With the Taylor series $\arctan x = \sum_{n=0}^{\infty}[(-1)^n x^{2n+1}/(2n+1)]$, we can write the expression in the brackets as a fast converging series

$$\frac{(3 + e'^2)\arctan e' - 3e'}{e'^5} = \sum_{n=1}^{\infty} \frac{4(-1)^{n+1} n}{(2n+1)(2n+3)} e'^{2(n-1)}, \tag{4.196}$$

which is more favorable for practical evaluation. In (4.196), neglecting all terms on the order of e' and smaller at the right-hand side results an approximate value of $4/15$. The substitution of this approximate value into (4.195) leads to an approximate initial value for the iteration, $e'^2 = 3J_2 + q$, which can also be inferred from (4.178). Furthermore, this implies that the numerator at the left-hand side of (4.196) is on the order of e'^5 in magnitude, where quantities of larger orders of magnitude cancel out. Therefore, the direct evaluation of it involves the computation of the difference between two nearly equal quantities, which is associated with a loss of significant digits. Our experience is that if (4.195) is solved without making use of (4.196), there is a need of quadruple precision if an accuracy of more than 10 significant digits is required, which confirms the advantage of the series in (4.196). The accurate result can be found in official GRS80 documents (e.g., Moritz, 2000). Here we provide the value of the flattening accurate to 15 digits computed in quadruple precision based on the above two formulae, $1/\alpha = 298.257222100883$, which can be used to compute all other parameters using the exact closed-form formulae without the need of iteration.

Before concluding this chapter, we particularly emphasize that a and U_0 are not independent; only one of them can be assigned to uniquely define the reference Earth ellipsoid.

- If the geoid used to define the reference Earth ellipsoid is chosen to be a specific equipotential surface of gravity, such as the one through the mean ocean surface at a particular tide gauge or the one through the average of the mean ocean surface at several tide gauges, the gravity potential on the geoid, W_0, is then known, and $U_0 = W_0$ should be assigned. In this case, the value of a should be computed from the value of U_0.
- If the value of a is assigned, e.g., to be its value obtained from the global sea level determined using satellite altimetry, the value of U_0 should then be computed from the value of a, implying that the geoid used to define the reference Earth ellipsoid is implicitly chosen to be the one with $W_0 = U_0$.

The geoid used to define the reference Earth ellipsoid and the one serving as vertical datum may be chosen differently; if this is the case, W_0 (the gravity potential on the geoid defining the vertical datum) and U_0 (the normal potential on the reference Earth ellipsoid or the gravity potential on the geoid defining the reference Earth ellipsoid) are different, and this difference must be taken into account in the interpretation of data.

References

Borkowski, K. M. (1987). Transformation of geocentric to geodetic coordinates without approximations. *Astrophysics and Space Science, 139*, 1–4.

He, S. J. (1957). *Gravimetry* (In Chinese). Press of Surveying and Mapping. Beijing, China.

Levallois, J. J. (1970). *Géodésie général, Vol 3: Le champ de la pesanteur* (In French). Eyrolles, Paris, France.

MacMillan W. D. (1958). *The theory of the potential*. Dover Publications.

Moritz, H. (2000). Geodetic Reference System 1980. *Journal of Geodesy, 74*, 128–133.

Stokes' Theory and Beyond

5

Abstract

This chapter formulates the solution of the disturbing gravity field, i.e., the difference between the real and normal gravity fields, by solving Laplace's equation outside a spherical boundary. The quantities describing the disturbing gravity field include the disturbing potential, geoidal height, deflection of the vertical, gravity disturbance, and gravity anomaly. In Stokes' classical theory, the gravity on the geoid is assumed to be known, which can be obtained from data of spirit leveling and gravimetry through gravity reduction. The gravity on the geoid is used to define the gravity anomaly, which serves as data in the boundary condition. In the solution of Laplace's equation, it is assumed that no mass is present above the geoid, and the geoid is approximated as a sphere. The boundary value problems with other kinds of data as boundary conditions—the gravity disturbance obtained by gravimetry positioned using GNSS and the deflection of the vertical or geoidal height obtained by satellite altimetry of ocean surface—are also formulated under the same assumptions. In particular, the formulation when errors are present in the data used to define the reference Earth ellipsoid is included. Finally, some characteristics of the Earth's gravity field are presented.

5.1 Stokes' Boundary Value Problem

5.1.1 Disturbing Potential and Gravity Disturbance

In practical determination, the Earth's gravity field is separated into a normal gravity field and a disturbing gravity field. The normal gravity field, which was formulated in the previous chapter, is assumed to be known. Therefore, the determination of the disturbing gravity field remains to be formulated.

By definition, the disturbing gravity field is the difference between the real and normal gravity fields and is represented by the disturbing potential T and gravity

© The Author(s), under exclusive license to Springer Nature Switzerland AG 2023
J.-Y. Guo, *Physical Geodesy*, Springer Textbooks in Earth Sciences, Geography and Environment, https://doi.org/10.1007/978-3-031-23320-3_5

disturbance $\delta \vec{g}$, which are defined as

$$T = W - U, \quad \delta \vec{g} = \vec{g} - \vec{\gamma}. \tag{5.1}$$

Furthermore, $\delta \vec{g}$ is related to T in the form of

$$\delta \vec{g} = \nabla W - \nabla U = \nabla T, \tag{5.2}$$

whose components are

$$\delta g_r = \frac{\partial T}{\partial r}, \quad \delta g_\theta = \frac{1}{r}\frac{\partial T}{\partial \theta}, \quad \delta g_\lambda = \frac{1}{r \sin \theta}\frac{\partial T}{\partial \lambda}. \tag{5.3}$$

As the same centrifugal potential is included in both the real and normal potentials, the disturbing potential is the difference between the real and normal gravitational potentials. Hence, the disturbing potential has all the properties of the gravitational potential, and consequently, the gravity disturbance has all the properties of the gravitational attraction. Furthermore, the disturbing potential and gravity disturbance are small quantities, since the normal gravity field is very close to the real gravity field.

In the literature, the gravity disturbance is also referred to as disturbing gravity.

As the disturbing gravity is small, the Earth's external gravity field defers little from the normal gravity field. Around the Earth's surface, the relative positions of the gravitational attraction, centrifugal acceleration, and gravity are similar to those of the normal gravity field shown in Fig. 4.4. Above the geoid, the equipotential surfaces of gravity and plumb lines are also similar to those of the normal gravity field shown in Fig. 4.5; the reference Earth ellipsoid becomes the geoid here, and an equipotential surface of gravity very close to the geoid may penetrate continents.

5.1.2   Geoidal Height and Deflection of the Vertical

The geoid, which is defined as an equipotential surface of gravity chosen as close as possible to the mean sea surface, is a geometrical representation of the gravity field. Its deviation from the reference Earth ellipsoid is then a representation of the disturbing gravity field. Two quantities are used to describe the deviation of the geoid from the reference Earth ellipsoid: geoidal height and deflection of the vertical. The deflection of the vertical is also referred to as vertical deflection in the literature.

The geoidal height N, as shown in Fig. 5.1, is a measure of the distance between the geoid and the reference Earth ellipsoid. We use Σ to denote the geoid and S the reference Earth ellipsoid. Let P be a point on the geoid, and its projection on the reference Earth ellipsoid is P_0. The geoidal height N at the point P is the distance from P_0 to P, positive upward.

Fig. 5.1 Geoidal height

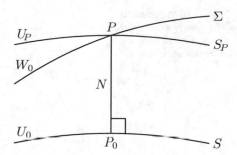

The relation between the geoidal height N and the disturbing potential T can be derived quite straightforwardly. We use U_0 to denote the normal potential on the reference Earth ellipsoid, W_0 the gravity potential on the geoid, and U_P the normal potential at P. Also shown in Fig. 5.1 is the equipotential surface of normal gravity through P, S_P. The geoidal height can be inferred by applying (1.49) to the normal gravity field, where S and S_P are two closely located equipotential surfaces of normal gravity

$$N = -\frac{U_P - U_{P_0}}{\gamma_{P_0}}.\tag{5.4}$$

Taking into account of (5.1) and with the facts $W = W_0$ on Σ and $U = U_0$ on S, we obtain

$$N = -\frac{(W_P - T_P) - U_{P_0}}{\gamma_{P_0}} = -\frac{W_0 - U_0}{\gamma_0} + \frac{T_P}{\gamma_{P_0}}.\tag{5.5}$$

As the reference Earth ellipsoid is defined such that $U_0 = W_0$, the first term at the right-hand side vanishes. After replacing the notations T_P and γ_{P_0} by T_0 and γ_0, respectively, we obtain a general relation between the geoidal height and the disturbing potential

$$N = \frac{T_0}{\gamma_0}.\tag{5.6}$$

The disturbing potential is introduced as a small quantity. Hence, replacing γ_0 by its global average $\bar{\gamma}_0$ would introduce an error of even smaller order of magnitude. Therefore, we have

$$N = \frac{T_0}{\bar{\gamma}_0 + \gamma_0 - \bar{\gamma}_0} = \frac{T_0}{\bar{\gamma}_0}\left[1 - \frac{\gamma_0 - \bar{\gamma}_0}{\bar{\gamma}_0}\right] = \frac{T_0}{\bar{\gamma}_0}.\tag{5.7}$$

The relation between the geoidal height and the disturbing potential, either (5.5), (5.6), or (5.7), is called Bruns' formula.

Fig. 5.2 Components of the
deflection of the vertical,
briefly referred to as
deflection components

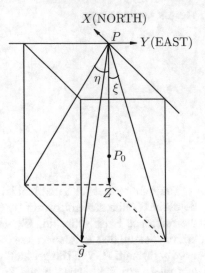

Evidently, the geoid serving as the vertical datum can be provided as a model
of the geoidal height, which can be used to convert the GNSS measured geodetic
height h to the orthometric height H

$$H = h - N. \tag{5.8}$$

The deflection of the vertical, or more specifically, gravimetric deflection of the
vertical here, is used to measure the angle between the geoid and the reference
Earth ellipsoid. For a point P on the geoid, the deflection of the vertical is defined
to be the angle between the normal to the geoid and that to the reference Earth
ellipsoid, or equivalently, the angle between the gravity at P and the normal gravity
at P_0. Evidently, besides magnitude, the direction should also be specified. Hence,
two components are necessary to completely describe the deflection of the vertical.
Customarily, the component to the south, ξ, and that to the west, η, are used, as
shown in Fig. 5.2, where the Z-axis is perpendicular to the reference Earth ellipsoid,
and the point P_0 is on it (see also Fig. 4.1).

As shown in Fig. 5.3, the projection of the deflection of the vertical in any plane
passing through the Z-axis is $\angle P_0 P Q$, where $P Q$ is the projection of the direction
of gravity. It is equal to the angle between the intersection lines of this plane with
the geoid and the reference Earth ellipsoid. We have

$$\epsilon = \lim_{\Delta l \to 0} \frac{\delta N}{\Delta l} = \frac{\partial N}{\partial l}. \tag{5.9}$$

The south and west components defined in Fig. 5.2 can be expressed as

$$\xi = -\frac{\partial N}{\partial X}, \quad \eta = -\frac{\partial N}{\partial Y}, \tag{5.10}$$

Fig. 5.3 Projection of the
deflection of the vertical in a
plane passing through the
Z-axis, which is
perpendicular to the reference
Earth ellipsoid

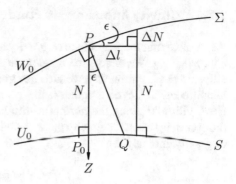

where the X- and Y-axes point to north and east, respectively, as shown in Fig. 5.2.
As the deflection of the vertical is a small quantity, approximating the reference
Earth ellipsoid by a sphere in (5.10) would introduce an error of even smaller order
of magnitude. Hence, these formulae can be approximated as

$$\xi = \frac{1}{R}\frac{\partial N}{\partial \theta}, \qquad \eta = -\frac{1}{R\sin\theta}\frac{\partial N}{\partial \lambda}, \tag{5.11}$$

where the radius R is chosen to be the equi-volumetric mean radius of the reference
Earth ellipsoid. Substitute (5.7) into this formula. We obtain an expression of the
deflection of the vertical expressed using disturbing potential

$$\xi = \frac{1}{R\bar{\gamma_0}}\frac{\partial T_0}{\partial \theta}, \qquad \eta = -\frac{1}{R\sin\theta\bar{\gamma_0}}\frac{\partial T_0}{\partial \lambda}. \tag{5.12}$$

According to (5.3), these formulae can also be written as

$$\xi = \frac{\delta g_{0\theta}}{\bar{\gamma_0}}, \qquad \eta = -\frac{\delta g_{0\lambda}}{\bar{\gamma_0}}, \tag{5.13}$$

where $\delta g_{0\theta}$ and $\delta g_{0\lambda}$ are the values of δg_θ and δg_λ, respectively, on the geoid.

Finally, we mention that the formula of the deflection of the vertical without
approximating the reference Earth ellipsoid as a sphere in (5.10) can be obtained
using the radius of curvature of the meridian, R_M, and that of the prime vertical,
R_N, derived in Appendix A.3. As the lengths of line segments toward north and east
are $\Delta X = R_M \Delta B$ and $\Delta Y = R_N \cos B \Delta\lambda$, respectively, we have

$$\xi = -\frac{1}{R_M}\frac{\partial N}{\partial B}, \qquad \eta = -\frac{1}{R_N\cos B}\frac{\partial N}{\partial \lambda}. \tag{5.14}$$

In this book, we seek solutions under spherical approximation.

5.1.3   Gravity Anomaly and Fundamental Gravimetric Equation

We assume that the magnitude of gravity at all points on the geoid is known and is denoted as g_0. We denote the magnitude of normal gravity on the reference Earth ellipsoid as γ_0. For brevity, we also use gravity and normal gravity to indicate their magnitudes. Let P be a point on the geoid and P_0 its projection on the reference Earth ellipsoid along the normal to the latter surface, as shown in Fig. 5.1. The difference between the gravity g_0 at P and the normal gravity γ_0 at P_0 is referred to as gravity anomaly Δg

$$\Delta g = g_0 - \gamma_0 . \tag{5.15}$$

As P and P_0 are separated by the geoidal height, the centrifugal accelerations in g_0 and γ_0 are different. Hence, Δg includes a contribution from the centrifugal acceleration.

In Stokes' theory, the gravity anomaly is assumed to be known. It can be computed from the gravity on the Earth's surface as well as the orthometric height of the Earth's surface, which will be the topic of the next chapter.

Here we first formulate a relation between Δg and T. We use $\hat{e}_{n_\Sigma}$ and $\hat{e}_{ns}$ to denote the upward normals of the geoid Σ and the reference Earth ellipsoid S. We have

$$g_0 = -\left(\frac{\partial W}{\partial n_\Sigma}\right)_\Sigma = -\left(\frac{\partial U}{\partial n_\Sigma}\right)_\Sigma - \left(\frac{\partial T}{\partial n_\Sigma}\right)_\Sigma , \tag{5.16}$$

$$\gamma_0 = -\left(\frac{\partial U}{\partial n_s}\right)_S . \tag{5.17}$$

Hence,

$$\Delta g = -\left(\frac{\partial U}{\partial n_\Sigma}\right)_\Sigma + \left(\frac{\partial U}{\partial n_s}\right)_S - \left(\frac{\partial T}{\partial n_\Sigma}\right)_\Sigma . \tag{5.18}$$

The first term can be further developed to

$$\left(\frac{\partial U}{\partial n_\Sigma}\right)_\Sigma = \left(\frac{\partial U}{\partial n_\Sigma}\right)_S + N\left(\frac{\partial^2 U}{\partial n_\Sigma^2}\right)_S = \left(\frac{\partial U}{\partial n_s}\right)_S \cos(\hat{e}_{n_\Sigma}, \hat{e}_{ns}) + N\left(\frac{\partial^2 U}{\partial n_\Sigma^2}\right)_S , \tag{5.19}$$

where $(\hat{e}_{n_\Sigma}, \hat{e}_{ns})$ is the magnitude of the deflection of the vertical. Accurate to only the first order terms of the quantities of the disturbing gravity field, we can approximate $\cos(\hat{e}_{n_\Sigma}, \hat{e}_{n_S})$ by 1 and $(\partial^2 U/\partial n_\Sigma^2)_S$ by $(\partial^2 U/\partial n_S^2)_S$. These lead to

$$\left(\frac{\partial U}{\partial n_\Sigma}\right)_\Sigma = \left(\frac{\partial U}{\partial n_S}\right)_S + N\left(\frac{\partial^2 U}{\partial n_S^2}\right)_S = \left(\frac{\partial U}{\partial n_S}\right)_S + N\left(\frac{\partial \gamma}{\partial n_S}\right)_S. \tag{5.20}$$

Finally, we obtain

$$\Delta g = N\left(\frac{\partial \gamma}{\partial n_S}\right)_S - \left(\frac{\partial T}{\partial n_\Sigma}\right)_\Sigma, \tag{5.21}$$

which can be written as, by substituting Bruns' formula into it,

$$\Delta g = \frac{T_0}{\gamma_0}\left(\frac{\partial \gamma}{\partial n_S}\right)_S - \left(\frac{\partial T}{\partial n_\Sigma}\right)_\Sigma. \tag{5.22}$$

This formula can also be expressed using the geodetic height h and the orthometric height H as

$$\Delta g = \frac{T_0}{\gamma_0}\left(\frac{\partial \gamma}{\partial h}\right)_{h=0} - \left(\frac{\partial T}{\partial H}\right)_{H=0}, \tag{5.23}$$

which is referred to as fundamental gravimetric equation.

As both Δg and T are small quantities, spherical approximation can be applied to the fundamental gravimetric equation with a reasonable precision, which includes two aspects. First, the reference Earth ellipsoid is approximated as a sphere when computing γ_0 and $(\partial \gamma/\partial h)_{h=0}$, meaning that

$$\gamma_0 = \frac{GM}{R^2}, \qquad \left(\frac{\partial \gamma}{\partial h}\right)_{h=0} = -2\frac{GM}{R^3}, \tag{5.24}$$

where R is the radius of the sphere. This is in fact the spherically symmetrical part, i.e., the degree 0 term of the spherical harmonic series, of the normal gravitational potential. Second, when the coordinate of a point is used as arguments of T, we approximate the geoid by the sphere of radius R, meaning that

$$T_0 = T_\Sigma \approx T_R, \qquad \left(\frac{\partial T}{\partial H}\right)_{H=0} \approx \left(\frac{\partial T}{\partial r}\right)_R, \tag{5.25}$$

where the subscript R stands for $r = R$, a convention to be used throughout this chapter. We thus obtain an approximate fundamental gravimetric equation on a spherical surface of radius R

$$\Delta g = -2\frac{T_0}{R} - \left(\frac{\partial T}{\partial r}\right)_R. \tag{5.26}$$

The accuracy of the approximations will be discussed in the next subsection. In this approximate formula, the small contribution of the centrifugal acceleration of Δg is neglected.

5.1.4 Stokes' Boundary Value Problem

Our purpose is to determine the Earth's external gravity potential, which is separated into a normal potential and a disturbing potential. As the normal potential is known, we have only to determine the disturbing potential, which satisfies Laplace's equation outside the Earth and satisfies the fundamental gravimetric equation on the geoid.

If there exists no mass above the geoid, the disturbing potential then satisfies Laplace's equation outside the geoid and satisfies the fundamental gravimetric equation on the geoid. This is a third-type boundary value problem of Laplace's equation with the geoid as boundary and the fundamental gravimetric equation as boundary condition and is referred to as Stokes' boundary value problem.

The boundary, i.e., the geoid, depends on the disturbing potential which is yet to be determined; it is not known. For this reason, Stokes' boundary value problem is a free boundary value problem.

Nevertheless, although the geoid as a boundary is not known, it deviates little from the reference Earth ellipsoid; the geoidal height seldom exceeds a hundred meters, and the deflection of the vertical seldom exceeds one arc minute. Hence, in solving Stokes' boundary value problem, replacing the free boundary, i.e., the geoid, by the fixed boundary, i.e., the reference Earth ellipsoid, leads to very accurate result. Furthermore, the flattening of the reference Earth ellipsoid is on the order of $1/300$, meaning that approximating the reference Earth ellipsoid with a sphere would introduce an error of the same level. This leads to the conclusion that spherical approximation is a reasonable first order approximation. In this book, we seek the solution of Stokes' boundary value problem under spherical approximation.

We emphasize that the spherical approximation is to be applied to small quantities like the disturbing potential or to parameters multiplied to a small quantity. In the computation of gravity anomaly according to (5.15), the normal gravity γ_0 must be computed using its formula on the reference Earth ellipsoid, rather than its spherical approximation understood as its global average.

As compared with the general form of the third-type boundary problem discussed in Sect. 2.3, α and β have the same sign in the fundamental gravimetric equation. We did not provide a proof that the solution of such a problem is unique for whatever shape the boundary has. It is indeed not unique. Nevertheless, the non-uniqueness can always be revealed in the formulation of the solution, and some physically meaningful constraint can always be applied to make the solution unique. Therefore, we leave the proof of uniqueness as a part of the derivation of the solution. Furthermore, the formulation of the solution also serves as a proof of the existence of the solution.

In the next section, Stokes' boundary value problem under spherical approximation is to be solved assuming that the gravity anomaly is known all over the geoid and that there exists no mass above the geoid.

5.2 Solution of Stokes' Boundary Value Problem

5.2.1 Solution of the Disturbing Potential and Geoidal Height

As the mass of the reference Earth ellipsoid is chosen to be identical to that of the Earth, the degree 0 term is missing from the spherical harmonic series of the disturbing potential. Furthermore, we assume that the center of the reference Earth ellipsoid is on the center of mass of the Earth. In this way, the degree 1 term is also missing. Hence, the spherical harmonic series of the disturbing potential is

$$
T = \frac{GM}{r} \sum_{n=2}^{\infty} \sum_{k=0}^{n} \left(\frac{R}{r}\right)^n (A_n^k \cos k\lambda + B_n^k \sin k\lambda) P_n^k (\cos \theta). \tag{5.27}
$$

This is a general form of the solution of Laplace's equation for our case. Our next task is to determine the coefficients. Here the scale length a is replaced by R under spherical approximation.

Substitute (5.27) into the fundamental gravimetric equation (5.26). We obtain an expression of Δg using the spherical harmonic coefficients of T

$$
\Delta g = \frac{GM}{R^2} \sum_{n=2}^{\infty} \sum_{k=0}^{n} (n-1)(A_n^k \cos k\lambda + B_n^k \sin k\lambda) P_n^k (\cos \theta). \tag{5.28}
$$

As Δg is assumed to be known all over the geoid, it can be expressed as a spherical harmonic series

$$
\Delta g = \sum_{n=2}^{\infty} \sum_{k=0}^{n} (c_n^k \cos k\lambda + s_n^k \sin k\lambda) P_n^k (\cos \theta), \tag{5.29}
$$

where the degrees 0 and 1 terms are known to be missing according to (5.28). The coefficients are, according to (3.374),

$$
\begin{cases}
c_n^k = \dfrac{1}{1+\delta_k} \dfrac{2n+1}{2\pi} \dfrac{(n-k)!}{(n+k)!} \displaystyle\int_\omega \Delta g(\theta', \lambda') P_n^k (\cos \theta') \cos k\lambda' d\omega, \\
s_n^k = \dfrac{1}{1+\delta_k} \dfrac{2n+1}{2\pi} \dfrac{(n-k)!}{(n+k)!} \displaystyle\int_\omega \Delta g(\theta', \lambda') P_n^k (\cos \theta') \sin k\lambda' d\omega,
\end{cases} \tag{5.30}
$$

where ω is a unit sphere, and θ' and λ' are the variables of integration, i.e., they are the coordinates of $d\omega$, and $d\omega = \sin \theta' d\theta' d\lambda'$. By comparing (5.28) and (5.29), we obtain the spherical harmonic coefficients of T as functions of those of Δg

$$A_n^k = \frac{R^2}{GM} \frac{c_n^k}{n-1}, \qquad B_n^k = \frac{R^2}{GM} \frac{s_n^k}{n-1}; \qquad n \geq 2. \tag{5.31}$$

The formulae (5.27), (5.30), and (5.31) constitute the solution of the problem in spherical harmonic series.

We can infer that the degree 1 term is missing in Δg even if it is present in T. Hence, strictly speaking, the solution of this boundary value problem is not unique, as an arbitrary degree 1 term can be added to T without altering the fact that T is a solution; the solution is made unique by assuming the degree 1 term of T to vanish. Physically, this is assured by setting the origin of the coordinate system on the center of mass of the Earth. Missing of the degree 0 term is not mathematically required to ensure the solution to be unique, but it is by its nature a physical requirement so that the reference Earth ellipsoid best represents the Earth in the average sense. The more general case with degrees 0 and 1 terms included will be formulated in Sect. 5.4.

In the following, we derive a closed-form expression of the spherical harmonic series.

Substitute (5.30) and (5.31) into (5.27). We obtain

$$T(r,\theta,\lambda) = \frac{R}{4\pi} \int_\omega \left[\Delta g(\theta',\lambda') \sum_{n=2}^{\infty} \frac{2n+1}{n-1} \left(\frac{R}{r}\right)^{n+1} \sum_{k=0}^{n} \frac{2}{1+\delta_k} \frac{(n-k)!}{(n+k)!} \right.$$

$$\left. \times P_n^k(\cos\theta) P_n^k(\cos\theta')(\cos k\lambda \cos k\lambda' + \sin k\lambda \sin k\lambda') \right] d\omega, \tag{5.32}$$

where the variables of integration are again θ' and λ', as in (5.30), and we have explicitly expressed T as a function of r, θ, and λ for clarity. Define

$$S(r,\theta,\lambda,\theta',\lambda') = \sum_{n=2}^{\infty} \frac{2n+1}{n-1} \left(\frac{R}{r}\right)^{n+1} \sum_{k=0}^{n} \frac{2}{1+\delta_k} \frac{(n-k)!}{(n+k)!} P_n^k(\cos\theta) P_n^k(\cos\theta')$$

$$\times (\cos k\lambda \cos k\lambda' + \sin k\lambda \sin k\lambda'). \tag{5.33}$$

It can be written as, according to the addition theorem (3.268),

$$S(r,\psi) = \sum_{n=2}^{\infty} \frac{2n+1}{n-1} \left(\frac{R}{r}\right)^{n+1} P_n(\cos\psi), \tag{5.34}$$

where $\psi = \psi(\theta,\lambda,\theta',\lambda')$ is the angle made by the two points on the sphere, (θ,λ) and (θ',λ'), with respect to the center O, or equally the angle made by the points

Fig. 5.4 Definition of ψ and l

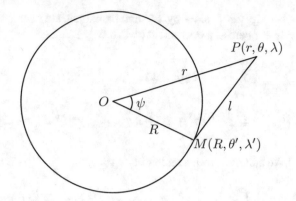

$P(r, \theta, \lambda)$ and $M(R, \theta', \lambda')$ with respect to O, which is given in (3.254), and is shown again in Fig. 5.4. The expression of T can then be written as an integral

$$T(r, \theta, \lambda) = \frac{R}{4\pi} \int_\omega \Delta g(\theta', \lambda') S(r, \psi) d\omega.$$ (5.35)

We emphasize again that the variables of integration are θ' and λ', meaning that they are the coordinates of $d\omega$, with $d\omega = \sin\theta' d\theta' d\lambda'$.

For brevity, we introduce the notation $x = R/r$. We then have $x \leq 1$, and

$$S(r, \psi) = \sum_{n=2}^{\infty} \left(2 + \frac{3}{n-1}\right) x^{n+1} P_n(\cos\psi).$$ (5.36)

In the relation

$$(1 + x^2 - 2x\cos\psi)^{-1/2} = \sum_{n=0}^{\infty} x^n P_n(\cos\psi),$$ (5.37)

we write the terms $P_0(\cos\psi) = 1$ and $P_1(\cos\psi) = \cos\psi$ separately to obtain

$$\sum_{n=2}^{\infty} x^n P_n(\cos\psi) = (1 + x^2 - 2x\cos\psi)^{-\frac{1}{2}} - 1 - x\cos\psi.$$ (5.38)

Multiply both sides by x^{-2} and then integrate over x from 0 to x,[1] the factor $1/(n-1)$ appears as a coefficient of $P_n(\cos\psi)$

$$\sum_{n=2}^{\infty} \frac{x^{n-1}}{n-1} P_n(\cos\psi)$$
$$= \int_0^x \left[x^{-2}(1+x^2-2x\cos\psi)^{-1/2} - x^{-2} - x^{-1}\cos\psi \right] dx . \tag{5.39}$$

We can find from a table of integral that

$$\int x^{-2}(1+x^2-2x\cos\psi)^{-1/2}dx = -\frac{1}{x}(1+x^2-2x\cos\psi)^{1/2}$$
$$-\cos\psi \ln\left[\frac{(1+x^2-2x\cos\psi)^{1/2}+1}{x} - \cos\psi \right] + C . \tag{5.40}$$

The substitution of it into (5.39) yields

$$\sum_{n=2}^{\infty} \frac{x^{n-1}}{n-1} P_n(\cos\psi) = \left\{ -\frac{1}{x}(1+x^2-2x\cos\psi)^{1/2} \right.$$
$$\left. -\cos\psi \ln\left[\frac{(1+x^2-2x\cos\psi)^{1/2}+1}{x} - \cos\psi \right] + \frac{1}{x} - \cos\psi \ln x \right\}_0^x$$
$$= \left\{ \frac{1}{x}\left[1 - (1+x^2-2x\cos\psi)^{1/2} \right] \right.$$
$$\left. -\cos\psi \ln\left[(1+x^2-2x\cos\psi)^{1/2} + 1 - x\cos\psi \right] \right\}_0^x . \tag{5.41}$$

Denote by I the limit of the right-hand side when $x \to 0$

$$I = \lim_{x\to 0} \frac{1-(1+x^2-2x\cos\psi)^{\frac{1}{2}}}{x} - \cos\psi \ln 2 . \tag{5.42}$$

The first term is a limit of the form $0/0$. According to l'Hopital's rule,

$$I = \lim_{x\to 0} \frac{-2x+2\cos\psi}{2(1+x^2-2x\cos\psi)^{\frac{1}{2}}} - \cos\psi \ln 2 = \cos\psi - \cos\psi \ln 2 . \tag{5.43}$$

---

[1] This is a brief statement. Strictly speaking, it includes replacing x by x' and integrating over x' from 0 to x, which avoids the use of x as both the variable of integration and the limit of integration. However, the brief statement is unlikely to bring confusion and will be used later as well.

Substitute it back into (5.41). We obtain

$$\sum_{n=2}^{\infty} \frac{x^{n-1}}{n-1} P_n(\cos\psi) = \frac{1}{x}\left[1 - x\cos\psi - (1 + x^2 - 2x\cos\psi)^{1/2}\right]$$

$$- \cos\psi \ln\left\{\frac{1}{2}\left[(1 + x^2 - 2x\cos\psi)^{1/2} + 1 - x\cos\psi\right]\right\}. \tag{5.44}$$

According to (5.38) and (5.44), (5.36) can be written as

$$S(r,\psi) = 2x\sum_{n=2}^{\infty} x^n P_n(\cos\psi) + 3x^2\sum_{n=2}^{\infty}\frac{x^{n-1}}{n-1} P_n(\cos\psi)$$

$$= 2x(1 + x^2 - 2x\cos\psi)^{-1/2} - 2x - 2x^2\cos\psi$$

$$+ 3x\left[1 - x\cos\psi - (1 + x^2 - 2x\cos\psi)^{1/2}\right]$$

$$- 3x^2\cos\psi \ln\left\{\frac{1}{2}\left[(1 + x^2 - 2x\cos\psi)^{1/2} + 1 + x\cos\psi\right]\right\}$$

$$= 2x(1 + x^2 - 2x\cos\psi)^{-1/2} + x - 5x^2\cos\psi - 3x(1 + x^2 - 2x\cos\psi)^{1/2}$$

$$- 3x^2\cos\psi \ln\left\{\frac{1}{2}\left[(1 + x^2 - 2x\cos\psi)^{1/2} + 1 + x\cos\psi\right]\right\}. \tag{5.45}$$

Finally, with $x = R/r$, we obtain

$$S(r,\psi) = 2R\left(r^2 + R^2 - 2rR\cos\psi\right)^{-1/2} + \frac{R}{r} - 5\frac{R^2}{r^2}\cos\psi$$

$$- 3\frac{R}{r^2}\left(r^2 + R^2 - 2rR\cos\psi\right)^{1/2}$$

$$- 3\frac{R^2}{r^2}\cos\psi \ln\left\{\frac{1}{2r}\left[\left(r^2 + R^2 - 2rR\cos\psi\right)^{1/2} + r - R\cos\psi\right]\right\}$$

$$= 2\frac{R}{l} + \frac{R}{r} - 5\frac{R^2}{r^2}\cos\psi - 3\frac{Rl}{r^2} - 3\frac{R^2}{r^2}\cos\psi \ln\frac{l + r - R\cos\psi}{2r}, \tag{5.46}$$

where, as shown in Fig. 5.4,

$$l = (r^2 + R^2 - 2rR\cos\psi)^{1/2} \tag{5.47}$$

is the distance between the point P, i.e., the point where T is computed, and the point M, i.e., the point on the sphere corresponding to $d\omega$.

The function $S(r, \psi)$ is referred to as generalized Stokes function and (5.35) generalized Stokes' formula, which can be used to compute the disturbing potential outside the geoid using data of Δg.

According to (5.7), the geoidal height can be considered as a degenerate case of the disturbing potential when $r = R$. The formulation of its spherical harmonic series is straightforward. Here we focus on the integral form.

When $r = R$, the degenerate form of the generalized Stokes' formula is

$$T_0(\theta, \lambda) = \frac{R}{4\pi} \int_\omega \Delta g(\theta', \lambda') S(\psi) d\omega, \tag{5.48}$$

where $S(\psi) = S(R, \psi)$ is the degenerate form of the generalized Stokes function (5.46). As $l = 2R \sin(\psi/2)$ when $r = R$, we have, by using also the formula $\cos \psi = 1 - 2 \sin^2(\psi/2)$

$$S(\psi) = \frac{1}{\sin(\psi/2)} + 1 - 5\cos\psi - 6\sin\frac{\psi}{2} - 3\cos\psi \ln\left(\sin\frac{\psi}{2} + \sin^2\frac{\psi}{2}\right), \tag{5.49}$$

which is referred to as Stokes function. The formula of geoidal height is, according to (5.7) and (5.48),

$$N(\theta, \lambda) = \frac{R}{4\pi \bar{\gamma}_0} \int_\omega \Delta g(\theta', \lambda') S(\psi) d\omega. \tag{5.50}$$

This formula is referred to as Stokes' formula and so is (5.48).

A graph of the Stokes function $S(\psi)$ is shown in Fig. 5.5, together with that of its leading term,

$$S(\psi) \approx \frac{1}{\sin(\psi/2)} \approx \frac{2}{\psi}. \tag{5.51}$$

As $S(\psi) \to \infty$ when $\psi \to 0$, the detail when $\psi \leq 10°$ is shown separately in the left panel in logarithm scale. When ψ is very small, the difference between $S(\psi)$ and $2/\psi$ is only a very small fraction of $S(\psi)$, and hence, the approximation (5.51) is often made.

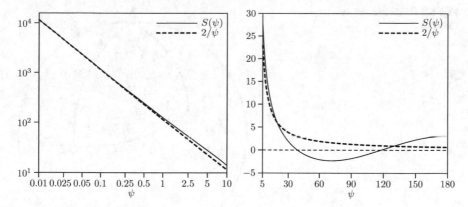

Fig. 5.5 The Stokes function

5.2.2 Solution of the Gravity Disturbance and Deflection of the Vertical

The spherical harmonic series of the gravity disturbance can be readily obtained by replacing the spherical harmonic coefficients in (3.303) to (3.305) by those of the disturbing potential. Here we focus on the integral form.

According to (5.3), we obtain from (5.35)

$$
\begin{cases}
\delta g_r(r, \theta, \lambda) = \dfrac{R}{4\pi} \displaystyle\int_\omega \Delta g \dfrac{\partial}{\partial r} S(r, \psi)\, d\omega\,, \\[2ex]
\delta g_\theta(r, \theta, \lambda) = \dfrac{R}{4\pi r} \displaystyle\int_\omega \Delta g \dfrac{\partial}{\partial \theta} S(r, \psi)\, d\omega = \dfrac{R}{4\pi r} \displaystyle\int_\omega \Delta g \dfrac{\partial}{\partial \psi} S(r, \psi) \dfrac{\partial \psi}{\partial \theta}\, d\omega\,, \\[2ex]
\delta g_\lambda(r, \theta, \lambda) = \dfrac{R}{4\pi r \sin\theta} \displaystyle\int_\omega \Delta g \dfrac{\partial}{\partial \lambda} S(r, \psi)\, d\omega \\[2ex]
\qquad\qquad\quad = \dfrac{R}{4\pi r \sin\theta} \displaystyle\int_\omega \Delta g \dfrac{\partial}{\partial \psi} S(r, \psi) \dfrac{\partial \psi}{\partial \lambda}\, d\omega\,,
\end{cases}
$$

$$(5.52)$$

where the left-hand side is understood as a function of r, θ, and λ, Δg is understood as a function of θ' and λ', and the integration is performed for the variables θ' and λ', i.e., $d\omega = \sin\theta' d\theta' d\lambda'$. Evidently, we have to derive the formulae for $\partial\psi/\partial\theta$, $\partial\psi/\partial\lambda$ and $(\partial/\partial r)S(r, \psi)$, $(\partial/\partial\psi)S(r, \psi)$.

First, we derive the formulae for $\partial\psi/\partial\theta$ and $\partial\psi/\partial\lambda$. As shown in Fig. 5.6, N is the north pole, P_0 is the projection of P on the geoid approximated as a sphere here, M is the point on the sphere corresponding to $d\omega$, and A is the azimuth of M with respect to P_0. We have

$$\sin\psi \sin A = \sin\theta' \sin(\lambda' - \lambda)\,, \tag{5.53}$$

$$\sin\psi \cos A = \cos\theta' \sin\theta - \sin\theta' \cos\theta \cos(\lambda' - \lambda)\,, \tag{5.54}$$

$$\cos\psi = \cos\theta' \cos\theta + \sin\theta' \sin\theta \cos(\lambda' - \lambda)\,. \tag{5.55}$$

Fig. 5.6 Definition of the azimuth A

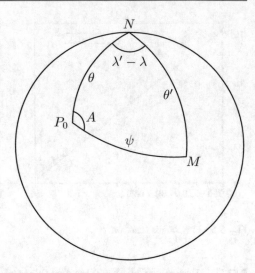

Take the derivatives of both sides of (5.55) with respect to θ and λ. We obtain

$$
\begin{cases}
- \sin \psi \dfrac{\partial \psi}{\partial \theta} = - \cos \theta' \sin \theta + \sin \theta' \cos \theta \cos(\lambda' - \lambda) , \\[2mm]
- \sin \psi \dfrac{\partial \psi}{\partial \lambda} = \sin \theta' \sin \theta \sin(\lambda' - \lambda) .
\end{cases}
\tag{5.56}
$$

By comparing these with (5.53) and (5.54), we obtain

$$
\frac{\partial \psi}{\partial \theta} = \cos A , \qquad \frac{\partial \psi}{\partial \lambda} = - \sin \theta \sin A .
\tag{5.57}
$$

Substitute these into (5.52). We obtain, written in a more explicit form,

$$
\begin{cases}
\delta g_r(r, \theta, \lambda) = \dfrac{R}{4\pi} \displaystyle\int_\omega \Delta g(\theta', \lambda') \dfrac{\partial}{\partial r} S(r, \psi) d\omega , \\[3mm]
\delta g_\theta(r, \theta, \lambda) = \dfrac{R}{4\pi r} \displaystyle\int_\omega \Delta g(\theta', \lambda') \dfrac{\partial}{\partial \psi} S(r, \psi) \cos A \, d\omega , \\[3mm]
\delta g_\lambda(r, \theta, \lambda) = - \dfrac{R}{4\pi r} \displaystyle\int_\omega \Delta g(\theta', \lambda') \dfrac{\partial}{\partial \psi} S(r, \psi) \sin A \, d\omega .
\end{cases}
\tag{5.58}
$$

According to (5.53) and (5.54), A could be computed from

$$
\tan A = \frac{\sin \theta' \sin(\lambda' - \lambda)}{\cos \theta' \sin \theta - \sin \theta' \cos \theta \cos(\lambda' - \lambda)} ,
\tag{5.59}
$$

where the numerator and denominator have the same signs as $\sin A$ and $\cos A$, respectively.

Finally, we derive the formulae of $(\partial/\partial r)S(r,\psi)$ and $(\partial/\partial\psi)S(r,\psi)$. According to (5.47), we have

$$\frac{\partial l}{\partial r}=\frac{r-R\cos\psi}{l}, \qquad \frac{\partial l}{\partial\psi}=\frac{Rr\sin\psi}{l}. \tag{5.60}$$

We can then obtain from (5.46), using (5.47) and the above formulae,

$$\frac{\partial}{\partial r}S(r,\psi)$$

$$=-2\frac{R}{l^2}\frac{r-R\cos\psi}{l}-\frac{R}{r^2}+10\frac{R^2}{r^3}\cos\psi+6\frac{Rl}{r^3}$$

$$-3\frac{R}{r^2}\frac{r-R\cos\psi}{l}+6\frac{R^2}{r^3}\cos\psi\ln\frac{l+r-R\cos\psi}{2r}$$

$$-3\frac{R^2}{r^2}\frac{2r\cos\psi}{l+r-R\cos\psi}\frac{2r[(r-R\cos\psi)/l+1]-2(l+r-R\cos\psi)}{4r^2}$$

$$=-\frac{2R(r-R\cos\psi)}{l^3}+\frac{6Rl-Rr}{r^3}-3\frac{R}{rl}$$

$$+\frac{R^2\cos\psi}{r^3}\left(13+6\ln\frac{l+r-R\cos\psi}{2r}\right).$$

$$=-\frac{R(r^2-R^2)}{rl^3}+\frac{6Rl-Rr}{r^3}-4\frac{R}{rl}$$

$$+\frac{R^2\cos\psi}{r^3}\left(13+6\ln\frac{l+r-R\cos\psi}{2r}\right). \tag{5.61}$$

$$\frac{\partial}{\partial\psi}S(r,\psi)$$

$$=-2\frac{R}{l^2}\frac{Rr\sin\psi}{l}+5\frac{R^2\sin\psi}{r^2}-3\frac{R}{r^2}\frac{Rr\sin\psi}{l}+3\frac{R^2}{r^2}\sin\psi\ln\frac{l+r-R\cos\psi}{2r}$$

$$-3\frac{R^2}{r^2}\cos\psi\frac{2r}{l+r-R\cos\psi}\frac{1}{2r}\left(\frac{Rr\sin\psi}{l}+R\sin\psi\right)$$

$$=\sin\psi\left\{-2\frac{R^2r}{l^3}+5\frac{R^2}{r^2}-3\frac{R^2}{rl}+3\frac{R^2}{r^2}\left[\ln\frac{l+r-R\cos\psi}{2r}\right.\right.$$

$$\left.\left.-\frac{R\cos\psi(r+l)}{l(l+r-R\cos\psi)}\right]\right\}$$

$$
= \sin \psi \left\{ -2\frac{R^2 r}{l^3} + 5\frac{R^2}{r^2} - 3\frac{R^2}{rl} + 3\frac{R^2}{r^2}\left[\ln \frac{l + r - R\cos\psi}{2r} \right.\right.
$$

$$
\left.\left. - \frac{R\cos\psi\,(l+r)(l-r+R\cos\psi)}{R^2 l \sin^2\psi} \right]\right\}
$$

$$
= \sin \psi \left[-2\frac{R^2 r}{l^3} + 8\frac{R^2}{r^2} - 6\frac{R^2}{rl} + 3\frac{R^2}{r^2}\left(\frac{r - l - R\cos\psi}{l\sin^2\psi} \right.\right.
$$

$$
\left.\left. + \ln \frac{l + r - R\cos\psi}{2r} \right)\right]. \tag{5.62}
$$

In summary, the integral formula for the gravity disturbance is (5.58) together with (5.59) and the two formulae above.

According to (5.13), the deflection of the vertical can be considered as a degenerate case of the horizontal components of the gravity disturbance when $r = R$. As in the case of geoidal height, the formulation of their spherical harmonic series is straightforward, and here we focus on the integral form.

When $r = R$, we have $l = 2R\sin(\psi/2)$. The degenerate form of $(\partial/\partial\psi)S(r, \psi)$ is, by using also the formulae $\cos\psi = 1 - 2\sin^2(\psi/2)$ and $\sin\psi = 2\sin(\psi/2)\cos(\psi/2)$,

$$
\frac{d}{d\psi}S(\psi) = -\frac{\cos(\psi/2)}{2\sin^2(\psi/2)} + 8\sin\psi - 6\cos\frac{\psi}{2} - 3\frac{1 - \sin(\psi/2)}{\sin\psi}
$$

$$
+ 3\sin\psi \ln\left(\sin\frac{\psi}{2} + \sin^2\frac{\psi}{2} \right), \tag{5.63}
$$

which is referred to as Vening Meinesz function. The formula of the deflection of the vertical can be inferred from (5.13) and (5.58) to be

$$
\begin{cases}
\xi(\theta, \lambda) = \dfrac{1}{4\pi\bar{\gamma}_0} \displaystyle\int_\omega \Delta g(\theta', \lambda')\dfrac{d}{d\psi}S(\psi)\cos A\, d\omega, \\[2mm]
\eta(\theta, \lambda) = \dfrac{1}{4\pi\bar{\gamma}_0} \displaystyle\int_\omega \Delta g(\theta', \lambda')\dfrac{d}{d\psi}S(\psi)\sin A\, d\omega,
\end{cases} \tag{5.64}
$$

which is referred to as Vening Meinesz' formula.

A graph of the Vening Meinesz function $(d/d\psi)S(\psi)$ is shown in Fig. 5.7, together with that of its leading term,

$$
\frac{d}{d\psi}S(\psi) \approx -\frac{\cos(\psi/2)}{2\sin^2(\psi/2)} \approx -\frac{1}{2\sin^2(\psi/2)} \approx -\frac{2}{\psi^2}. \tag{5.65}
$$

As compared to the case of Stokes function in (5.51), when ψ is small, $(d/d\psi)S(\psi)$ is negative, and the inverse proportion to ψ^2 renders the graph steeper. Hence, in the figure, the magnitude decreases from bottom to top. Like the case of Stokes function,

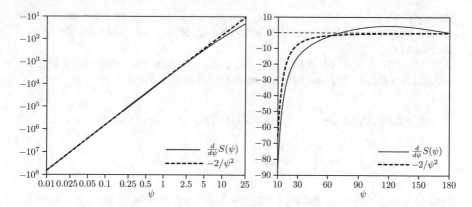

Fig. 5.7 Vening Meinesz function. As the function is negative when ψ is small, the lower the larger in magnitude

the difference between $(d/d\psi)S(\psi)$ and $-2/\psi^2$ is only a very small fraction of $(d/d\psi)S(\psi)$ when ψ is small, and hence, the approximation (5.65) is often made.

We also use the term *gravity disturbance* to indicate its downward component, i.e., $\delta g(r, \theta, \lambda) = -\delta g_r(r, \theta, \lambda)$. In spherical approximation, the value of δg on the geoid, δg_0, can be computed using Δg and T_0 according to the fundamental gravimetric equation (5.26)

$$\delta g_0(\theta, \lambda) = \Delta g(\theta, \lambda) + \frac{2}{R}T_0(\theta, \lambda) = \Delta g(\theta, \lambda) + \frac{1}{2\pi}\int_\omega \Delta g(\theta', \lambda')S(\psi)d\omega.$$

$$(5.66)$$

It is not evident to compute $\delta g_0(\theta, \lambda) = -g_r(R, \theta, \lambda)$ by setting $r = R$ in the first formula of (5.58) and in (5.61), which will be discussed in the next subsection.

5.2.3   Gravity Anomaly Outside the Geoid

Gravity anomaly is defined using the gravity on the geoid and the normal gravity on the reference Earth ellipsoid and is considered to be a function defined on the geoid. In spherical approximation, it is related to the disturbing potential according to the fundamental gravimetric equation (5.26), where the geoid is approximately considered as a sphere. Now we extend the definition of gravity anomaly to outside the geoid by extending the fundamental gravimetric equation (5.26) to any $r \geq R$

$$\Delta g = -2\frac{T}{r} - \frac{\partial T}{\partial r}.$$

$$(5.67)$$

Although still bearing the name gravity anomaly, it is no longer related to the geoid and the reference Earth ellipsoid when $r > R$. Hence, we just consider it as a function of r.

Multiply both sides of (5.67) by r, and then substitute the solutions of T given in (5.35) and that of $\delta g_r = \partial T/\partial r$ given in (5.52) into it. We obtain

$$
r\Delta g(r, \theta, \lambda) = \frac{1}{2\pi} \int_\omega \Delta g(\theta', \lambda') \left[-2S(r, \psi) - r\frac{\partial}{\partial r} S(r, \psi) \right] d\omega
$$

$$
= \frac{1}{2\pi} \int_\omega [R\Delta g(\theta', \lambda')] G'(r, \psi) d\omega , \tag{5.68}
$$

which is a formula for computing $r\Delta g$ outside the geoid using its value on the geoid, $R\Delta g$. The kernel $G'(r, \psi)$ can be obtained from (5.46) and (5.61) to be

$$
G'(r, \psi) = -2S(r, \psi) - r\frac{\partial}{\partial r} S(r, \psi) = \frac{R(r^2 - R^2)}{l^3} - \frac{R}{r} - 3\frac{R^2}{r^2} \cos \psi
$$

$$
= G(r, \psi) - \frac{R}{r} - 3\frac{R^2}{r^2} \cos \psi , \tag{5.69}
$$

where $G(r, \psi) = R(r^2 - R^2)/l^3$ is the kernel of the Poisson integral given in (3.387). According to the definition of $G(r, \psi)$ using spherical harmonics given in (3.384), $G'(r, \psi)$ is different from $G(r, \psi)$ just by omitting the degree 0 term R/r and degree 1 term $3(R/r)^2 \cos \psi$.

In fact, it is expected to obtain (5.68) and (5.69). First, it can be verified that $r(\partial T/\partial r)$ is harmonic based on the fact that T is harmonic, and furthermore, $r\Delta g$ is harmonic based on its definition given in (5.67). Hence, the Poisson integral is applicable to it. Second, the degrees 0 and 1 terms are removed throughout the formulation.

However, it is not really necessary to remove the degrees 0 and 1 terms in (5.68) and (5.69) even if Δg does not include them at all, which could be best explained based on the solution expressed as a spherical harmonic series provided in Sect. 3.6.6. We first rewrite (5.68) including degrees 0 and 1 as

$$
r\Delta g(r, \theta, \lambda) = \frac{1}{2\pi} \int_\omega [R\Delta g(\theta', \lambda')] G(r, \psi) d\omega . \tag{5.70}
$$

Due to the orthogonality among spherical harmonics of different degrees and orders, the degrees 0 and 1 terms in $G(r, \psi)$ would yield non-zero results only in their products with the degrees 0 and 1 terms of $\Delta g(\theta', \lambda')$, respectively. Therefore, if the degrees 0 and 1 parts are missing in $\Delta g(\theta', \lambda')$, they would be missing in $\Delta g(r, \theta, \lambda)$. Consequently, (5.70) can be used without modification.

Another remark is that unlike the case of disturbing potential, it is impossible to obtain a degenerate kernel $G(\psi)$ by setting $r = R$, so that the right-hand side of (5.70) degenerates to an integral in the form of $[R\Delta g(\theta', \lambda')]G(\psi)$ on ω. This is due to the fact that if we set $r \to R$, $\lim_{r \to R} G(r, \psi) = 0$ for any $\psi \neq 0$, but $\lim_{r \to R} G(r, 0) = \infty$ (set $\psi = 0$ first, and then seek the limit $r \to R$).

We mentioned in the previous subsection that it is not evident to compute $\delta g_0(\theta, \lambda) = -g_r(R, \theta, \lambda)$ by setting $r = R$ in the first formula of (5.58) and in (5.61). This is due to the fact that the first term in (5.61) is the kernel of the Poisson integral (with an extra factor $-1/r$), to which the discussion in the last paragraph applies.

5.3   Boundary Value Theories Beyond That of Stokes

In the previous section, we formulated the problem to compute all other quantities of the disturbing gravity field using data of gravity anomaly. In the present era of satellite geodesy, various other types of data become available. Satellite altimetry provides measurements of geoidal height and/or deflection of the vertical over oceans. When GNSS[2] is used for positioning in gravimetry, the gravity disturbance is obtained by gravity reduction. Naturally, it is now of both theoretical and practical interests to determine the disturbing gravity field using data of any of the quantities mentioned above, which is the subjects of this section. The same as in Stokes' theory, spherical approximation will be assumed.

Before proceeding to the formulation, we make some remarks on the degrees 0 and 1 terms. In the formulation of Stokes' theory, we have imposed several requirements for the reference Earth ellipsoid. The angular velocity of the reference Earth ellipsoid is the same as that of the Earth, so that T is harmonic. The value of GM of the reference Earth ellipsoid equals that of the Earth, so that the degree 0 term of T vanishes, which is implied by expressing T in the form of (5.27). The normal potential on the reference Earth ellipsoid, U_0, is the same as the real gravity potential of the Earth on the geoid, W_0, which was introduced in deriving Bruns' formula of geoidal height. The fact that the degree 1 term has to be removed is due to the definition of gravity anomaly. Taking reference to (5.28), even if T includes the degree 1 term, it would not be present in Δg, i.e., the degree 1 term of T does not contribute to Δg at all. Therefore, as already mentioned in Sect. 5.2.1, T is not uniquely determined by Δg. The reverse, i.e., determining Δg using T, does not have this kind of problem; Δg can be computed from data of T according the fundamental gravimetric equation without taking extra care, and the result of Δg does not have the degree 1 term.

To understand why Δg does not include any degree 1 information based on physical inference, we consider a special case when the Earth is identical to the reference Earth ellipsoid and the center of the reference Earth ellipsoid is offset slightly with respect to the Earth's center of mass. In this case, only the degree 1 term is present in T. However, as the position of P with respect to the Earth is the

---

[2] The term *GNSS* is an acronym of *Global Navigation Satellite System*. It is well know even to the public, and we use it as a well-defined nomenclature.

same as that of P_0 with respect to the reference Earth ellipsoid, the gravity at P
equals the normal gravity at P_0, which leads to $\Delta g = 0$.

The non-uniqueness of solution of T is a characteristic of Stokes' theory when
Δg is known, which is the reason why the degree 1 term has to be excluded in
the formulation. As for the degree 0 term, we have assumed it to be missing by
setting GM of the reference Earth ellipsoid to be that of the Earth and $U_0 = W_0$.
In the problems to be formulated in this section, non-uniqueness of solution does
not happen with only one simple exception, and we will first formulate the solution
with the degrees 0 and 1 terms included and then mention how to remove them.

5.3.1   Formulation When the Gravity Disturbance Is Known

The background theory is the solution of the second-type boundary value problem.
We first derive the formula using more general notations: let S be a sphere of radius
R and V a harmonic function on and outside the sphere, and $\partial V / \partial r$ is known on
the sphere. Derive the formula for computing V outside the sphere.

We use spherical harmonic series to solve the problem, as in the case of solving
Stokes's problem. We express $\partial V / \partial r$ on the sphere as a spherical harmonic series

$$\left(\frac{\partial V}{\partial r} \right)_R = \sum_{n=0}^{\infty} \sum_{k=0}^{n} (A_n^k \cos k\lambda + B_n^k \sin k\lambda) P_n^k (\cos \theta) , \tag{5.71}$$

where

$$\begin{cases} A_n^k = \dfrac{1}{1 + \delta_k^0} \dfrac{2n + 1}{2\pi} \dfrac{(n - k)!}{(n + k)!} \displaystyle\int_{\omega} \left(\frac{\partial V}{\partial r} \right)_R P_n^k (\cos \theta') \cos k\lambda' d\omega , \\[4mm] B_n^k = \dfrac{1}{1 + \delta_k^0} \dfrac{2n + 1}{2\pi} \dfrac{(n - k)!}{(n + k)!} \displaystyle\int_{\omega} \left(\frac{\partial V}{\partial r} \right)_R P_n^k (\cos \theta') \sin k\lambda' d\omega . \end{cases} \tag{5.72}$$

As V is harmonic on and outside the sphere, it can be expressed as a spherical
harmonic series in the form of

$$V = \sum_{n=0}^{\infty} \sum_{k=0}^{n} \left(\frac{R}{r} \right)^{n+1} (C_n^k \cos k\lambda + D_n^k \sin k\lambda) P_n^k (\cos \theta) . \tag{5.73}$$

Take the derivative of both sides with respect to r, and then set $r = R$. We obtain

$$\left(\frac{\partial V}{\partial r} \right)_R = \sum_{n=0}^{\infty} \sum_{k=0}^{n} \left(-\frac{n + 1}{R} \right) (C_n^k \cos k\lambda + D_n^k \sin k\lambda) P_n^k (\cos \theta) . \tag{5.74}$$

By comparing (5.71) and (5.74), we obtain

$$C_n^k = -\frac{R}{n+1}A_n^k, \qquad D_n^k = -\frac{R}{n+1}B_n^k. \tag{5.75}$$

Substitute this formula into (5.73), and then use (5.72). We obtain

$$V = -\sum_{n=0}^{\infty}\sum_{k=0}^{n}\left(\frac{R}{r}\right)^{n+1}\frac{R}{n+1}(C_n^k\cos k\lambda + D_n^k\sin k\lambda)P_n^k(\cos\theta)$$

$$= -\sum_{n=0}^{\infty}\sum_{k=-n}^{n}\left(\frac{R}{r}\right)^{n+1}\frac{R}{n+1}\frac{1}{1+\delta_k^0}\frac{2n+1}{2\pi}\frac{(n-k)!}{(n+k)!}\int_{\omega}\left(\frac{\partial V}{\partial r}\right)_R$$

$$\times P_n^k(\cos\theta)P_n^k(\cos\theta')(\cos k\lambda\cos k\lambda' + \sin k\lambda\sin k\lambda')d\omega, \tag{5.76}$$

where $(\partial V/\partial r)_R$ should be considered as a function of the variables of integration θ' and λ'. According to the addition theorem, this formula can be written as

$$V = -\sum_{n=0}^{\infty}\frac{R}{n+1}\frac{2n+1}{4\pi}\left(\frac{R}{r}\right)^{n+1}\int_{\omega}\left(\frac{\partial V}{\partial r}\right)_R P_n(\cos\psi)d\omega, \tag{5.77}$$

where ψ is the angle between the two points (θ, λ) and (θ', λ') with respect to the center of the sphere. In the following, we proceed to express V in the form of

$$V = -\frac{R}{4\pi}\int_{\omega}\left(\frac{\partial V}{\partial r}\right)_R K(r, \psi)d\omega, \tag{5.78}$$

where

$$K(r, \psi) = \sum_{n=0}^{\infty}\frac{2n+1}{n+1}\left(\frac{R}{r}\right)^{n+1}P_n(\cos\psi). \tag{5.79}$$

Now we derive a closed-form formula of the Legendre series (5.79). For brevity, we adopt $x = R/r$. First of all, we write (5.79) as

$$K(r, \psi) = \sum_{n=0}^{\infty}\left(2 - \frac{1}{n+1}\right)x^{n+1}P_n(\cos\psi). \tag{5.80}$$

Multiply both sides of the formula $(1 + x^2 - 2x\cos\psi)^{-1/2} = \sum_{n=0}^{\infty}x^n P_n(\cos\psi)$ by x. We obtain

$$\sum_{n=0}^{\infty}x^{n+1}P_n(\cos\psi) = \frac{x}{(1 + x^2 - 2x\cos\psi)^{1/2}}. \tag{5.81}$$

Integrate the formula $(1 + x^2 - 2x \cos \psi)^{-1/2} = \sum_{n=0}^{\infty} x^n P_n(\cos \psi)$ over x from 0 to x. We obtain

$$\sum_{n=0}^{\infty} \frac{1}{n+1} x^{n+1} P_n(\cos \psi) = \int_0^x \frac{dx}{(1 + x^2 - 2x \cos \psi)^{1/2}} . \tag{5.82}$$

It can be found from a table of integrals that

$$\int \frac{dx}{(1 + x^2 - 2x \cos \psi)^{1/2}} = \ln \left[2x - 2\cos \psi + 2(1 + x^2 - 2x \cos \psi)^{1/2} \right] + C . \tag{5.83}$$

The substitution of it into (5.82) yields

$$\sum_{n=0}^{\infty} \frac{1}{n+1} x^{n+1} P_n(\cos \psi) = \left\{ \ln \left[2x - 2\cos \psi + 2(1 + x^2 - 2x \cos \psi)^{1/2} \right] \right\}_0^x$$

$$= \ln \left[2x - 2\cos \psi + 2(1 + x^2 - 2x \cos \psi)^{1/2} \right]$$

$$- \ln(2 - 2\cos \psi)$$

$$= \ln \frac{x - \cos \psi + (1 + x^2 - 2x \cos \psi)^{1/2}}{1 - \cos \psi} . \tag{5.84}$$

Substitute (5.81) and (5.84) into (5.79). We obtain

$$K(r, \psi) = \frac{2x}{(1 + x^2 - 2x \cos \psi)^{1/2}} - \ln \frac{x - \cos \psi + (1 + x^2 - 2x \cos \psi)^{1/2}}{1 - \cos \psi}$$

$$= \frac{2R}{l} - \ln \frac{l + R - r \cos \psi}{r(1 - \cos \psi)} , \tag{5.85}$$

where l is defined as usual

$$l = (r^2 + R^2 - 2rR \cos \psi)^{1/2} . \tag{5.86}$$

The kernel $K(r, \psi)$ is a generalization of the Hotine function to be defined in the next paragraph.

Setting $V = T$ and $r = R$ in (5.78) and (5.85) leads to Hotine's formula

$$T_0 = -\frac{R}{4\pi} \int_\omega \left(\frac{\partial T}{\partial r} \right)_R K(\psi) d\omega = \frac{R}{4\pi} \int_\omega \delta g(\theta', \lambda') K(\psi) d\omega , \tag{5.87}$$

where $\delta g = -(\partial T/\partial r)_{r=R}$ is the gravity disturbance, and

$$K(\psi) = \frac{1}{\sin(\psi/2)} - \ln\left[1 + \frac{1}{\sin(\psi/2)}\right] \tag{5.88}$$

is referred to as Hotine function. We have chosen to use K instead of H to denote this function as it is also referred to as Neumann–Koch function.

According to Bruns' formula, we can write (5.87) in the form for computing geoidal height using data of gravity disturbance

$$N(\theta, \lambda) = \frac{R}{4\pi \bar{\gamma}_0} \int_\omega \delta g(\theta', \lambda') K(\psi) d\omega, \tag{5.89}$$

which is also referred to as Hotine's formula. The deflection of the vertical can be easily formulated to be

$$\begin{cases} \xi(\theta, \lambda) = \dfrac{1}{4\pi \bar{\gamma}_0} \displaystyle\int_\omega \delta g(\theta', \lambda') \dfrac{d}{d\psi} K(\psi) \cos A \, d\omega, \\[4mm] \eta(\theta, \lambda) = \dfrac{1}{4\pi \bar{\gamma}_0} \displaystyle\int_\omega \delta g(\theta', \lambda') \dfrac{d}{d\psi} K(\psi) \sin A \, d\omega, \end{cases} \tag{5.90}$$

where

$$\frac{d}{d\psi} K(\psi) = -\frac{\cos(\psi/2)}{2\sin^2(\psi/2)[1 + \sin(\psi/2)]}. \tag{5.91}$$

The formula for computing gravity anomaly using data of gravity disturbance can be obtained readily by substituting (5.87) into the fundamental gravimetric equation (5.26)

$$\Delta g(\theta, \lambda) = \delta g(\theta, \lambda) - \frac{1}{2\pi} \int_\omega \delta g(\theta', \lambda') K(\psi) d\omega. \tag{5.92}$$

Removing the degrees 0 and 1 terms in the formulation is straightforward. Without the degrees 0 and 1 terms, the summation in (5.77) and (5.79) would start from $n = 2$. Hence, it concerns removing the degrees 0 and 1 terms in $K(r, \psi)$, which are $K_0(r, \psi) = 1$ and $K_1(r, \psi) = (3/2)(R/r)\cos\psi$. In the degenerate case $r = R$, they are $K_0(\psi) = 1$ and $K_1(\psi) = (3/2)\cos\psi$, and further, $(d/d\psi)K_0(\psi) = 0$ and $(d/d\psi)K_1(\psi) = -(3/2)\sin\psi$. We mention, however, that it is not necessary to remove any of these terms in the formulation to ensure the solution to be unique; the solution is proven to be unique, which can also be inferred from the derivation of the solution. Furthermore, if the effects of these terms are not present in the data of δg, their effects will not appear in the result either even when they are included in the formulation. Removal of these terms in the formulation is necessary only if their effects are present in the data of δg, but their effects are to be excluded in the result.

5.3.2 Formulation When the Deflection of the Vertical Is Known

Considering that the deflection of the vertical is related to the disturbing potential in the form of (5.12), here we start from guessing that the solution for T is in the form

$$
T(r, \theta, \lambda) = \frac{R\bar{\gamma}_0}{4\pi} \int_\omega \left[\frac{\partial}{\partial \theta'} C(r, \psi) \xi(\theta', \lambda') - \frac{1}{\sin\theta'} \frac{\partial}{\partial \lambda'} C(r, \psi) \eta(\theta', \lambda') \right] d\omega .
$$

(5.93)

The coefficient $R\bar{\gamma}_0/(4\pi)$ is chosen so that the kernel $C(r, \psi)$ be dimensionless. In the subsequent development, we will see that this guess is well educated based on the understanding of spherical harmonics. We express the kernel in the form of

$$
C(r, \psi) = \sum_{n=1}^{\infty} \alpha_n \left(\frac{R}{r} \right)^{n+1} P_n(\cos\psi) ,
$$

(5.94)

where the coefficient α_n is to be determined according to (5.93). The dependence on r in the form $(R/r)^{n+1}$ is evident as T is harmonic. The degree 0 term is missing as it does not contribute to the derivatives in (5.93). In other words, data of the deflection of the vertical have no information of the degree 0 term of T, and the solution of T is made unique by setting its degree 0 term to vanish.

According to the addition theorem (3.268), we write $C(r, \psi)$ as

$$
C(r, \psi) = \sum_{n=1}^{\infty} \sum_{k=0}^{n} \alpha_n \left(\frac{R}{r} \right)^{n+1} \frac{2}{1+\delta_k} \frac{(n-k)!}{(n+k)!} P_n^k(\cos\theta) P_n^k(\cos\theta')
$$

$$
\times (\cos k\lambda \cos k\lambda' + \sin k\lambda \sin k\lambda') .
$$

(5.95)

We then have

$$
\begin{cases}
\dfrac{\partial}{\partial \theta'} C(r, \psi) = \displaystyle\sum_{n=1}^{\infty} \sum_{k=0}^{n} \alpha_n \left(\frac{R}{r} \right)^{n+1} \frac{2}{1+\delta_k} \frac{(n-k)!}{(n+k)!} P_n^k(\cos\theta) \\
\qquad\qquad\qquad \times \dfrac{\partial}{\partial \theta'} P_n^k(\cos\theta')(\cos k\lambda \cos k\lambda' + \sin k\lambda \sin k\lambda') , \\[4mm]
-\dfrac{1}{\sin\theta'} \dfrac{\partial}{\partial \lambda'} C(r, \psi) = -\displaystyle\sum_{n=1}^{\infty} \sum_{k=0}^{n} \alpha_n \left(\frac{R}{r} \right)^{n+1} \frac{2}{1+\delta_k} \frac{(n-k)!}{(n+k)!} P_n^k(\cos\theta) \\
\qquad\qquad\qquad \times \dfrac{k}{\sin\theta'} P_n^k(\cos\theta')(-\cos k\lambda \sin k\lambda' + \sin k\lambda \cos k\lambda') .
\end{cases}
$$

(5.96)

According to (5.27) and (5.12), the deflection of the vertical is (here using different summation indexes to avoid confusion)

$$
\begin{cases}
\xi(\theta', \lambda') = \dfrac{GM}{R^2 \bar{\gamma}_0} \displaystyle\sum_{m=1}^{\infty} \sum_{l=0}^{m} \dfrac{\partial}{\partial \theta'} P_m^l(\cos\theta')(A_m^l \cos l\lambda' + B_m^l \sin l\lambda'), \\[4mm]
\eta(\theta', \lambda') = -\dfrac{GM}{R^2 \bar{\gamma}_0} \displaystyle\sum_{m=1}^{\infty} \sum_{l=0}^{m} \dfrac{l}{\sin\theta'} P_m^l(\cos\theta')(-A_m^l \sin l\lambda' + B_m^l \cos l\lambda'),
\end{cases}
$$
$$(5.97)$$

where we emphasize that A_m^l and B_m^l are the spherical harmonic coefficients of the disturbing potential T including degree 1, but not degree 0 as it does not contribute the deflection of the vertical. Substitute (5.96) and (5.97) into (5.93). We obtain, with $d\omega = \sin\theta' d\theta' d\lambda'$ and the trigonometric orthogonality relations (3.243) to (3.245),

$$
T(r, \theta, \lambda) = \frac{GM}{2R} \sum_{n=1}^{\infty} \sum_{m=1}^{\infty} \sum_{k=0}^{n} \alpha_n \left(\frac{R}{r}\right)^{n+1} \frac{(n-k)!}{(n+k)!} P_n^k(\cos\theta)
$$
$$
\times (A_m^k \cos k\lambda + B_m^k \sin k\lambda) \int_0^{\pi} \left\{ \frac{\partial}{\partial \theta'} P_n^k(\cos\theta') \frac{\partial}{\partial \theta'} P_m^k(\cos\theta') \right.
$$
$$
\left. + \frac{k^2}{\sin^2\theta'} P_n^k(\cos\theta') P_m^k(\cos\theta') \right\} \sin\theta' d\theta', \tag{5.98}
$$

where the appearance of the integral was predicted while guessing (5.93). By applying the orthogonality relation of the Legendre function (3.227) to it, we obtain

$$
T(r, \theta, \lambda) = \frac{GM}{R} \sum_{n=1}^{\infty} \sum_{k=0}^{n} \alpha_n \frac{n(n+1)}{2n+1} \left(\frac{R}{r}\right)^{n+1} P_n^k(\cos\theta)(A_n^k \cos k\lambda + B_n^k \sin k\lambda). \tag{5.99}
$$

By comparing this expression with (5.27) with the degree 1 term added in, we obtain the coefficient α_n

$$
\alpha_n = \frac{2n+1}{n(n+1)}. \tag{5.100}
$$

Hence,

$$
C(r, \psi) = \sum_{n=1}^{\infty} \frac{2n+1}{n(n+1)} \left(\frac{R}{r}\right)^{n+1} P_n(\cos\psi). \tag{5.101}
$$

We mention again that as the degree 0 term does not contribute to the deflection of the vertical, it cannot be determined from data of the deflection of the vertical.

In the following, we first derive a closed-form formula of $C(r, \psi)$ and then develop (5.93) to a more explicit form for computation.

For brevity, we substitute $x = R/r$ and write (5.101) as

$$
C(r, \psi) = \sum_{n=1}^{\infty} \frac{2n+1}{n(n+1)} x^{n+1} P_n(\cos \psi)
$$

$$
= \sum_{n=1}^{\infty} \frac{1}{n+1} x^{n+1} P_n(\cos \psi) + \sum_{n=1}^{\infty} \frac{1}{n} x^{n+1} P_n(\cos \psi). \quad (5.102)
$$

The first term has already been given in (5.85) with the degree 0 term as extra. Here we move the degree 0 term to the right-hand side to obtain

$$
\sum_{n=1}^{\infty} \frac{1}{n+1} x^{n+1} P_n(\cos \psi) = \ln \frac{x - \cos \psi + (1 + x^2 - 2x \cos \psi)^{1/2}}{1 - \cos \psi} - x.
$$

$$
(5.103)
$$

The second term can be written as

$$
\sum_{n=1}^{\infty} \frac{1}{n} x^{n+1} P_n(\cos \psi) = x \int_0^x \left[\sum_{n=1}^{\infty} x^{n-1} P_n(\cos \psi) \right] dx. \quad (5.104)
$$

According to the formula $(1 + x^2 - 2x \cos \psi)^{-1/2} = \sum_{n=0}^{\infty} x^n P_n(\cos \psi)$, it can be written as

$$
\sum_{n=1}^{\infty} \frac{1}{n} x^{n+1} P_n(\cos \psi) = x \int_0^x \frac{(1 + x^2 - 2x \cos \psi)^{-1/2} - 1}{x} dx
$$

$$
= x \left[\int_0^x \frac{1}{x(1 + x^2 - 2x \cos \psi)^{1/2}} dx - \int_0^x \frac{1}{x} dx \right].
$$

$$
(5.105)
$$

We can find, from a table of integrals,

$$
\int \frac{1}{x(1 + x^2 - 2x \cos \psi)^{1/2}} dx
$$

$$
= -\ln \frac{2(1 + x^2 - 2x \cos \psi)^{1/2} - 2x \cos \psi + 2}{x} + C, \quad (5.106)
$$

where C is a constant of integral. Substitute it into (5.105). We obtain

$$\sum_{n=1}^{\infty} \frac{1}{n} x^{n+1} P_n(\cos \psi) = x \left[-\ln \frac{2(1 + x^2 - 2x \cos \psi)^{1/2} - 2x \cos \psi + 2}{x} - \ln x \right]_0^x$$

$$= x \left\{ -\ln[2(1 + x^2 - 2x \cos \psi)^{1/2} - 2x \cos \psi + 2] \right\}_0^x$$

$$= x \{ -\ln[2(1 + x^2 - 2x \cos \psi)^{1/2} - 2x \cos \psi + 2] + (\ln 4) \}$$

$$= -x \ln \frac{(1 + x^2 - 2x \cos \psi)^{1/2} - x \cos \psi + 1}{2}.$$

$$(5.107)$$

Substitute (5.103) and (5.107) into (5.102). We obtain

$$C(r, \psi) = \ln \frac{x - \cos \psi + (1 + x^2 - 2x \cos \psi)^{1/2}}{1 - \cos \psi}$$

$$- x \left[1 + \ln \frac{(1 + x^2 - 2x \cos \psi)^{1/2} - x \cos \psi + 1}{2} \right]$$

$$= \ln \frac{R - r \cos \psi + l}{r(1 - \cos \psi)} - \frac{R}{r} \left[1 + \ln \frac{l - R \cos \psi + r}{2r} \right]. \qquad (5.108)$$

A more practical expression of T can be obtained by substituting (5.101) or (5.108) into (5.93)

$$T(r, \theta, \lambda) = \frac{R \bar{\gamma}_0}{4\pi} \int_\omega \frac{\partial}{\partial \psi} C(r, \psi) \left[\xi(\theta', \lambda') \frac{\partial \psi}{\partial \theta'} - \frac{1}{\sin \theta'} \eta(\theta', \lambda') \frac{\partial \psi}{\partial \lambda'} \right] d\omega.$$

$$(5.109)$$

To find an explicit expression of $\partial \psi / \partial \theta'$ and $\partial \psi / \partial \lambda'$, we define A' as shown in Fig. 5.8, where N is the north pole, P_0 the projection of P on the geoid approximated as a sphere, and M the point on the sphere corresponding to $d\omega$. We see that A' is the azimuth of P_0 with respect to M. We have

$$-\sin \psi \sin A' = \sin \theta \sin(\lambda' - \lambda), \qquad (5.110)$$

$$\sin \psi \cos A' = \cos \theta \sin \theta' - \sin \theta \cos \theta' \cos(\lambda' - \lambda), \qquad (5.111)$$

$$\cos \psi = \cos \theta' \cos \theta + \sin \theta' \sin \theta \cos(\lambda' - \lambda). \qquad (5.112)$$

Take the derivatives of both sides of (5.55) with respect to θ' and λ'. We obtain

$$\begin{cases} -\sin \psi \dfrac{\partial \psi}{\partial \theta'} = -\sin \theta' \cos \theta + \cos \theta' \sin \theta \cos(\lambda' - \lambda), \\ -\sin \psi \dfrac{\partial \psi}{\partial \lambda'} = -\sin \theta' \sin \theta \sin(\lambda' - \lambda). \end{cases} \qquad (5.113)$$

Fig. 5.8 Definition of azimuth A'

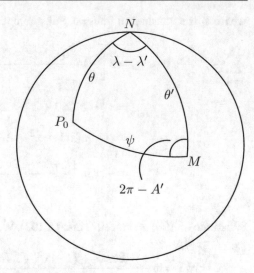

By comparing these with (5.53) and (5.54), we obtain

$$\frac{\partial \psi}{\partial \theta'} = \cos A', \qquad \frac{\partial \psi}{\partial \lambda'} = -\sin \theta' \sin A'. \qquad (5.114)$$

The azimuth A' can be computed from

$$\tan A' = \frac{-\sin \theta \sin(\lambda' - \lambda)}{\cos \theta \sin \theta' - \sin \theta \cos \theta' \cos(\lambda' - \lambda)}, \qquad (5.115)$$

where the numerator and denominator have the same signs as $\sin A'$ and $\cos A'$, respectively. Substitute (5.114) into (5.109). We obtain

$$T(r, \theta, \lambda) = \frac{R\bar{\gamma}_0}{4\pi} \int_\omega C'(r, \psi) \left[\xi(\theta', \lambda') \cos A' + \eta(\theta', \lambda') \sin A' \right] d\omega, \qquad (5.116)$$

where $C'(r, \psi)$ is used as a short form of $(\partial / \partial \psi) C(r, \psi)$.

The rest is to further formulate $C'(r, \psi)$. The formula in terms of spherical harmonic series can be readily derived based on (5.101). Here we provide the formula in closed-form, which is derived based on (5.108) by using (5.60),

$$C'(r, \psi) = \frac{\partial}{\partial \psi} C(r, \psi)$$

$$= \frac{\sin \psi [(r - R)l + Rr(1 - \cos \psi) - l^2]}{l(1 - \cos \psi)(R - r \cos \psi + l)} - \frac{R^2 \sin \psi (r + l)}{rl(l - R \cos \psi + r)}. \qquad (5.117)$$

With the expression of T, we can derive the formula for the gravity disturbance. The procedure is similar to the case of Stokes' boundary value problem in Sect. 5.2. Here we derive the formula of δg_r and leave case of δg_θ and δg_λ to the readers. We have

$$
\begin{aligned}
\delta g_r(r, \theta, \lambda) &= \frac{\partial}{\partial r} T(\theta, \lambda) \\
&= \frac{\bar{\gamma}_0}{4\pi} \int_\omega \left[R \frac{\partial}{\partial r} C'(r, \psi) \right] \left[\xi(\theta', \lambda') \cos A' + \eta(\theta', \lambda') \sin A' \right] d\omega ,
\end{aligned}
$$

$$(5.118)$$

where the dimensionless kernel, i.e., the term in the first pair of brackets, can be obtained in the form of spherical harmonic series based on (5.101) or in closed-form based on (5.108). Here we provide the result in closed-form

$$
\begin{aligned}
R\frac{\partial}{\partial r} & C'(r, \psi) \\
&= \frac{R \sin \psi [l^2 + (r - R - 2l)(r - R \cos \psi) + Rl(1 - \cos \psi)]}{l^2(1 - \cos \psi)(R - r \cos \psi + l)} \\
& - \frac{\left[\begin{array}{l} R \sin \psi [(r - R)l + Rr(1 - \cos \psi) - l^2] \\ \times [(r - R \cos \psi)(R - r \cos \psi + 2l) - l^2 \cos \psi] \end{array} \right]}{l^3(1 - \cos \psi)(R - r \cos \psi + l)^2} \\
& + \frac{R^3 \sin \psi}{l^3} + \frac{R^3 \sin \psi (r + l)}{r^2 l (l - R \cos \psi + r)} ,
\end{aligned}
$$

$$(5.119)$$

where (5.60) is used in the derivation.

With (5.116) to (5.119), we can derive the formulae of the geoidal height, the gravity disturbance on the geoid, and the gravity anomaly. The geoidal height is obtained by setting $r = R$ in (5.116) and (5.117) and then dividing by $\bar{\gamma}_0$

$$
N(\theta, \lambda) = \frac{R}{4\pi} \int_\omega C'(\psi) \left[\xi(\theta', \lambda') \cos A' + \eta(\theta', \lambda') \sin A' \right] d\omega ,
$$

$$(5.120)$$

where $C'(\psi)$ is the degenerate form of $C'(r, \psi)$ when $r = R$, which can be straightforwardly derived,

$$
C'(\psi) = - \cot \frac{\psi}{2} .
$$

$$(5.121)$$

The gravity disturbance is obtained by setting $r = R$ in (5.118) and (5.119)

$$\delta g(\theta, \lambda) = -\delta g_r(R, \theta, \lambda)$$

$$= \frac{\bar{\gamma_0}}{4\pi} \int_\omega \left[-R \frac{\partial}{\partial r} C'(r, \psi)_R \right] \left[\xi(\theta', \lambda') \cos A' + \eta(\theta', \lambda') \sin A' \right] d\omega,$$

$$(5.122)$$

where $R(\partial/\partial r)C'(r, \psi)_R$ is the degenerate form of $R(\partial/\partial r)C'(r, \psi)$ when $r = R$, which can also be straightforwardly derived,

$$-R \frac{\partial}{\partial r} C'(r, \psi)_R = -\frac{\cos(\psi/2)}{2\sin^2(\psi/2)} - \frac{\cos(\psi/2)(1 + 2\sin(\psi/2))}{2\sin(\psi/2)[1 + \sin(\psi/2)]}. \qquad (5.123)$$

Taking reference to Bruns' formula (5.7) and the fundamental gravimetric equation (5.26), the gravity anomaly can be obtained using the formulae of geoidal height and gravity disturbance

$$\Delta g(\theta, \lambda) = \frac{\bar{\gamma_0}}{4\pi} \int_\omega H'(\psi) \left[\xi(\theta', \lambda') \cos A' + \eta(\theta', \lambda') \sin A' \right] d\omega, \qquad (5.124)$$

which is the inverse Vening Meinesz formula, where

$$H'(\psi) = -2C'(\psi) - R \frac{\partial}{\partial r} C'(r, \psi)_R$$

$$= -\frac{\cos(\psi/2)}{2\sin^2(\psi/2)} + \frac{\cos(\psi/2)[3 + 2\sin(\psi/2)]}{2\sin(\psi/2)[1 + \sin(\psi/2)]}. \qquad (5.125)$$

The degree 0 term has already been excluded in the formulation, as the deflection of the vertical does not include its effect. The degree 1 term is $C_1(r, \psi) = (3/2)(R/r)^2 \cos\psi$, $C_1'(r, \psi) = -(3/2)(R/r)^2 \sin\psi$, $R(\partial/\partial r)C_1'(r, \psi) = 3(R/r)^3 \sin\psi$, $C_1'(\psi) = -(3/2)\sin\psi$, $R(\partial/\partial r)C_1'(r, \psi)_R = 3\sin\psi$, and $H_1'(\psi) = 0$. The vanishing of the last one is expected, as Δg does not include the degree 1 term at all. These can be subtracted from the corresponding functions for removing the degree 1 term in the formulation.

We refer to (5.120) and (5.124) as Hwang's formulae and the kernels $C'(\psi)$ and $H'(\psi)$ Hwang functions, to credit the author who first derived them (Hwang, 1998). Actually this is the last piece of the puzzle of the general formulation for the conversion from any kind of data to all others of the disturbing gravity field assuming no mass exists above the geoid and approximating the geoid as a sphere.

5.3.3 Formulation When the Geoidal Height Is Known

We first derive the integral formula for computing gravity disturbance. Based on the characteristic of the problem, we assume that V is a harmonic function outside a sphere S of radius R and derive the formula for computing $\partial V/\partial r$ on S when V on S is known.

We start from the Poisson integral (3.388). Take the derivative of (3.388) with respect to r. We obtain

$$\frac{\partial}{\partial r} V(r, \theta, \lambda) = \frac{1}{4\pi} \int_\omega V(\theta', \lambda') \frac{\partial}{\partial r} \left[\frac{R(r^2 - R^2)}{l^3} \right] d\omega. \tag{5.126}$$

Apply it to the particular harmonic function $V = R/r$, $\partial V/\partial r = -R/r^2$. We obtain

$$-\frac{R}{r^2} = \frac{1}{4\pi} \int_\omega \frac{\partial}{\partial r} \left[\frac{R(r^2 - R^2)}{l^3} \right] d\omega. \tag{5.127}$$

Multiply (5.127) with $V(\theta, \lambda) = V(R, \theta, \lambda)$, and then subtract the result from (5.126). We obtain

$$\frac{\partial}{\partial r} V(r, \theta, \lambda) = -\frac{R}{r^2} V(\theta, \lambda) + \frac{1}{4\pi} \int_\omega [V(\theta', \lambda') - V(\theta, \lambda)] \frac{\partial}{\partial r} \left[\frac{R(r^2 - R^2)}{l^3} \right] d\omega, \tag{5.128}$$

where the derivative in the integrand can be readily derived out to be

$$\frac{\partial}{\partial r} \left[\frac{R(r^2 - R^2)}{l^3} \right] = \frac{R(5rR^2 - r^3 - r^2 R \cos \psi - 3R^3 \cos \psi)}{l^5}. \tag{5.129}$$

On the sphere S of radius R, we set $r = R$ to obtain

$$l = l_0 = R(2 - 2 \cos \psi)^{1/2} = 2R \sin \frac{\psi}{2}, \tag{5.130}$$

where l_0 is used to denote l when P is on the sphere. In this case, (5.129) degenerates to

$$\frac{\partial}{\partial r} \left[\frac{R(r^2 - R^2)}{l^3} \right]_R = \frac{2R^2}{l_0^3}. \tag{5.131}$$

With these two formulae, we obtain, by setting $r = R$ in (5.128),

$$\frac{\partial}{\partial r} V(r, \theta, \lambda)_R = -\frac{1}{R} V(\theta, \lambda) + \frac{1}{4\pi} \int_\omega [V(\theta', \lambda') - V(\theta, \lambda)] \frac{2R^2}{l_0^3} d\omega. \tag{5.132}$$

This is the formula to compute the radial derivative of a harmonic function on a sphere using its value on the sphere. The appearance of $V(\theta', \lambda') - V(\theta, \lambda)$ in the integrand at the right-hand side is arranged intentionally to favorite numerical evaluation of the integral.

Using (5.132) to compute gravity disturbance from geoidal height is straightforward. We just have to replace V by T and then apply Bruns' formula. Its magnitude in the downward direction is

$$\delta g(\theta, \lambda) = \frac{\bar{\gamma_0}}{R} N(\theta, \lambda) - \frac{\bar{\gamma_0}}{4\pi} \int_\omega [N(\theta', \lambda') - N(\theta, \lambda)] \frac{2R^2}{l_0^3} d\omega, \qquad (5.133)$$

which is referred to as inverse Hotine formula.

In order to apply (5.132) to compute gravity anomaly, we replace V by T and then substitute it into the fundamental gravimetric equation (5.26) to obtain

$$\Delta g(\theta, \lambda) = -\frac{1}{R} T(\theta, \lambda) - \frac{1}{4\pi} \int_\omega [T(\theta', \lambda') - T(\theta, \lambda)] \frac{2R^2}{l_0^3} d\omega. \qquad (5.134)$$

In terms of geoidal height, this formula can be written as

$$\Delta g(\theta, \lambda) = -\frac{\bar{\gamma_0}}{R} N(\theta, \lambda) - \frac{\bar{\gamma_0}}{4\pi} \int_\omega [N(\theta', \lambda') - N(\theta, \lambda)] \frac{2R^2}{l_0^3} d\omega, \qquad (5.135)$$

which is referred to as inverse Stokes formula. We have intentionally included its alternative form (5.134) for reference in a later chapter.

The removal of degrees 0 and 1 terms in the formulation is more complicated in this case. The formulations relied on the kernel of the Poisson integral, $G(r, \psi) = R(r^2 - R^2)/l^3$, with its spherical harmonic series defined in (3.387), where the degrees 0 and 1 terms are $G_0(r, \psi) = R/r$ and $G_1(r, \psi) = 3(R^2/r^2) \cos \psi$, and hence $(\partial/\partial r) G_0(r, \psi) = -R/r^2$ and $(\partial/\partial r) G_1(r, \psi) = -6(R^2/r^3) \cos \psi$. It is not just a matter to take these away from the kernels. It is alright to remove them in (5.126) to exclude the degrees 0 and 1 terms. However, (5.127) is the degree 0 part of (5.126), and removing the degree 0 term in the kernel of the integral makes the integral to vanish. This could be explained by separating the degree 0 with the rest part in (5.127)

$$-\frac{R}{r^2} = \frac{1}{4\pi} \int_\omega \frac{\partial}{\partial r} \left[\frac{R(r^2 - R^2)}{l^3} - G_0(r, \psi) \right] d\omega + \frac{1}{4\pi} \int_\omega \frac{\partial}{\partial r} G_0(r, \psi) d\omega, \qquad (5.136)$$

where the second term at right-hand side is equal to the left-hand side. Hence, in (5.128), the first term at the right-hand side should be omitted if the degree 0 term is removed from the kernel of the integral, $(\partial/\partial r)[R(r^2 - R^2)/l^3]$. The same is for (5.132) and (5.133) with the kernel $2R^2/l_0^3$. As a result, in (5.134) and (5.135), a factor 2 should be added to the first term at the right-hand side when the degree 0 term is removed from the kernels of the integral, $2R^2/l_0^3$. In all these formulae, removing the degree 1 term requires only to remove the degree 1 term of the kernel of the integral. We mention again that the solution is unique. If the effects of the degrees 0 and 1 terms are removed in the data of N, their influence to the result would be missing too, no matter whether the degree 0 or 1 terms are removed in the formulation. Hence, it is not really necessary to remove these terms from the formulation, unless their effects are present in the data and their effects in the result are to be excluded.

5.4 Inclusion of Errors of the Reference Earth Ellipsoid in Stokes' Theory

5.4.1 Fundamental Relations

We have made the following assumptions to the reference Earth ellipsoid:

1. Its value of GM is equal to that of the Earth.
2. Its center coincides with the center of mass of the Earth, and serves as the origin of the coordinate system used to express positions.
3. It rotates around its minor axis, which coincides with the Earth's rotation axis and the rate is the same as that of the Earth.
4. Its gravity potential, i.e., the normal potential, on its surface, U_0, is equal to the gravity potential of the Earth on the geoid, W_0.
5. Its dynamical form factor is equal to that of the Earth.

In the formulation of Stokes's boundary value problem in Sect. 5.1, only conditions (3) and (4) are enforced. In the solution of the problem in Sect. 5.2, condition (1) is enforced by expressing T in the form (5.27). Condition (2) is enforced due to the fact that Δg does not have degree 1 at all by definition, as can be inferred from (5.28), which is also the reason why the degree 1 is omitted in the expression of T in (5.27).

Here we reformulate Stokes' boundary value problem allowing errors in GM and W_0, and also assuming the center of the reference Earth ellipsoid, which serves as coordinate origin, does not coincide with the center of mass of the Earth.

First, we assume that there is an error in GM such that its true value is $GM + \delta(GM)$, and the coordinates of the center of mass of the Earth are (X_0, Y_0, Z_0). The expression of the total gravity potential of the Earth is

$$
W = \left[\frac{GM}{r} + \frac{\delta(GM)}{r} \right] \sum_{n=0}^{\infty} \sum_{k=0}^{n} \left(\frac{R}{r} \right)^n \left(C_n^k \cos k\lambda + S_n^k \sin k\lambda \right) P_n^k (\cos \theta)
$$

$$
+ \frac{1}{2} \Omega^2 r^2 \sin^2 \theta, \tag{5.137}
$$

where the degrees 0 and 1 coefficients are

$$C_0 = 1, \quad C_1 = \frac{Z_0}{R}, \quad C_1^1 = \frac{X_0}{R}, \quad S_1^1 = \frac{Y_0}{R}. \tag{5.138}$$

The normal potential is based on the known value of GM. Hence,

$$U = \frac{GM}{r} \left[1 - \sum_{n=0}^{\infty} J_{2n} \left(\frac{R}{r} \right)^n P_{2n}(\cos\theta) \right] + \frac{1}{2}\Omega^2 r^2 \sin^2\theta. \tag{5.139}$$

By definition, the disturbing potential is

$$T = \frac{\delta(GM)}{r} + \frac{GM}{r} \sum_{n=1}^{\infty} \sum_{k=0}^{n} \left(\frac{R}{r} \right)^n \left(A_n^k \cos k\lambda + B_n^k \sin k\lambda \right) P_n^k(\cos\theta), \tag{5.140}$$

where the degree 1 coefficients are

$$A_1 = \frac{Z_0}{R}, \quad A_1^1 = \frac{X_0}{R}, \quad B_1^1 = \frac{Y_0}{R}. \tag{5.141}$$

We have neglected the products between $\delta(GM)$ and X_0, Y_0, and Z_0, which are second order small quantities.

We further assume that there is an error in W_0, such that its true value is $W_0 + \delta W_0$. In this case, U_0 is chosen to be the value known, i.e., $U_0 = W_0$, and the geoidal height can be obtained from (5.5) to be

$$N = -\frac{\delta W_0}{\gamma_0} + \frac{T_0}{\gamma_0}. \tag{5.142}$$

Its spherical approximation is

$$N = -\frac{\delta W_0}{\bar{\gamma}_0} + \frac{T_0}{\bar{\gamma}_0}. \tag{5.143}$$

By substituting (5.142) into (5.21) and referencing the procedure to derive (5.26), we obtain the fundamental gravimetric equation under spherical approximation for the present case

$$\Delta g = 2\frac{\delta W_0}{R} - 2\frac{T_0}{R} - \left(\frac{\partial T}{\partial r} \right)_R. \tag{5.144}$$

As we have already mentioned, the geoid defining the reference Earth ellipsoid may be different from the one serving as the vertical datum. If this is the case, the formulation of this subsection and the next may apply by understanding δW_0 as the difference between the gravity potentials of these two geoids.

5.4.2   Solution of Disturbing Potential, Geoidal Height, and Deflection of the Vertical

The determination of the disturbing potential using data of Δg can be formulated similar to the solution of Stokes' boundary value problem in Sect. 5.2. The substitution of (5.140) into (5.144) yields a relation between Δg and the spherical harmonic coefficients of T

$$\Delta g = 2\frac{\delta W_0}{R} - \frac{\delta(GM)}{R^2} + \frac{GM}{R^2}\sum_{n=1}^{\infty}\sum_{k=0}^{n}(n-1)\left(A_n^k\cos k\lambda + B_n^k\sin k\lambda\right)P_n^k(\cos\theta).$$

$$(5.145)$$

The appearance of the factor $n-1$ implies that the degree 1 term is excluded in Δg. Hence, the spherical harmonic series of Δg is

$$\Delta g = \Delta g_0 + \sum_{n=2}^{\infty}\sum_{k=0}^{n}\left(a_n^k\cos k\lambda + a_n^k\sin k\lambda\right)P_n^k(\cos\theta),\qquad(5.146)$$

where a_n^k and b_n^k are the same as in (5.30), and

$$\Delta g_0 = \frac{1}{4\pi}\int_{\omega}\Delta g\,d\omega.\qquad(5.147)$$

By comparing (5.145) and (5.146), we obtain the same formulae as (5.32) for A_n^k and B_n^k, and

$$2\frac{\delta W_0}{R} - \frac{\delta(GM)}{R^2} = \Delta g_0.\qquad(5.148)$$

We thus obtain T

$$T = \frac{\delta(GM)}{r} + \frac{GM}{r^2}(Z_0\cos\theta + X_0\sin\theta\cos\lambda + Y_0\sin\theta\sin\lambda)$$

$$+ \frac{GM}{r}\sum_{n=2}^{\infty}\sum_{k=0}^{n}\left(\frac{R}{r}\right)^n\left(A_n^k\cos k\lambda + B_n^k\sin k\lambda\right)P_n^k(\cos\theta).\qquad(5.149)$$

The sum can be expressed as an integral, as proven in Sect. 5.2. Hence,

$$T = \frac{\delta(GM)}{r} + \frac{GM}{r^2}(Z_0 \cos\theta + X_0 \sin\theta \cos\lambda + Y_0 \sin\theta \sin\lambda)$$

$$+ \frac{R}{4\pi}\int_\omega \Delta g S(r, \psi) d\omega . \tag{5.150}$$

We mention that the degrees 0 and 1 terms are excluded in the integral according to the definition of $S(r, \psi)$.

Expressions of the geoidal height and deflection of the vertical are straightforward to derive. Here we write out their integral forms with the approximation $\bar{\gamma}_0 = GM/R^2$

$$N = -\frac{\delta W_0}{\bar{\gamma}_0} + \frac{\delta(GM)}{R\bar{\gamma}_0} + (Z_0 \cos\theta + X_0 \sin\theta \cos\lambda + Y_0 \sin\theta \sin\lambda)$$

$$+ \frac{R}{4\pi\bar{\gamma}_0}\int_\omega \Delta g S(\psi) d\omega ; \tag{5.151}$$

$$\begin{cases} \xi = \frac{1}{R}(-Z_0 \sin\theta + X_0 \cos\theta \cos\lambda + Y_0 \cos\theta \sin\lambda) \\ \quad + \frac{1}{4\pi\bar{\gamma}_0}\int_\omega \Delta g \frac{d}{d\psi}S(\psi)\cos A d\omega , \\ \eta = -\frac{1}{R\sin\theta}(-X_0 \sin\theta \sin\lambda + Y_0 \sin\theta \cos\lambda) + \frac{1}{4\pi\bar{\gamma}_0}\int_\omega \Delta g \frac{d}{d\psi}S(\psi)\sin A d\omega . \end{cases} \tag{5.152}$$

The formulation in terms of spherical harmonic series is left to the readers.

The generalized Stokes' formula (5.151) can be written in two alternative forms. The degree 0 term of N is

$$N_0 = -\frac{\delta W_0}{\bar{\gamma}_0} + \frac{\delta(GM)}{R\bar{\gamma}_0} . \tag{5.153}$$

With (5.148) and (5.147), it can be written in any of the two forms:

$$N_0 = \frac{\delta W_0}{\bar{\gamma}_0} - \frac{R}{\bar{\gamma}_0}\Delta g_0 = \frac{\delta W_0}{\bar{\gamma}_0} - \frac{R}{4\pi\gamma_0}\int_\omega \Delta g d\omega , \tag{5.154}$$

$$N_0 = \frac{\delta(GM)}{2R\bar{\gamma}_0} - \frac{R}{2\bar{\gamma}_0}\Delta g_0 = \frac{\delta(GM)}{2R\bar{\gamma}_0} - \frac{R}{8\pi\gamma_0}\int_\omega \Delta g d\omega . \tag{5.155}$$

Accordingly, the two alternative expressions of N are

$$
N = \frac{\delta W_0}{\bar{\gamma}_0} + (Z_0 \cos\theta + X_0 \sin\theta \cos\lambda + Y_0 \sin\theta \sin\lambda)
$$

$$
+ \frac{R}{4\pi\bar{\gamma}_0} \int_\omega \Delta g[S(\psi) - 1]d\omega, \tag{5.156}
$$

$$
N = \frac{\delta(GM)}{2R\bar{\gamma}_0} + (Z_0 \cos\theta + X_0 \sin\theta \cos\lambda + Y_0 \sin\theta \sin\lambda)
$$

$$
+ \frac{R}{4\pi\bar{\gamma}_0} \int_\omega \Delta g \left[S(\psi) - \frac{1}{2} \right] d\omega. \tag{5.157}
$$

5.4.3 Relation Between the Reference Earth Ellipsoid and the Geoid

Relation Between the Reference Earth Ellipsoid and the Geoid
In the present space geodesy era, GM and J_2 are determined very accurately by observing satellite orbits, and Ω is determined very accurately by observing very distant celestial bodies. Furthermore, the use of GNSS provides very accurate coordinates with respect to the center of mass of the Earth; hence, by understanding in the reverse sense, the center of mass of the Earth is very accurately determined. Based on these, we show an example to determine a reference Earth ellipsoid assuming that GM, J_2, Ω, and the location of the center of mass of the Earth are accurately known. We assume to have an approximate reference Earth ellipsoid. The degree 0 term of the geoidal height with respect to this approximate reference Earth ellipsoid is, according to (5.153), (5.147), and (5.148),

$$
N_0 = -\frac{\delta W_0}{\bar{\gamma}_0} = -\frac{R}{8\pi\bar{\gamma}_0} \int_\omega \Delta g\, d\omega, \tag{5.158}
$$

which can also be obtained from (5.155) directly. This formula implies that the degree 0 term of the geoidal height, N_0, and so is the error of the gravity potential on the geoid, δW_0, can be determined by gravity measurements. The gravity potential on the geoid can be corrected to $W_0 + \delta W_0$.

In the following, we derive a general relation between the reference Earth ellipsoid and the geoid. We use the spherical harmonic series of the geoidal height, which is straightforward to obtain from (5.149) and (5.143). With the approximation $\bar{\gamma}_0 = GM/R^2$, we have

$$
N = -\frac{\delta W_0}{\bar{\gamma}_0} + \frac{\delta(GM)}{R\bar{\gamma}_0} + (Z_0 \cos\theta + X_0 \sin\theta \cos\lambda + Y_0 \sin\theta \sin\lambda)
$$

$$
+ R \sum_{n=2}^{\infty} \sum_{k=0}^{n} \left(A_n^k \cos k\lambda + B_n^k \sin k\lambda \right) P_n^k(\cos\theta). \tag{5.159}
$$

According to the orthogonality of the spherical harmonics, (3.247) to (3.249), we can obtain

$$\frac{1}{4\pi} \int_\omega N^2 d\omega = \left[-\frac{\delta W_0}{\bar{\gamma}_0} + \frac{\delta(GM)}{R\bar{\gamma}_0} \right]^2 + \frac{1}{3}(Z_0^2 + X_0^2 + Y_0^2)$$

$$+ R^2 \sum_{n=2}^{\infty} \sum_{k=0}^{n} \frac{1+\delta_k}{2(2n+1)} \frac{(n+k)!}{(n-k)!} \left[(A_n^k)^2 + (B_n^k)^2 \right], \qquad (5.160)$$

where we set the meaningless coefficient $B_n^0 = 0$. We now update the position of the center and the geometrical parameters a and α of the reference Earth ellipsoid so that

$$\frac{1}{4\pi} \int_\omega N^2 d\omega = \min \qquad (5.161)$$

for the updated reference Earth ellipsoid. As the three terms at the right-hand side of (5.160) are independent of one another, the result implies that

$$-\frac{\delta W_0}{\bar{\gamma}_0} + \frac{\delta(GM)}{R\bar{\gamma}_0} = 0, \qquad (5.162)$$

$$Z_0 = X_0 = Y_0 = 0. \qquad (5.163)$$

As $J_2, J_4, J_6, \cdots$ are related, $A_2, A_4, A_6, \cdots$ are also related. Hence, the result implies only $A_2 \approx 0$, meaning that $J_2 = C_2$ as defined in (4.106) is achieved only approximately. We thus conclude that if N can be measured all over the globe, a reference Earth ellipsoid can be determined based on (5.161) alone, provided that GM is determined by observing satellite orbit very accurately and that Ω and the direction of Earth's rotation axis are also determined accurately.

We note that we have assumed the exact value of the gravity potential on the geoid to be $W_0 + \delta W_0$, where W_0 is the known value, and δW_0 is its error, and defined the reference Earth ellipsoid by setting $U_0 = W_0$. The result of δW_0 is to be used as a correction to W_0 or U_0.

The Geoid Best Fitting the Global Ocean Surface

Nowadays, the mean ocean surface height above the reference Earth ellipsoid, h, is very accurately measured by satellite altimetry. We know that the mean ocean surface is very close to the geoid, $h = N + d$, where the small quantity d is the height of the mean ocean surface above the geoid and is referred to as mean dynamic ocean surface topography, which is a very important parameter for modeling global ocean circulations. On the other hand, as already mentioned, GM, J_2, Ω, and the direction of Earth's rotation axis can be assumed to be error free, and $Z_0 = X_0 = Y_0 = 0$ can also be assumed. Furthermore, the gravitational potential is also determined using satellites very accurately up to a certain degree, say, n_{max}. Hence, we have the relation

$$d = h - N = h + \frac{\delta W_0}{\bar{\gamma}_0} - N_{2 \leq n \leq n_{max}} - N_{n > n_{max}} , \tag{5.164}$$

where

$$N_{2 \leq n \leq n_{max}} = R \sum_{n=2}^{n_{max}} \sum_{k=0}^{n} \left(A_n^k \cos k\lambda + B_n^k \sin k\lambda \right) P_n^k (\cos \theta) , \tag{5.165}$$

$$N_{n > n_{max}} = R \sum_{n=n_{max}+1}^{\infty} \sum_{k=0}^{n} \left(A_n^k \cos k\lambda + B_n^k \sin k\lambda \right) P_n^k (\cos \theta) . \tag{5.166}$$

These relations imply that the mean dynamic ocean surface topography can be partly obtained from satellite altimetry and satellite gravimetry data. We said partly because the result is limited by the value of n_{max} reached and the coverage of the data of h.

We now define a new geoid, which is the equipotential surface of gravity that best fits the global mean ocean surface, i.e., to minimize d^2 globally. However, as $N_{n > n_{max}}$ is not measured, it cannot be subtracted as appeared in the formula, meaning that the computation can be done only without it. This is equivalent to treat it as an error. Now we write (5.164) in the alternative form

$$\frac{\delta W_0}{\bar{\gamma}_0} + \left(h - N_{2 \leq n \leq n_{max}} \right) = d + N_{n > n_{max}} . \tag{5.167}$$

Although it is ideal to define the geoid by minimize d^2 globally, it is not practical as the oceans cover only a portion of the Earth's surface. The realizable compromise is to minimize $(d + N_{n > n_{max}})^2$ over the oceans. In the least squares sense, we have

$$\int_{\omega_{oceans}} \left[\frac{\delta W_0}{\bar{\gamma}_0} + \left(h - N_{2 \leq n \leq n_{max}} \right) \right]^2 d\omega = \min . \tag{5.168}$$

This criterion requires that data of h are available all over the oceans. It should be modified if this data requirement is not fulfilled. The result of δW_0 obtained is to be used to update the value of W_0 or U_0, so that the geoid defining the reference Earth ellipsoid be updated to the one best fitting the global mean ocean surface.

5.5   Some Characteristics of the Earth's Gravitational Field

Although we are not yet at the point of knowledge to compute a model of the gravitational field using data, we show a global model here for a general impression of how the Earth's gravitational field behaves. A global model of the gravitational potential is customarily referred to as an Earth gravitational model (EGM), which is provided in the form of a spherical harmonic series.

5.5.1 A Global Model of the Gravitational Potential

We chose to show an EGM named GGM05C[3] (Ries et al., 2016), which is computed by combining data from the Gravity Recovery and Climate Experiment (GRACE) satellite mission, the Gravity Field and Steady-State Ocean Circulation Explorer (GOCE) satellite mission, satellite altimetry, and ground gravimetry. The GRACE consists of two satellites flying one after another. Due to gravity changes along the trajectory, the distance between the two satellites changes, and the rate of change of this distance, which is referred to as range rate in the GRACE community, is measured very precisely. The gravitational potential is determined mainly from the range rate data, together with other data measured by on-board sensors, including particularly the acceleration due to non-gravitational forces measured by accelerometers, and the position and velocity of the satellites measured by GPS receivers. The main equipment on-board the GOCE is a gradiometer measuring gravity gradients in all directions, together with a set of accelerometers and a GPS receiver. The gravitational potential is determined mainly from the gravity gradient data. Satellite altimetry provides mean ocean surface height above the reference Earth ellipsoid, i.e., geodetic height. In order to convert geodetic height to geoidal height, a model of mean dynamic ocean surface topography is required. An alternative is to use the rate of change of geoidal height along the satellite's ground track as the deflection of the vertical for alleviating the influence of satellite orbit error. The geoidal height or deflection of the vertical is converted to gravity anomaly, which is the same type of data as that deduced from ground gravity measurements. The determination of the gravitational potential using observations of satellite gravimetry missions (GRACE and GOCE) is beyond the scope of this book. Here we just take the model as an example.

This model is complete up to degree and order $n_{\max} = 360$, with the reference length chosen to be $a = 6378136.3$ meters. All spherical harmonic coefficients are fully normalized, meaning that the gravitational potential is expressed in the form of (3.314), which is copied to here for convenience of reference

$$V = \frac{GM}{r}\left[1 + \sum_{n=2}^{n_{\max}}\left(\frac{a}{r}\right)^n \sum_{k=0}^{n}(\bar{C}_n^k \cos k\lambda + \bar{S}_n^k \sin k\lambda)\bar{P}_n^k(\cos\theta)\right]. \tag{5.169}$$

---

[3] As noted by the authors, these data are freely available under a Creative Commons Attribution 4.0 International License (CC BY 4.0).

In Table 5.1, we list only the coefficients up to degree and order 10, which are all multiplied by a factor 10^7. We see that the coefficient $\bar{C}_2 = -J_2/\sqrt{5}$ is mostly at least 3 orders of magnitude larger than the others.

A map of the geoidal height based on this model and the reference Earth ellipsoid GRS80 is shown in Fig. 5.9. The normal gravitational potential is computed rigorously according to (4.102) and (4.115) up to degree 20, which is as small as on the order of 1.6×10^{-25}, far below the magnitude of the coefficients listed in Table 5.1. It is then removed from the gravitational potential to obtain the disturbing potential. In the computation of the geoidal height using the disturbing potential, the reference Earth ellipsoid is approximated as a sphere, which would not bring an evident visible effect to such a small-scale map. The magnitude of the geoidal height is mostly below 40 meters, and it rarely exceeds 80 meters. Considering the size of the Earth, the geoid is really very close to an ellipsoid of revolution.

Table 5.1 Coefficients ($\times 10^7$) of the EGM GGM05C up to degree and order 10

n	k	$\bar{C}_n^k$	$\bar{S}_n^k$	n	k	$\bar{C}_n^k$	$\bar{S}_n^k$	n	k	$\bar{C}_n^k$	$\bar{S}_n^k$
2	0	$\bar{C}_2$	0	3	0	9.57165	0	4	0	5.39982	0
5	0	0.68650	0	6	0	−1.49976	0	7	0	0.90501	0
8	0	0.49478	0	9	0	0.28023	0	10	0	0.53345	0
2	1	−0.00310	0.01411	3	1	20.30447	2.48241	4	1	−5.36181	−4.73577
5	1	−0.62915	−0.94343	6	1	−0.75943	0.26526	7	1	2.80886	0.95160
8	1	0.23141	0.58905	9	1	1.42149	0.21416	10	1	0.83754	−1.31094
2	2	24.39373	−14.00294	3	2	9.04765	−6.19007	4	2	3.50492	6.62505
5	2	6.52059	−3.23343	6	2	0.48636	−3.73769	7	2	3.30396	0.93020
8	2	0.80020	0.65300	9	2	0.21412	−0.31680	10	2	−0.93978	−0.51265
3	3	7.21285	14.14400	4	3	9.90861	−2.00951	5	3	−4.51831	−2.14942
6	3	0.57251	0.08973	7	3	2.50463	−2.17097	8	3	−0.19359	−0.85942
9	3	−1.60602	−0.74250	10	3	−0.06991	−1.54128	4	4	−1.88492	3.08819
5	4	−2.95323	0.49811	6	4	−0.85997	−4.71427	7	4	−2.74986	−1.24051
8	4	−2.44345	0.69812	9	4	−0.09349	0.19899	10	4	−0.84459	−0.79024
5	5	1.74814	−6.69355	6	5	−2.67163	−5.36496	7	5	0.01647	0.17931
8	5	−0.25702	0.89206	9	5	−0.16315	−0.54044	10	5	−0.49286	−0.50617
6	6	0.09480	−2.37384	7	6	−3.58799	1.51793	8	6	−0.65971	3.08945
9	6	0.62784	2.22964	10	6	−0.37587	−0.79771	7	7	0.01523	0.24103
8	7	0.67257	0.74864	9	7	−1.17981	−0.96921	10	7	0.08257	−0.03049
8	8	−1.24033	1.20540	9	8	1.88132	−0.03002	10	8	0.40590	−0.91711
9	9	−0.47558	0.96877	10	9	1.25385	−0.37942	10	10	1.00423	−0.23864

$\bar{C}_2 = -4841.69457$ (after multiplication by 10^7)

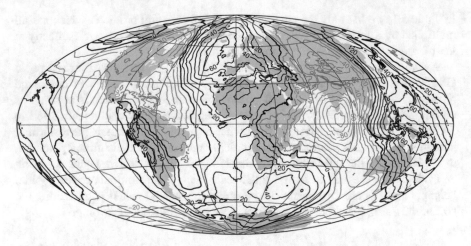

Fig. 5.9 The geoidal height based on GGM05C and GRS80. Thick black line: 0 geoidal height; black line: positive geoidal height; and gray line: negative geoidal height

5.5.2 Degree Power Spectrum of the Gravitational Potential and Geoidal Height

An immediate observation is that the larger higher degree coefficients are, the richer finer features are, as higher degree spherical harmonics represent variations in shorter space scale. The resolution of degree n spherical harmonics is $180°/n$ half wavelength on the Earth's surface, with $1°$ corresponding to $2\pi R/360 = 111$ km. This could be understood in the following ways:

1. At the equator, the spherical harmonics of degree n degenerates to a Fourier series of order n, meaning that the highest resolution is n cycles of the sine or cosine function from $0°$ to $360°$. The wavelength of each cycle is $360°/n$, and the half wavelength is $180°/n$.
2. According to the discussion in Sect. 3.4.4, where (3.256) and (3.258) represent the same function, we know that if a function is a degree n spherical harmonics in one coordinate system, it remains so in all coordinates systems with the same origin. In the Cartesian coordinate system $Oxyz$ associated with the spherical coordinate system $Or\theta\lambda$, the equator is just a great arc in the Oxy plane. By rotating the coordinate system, any great arc can be arranged to lie in the Oxy plane. Hence, the conclusion (1) holds for any great arc.

In order to understand how large the signal is represented by the spherical harmonic coefficients of each degree, we formulate

$$\frac{1}{4\pi} \int_\omega V^2 d\omega, \tag{5.170}$$

which is the mean square of the gravitational potential over a sphere. It can be understand as a measure of the overall magnitude of V over the sphere of radius r. We use the normalized spherical harmonic coefficients as appeared in (5.169). By substituting $\bar{P}_n^k(\cos\theta)$ by $P_n^k(\cos\theta)$ according to (3.313), we obtain an expression of V using fully normalized coefficients and un-normalized Legendre function, which we write in the form of

$$V = \sum_{n=0}^{n_{max}} V_n, \tag{5.171}$$

$$V_n = \frac{GM}{R} \sum_{n=0}^{n_{max}} \left(\frac{R}{r}\right)^{n+1} \sum_{k=0}^{n} \left[\frac{2(2n+1)}{1+\delta_k} \frac{(n-k)!}{(n+k)!}\right]^{1/2}$$

$$\times (\bar{C}_n^k \cos k\lambda + \bar{S}_n^k \sin k\lambda) P_n^k(\cos\theta), \tag{5.172}$$

where we followed the convention of this chapter to use the mean radius R as scale length. First, for V_n, according to the orthogonality relations (3.247) to (3.249), we have

$$\frac{1}{4\pi} \int_\omega (V_n)^2 d\omega = \left(\frac{GM}{R}\right)^2 \left(\frac{R}{r}\right)^{2n+2} \sum_{k=0}^{n} \left[\left(\bar{C}_n^k\right)^2 + \left(\bar{S}_n^k\right)^2\right]. \tag{5.173}$$

When rotating the coordinate system, the spherical harmonic coefficients of V_n would change, but the value of V_n remains the same. This implies that the right-hand side of (5.173) is independent of coordinate system, although the spherical harmonic coefficients do.

Second, substitute (5.171) and (5.172) into (5.170), and then use the orthogonality relations (3.247) to (3.249). We obtain

$$\frac{1}{4\pi} \int_\omega V^2 d\omega = \left(\frac{GM}{R}\right)^2 \sum_{n=0}^{n_{max}} \left(\frac{R}{r}\right)^{2n+2} \sum_{k=0}^{n} \left[\left(\bar{C}_n^k\right)^2 + \left(\bar{S}_n^k\right)^2\right]$$

$$= \sum_{n=0}^{n_{max}} \left(\frac{1}{4\pi} \int_\sigma (V_n)^2 d\sigma\right). \tag{5.174}$$

We see that there is no cross product among the coefficients of different degrees. Hence, if we understand the mean square as the power of the signal, it can be decomposed degree by degree according to (5.174), which is called degree variance power spectrum.

In the geodetic community, particular attention is given to

$$\left(\sigma_n^V\right)^2 = \sum_{k=0}^{n} \left[\left(\bar{C}_n^k\right)^2 + \left(\bar{S}_n^k\right)^2\right], \tag{5.175}$$

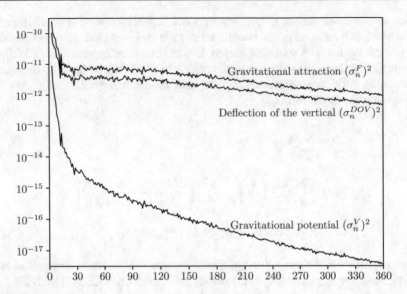

Fig. 5.10 Degree power spectral variance of the EGM GGM05C from degree 3 to degree 360

which is a degenerate form of (5.173) when $r = R$ omitting the constant factor $(GM/R)^2$. A graph of it is shown in Fig. 5.10 for the EGM GGM05C. We see that $(\sigma_n^V)^2$ decreases rapidly when n increases. An early fitting of a relation between $(\sigma_n)^2$ and n is (Kaula, 1966)

$$\left(\sigma_n^V\right)^2 \approx (2n+1)\left(\frac{10^{-5}}{n^2}\right)^2, \tag{5.176}$$

meaning that $\left(\sigma_n^V\right)^2$ is approximately inverse proportional to n^3, which is referred to as Kaula's rule. Later on, alternative forms have also been used, e.g., (Rapp, 1989),

$$\left(\sigma_n^V\right)^2 \approx \frac{\alpha}{n^\beta}. \tag{5.177}$$

The degree variance power spectrum can be defined for other quantities as well. For example, if we replace V by N as given in (5.159), we obtain the degree variance power spectrum of the geoidal height (considering only the $2 \leq n \leq n_{\max}$ part)

$$\frac{1}{4\pi}\int_\omega (N)^2 d\omega = \sum_{n=2}^{n_{\max}} \frac{1}{4\pi}\int_\omega (N_n)^2 d\omega, \tag{5.178}$$

$$\frac{1}{4\pi}\int_\omega (N_n)^2 d\omega = R^2 \sum_{k=0}^n \left[\left(\bar{A}_n^k\right)^2 + \left(\bar{B}_n^k\right)^2\right] = R^2 (\sigma_n^T)^2, \tag{5.179}$$

where $\bar{A}_n^k$ and $\bar{B}_n^k$ are the fully normalized spherical harmonic coefficients of the disturbing potential, which are practically indifferent from $\bar{C}_n^k$ and $\bar{S}_n^k$ beyond degree 4. The normalized coefficient of the normal potential, $\bar{J}_6$, is already $\sim 1\%$ of $\bar{C}_6$ in magnitude. Hence, no visible difference between the graphs of $\left(\sigma_n^V\right)^2$ and $\left(\sigma_n^T\right)^2$ can be expected. We mention that (5.178) and (5.179) have already been provided in (5.160) using un-normalized coefficients.

5.5.3 Degree Power Spectrum of the Gravitational Attraction and Deflection of the Vertical

For the gravitational attraction $\vec{F}$, a relevant measure of its overall magnitude over a sphere similar to (5.170) is

$$\frac{1}{4\pi} \int_\omega \vec{F} \cdot \vec{F}\, d\omega = \frac{1}{4\pi} \int_\omega F_r^2 d\omega + \frac{1}{4\pi} \int_\omega (F_\theta^2 + F_\lambda^2) d\omega. \tag{5.180}$$

We have intentionally separated it into two parts to formulate individually, since a degenerate form of the second term can be used as a measure of the overall magnitude of the deflection of the vertical over a sphere.

The result of degree variance power spectrum can be obtained by converting the expressions of $\vec{F}$, i.e., (3.303) to (3.305), into those expressed using fully normalized coefficients similar to (5.171) and (5.172) and then substituting into (5.180). As expected, no cross degree term would appear. Hence, we only write the per degree result. The formulation of the first term is almost identical to that of V. The result is

$$\frac{1}{4\pi} \int_\omega (F_{rn})^2 d\omega = \left(\frac{GM}{R^2}\right)^2 \left(\frac{R}{r}\right)^{2n+4} \left\{(n+1)^2 \sum_{k=0}^n \left[\left(\bar{C}_n^k\right)^2 + \left(\bar{S}_n^k\right)^2\right]\right\}. \tag{5.181}$$

The formulation of the second term is similar to that from (5.98) to (5.99). The result is

$$\frac{1}{4\pi} \int_\omega \left[(F_{\theta n})^2 + (F_{\lambda n})^2\right] d\omega$$
$$= \left(\frac{GM}{R^2}\right)^2 \left(\frac{R}{r}\right)^{2n+4} \left\{n(n+1) \sum_{k=0}^n \left[\left(\bar{C}_n^k\right)^2 + \left(\bar{S}_n^k\right)^2\right]\right\}. \tag{5.182}$$

Their sum is, with $F^2 = (F_{rn})^2 + (F_{\theta n})^2 + (F_{\lambda n})^2$,

$$\frac{1}{4\pi} \int_\omega F^2 d\omega = \left(\frac{GM}{R^2}\right)^2 \left(\frac{R}{r}\right)^{2n+4} \left\{(n+1)(2n+1) \sum_{k=0}^n \left[\left(\bar{C}_n^k\right)^2 + \left(\bar{S}_n^k\right)^2\right]\right\}. \tag{5.183}$$

Similar to the case of V, we consider particularly

$$\left(\sigma_n^F\right)^2 = (n+1)(2n+1) \sum_{k=0}^{n} \left[\left(\bar{C}_n^k\right)^2 + \left(\bar{S}_n^k\right)^2\right], \tag{5.184}$$

which is a degenerate form of (5.183) when $r = R$ and omitting the constant factor $(GM/R^2)^2$. A graph of it is also shown in Fig. 5.10 for the EGM GGM05C. We see that when n increases, it decreases much slower than $\left(\sigma_n^V\right)^2$ shown in the same figure. Actually, if $\left(\sigma_n^V\right)^2$ obeys Kaula's rule (5.176), $\left(\sigma_n^F\right)^2$ obeys

$$\left(\sigma_n^F\right)^2 \approx (n+1)(2n+1)^2 \left(\frac{10^{-5}}{n^2}\right)^2, \tag{5.185}$$

which means that $\left(\sigma_n^F\right)^2$ is approximately inverse proportional to n. It decreases two order of magnitude slower than $\left(\sigma_n^V\right)^2$ as n increases.

Taking reference to (5.13), the degree variance power spectrum of the deflection of the vertical can be obtained from (5.182) by setting $r = R$, dividing by $\bar{\gamma}_0^2 = (GM/R^2)^2$, and replacing the coefficients by those of the disturbing potential

$$\left(\sigma_n^{DOV}\right)^2 = \frac{1}{4\pi} \int_{\omega} \left[(\xi_n)^2 + (\eta_n)^2\right] d\omega = n(n+1) \sum_{k=0}^{n} \left[\left(\bar{A}_n^k\right)^2 + \left(\bar{B}_n^k\right)^2\right], \tag{5.186}$$

where $\bar{A}_n^k$ and $\bar{B}_n^k$ are the fully normalized spherical harmonic coefficients of the disturbing potential, which are practically indifferent from $\bar{C}_n^k$ and $\bar{S}_n^k$ beyond degree 4. As already mentioned, the normalized coefficient of the normal potential, $\bar{J}_6$, is already $\sim 1\%$ of $\bar{C}_6$ in magnitude. By comparing (5.184) and (5.186), we see that $\left(\sigma_n^F\right)^2 \approx 2(\sigma_n^{DOV})^2$, which means that $\left(\sigma_n^{DOV}\right)^2$ is about of the same order of magnitude as $\left(\sigma_n^F\right)^2$. A graph of $\left(\sigma_n^{DOV}\right)^2$ is also shown in Fig. 5.10 for the EGM GGM05C.

The gravitational attraction and the deflection of the vertical are on the order of magnitude of the first order derivatives of the gravitational potential. We have approximately $\left(\sigma_n^F\right)^2 \sim 2\left(\sigma_n^{DOV}\right)^2 \sim 2n^2 \left(\sigma_n^V\right)^2$. It can be readily inferred that the degree variance power spectrum of the second order gradient of the gravitational potential, i.e., the gradient of the gravitational attraction, would be larger than $\left(\sigma_n^F\right)^2$ or $\left(\sigma_n^{DOV}\right)^2$ by a factor of about n^2. The readers are suggested to verify this for the component $F_{rrn} = \partial F_{rn}/\partial r$. Subsequently, when the order of the derivative increases by 1, the order of magnitude of the degree variance power spectrum would increase by a factor of about n^2.

References

Hwang, C. (1998). Inverse Vening Meinesz formula and deflection-geoid formula: applications to the predictions of gravity and geoid over the South China Sea. *Journal of Geodesy, 72*, 304–312. http://dx.doi.org/10.1007/s001900050169.

Kaula, W. M. (1966). *Theory of satellite geodesy: Application of satellites to geodesy*. Dover publications Incorporated.

Rapp, R. H. (1989). The decay of the spectrum of the gravitational potential and the topography for the Earth. *Geophysical Journal International, 99*, 449–455. https://doi.org/10.1111/j.1365-246X.1989.tb02031.x.

Ries, J., Bettadpur, S., Eanes, R., Kang, Z., Ko, U., McCullough, C., Nagel, P., Pie, N., Poole, S., Richter, T., Save, H., & Tapley, B. (2016). The Combined Gravity Model GGM05C. GFZ Data Services. http://doi.org/10.5880/icgem.2016.002.

Gravity Reduction

6

Abstract

This chapter formulates gravity reduction, which concerns the computation of the gravity on the geoid for the gravity anomaly as required by Stokes' theory. Several kinds of gravity anomaly—free air anomaly, incomplete Bouguer anomaly, complete Bouguer anomaly, Helmert anomaly, and isostatic anomaly—are introduced by correcting the gravity on the ground. The Helmert orthometric height, which is widely used as an approximation of the orthometric height, is formulated in this chapter, as it relies on the formulation of gravity reduction. The isostatic models customarily adopted in gravity reduction—the Pratt–Hayford model, the Airy–Heiskanen model, and the Vening Meinesz model—are all formulated. In particular, a systematic formulation of gravity reduction based on a spherical Earth model is included. Gravity reduction involves adjusting the Earth's mass distribution for removing or relocating the mass above the geoid for the need of Stokes' theory, i.e., no mass is present above the geoid. Hence, the Earth's gravity field is adjusted, and a correction called the indirect effect should be applied to obtain the Earth's real gravity field. Therefore, a systematic formulation of the indirect effect based on a spherical Earth model is also included.

6.1   Basic Corrections and Gravity Anomalies

In order to determine the Earth's gravity field using Stokes' theory, we have to know the gravity on the geoid, and meanwhile, there must be no mass above the geoid. However, gravity measurements are performed on the Earth's surface, on either the ground or ocean surface, and the geoid is below the ground in virtually all continental regions. Gravity reduction is to convert the gravity on the Earth's surface to the geoid, and meanwhile to adjust the mass distribution of the Earth so that the mass above the geoid be removed or relocated. Mass adjustment causes change

© The Author(s), under exclusive license to Springer Nature Switzerland AG 2023
J.-Y. Guo, *Physical Geodesy*, Springer Textbooks in Earth Sciences, Geography and Environment, https://doi.org/10.1007/978-3-031-23320-3_6

of the Earth's gravity field, which should be either negligibly small or restored afterward.

6.1.1 Free Air Correction and Free Air Anomaly

The most primitive gravity reduction is to ignore the mass above the geoid and compute the gravity on the geoid, g_0, from that on the ground, g, using the gradient of the gravity

$$g_0 = g - \left(\frac{\partial g}{\partial H}\right)_\Sigma H - \frac{1}{2}\left(\frac{\partial^2 g}{\partial H^2}\right)_\Sigma H^2, \tag{6.1}$$

where H is the orthometric height, and Σ is still used to denote the geoid as in the previous chapter. This formula can be understood more straightforwardly if the second and third terms at the right-hand side are moved to the left-hand side with sign reversed. We refer the second term at the right-hand side as first order gradient correction, and the third term second order gradient correction.

However, $(\partial g/\partial H)_\Sigma$ and $(\partial^2 g/\partial H^2)_\Sigma$ are not known. In practical computations, they are approximated by the normal gravity gradients as appeared in (4.174)

$$\left(\frac{\partial g}{\partial H}\right)_\Sigma \approx \left(\frac{\partial \gamma}{\partial h}\right)_S = -2\frac{\gamma_e}{a}\left[1 + \alpha + m + \left(-3\alpha + \frac{5}{2}m\right)\cos^2\theta\right], \tag{6.2}$$

$$\frac{1}{2}\left(\frac{\partial^2 g}{\partial H^2}\right)_\Sigma \approx \frac{1}{2}\left(\frac{\partial^2 \gamma}{\partial h^2}\right)_S = 3\frac{\gamma_e}{a^2}, \tag{6.3}$$

where, the same as in the previous chapter, S is used to denote the reference Earth ellipsoid, h the geodetic height, and a the semi-major axis of the reference Earth ellipsoid.

Upon substitution of (6.2) and (6.3) into (6.1), it can be seen that H/a and $(H/a)^2$ appear in the first and second order gradient corrections, respectively, which are multiplied by a quantity in the same order of magnitude as the gravity on the Earth's surface (by definition, the order of magnitude of the normal gravity on the reference Earth ellipsoid is the same as that of the Earth on the surface), thus representing the order of magnitude of the respective corrections. In reality, H rarely exceeds 2000 m, where $H/a \sim 3.1 \times 10^{-4}$ and $(H/a)^2 = 9.9 \times 10^{-8}$, and the highest point is $H = 8848$ m, where $H/a \sim 1.3 \times 10^{-3}$ and $(H/a)^2 = 1.9 \times 10^{-6}$. This is the reason why terms on the order of magnitude of α are retained in the first order gradient correction and omitted in the second order gradient correction.

It can be obtained using data provided in Chap. 4 that $(\partial\gamma/\partial h)_S$ is in between -0.3083 and -0.3088 mGal·m^{-1}, and $(1/2)(\partial^2\gamma/\partial h^2)_S = 7.2 \times 10^{-8}$ mGal·m^{-2}.

Under spherical approximation, we replace $(\partial \gamma / \partial h)_S$ by its global average

$$\left(\overline{\frac{\partial \gamma}{\partial h}} \right)_S = \frac{1}{4\pi} \int_\omega \left(\frac{\partial \gamma}{\partial h} \right)_S d\omega = -2 \frac{\gamma_e}{a} \left(1 + \frac{11}{6} m \right) = -0.3086 \, \mathrm{mGal \cdot m^{-1}} .$$

(6.4)

When $H = 2000$ m, the difference between the first order gradient corrections computed using $(\partial \gamma / \partial h)_S$ and $(\partial \bar{\gamma} / \partial h)_S$ is 0.6 mGal, and the second order gradient correction is 0.29 mGal. When $H = 8848$ m, the difference between the first order gradient corrections computed using $(\partial \gamma / \partial h)_S$ and $(\partial \bar{\gamma} / \partial h)_S$ is 2.7 mGal, and the second order gradient correction is 5.6 mGal.

Hence, when H is not very large, a practical formula customarily used is

$$g_0 = g + \Delta_1 g .$$

(6.5)

The correction is

$$\Delta_1 g = 0.3086 H ,$$

(6.6)

which is referred to as free air correction, where the unit of H is m, and the unit of the correction $\Delta_1 g$ is mGal. The gravity anomaly obtained after applying the free air correction is

$$g_0 - \gamma_0 = g + \Delta_1 g - \gamma_0 ,$$

(6.7)

which is referred to as free air gravity anomaly, or shortly free air anomaly.

6.1.2 Plate Correction and Incomplete Bouguer Anomaly

In the free air correction formulated in the previous subsection, we assumed that mass above the geoid vanishes, which is not the case in reality. In this subsection, we remove the effect of the mass above the geoid under a simple assumption: the geoid and the Earth's surface are planes parallel to each other. The correction is called Bouguer plate correction or simply plate correction and is denoted as $\Delta_2 g$.

The plate can be considered as a cylinder with an extremely large radius as compared to its height, e.g., a radius of 100 km with a height of 2 km, and the gravity is measured at the center of the top. The formula of gravitational attraction of a cylinder is provided in (1.95). As removing the effect of the plate always brings in a negative correction, we obtain, by setting $z = 0$ and $h = H$ in (1.95),

$$\Delta_2 g = -2\pi G \rho \left[(a + H) - \left(a^2 + H^2 \right)^{1/2} \right] .$$

(6.8)

This is the exact formula for a cylindrical plate of radius a and thickness H. When H/a is small enough, we can derive an approximate formula

$$\Delta_2 g = -2\pi G \rho a \left[1 + \frac{H}{a} - \left(1 + \frac{H^2}{a^2} \right)^{1/2} \right]$$

$$\approx -2\pi G \rho a \left[1 + \frac{H}{a} - \left(1 + \frac{H^2}{2a^2} \right) \right]$$

$$= -2\pi G \rho H \left(1 - \frac{H}{2a} \right). \tag{6.9}$$

A further approximation is to assume that H/a is extremely small, so that the term $H/(2a)$ can be neglected. This leads to

$$\Delta_2 g = -2\pi G \rho H. \tag{6.10}$$

This is the most known Bouguer plate correction formula, where quantities of relative magnitude on the order of H/a or smaller are neglected. It is independent of a and is in fact the exact formula if the radius of the cylinder is infinitely large, i.e., $a = \infty$. In practical applications, it should be understood as an approximate formula for a cylinder of finite radius rather than as the exact formula for a cylinder of infinite radius. If the term neglected, $H/(2a)$, is not significantly small, the approximate formula (6.9) including this term should be used. In special applications where H/a is not a small quantity, the exact formula (6.8) should be used. This will be further discussed in the formulation of gravity reduction for the spherical Earth model in Sect. 6.4.1.

The average density of crust is $\rho = 2.67$ g cm^{-3}, which leads to

$$\Delta_2 g = -0.1119 H, \tag{6.11}$$

where, the same as the free air correction, the unit of H is m, and the unit of the correction $\Delta_2 g$ is mGal. The gravity anomaly obtained by applying both the corrections $\Delta_1 g$ and $\Delta_2 g$ is

$$g_0 - \gamma_0 = g + \Delta_1 g + \Delta_2 g - \gamma_0, \tag{6.12}$$

which is referred to as incomplete Bouguer gravity anomaly, or simply incomplete Bouguer anomaly. It is computed by removing the effect of the mass above the geoid assuming the ground to be a plane parallel to the geoid, and then project the gravity to the geoid using free air correction. The correction $\Delta_1 g + \Delta_2 g$ is referred to as incomplete Bouguer correction. In the literature, the word *incomplete* is sometimes omitted. In this book, we keep it to distinguish from the complete Bouguer correction and anomaly to be discussed in the next subsection.

6.1.3 Terrain Correction and Complete Bouguer Anomaly

The plate correction removes the effect of a layer of mass between the local horizontal plane (that passes through the ground point where gravity is measured) and the geoid. Evidently, it is more accurate to consider the actual topography. Therefore, we add a correction, called terrain correction denoted as $\Delta_3 g$, on top of the plate correction $\Delta_2 g$, such that their sum $\Delta_2 g + \Delta_3 g$, which is referred to as topographic correction, removes the effect of all masses above the geoid and below the Earth's surface.

As shown in Fig. 6.1, the terrain correction at a point P includes two parts. One part is from places where the ground is above the horizontal plane through P, and the mass between the plane and the ground is to be removed. The other part is from places where the ground is below the horizontal plane through P, and the hollow between the plane and the ground should be filled with mass; this is due to the fact that mass is assumed to exist there, and its effect is removed in the plate correction.

The contribution to $\Delta_3 g$ from any place is positive, also as shown in Fig. 6.1. For a place above the horizontal plain through P, the contribution of the mass above the horizontal plane to the gravitational attraction at P points upward. Hence, the gravity at P would become larger after removing this contribution, and thus, the correction is positive. For a place below the horizontal plain through P, it is evident that adding mass to fill the hollow makes the gravitational attraction at P larger, so the correction is also positive.

In order to formulate $\Delta_3 g$, we use a local cylindrical coordinate system $Pr\alpha z$, as shown in Fig. 6.2, where the gravity correction at P is to be computed. Consider a volume element dv. Its vertical projection on the ground is M, and that on the horizontal plane is M_0. The contribution of the volume element to the gravitational attraction at P along the upward direction is $(\rho dv/l^2) \sin \beta$, where ρ is the density assumed to be constant, and other symbols are defined in Fig. 6.2. Evidently, $l^2 = r^2 + z^2$, $\sin \beta = z/(r^2 + z^2)^{1/2}$, and $dv = rd\alpha drdz$. We consider only the total contribution of the region within the distance a measured in the horizontal plane.

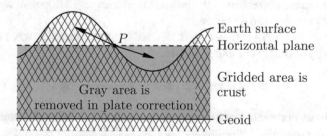

Fig. 6.1 Terrain correction

Fig. 6.2 Local coordinate system $Or\alpha z$

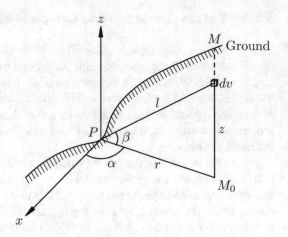

Hence,

$$\Delta_3 g = G\rho \int_0^{2\pi} d\alpha \int_0^a dr \int_0^{\Delta H} \frac{zr}{(r^2 + z^2)^{2/3}} dz , \qquad (6.13)$$

where $\Delta H = M_0 M$ is the difference between the heights of the ground points M and P, $\Delta H = H_M - H_P$. The integral over z can be easily obtained analytically. We obtain

$$\Delta_3 g = G\rho \int_0^{2\pi} d\alpha \int_0^a \left[1 - \frac{r}{(r^2 + \Delta H^2)^{1/2}} \right] dr , \qquad (6.14)$$

which can be evaluated numerically using a digital ground elevation model.

The formula (6.14) is the projection of the gravitational attraction of the terrain mass in upward direction at P assuming M to be above P; hence, it is positive. It can be readily shown that the formula of terrain correction is the same when M is below P, where the correction is the projection of the gravitational attraction of the terrain mass in downward direction.

The correction $\Delta_1 g + \Delta_2 g + \Delta_3 g$ is referred to as complete Bouguer correction, and the gravity anomaly computed according to

$$g_0 - \gamma_0 = g + \Delta_1 g + \Delta_2 g + \Delta_3 g - \gamma_0 \qquad (6.15)$$

is referred to as complete Bouguer gravity anomaly, or briefly complete Bouguer anomaly. It is computed by removing the effect of all masses above the geoid and then projecting the gravity to the geoid using the free air correction.

The gravity anomaly obtained by applying the free air and terrain corrections,

$$g_0 - \gamma_0 = g + \Delta_1 g + \Delta_3 g - \gamma_0 , \qquad (6.16)$$

is called Faye gravity anomaly, or briefly Faye anomaly. It is computed by adjusting the ground to a horizontal plane through the point where gravity is measured and then projecting the gravity to the geoid using the free air correction.[1]

6.1.4 A Practical Formula for Computing Orthometric Height Using Spirit Leveling Data

In Sect. 1.3.2, we derived a primitive formula for computing the orthometric height using data of spirit leveling, which we summarize here. We assume that spirit leveling is performed from a point A on the geoid to a point P where the height is to be determined. The basic quantity representing the height of P is the potential number computed using data of spirit leveling and gravity

$$C = \int_A^P g \, dH_L \,, \tag{6.17}$$

which is an integral along the path of spirit leveling. The formula for the orthometric height is

$$H = \frac{C}{g_m} \,, \tag{6.18}$$

where

$$g_m = \frac{1}{H} \int_{P_0}^P g(H') \, dH' = \frac{1}{H} \int_0^H g(H') \, dH' \tag{6.19}$$

is the average gravity between P and its projection on the geoid, P_0, as shown in Fig. 1.7.

In (6.19), $g(H')$ is the gravity inside the crust, which is not known. However, an approximate formula for it can be derived based on gravity reduction discussed in the previous subsections. Assume the ground to be a plane parallel to the geoid, as in the case of plate correction. The gravity between the ground and the geoid can be computed according to the following procedure, as shown in Fig. 6.3. Three corrections should be applied to obtain the gravity $g(H')$ at P' from the gravity g at P, which is referred to as Poincaré and Prey gravity reduction. (1) A plate correction at P to remove the effect of the layer between P and P': $-0.1119(H - H')$. (2) A free air correction to project the gravity at P to P': $+0.3086(H - H')$. (3) A plate correction at P' to place back the effect of the layer between P and P':

---

[1] In the literature, Faye correction and anomaly are sometimes used as synonyms of free air correction and anomaly. Hence, there is always a need to verify or clarify how it is defined.

Fig. 6.3 Computation of
gravity between ground and
geoid

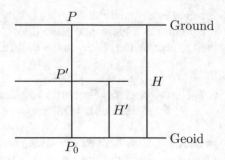

$-0.1119(H - H')$. After adding these three corrections together, we obtain

$$g(H') = g + 0.0848(H - H'). \tag{6.20}$$

This leads to

$$g_m = \frac{1}{H} \int_0^H [g + 0.0848(H - H')]dH' = g + 0.0424H. \tag{6.21}$$

In practical computations, H could be replaced by an approximate value, e.g., the
height obtained from spirit leveling, H_L. The other way is to use (6.18) and (6.21)
iteratively.

The height computed using (6.18) and (6.21) is referred to as Helmert orthometric height. We remark that more accurate gravity reduction for this purpose may
include terrain correction, variation of density, or even consider the curvature of
the Earth's surface. However, consideration of these factors would require much
more computations. Actually, Helmert orthometric height is widely used in practical
applications.

6.1.5 Helmert Condensation and Helmert Anomaly

In the computation of the complete Bouguer anomaly, the masses above the geoid
are all removed. However, removing these masses changes the Earth's total mass.
A simple way to reserve the Earth's total mass is to put back the masses removed
as a material surface on the geoid, which is referred to as Helmert condensation. In
this way, the no mass above the geoid condition to use Stokes' theory is satisfied by
relocating the masses above the geoid to the geoid.

As shown in Fig. 6.4, gravity reduction is performed for the point P, and P_0 is
its projection on the geoid. The density of the condensed material surface, called
Helmert condensation layer, at a point (r, α) is ρH, where ρ is the crustal density
considered as constant, and H the orthometric height of the Earth's surface. The
Helmert condensation correction due to a surface element dS as shown in Fig. 6.4
is $(G\rho H dS/l^2) \cos \beta$. We use H_P to denote the orthometric height of the point P.

Fig. 6.4 Gravity of a Helmert condensation layer

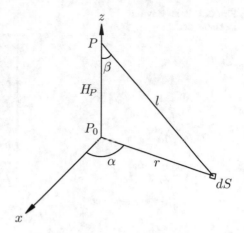

We thus have $l = (r^2 + H_P^2)^{1/2}$ and $\cos \beta = H_P/(r^2 + H_P^2)^{1/2}$. As $dS = rdrd\alpha$, the total contribution from within the distance a can be written as

$$\Delta_4 g = G\rho H_P \int_0^{2\pi} d\alpha \int_0^a \frac{Hr}{(r^2 + H_P^2)^{3/2}} dr ,\qquad (6.22)$$

which can be computed using a numerical integration technique. The gravity anomaly defined by

$$g_0 - \gamma_0 = g + \Delta_1 g + \Delta_2 g + \Delta_3 g + \Delta_4 g - \gamma_0 \qquad (6.23)$$

is referred to as Helmert condensation gravity anomaly, or briefly Helmert condensation anomaly, or Helmert anomaly.

We remark that this method is often referred to as Helmert's second method of condensation in the literature. In this book, we use the more brief term Helmert condensation for it, as his first condensation method is seldom used contemporarily and is not to be introduced in this book.

6.2 Isostasy, Isostatic Correction, and Isostatic Gravity Anomaly

6.2.1 Background Fact and the Pratt–Hayford Model

If the mass above the geoid is the main cause of the gravity anomaly, the complete Bouguer anomaly should be very small, as the effect of the mass above the geoid on gravity is completely removed. However, the reality is the reverse. The complete Bouguer anomaly obtained is very large in magnitude in mountainous regions, and the sign is negative. This implies that there may be some kind of mass deficit under mountains. Observations of the deflection of the vertical demonstrate the same

Fig. 6.5 Pratt–Hayford
isostatic model

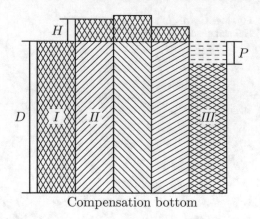

Compensation bottom

phenomenon; the magnitude of the deflection of the vertical near a mountain is far smaller than what is caused by the visible topographic masses above the geoid, which may also imply a mass deficit under mountains. These observed phenomena have led to the hypothesis of isostasy: the mass above the geoid is compensated by a mass deficit under the geoid, called isostatic compensation. Several models of isostasy have been proposed, which may be considered as models of crustal density based on gravity measurements.

Pratt–Hayford used a model as shown in Fig. 6.5. This model is based on an ideal assumption that the crustal density above the geoid is constant (to agree with that in the complete Bouguer correction), $\rho = 2.67$ g cm^{-3}, but the crustal density below the geoid and above a compensation bottom of depth D varies, such that all columns of equal cross section area from the compensation bottom to the Earth's surface have the same mass. With this assumption, the pressure at the compensation bottom is identical under all columns, meaning that it is position independent. We still use H to denote the orthometric height over land and use ρ' to denote the density between the geoid and the compensation bottom. When $H = 0$, such as the column I in Fig. 6.5, there is no compensation under the geoid, and the density of the column under the geoid is assumed to be $\rho' = \rho$. When $H > 0$, such as the column II in Fig. 6.5, the density above the geoid is ρ, and that below is ρ'. As the cross section areas of the columns I and II are equal, and their masses from the compensation bottom to the Earth's surface are also equal, we have

$$\rho D = \rho H + \rho' D . \tag{6.24}$$

This leads to a density deficit of

$$\Delta \rho = \rho - \rho' = \frac{H}{D} \rho . \tag{6.25}$$

Over oceans, such as the column *III* in Fig. 6.5, we have

$$\rho D = \rho_w P + \rho'(D - P),\qquad(6.26)$$

where ρ_w is the density of ocean water, and P the ocean depth. The density deficit is

$$\Delta\rho = \rho - \rho' = -\frac{(\rho - \rho_w)P}{D - P}\rho.\qquad(6.27)$$

This deficit has a negative sign, meaning that it is indeed a surplus of mass in compensation to the deficit of mass in the ocean.

It is to be noted that the compensation region under the ground is from the geoid to the compensation bottom, but that under oceans is from the ocean bottom to the compensation bottom. The compensation depth is mostly $D = 100\,\text{km}$.

6.2.2   The Airy–Heiskanen Model

Airy–Heiskanen used a model as shown in Fig. 6.6. A homogeneous crust of density ρ floats on a homogeneous mantle of larger density ρ_1, as if all crustal columns in the figure are detached from each other, and the weight of each column is balanced by the buoyancy. Normally, the densities $\rho = 2.67\ \text{g cm}^{-3}$ and $\rho_1 = 3.27\ \text{g cm}^{-3}$ are adopted. We denote the crustal thickness when $H = 0$ as T, which is referred to as the normal thickness of the crust. In most regions, $T \approx 30$ km. In this model, the topographic mass above the geoid is compensated by the deficit of mass within the

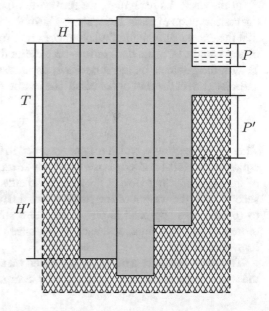

Fig. 6.6 Airy–Heiskanen isostatic model

crust below the depth T in contrast to the larger density of the mantle. We use H' to denote the depth of the crustal bottom below the surface of depth T, as shown in Fig. 6.6. We then have, over land,

$$\rho H = (\rho_1 - \rho)H'. \tag{6.28}$$

Hence,

$$H' = \frac{\rho}{\rho_1 - \rho} H. \tag{6.29}$$

As $\rho/(\rho_1 - \rho) = 4.45$, this formula implies that a 1 km high mountain should have a 4.45 km deep root. In oceans, the balance between the mass and bouncy of a column leads to

$$(\rho - \rho_w)P = (\rho_1 - \rho)P', \tag{6.30}$$

where P is the ocean depth, and P' the height of anti-root of the ocean, as shown in Fig. 6.6. We obtain

$$P' = \frac{\rho - \rho_w}{\rho_1 - \rho} P. \tag{6.31}$$

In practical computations, oceans and lands can be handled altogether with the formulation for lands. The idea is to "compress water" of depth P to a "rock-equivalent" layer of thickness t with the density of crust, i.e., $\rho t = \rho_w P$, $t = (\rho_w/\rho)P$. In this way, oceans behave the same as lands with a height of $-(P - t)$.

In the previous paragraph, we formulated the Airy–Heiskanen model based on the conservation of mass. It can also be formulated based on the physical assumption that the excess of weight of a column of crust from the bottom to the top in access of the weight of a standard crust—here defined as the crust of constant thickness T with the geoid as its top surface—is balanced by the buoyancy from the mantle associated with the root of the crust. The mathematical expression of this balance is

$$\rho_1 H'g = \rho(H + T + H')g - \rho Tg, \tag{6.32}$$

where g is the gravity, which may be assumed to be constant for this problem. In this equation, the left-hand side is the buoyancy from the mantle associated with the root of the crust, the first term at the right-hand side is the weight of the crust, and the second term the weight of the standard crust. Evidently, this formula is equivalent to (6.28). This approach is physically more straightforward to be generalized to more complicated situations, such as the case when the density of the crust is no longer constant.

The Airy–Heiskanen model is also based on an idealized assumption: the crustal mass above the geoid at any place is fully compensated by a deficit of mass strictly below. This is in fact based on an assumption that all the columns as shown in Fig. 6.6 can move independently in vertical direction, meaning that no friction exists between the vertical walls of adjacent columns. This assumption appears totally unrealistic at a first glance, as both the crust and mantle are solid. However, the topographic mass is a long term load of millions of years. Under the action of such a long time scale forcing, both the crust and mantle may flow slowly to reach an ultimate balanced state. The Airy–Heiskanen model is in fact the simplest model of such a balance.

6.2.3 The Vening Meinesz Model

Vening Meinesz generalized the idea of the Airy–Heiskanen model by taking into account of more geophysical considerations. In Fig. 6.6, imagine a crustal column that is very high above the geoid, but its bottom is not as deep as that of the root required to achieve isostatic compensation in the sense of the Airy–Heiskanen model. An evident fact would be that this column tends to sink under the action of its weight. The reality would not be as ideal as the Airy–Heiskanen model where adjacent columns simply let this column slide down by the walls. Instead, adjacent columns are attached to this column and would yield and also sink under the downward dragging of this column. As a result, adjacent columns would also contribute to the final balance. A conceptual comparison between the Airy–Heiskanen and Vening Meinesz models is demonstrated in Fig. 6.7.

In order to develop a mathematical theory for the Vening Meinesz model, we consider first the load of a unit point mass as shown in Fig. 6.8. Under the action of

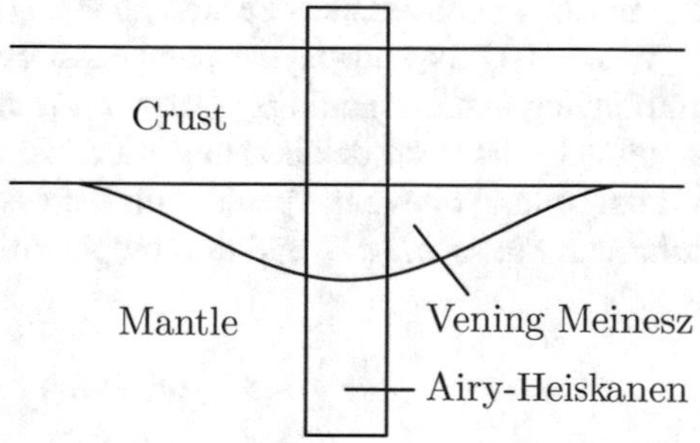

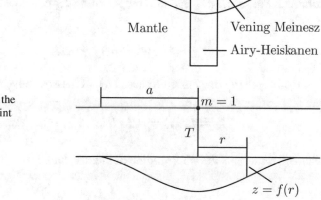

Fig. 6.7 A conceptual comparison between Airy–Heiskanen and Vening Meinesz isostatic models

Fig. 6.8 Definition of compensation function as the compensation to a unit point mass

this load, both the top and bottom of the crust would yield and curve, leading to a concave on the top and a convex at the bottom. Here we assume that the concave on the top is refilled with crustal material, such that the top remains a plane. As a result, the convex at bottom would be larger, which is to be also included in the mathematical model. We use T to denote the normal thickness of the crust, i.e., the thickness of the crust when $H = 0$, and express the depth of the convex below T as

$$z = f(r), \tag{6.33}$$

called compensation function, where r is the horizontal distance to the unit point mass. This implies that the compensation is assumed to be symmetrical about the load, which is rational. We further assume that the compensation is limited to the region within the distance a and denote the region as S. The requirement of mass conservation (deficit of mass below the geoid is equal to the mass above the geoid) is then

$$(\rho_1 - \rho) \int_S f(r)dS = 1. \tag{6.34}$$

We assume that the lower convex is the result of bending (without breaking) of the crust. An additional constraint to $f(r)$ would then be

$$\left[\frac{d}{dr}f(r)\right]_{r=a} = 0, \quad \left[\frac{d}{dr}f(r)\right]_{r=0} = 0. \tag{6.35}$$

The first identity implies that the lower convex is tangential to the horizontal plane at its boundary where $r = a$, and the second one is a consequence of an evident fact that the lower convex must be smooth and symmetrical around the point mass.

When $f(r)$ is known, the compensation depth below T can be obtained by multiplying and integrating $f(r)$ with the mass above the geoid computed using topography, as to be detailed hereafter. We replace the point mass by the mass of a crust column above the geoid with a cross section area of $d\Sigma$. The mass of this column is $dm = \rho H d\Sigma$, and the compensation depth below T caused by this mass is

$$dH' = \rho H f(r)d\Sigma. \tag{6.36}$$

The total depth of compensation caused by all masses above the geoid is

$$H'(x, y) = \rho \int_\Sigma H(x', y')f(r)dx'dy', \tag{6.37}$$

where Σ represents the whole region of interest, and

$$r = \left[(x' - x)^2 + (y' - y)^2\right]^{1/2}. \tag{6.38}$$

The compensation region is in between the depth of T and $T + H'$. We see that this model gets closer and closer to the Airy–Heiskanen model when a is chosen smaller and smaller.

There exist crustal flexure models to compute $f(r)$. However, models have to be based on assumptions of how the crust and mantle react to forcing at a time scale of millions of years, which is in fact quite hypothetical. Here, we take $f(r)$ itself as an empirical function expressed using a few parameters in addition to T. As an example, we take

$$f(r) = A\left[1 + \cos\left(\frac{\pi r}{a}\right)\right] + B\left[1 - \cos\left(\frac{2\pi r}{a}\right)\right], \tag{6.39}$$

which satisfies the prerequisite (6.35). The sharpness of the convex can be adjusted by adjusting the parameters A and B. However, A and B are not free of choice, as (6.34) must be satisfied also. Substitute (6.39) into (6.34). We obtain

$$(\rho_1 - \rho)\int_S f(r)dS = \pi a^2(\rho_1 - \rho)(A + B) = 1, \tag{6.40}$$

which leads to

$$A + B = \frac{a}{\pi a^3(\rho_1 - \rho)}. \tag{6.41}$$

Considering that $f(r)$ represents the compensation of a unit point mass, we have intentionally arranged the dimension of the numerator to be that of length, and that of the denominator to be that of mass. As A should not vanish for the curve of $f(r)$ to look like that in Fig. 6.8, we set $B = \alpha A$ and modify (6.41) to

$$A = \frac{a}{\pi(1+\alpha)a^3(\rho_1 - \rho)}, \qquad B = \frac{\alpha a}{\pi(1+\alpha)a^3(\rho_1 - \rho)}. \tag{6.42}$$

In this way, the parameters defining $f(r)$ are a and α. Some cases with different values of α are shown in Fig. 6.9, where we see how the shape of $f(r)$ depends on α. We remark that a unit mass does not mean a mass of 1 kg or 1 ton. Any mass could be chosen as unit. Here, for illustrative purpose, we chose $a = 100$ km and the unit mass to be the mass of a crust column 5 km high above the geoid with a cross section area of $5 \times 5 = 25$ km^2. It can be seen that the shape of $f(r)$ is adjusted reasonably well by choosing $\alpha \in [-0.25, 0.25]$. In this figure, while the case $\alpha = 0$ represents solely the first term at right-hand side of (6.39), we have intentionally drawn in dashed line the second term with an amplitude of $B = 0.25A_{\alpha=0}$, which gives an idea of how the second term is added to or subtracted from the first term for widening or sharpening the curve. We remark that making the curve too wide or too sharp by adjusting the value of α is pointless, as the effect would be too close to choosing a larger or smaller value for a.

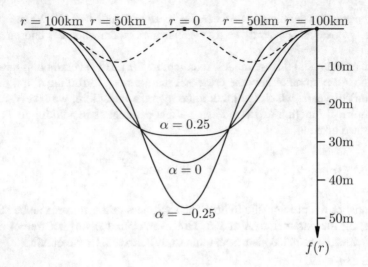

Fig. 6.9 $f(r)$ according to (6.39)

At this stage, we see that, if (6.39) is empirically chosen to be the compensation function, the Vening Meinesz model has only three parameters, the normal crust bottom depth T, the radius of compensation extent a, and a parameter α defining the shape of the compensation function. In contrast, both Pratt–Hayford and Airy–Heiskanen models have only one parameter, which are D and T, respectively.

The same as in the case of Airy–Heiskanen model, oceans can be handled in the same way as land by compressing the water layer to a rock-equivalent layer. However, we mention that the oceanic crust is basically different from the continental crust. The oceanic crust is formed as a result of sea floor spreading and cooling, and both its thickness and density are functions of its age. Therefore, special care should be taken if the Vening Meinesz model is to be applied, as the basis of this model is to take into account of the long term bending/yielding of the crust as well as the mantle under the load of topographic mass.

6.2.4 Isostatic Correction and Isostatic Anomaly

The formulation of the isostatic correction is almost identical to that of the terrain correction (6.13). We still consider only the total contribution of a region within the distance a (to be distinguished from the parameter a in Vening Meinesz model, which is specific to the model only) measured in the horizontal plane. We take the Airy–Heiskanen model as example. What we need is the effect of gravity in the downward direction by filling the mass deficit region with the density $\Delta\rho = \rho_1 - \rho$. Hence, we need to replace ρ by $\Delta\rho$ in (6.13), and the limits of integration for z should be from $-(H_P + T + H')$ to $-(H_P + T)$, H_P being the orthometric height of the point P where the correction is computed. We have

$$\Delta_4 g = -G\Delta\rho \int_0^{2\pi} d\alpha \int_0^a dr \int_{-(H_P+T+H')}^{-(H_P+T)} \frac{zr}{(r^2+z^2)^{2/3}} dz, \qquad (6.43)$$

where a negative sign is present as compared to (6.13), since this is the projection of the gravitational attraction along the downward direction. We use the same symbol as the Helmert condensation correction, $\Delta_4 g$, for the isostatic correction, as they are both a correction to put back the mass removed in the complete Bouguer correction. After integrating over z, we obtain

$$\Delta_4 g = G\rho \int_0^{2\pi} d\alpha \int_0^a \left\{ \frac{r}{[r^2+(H_P+T)^2]^{1/2}} - \frac{r}{[r^2+(H_P+T+H')^2]^{1/2}} \right\} dr,$$
$$(6.44)$$

which can be evaluated using a numerical integration technique. This formula applies equally to the Vening Meinesz model, which is different from the Airy–Heiskanen model only in the values of T and H'.

The gravity anomaly defined by

$$g_0 - \gamma_0 = g + \Delta_1 g + \Delta_2 g + \Delta_3 g + \Delta_4 g - \gamma_0 \qquad (6.45)$$

is referred to as isostatic gravity anomaly, or briefly isostatic anomaly. It is computed by removing all the masses above the geoid, refilling the masses removed to a region below the geoid where mass deficit is supposed to be present, and then projecting the gravity to the geoid using the free air correction.

Imagine an extreme Pratt–Hayford model whose compensation depth D is infinitesimally small. In this case, the isostatic correction is identical to the Helmert condensation correction. For this reason, we take Helmert condensation as an extreme Pratt–Hayford model with $D = 0$.

6.2.5   Determination of Isostatic Models: The Concept

First, we consider only the application in the determination of the Earth's external gravity field. In this case, we need to build an isostatic model using data of gravity and topography for obtaining an isostatic anomaly that varies as smoothly as possible for interpolating gravity data, as the gravity is often measured only in sparse stations.

As the isostatic anomaly is computed by adjusting the crustal mass to the most possible homogeneous distribution and is projected to the geoid, it should be extremely small if the adjustment is perfect, and other heterogeneity in mass distribution is minimal. Hence, we may expect

$$g + \Delta_1 g + \Delta_2 g + \Delta_3 g + \Delta_4 g - \gamma_0 = 0. \qquad (6.46)$$

However, this is a fictitious ideal circumstance, and it is not possible for this relation to hold in reality. A manageable relation for determining an isostatic model using gravity measurements and topography data is

$$\sum_i (g_i + \Delta_1 g_i + \Delta_2 g_i + \Delta_3 g_i + \Delta_4 g_i - \gamma_{0i})^2 = \min, \qquad (6.47)$$

where the subscript i indicates the ith gravity measurement station. The logic is using this relation to estimate the parameters of the isostatic model, e.g., T for the Airy–Heiskanen model, and T, a, and α for the Vening Meinesz model if the function $f(r)$ is assumed to be in the form of (6.39).

In all the isostatic models discussed so far, the geoid is taken as the upper surface of compensation. This is mostly due to the fact that our objective here is gravity reduction to remove or relocate the mass above the geoid while maintaining the Earth's total mass unchanged. For this reason, the perfect excess-deficit compensation of the mass has been assumed. A particular case is the Helmert condensation that does not mention isostatic mass compensation at all, where the topographic mass is relocated to the geoid as a material surface for using Stokes' theory. It is evident that relocating the topographic mass is superior than removing it as in the complete Bouguer anomaly, since the former alternates less mass distribution than the latter. Furthermore, it is also expected that the isostatic anomaly (and the Helmert anomaly) be smaller in magnitude and smoother than the complete Bouguer anomaly, which is a consequence of isostatic mass compensation.

However, in the context of geophysics where the reality is sought other than making models in favor of gravity reduction, the constraints in establishing isostatic models could be loosened. For example, instead of fixing on the geoid, the top surface of compensation could be above or below the geoid depending on the region studied. Furthermore, the full excess-deficit mass compensation could also be compromised, e.g., in the case of Airy–Heiskanen model, a parameter such as percentage of compensation could be included at the right-hand side of (6.29).

The Mohorovičić discontinuity, or briefly Moho, which is the bottom of the crust, is not a surface purely determined using gravity data; the crust is also mapped using seismic tomography with finer and finer resolution, from averages over $5° \times 5°$ grids to $2° \times 2°$, and then to $1° \times 1°$, which are provided as crust models named CRUST5.1, CRUST2.0, and CRUST1.0 (Laske et al., 2013), respectively. The last one is extended to include the lithospheric mantle in a model referred to as LITHO1.0 (Pasyanos el al., 2014), where the lithospheric mantle is understood as the mantle right below the crust. We remark that the lithosphere is roughly defined as the Earth's cold and hard top layer including the crust and a portion of the mantle, which could be different for different geophysical phenomena, and it is beyond the scope of this book to explore further into this subject. Data of all the models mentioned above are available over the Internet, including the density that we need. We have compared the LITHO1.0 crust with the Airy–Heiskanen style isostatic model, which can be regarded as a verification using independent data whether isostasy is realistic. The result confirms that the LITHO1.0 is indeed in reasonably good agreement with the Airy–Heiskanen style isostasy, except some particular regions such as tectonic subduction zones. However, we must keep in mind that isostatic compensation is an idealized model, but not the exact reality. Considering that the LITHO1.0 is provided as averages over $1° \times 1°$ grids, it confirms isostasy only down to this spatial scale, i.e., the order of $\sim$100 km or above (a $1°$ great arc on the Earth's surface corresponds to 111 km). This is the lower limit of spatial scale that the crust is supposed to be in isostatic compensation; the topographic mass in shorter spatial scale is supposed to be supported by the rigidity of the lithosphere.

Finally, we mention again that the geophysical reality is one thing, but a model for gravity reduction is another thing. For example, the Helmert condensation, which could be understood as the simplest and most extreme isostatic model in gravity reduction, serves reasonably well the purpose of studying the Earth's external gravity field.

6.3 Gravitational Field of a Layer of Mass Around the Earth's Surface

In the previous sections, we have assumed the geoid to be a plane when removal or relocation of mass is concerned. We refer such a model of the Earth as flat Earth model. Starting from this section, we assume the geoid to be a sphere and refer such a model of the Earth as spherical Earth model.

This section is a theoretical preparation for the next two sections in gravity reduction for the spherical Earth model. The model considered here is general enough, so that the formulae derived can be applied to compute the Earth's gravitational field using a model of density within the Earth like the LITHO1.0, which, when combined with a higher resolution topography model, may be used to predict or interpolate gravity.

6.3.1 A Laterally Heterogeneous Layer of Mass Around the Earth's Surface

Basic Equations
As shown in Fig. 6.10, we consider a laterally heterogeneous layer of mass around a sphere of radius R, which, in our case here, is an approximation of the geoid. The radial distances of the bottom and top are R_B and R_T, respectively, both varying laterally, i.e., they are functions of position on the sphere. We assume the density of the layer, ρ, to be independent of vertical position but varies laterally. We compute the gravitational potential and attraction at a point P of coordinate (r, θ, λ).

By definition, the gravitational potential of this layer of mass is

$$V = G \int_v \frac{\rho}{l} dv. \tag{6.48}$$

Fig. 6.10 A layer of mass around the Earth's surface

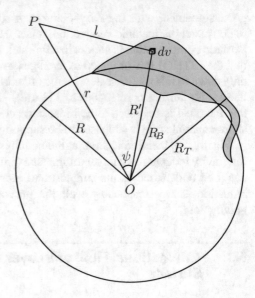

In this formula, the variables of integration, i.e., the coordinates of the volume element dv, are assumed to be R', θ', and λ', ρ is a function of θ' and λ', and l is the distance between the position of the volume element dv and the point P,

$$l = (r^2 + R'^2 - 2rR'\cos\psi)^{1/2}, \qquad (6.49)$$

where ψ is the geocentric angle between the position of dv and the point P

$$\cos\psi = \cos\theta\cos\theta' + \sin\theta\sin\theta'\cos(\lambda - \lambda'). \qquad (6.50)$$

Taking into account of the assumptions that ρ is independent of R', (6.48) can be written as

$$V = G \int_0^{2\pi} \int_0^{\pi} \rho \left(\int_{R_B}^{R_T} \frac{1}{l} R'^2 dR' \right) \sin\theta' d\theta' d\lambda' = G \int_{\omega} \rho \left(\int_{R_B}^{R_T} \frac{1}{l} R'^2 dR' \right) d\omega, \qquad (6.51)$$

where ω is a unit sphere, thus $d\omega = \sin\theta' d\theta' d\lambda'$. The gravitational attraction defined positive downward, $F = -\partial V/\partial r$, is

$$F = -G \int_{\omega} \rho \left[\int_{R_B}^{R_T} \frac{\partial}{\partial r} \left(\frac{1}{l} \right) R'^2 dR' \right] d\omega$$

$$= G \int_{\omega} \rho \left(\int_{R_B}^{R_T} \frac{r - R'\cos\psi}{l^3} R'^2 dR' \right) d\omega. \qquad (6.52)$$

These are the starting formulae for subsequent formulations of this subsection.

Gravitational Potential in Spherical Harmonic Series

Now we assume the point P to be outside the Brillouin sphere, and thus $1/l$ can be expressed as a spherical harmonic series as provided in (3.276), which we copy to here,

$$\frac{1}{l} = \sum_{k=0}^{n} \frac{2}{1 + \delta_k} \frac{(n-k)!}{(n+k)!} \frac{R'^n}{r^{n+1}} P_n^k(\cos\theta) P_n^k(\cos\theta')(\cos k\lambda \cos k\lambda' + \sin k\lambda \sin k\lambda') .$$

(6.53)

The substitution of (6.53) into (6.51) yields the spherical harmonic series of V. Similar to the case in Sect. 3.5, we write V in the form

$$V = \frac{GM}{r} \sum_{n=0}^{\infty} \left(\frac{R}{r}\right)^n \sum_{k=0}^{n} (C_n^k \cos k\lambda + S_n^k \sin k\lambda) P_n^k(\cos\theta) .$$

(6.54)

The coefficients C_n^k and S_n^k are

$$\begin{cases} C_n^k = \dfrac{2}{1 + \delta_k} \dfrac{(n-k)!}{(n+k)!} \displaystyle\int_\omega I_n^k P_n^k(\cos\theta') \cos k\lambda' d\omega , \\ S_n^k = \dfrac{2}{1 + \delta_k} \dfrac{(n-k)!}{(n+k)!} \displaystyle\int_\omega I_n^k P_n^k(\cos\theta') \sin k\lambda' d\omega , \end{cases}$$

(6.55)

where I_n^k is

$$I_n^k = \frac{R^3 \rho}{M} \int_{R_B}^{R_T} \frac{R'^{n+2}}{R^{n+3}} dR' = \frac{R^3 \rho}{(n+3)M} \left[\left(\frac{R_T}{R}\right)^{n+3} - \left(\frac{R_B}{R}\right)^{n+3} \right] .$$

(6.56)

In the result provided by (6.54) to (6.56), we were able to integrate analytically over height, as ρ is assumed to be independent of vertical position. However, as ρ, R_T, R_B, and I_n^k depend on θ' and λ', the integrals in (6.55) have to be computed using a numerical method.

Gravitational Potential in Integral Form

Exact Formula

An integral formula for the gravitational potential can be obtained by analytically evaluating the integral over R' in (6.51). Substitute (6.49) into (6.51). We obtain

$$V = G \int_\omega \rho \left(\int_{R_B}^{R_T} \frac{R'^2 dR'}{(r^2 + R'^2 - 2rR'\cos\psi)^{1/2}} \right) d\omega .$$

(6.57)

With the variable substitution $x = R'/r$, we can write it in the form of

$$V = G \int_\omega \rho f(R_B, R_T, r, \psi) d\omega, \tag{6.58}$$

where the kernel $f(R_B, R_T, r, \psi)$ is defined as

$$f(R_B, R_T, r, \psi) = r^2 \int_{R_B/r}^{R_T/r} \frac{x^2 dx}{(1 + x^2 - 2x \cos \psi)^{1/2}}. \tag{6.59}$$

By consulting a table of integrals, we have the indefinite integral formula

$$\int \frac{x^2 dx}{(1 + x^2 - 2x \cos \psi)^{1/2}} = \frac{1}{2}(x + 3 \cos \psi)(1 + x^2 - 2x \cos \psi)^{1/2}$$

$$+ \frac{1}{2}(3 \cos^2 \psi - 1) \ln \left[(1 + x^2 - 2x \cos \psi)^{1/2} + x - \cos \psi \right] + C, \tag{6.60}$$

where C is a constant of integration. Hence, the kernel (6.59) is

$$f(R_B, R_T, r, \psi) = \frac{r^2}{2} \left\{ (x + 3 \cos \psi)(1 + x^2 - 2x \cos \psi)^{1/2} \right.$$

$$\left. + (3 \cos^2 \psi - 1) \ln \left[(1 + x^2 - 2x \cos \psi)^{1/2} + x - \cos \psi \right] \right\}_{R_B/r}^{R_T/r}. \tag{6.61}$$

The result is (6.58) and (6.61).

An Approximate Formula

An approximate formula can be obtained based on (6.58) and (6.61) assuming P to be very close to the Earth's surface. In this case, we denote $H_{Br} = R_B - r$ and $H_{Tr} = R_T - r$, and assume $| H_{Br}/r | \ll 1$ and $| H_{Tr}/r | \ll 1$. We use $Q(x)$ to denote the function within the braces in (6.61). We can thus write (6.61) as

$$f(R_B, R_T, r, \psi) = \frac{r^2}{2} [Q(x)]_{1+H_{Br}/r}^{1+H_{Tr}/r}. \tag{6.62}$$

Following Sjöberg and Nahavandchi (1999), an approximate form of it can be derived based on the Taylor series expansion

$$Q(x) = Q(1) + \left(\frac{dQ}{dx} \right)_{x=1} (x - 1) + \frac{1}{2} \left(\frac{d^2 Q}{dx^2} \right)_{x=1} (x - 1)^2$$

$$+ \frac{1}{6} \left(\frac{d^3 Q}{dx^3} \right)_{x=1} (x - 1)^3, \tag{6.63}$$

which leads to

$$f(R_B, R_T, r, \psi) = \frac{r^2}{2} \left[\left(\frac{dQ}{dx} \right)_{x=1} \frac{H_{Tr} - H_{Br}}{r} \right.$$
$$+ \left(\frac{d^2 Q}{dx^2} \right)_{x=1} \frac{H_{Tr}^2 - H_{Br}^2}{2r^2} + \left(\frac{d^3 Q}{dx^3} \right)_{x=1} \left. \frac{H_{Tr}^3 - H_{Br}^3}{6r^3} \right].$$

$$(6.64)$$

The derivation of the formulae for dQ/dx, d^2Q/dx^2 and d^3Q/dx^3 requires some tedious algebraic manipulation. Here we write out the result directly:

$$\frac{dQ}{dx} = \frac{2x^2}{(1 + x^2 - 2x \cos \psi)^{1/2}}, \tag{6.65}$$

$$\frac{d^2 Q}{dx^2} = \frac{4x + 2x^3 - 6x^2 \cos \psi}{(1 + x^2 - 2x \cos \psi)^{3/2}}, \tag{6.66}$$

$$\frac{d^3 Q}{dx^3} = \frac{3}{2(1 + x^2 - 2x \cos \psi)^{1/2}} + \frac{5 - 9x^2}{2(1 + x^2 - 2x \cos \psi)^{3/2}}$$
$$- \frac{3x(1 - x^2)(x - \cos \psi)}{2(1 + x^2 - 2x \cos \psi)^{5/2}}. \tag{6.67}$$

We further obtain, by making use of $1 - \cos \psi = 2 \sin^2 (\psi/2)$,

$$\left(\frac{dQ}{dx} \right)_{x=1} = \frac{1}{\sin(\psi/2)} = \frac{2r}{l_r}, \tag{6.68}$$

$$\left(\frac{d^2 Q}{dx^2} \right)_{x=1} = \frac{3}{2 \sin(\psi/2)} = \frac{3r}{l_r}, \tag{6.69}$$

$$\left(\frac{d^3 Q}{dx^3} \right)_{x=1} = \frac{3}{4 \sin(\psi/2)} - \frac{1}{4 \sin^3(\psi/2)} = \frac{3r}{2l_r} - \frac{2r^3}{l_r^3} = -\frac{2r^3}{l_r^3} \left(1 - \frac{3l_r^2}{4r^2} \right), \tag{6.70}$$

where

$$l_r = r(2 - 2 \cos \psi)^{1/2} = 2r \sin \frac{\psi}{2} \tag{6.71}$$

is the length of the chord of a sphere of radius r defined by the angle ψ with respect to the center. The substitution of (6.68) to (6.70) into (6.64) and then into (6.58) yields the result we seek

$$V = Gr^2 \int_\omega \rho \frac{H_{Tr} - H_{Br}}{l_r} d\omega + \frac{3}{4} Gr \int_\omega \rho \frac{H_{Tr}^2 - H_{Br}^2}{l_r} d\omega$$

$$- \frac{1}{6} Gr^2 \int_\omega \rho \frac{H_{Tr}^3 - H_{Br}^3}{l_r^3} \left(1 - \frac{3l_r^2}{4r^2}\right) d\omega . \tag{6.72}$$

We use H to denote the order of magnitude of H_{Tr} and H_{Br}. An immediate observation is that, as compared to the first term, the second term is on the relative order of H/r, and that the expression $(3l_r^2)/(4r^2)$ within the parentheses of the third term represents a quantity on the relative order of H^2/r^2. Hence, they could be neglected as further approximations.

An Additional Approximate Formula

An additional approximate formula can be derived by approximating the integral kernel in (6.59). We denote again $H_{Br} = R_B - r$ and $H_{Tr} = R_T - r$ and assume $| H_{Br}/r | \ll 1$ and $| H_{Tr}/r | \ll 1$. Substitute x according to $x = 1 + \delta x$ in (6.59). We obtain, by neglecting terms on the order of δx^2 and higher,

$$f(R_B, R_T, r, \psi)$$

$$= r^2 \int_{H_{Br}/r}^{H_{Tr}/r} \frac{(1 + 2\delta x + \delta x^2)d\delta x}{\left[\begin{array}{c} (2 - 2\cos\psi + \delta x^2)^{1/2} \\ \times \left[1 + 2\delta x(1 - \cos\psi)/(2 - 2\cos\psi + \delta x^2)\right]^{1/2} \end{array} \right]}$$

$$\approx r^2 \int_{H_{Br}/r}^{H_{Tr}/r} \frac{1 + 2\delta x}{(2 - 2\cos\psi + \delta x^2)^{1/2}} \left[1 - \frac{\delta x(1 - \cos\psi)}{2 - 2\cos\psi + \delta x^2}\right] d\delta x$$

$$= r^2 \int_{H_{Br}/r}^{H_{Tr}/r} \left[\frac{1 + 2\delta x}{(2 - 2\cos\psi + \delta x^2)^{1/2}} - \frac{\delta x(1 - \cos\psi)}{(2 - 2\cos\psi + \delta x^2)^{3/2}}\right] d\delta x . \tag{6.73}$$

The following indefinite integrals can be obtained by consulting a table of integrals

$$\int \frac{d\delta x}{(2 - 2\cos\psi + \delta x^2)^{1/2}} = \ln\left[\delta x + (2 - 2\cos\psi + \delta x^2)^{1/2}\right] + C , \tag{6.74}$$

$$\int \frac{\delta x d\delta x}{(2 - 2\cos\psi + \delta x^2)^{1/2}} = (2 - 2\cos\psi + \delta x^2)^{1/2} + C, \tag{6.75}$$

$$\int \frac{\delta x d\delta x}{(2 - 2\cos\psi + \delta x^2)^{3/2}} = -\frac{1}{(2 - 2\cos\psi + \delta x^2)^{1/2}} + C , \tag{6.76}$$

where C is a constant of integration. The substitution of (6.74) to (6.76) into (6.73) results in

$$f(R_B, R_T, r, \psi) = r^2 \left\{ \ln \left[\delta x + (2 - 2\cos\psi + \delta x^2)^{1/2} \right] \right.$$

$$\left. + 2(2 - 2\cos\psi + \delta x^2)^{1/2} + \frac{1 - \cos\psi}{(2 - 2\cos\psi + \delta x^2)^{1/2}} \right\}_{H_{Br}/r}^{H_{Tr}/r}.$$

$$(6.77)$$

Define $H' = R' - R$. We have $\delta x = H'/r$ and can write the above formula as

$$f(R_B, R_T, r, \psi) = r^2 \left[\ln \frac{H' + (l_r^2 + H'^2)^{1/2}}{r} + \frac{2(l_r^2 + H'^2)^{1/2}}{r} \right.$$

$$\left. + \frac{l_r^2}{2r(l_r^2 + H'^2)^{1/2}} \right]_{H_{Br}}^{H_{Tr}},$$

$$(6.78)$$

where l_r is defined in (6.71). The formula for the potential is then

$$V = Gr^2 \int_\omega \rho \left[\ln \frac{H' + (l_r^2 + H'^2)^{1/2}}{r} + \frac{2(l_r^2 + H'^2)^{1/2}}{r} \right.$$

$$\left. + \frac{l_r^2}{2r(l_r^2 + H'^2)^{1/2}} \right]_{H_{Br}}^{H_{Tr}} d\omega,$$

$$(6.79)$$

where we emphasize that the quantity to be substituted by the limits H_{Tr} and H_{Br} is only H', but not r and l_r, which depend solely on the radial distance of the position where V is computed. This formula is that for the flat Earth model with spherical corrections. It is left to readers to simplify it to the formula for the flat Earth model (Hints: $r^2 d\omega = dS$, the denominator r under the ln sign can be omitted without affecting the result, and set $r \to \infty$ while restricting to a finite region, i.e., keeping l_r finite). We mention that this formula can also be further developed to a form as (6.72) by approximating the kernel using a Taylor series of H', which is also left to readers.

Gravitational Attraction in Integral Form
Exact Formula
We further develop (6.52). Substitute (6.49) into (6.52). We obtain

$$F = G \int_\omega \rho \left(\int_{R_B}^{R_T} \frac{(r R'^2 - R'^3 \cos\psi) dR'}{(r^2 + R'^2 - 2r R' \cos\psi)^{3/2}} \right) d\omega.$$

$$(6.80)$$

With the variable substitution $x = R'/r$, we can write it in the form of

$$F = G \int_\omega \rho f'(R_B, R_T, r, \psi) d\omega, \tag{6.81}$$

where the kernel $f'(R_B, R_T, r, \psi)$ is defined as

$$f'(R_B, R_T, r, \psi) = r \int_{R_B/r}^{R_T/r} \frac{(x^2 - x^3 \cos \psi) dx}{(1 + x^2 - 2x \cos \psi)^{3/2}}. \tag{6.82}$$

By consulting a table of integrals, we can obtain the following indefinite integrals:

$$\int \frac{x^2 dx}{(1 + x^2 - 2x \cos \psi)^{3/2}}$$

$$= -\frac{(1 - 2 \cos^2 \psi)x + \cos \psi}{(1 - \cos^2 \psi)(1 + x^2 - 2x \cos \psi)^{1/2}}$$

$$+ \ln \left[(1 + x^2 - 2x \cos \psi)^{1/2} + x - \cos \psi \right] + C, \tag{6.83}$$

$$\int \frac{x^3 dx}{(1 + x^2 - 2x \cos \psi)^{3/2}}$$

$$= -\frac{(1 - \cos^2 \psi)x^2 - \cos \psi (5 - 6 \cos^2 \psi)x + 2 - 3 \cos^2 \psi}{(1 - \cos^2 \psi)(1 + x^2 - 2x \cos \psi)^{1/2}}$$

$$+ 3 \cos \psi \ln \left[(1 + x^2 - 2x \cos \psi)^{1/2} + x - \cos \psi \right] + C, \tag{6.84}$$

where C is a constant of integration. After some tedious algebraic operation, we obtain

$$\int \frac{(x^2 - x^3 \cos \psi) dx}{(1 + x^2 - 2x \cos \psi)^{3/2}}$$

$$= \frac{x^3}{(1 + x^2 - 2x \cos \psi)^{1/2}} - (x + 3 \cos \psi)(1 + x^2 - 2x \cos \psi)^{1/2}$$

$$- (3 \cos^2 \psi - 1) \ln \left[(1 + x^2 - 2x \cos \psi)^{1/2} + x - \cos \psi \right] + C. \tag{6.85}$$

The kernel $f'(R_B, R_T, \psi)$ can be obtained to be

$$f'(R_B, R_T, r, \psi)$$

$$= r \left\{ \frac{x^3}{(1 + x^2 - 2x \cos \psi)^{1/2}} - (x + 3 \cos \psi)(1 + x^2 - 2x \cos \psi)^{1/2} \right.$$

$$\left. - (3 \cos^2 \psi - 1) \ln \left[(1 + x^2 - 2x \cos \psi)^{1/2} + x - \cos \psi \right] \right\}_{R_B/r}^{R_T/r}. \tag{6.86}$$

The result of the gravitational potential is (6.81) and (6.86). By comparing with (6.61), we find the following relation:

$$f'(R_B, R_T, r, \psi) = r\left[\frac{x^3}{(1 + x^2 - 2x\cos\psi)^{1/2}}\right]_{R_B/r}^{R_T/r} - \frac{2}{r}f(R_B, R_T, r, \psi).$$

(6.87)

An Approximate Formula

As in the case of potential, an approximate formula can be obtained based on (6.81) and (6.86) assuming P to be very close to the Earth's surface. We still denote $H_{Br} = R_B - r$ and $H_{Tr} = R_T - r$ and assume $|H_{Br}/r| \ll 1$ and $|H_{Tr}/r| \ll 1$. We also use $Q(x)$ to denote the function within the braces in (6.86). We can thus write (6.86) as

$$f'(R_B, R_T, r, \psi) = r\left[Q(x)\right]_{1+H_{Br}/r}^{1+H_{Tr}/r}.$$

(6.88)

An approximate form of it can be derived based on the Taylor series expansion (6.63)

$$f'(R_B, R_T, r, \psi) = r\left[\left(\frac{dQ}{dx}\right)_{x=1}\frac{H_{Tr} - H_{Br}}{r}\right.$$
$$\left. + \left(\frac{d^2Q}{dx^2}\right)_{x=1}\frac{H_{Tr}^2 - H_{Br}^2}{2r^2} + \left(\frac{d^3Q}{dx^3}\right)_{x=1}\frac{H_{Tr}^3 - H_{Br}^3}{6r^3}\right].$$

(6.89)

The derivation of formulae for dQ/dx, d^2Q/dx^2, and d^3Q/dx^3 requires some tedious algebraic manipulation. Here we write out the result directly:

$$\frac{dQ}{dx} = \frac{x^2}{(1 + x^2 - 2x\cos\psi)^{1/2}} - \frac{x^3(x - \cos\psi)}{(1 + x^2 - 2x\cos\psi)^{3/2}},$$

(6.90)

$$\frac{d^2Q}{dx^2} = \frac{2x}{(1 + x^2 - 2x\cos\psi)^{1/2}} - \frac{4x^2(x - \cos\psi) + x^3}{(1 + x^2 - 2x\cos\psi)^{3/2}}$$
$$+ \frac{3x^3(x - \cos\psi)^2}{(1 + x^2 - 2x\cos\psi)^{5/2}},$$

(6.91)

$$\frac{d^3Q}{dx^3} = \frac{2}{(1 + x^2 - 2x\cos\psi)^{1/2}} - \frac{10x(x - \cos\psi) + 7x^2}{(1 + x^2 - 2x\cos\psi)^{3/2}}$$
$$+ \frac{21x^2(x - \cos\psi)^2 + 9x^3(x - \cos\psi)}{(1 + x^2 - 2x\cos\psi)^{5/2}} - \frac{15x^3(x - \cos\psi)^3}{(1 + x^2 - 2x\cos\psi)^{7/2}}.$$

(6.92)

We can further obtain, by making use of $1 - \cos \psi = 2 \sin^2(\psi/2)$,

$$\left(\frac{dQ}{dx}\right)_{x=1} = \frac{1}{4\sin(\psi/2)} = \frac{r}{2l_r}, \tag{6.93}$$

$$\left(\frac{d^2Q}{dx^2}\right)_{x=1} = \frac{3}{8\sin(\psi/2)} - \frac{1}{8\sin^3(\psi/2)} = \frac{3r}{4l_r} - \frac{r^3}{l_r^3} = -\frac{r^3}{l_r^3}\left(1 - \frac{3l_r^2}{4r^2}\right), \tag{6.94}$$

$$\left(\frac{d^3Q}{dx^3}\right)_{x=1} = \frac{3}{16\sin(\psi/2)} - \frac{5}{16\sin^3(\psi/2)} = \frac{3r}{8l_r} - \frac{5r^3}{2l_r^3} = -\frac{r^3}{2l_r^3}\left(5 - \frac{3l_r^2}{4r^2}\right), \tag{6.95}$$

where l_r is defined in (6.71). The substitution of (6.93) to (6.95) into (6.89) and then into (6.81) yields the result we seek

$$F = \frac{1}{2}Gr\int_\omega \rho\frac{H_{Tr} - H_{Br}}{l_r}d\omega - \frac{1}{2}Gr^2\int_\omega \rho\frac{H_{Tr}^2 - H_{Br}^2}{l_r^3}\left(1 - \frac{3l_r^2}{4r^2}\right)d\omega$$

$$- \frac{1}{12}Gr\int_\omega \rho\frac{H_{Tr}^3 - H_{Br}^3}{l_r^3}\left(5 - \frac{3l_r^2}{4r^2}\right)d\omega. \tag{6.96}$$

Again, we use H to denote the order of magnitude of H_{Tr} and H_{Br}. We see that, as compared to the first term, the expression $(3l_r^2)/(4r^2)$ in the parentheses of the second and third terms represents a quantity on the relative order of H/r and H^2/r^2, respectively. Therefore, they could be neglected as further approximations. Actually, the third term is on the relative order of H/r as compared to the second term and can also be neglected as an approximation.

By comparing (6.96) with the approximate formula of V, (6.72), we find the following relation between F and V:

$$F = \frac{1}{2r}V - \frac{1}{2}Gr^2\int_\omega \rho\frac{H_{Tr}^2 - H_{Br}^2}{l_r^3}d\omega - \frac{1}{3}Gr\int_\omega \rho\frac{H_{Tr}^3 - H_{Br}^3}{l_r^3}d\omega. \tag{6.97}$$

As already mentioned, the third term is on the relative order of H/r as compared to the second term and can be neglected as a further approximation.

An Additional Approximate Formula

Again, as in the case of gravitational potential, an additional approximate formula can be derived by approximating the integral kernel in (6.82). We also denote $H_{Br} = R_B - r$ and $H_{Tr} = R_T - r$ and assume $| H_{Br}/r | \ll 1$ and $| H_{Tr}/r | \ll 1$. Substitute x according to $x = 1 + \delta x$ in (6.82). We obtain, by neglecting terms on the order of

δx^2 and higher,

$$f'(R_B, R_T, r, \psi)$$

$$= r \int_{H_{Br}/r}^{H_{Tr}/r} \frac{[(1 + 2\delta x + \delta x^2) - (1 + 3\delta x + 3\delta x^2 + \delta x^3)\cos\psi]d\delta x}{(2 - 2\cos\psi + \delta x^2)^{3/2}}$$
$$\times \left[1 + 2\delta x(1 - \cos\psi)/(2 - 2\cos\psi + \delta x^2)\right]^{3/2}$$

$$\approx r \int_{H_{Br}/r}^{H_{Tr}/r} \frac{-\delta x + (1 - \cos\psi)(1 + 3\delta x)}{(2 - 2\cos\psi + \delta x^2)^{3/2}}\left[1 - \frac{3\delta x(1 - \cos\psi)}{2 - 2\cos\psi + \delta x^2}\right]d\delta x$$

$$= r \int_{H_{Br}/r}^{H_{Tr}/r} \left[\frac{-\delta x + (1 - \cos\psi)(1 + 3\delta x)}{(2 - 2\cos\psi + \delta x^2)^{3/2}} - \frac{3\delta x(1 - \cos\psi)^2}{(2 - 2\cos\psi + \delta x^2)^{5/2}}\right]d\delta x \,.$$

$$(6.98)$$

The following indefinite integrals can be obtained by consulting a table of integrals

$$\int \frac{\delta x d\delta x}{(2 - 2\cos\psi + \delta x^2)^{3/2}} = -\frac{1}{(2 - 2\cos\psi + \delta x^2)^{1/2}} + C\,, \qquad (6.99)$$

$$\int \frac{d\delta x}{(2 - 2\cos\psi + \delta x^2)^{3/2}} = \frac{\delta x}{(2 - 2\cos\psi)(2 - 2\cos\psi + \delta x^2)^{1/2}} + C\,,$$

$$(6.100)$$

$$\int \frac{\delta x d\delta x}{(2 - 2\cos\psi + \delta x^2)^{5/2}} = -\frac{1}{3(2 - 2\cos\psi + \delta x^2)^{3/2}} + C\,, \qquad (6.101)$$

where C is a constant of integration. By substituting (6.99) to (6.101) into (6.98), we obtain, after some algebraic manipulation,

$$f'(R_B, R_T, r, \psi) = r\left[\frac{1}{(2 - 2\cos\psi + \delta x^2)^{1/2}} + \frac{\delta x}{2(2 - 2\cos\psi + \delta x^2)^{1/2}}\right.$$

$$\left. - \frac{3(1 - \cos\psi)}{(2 - 2\cos\psi + \delta x^2)^{1/2}} + \frac{(1 - \cos\psi)^2}{(2 - 2\cos\psi + \delta x^2)^{3/2}}\right]_{H_{Br}/r}^{H_{Tr}/r}\,.$$

$$(6.102)$$

Define $H' = R' - R$. We have $\delta x = H'/r$ and can write the above formula as

$$f'(R_B, R_T, r, \psi) = r\left[\frac{r}{(l_r^2 + H'^2)^{1/2}} + \frac{H'}{2(l_r^2 + H'^2)^{1/2}} - \frac{3l_r^2}{2r(l_r^2 + H'^2)^{1/2}}\right.$$

$$\left. + \frac{l_r^4}{4r(l_r^2 + H'^2)^{3/2}}\right]_{H_{Br}}^{H_{Tr}}\,,$$

$$(6.103)$$

where l_r is defined in (6.71). The formula for the gravitational attraction is

$$F = Gr^2 \int_\omega \rho \left[\frac{1}{(l_r^2 + H'^2)^{1/2}} + \frac{H'}{2r(l_r^2 + H'^2)^{1/2}} - \frac{3l_r^2}{2r^2(l_r^2 + H'^2)^{1/2}} \right.$$

$$\left. + \frac{l_r^4}{4r^2(l_r^2 + H'^2)^{3/2}} \right]_{H_{Br}}^{H_{Tr}} d\omega, \tag{6.104}$$

where, as in the case of gravitational potential, we emphasize that the quantity to be substituted by the limits H_{Tr} and H_{Br} is only H', but not r and l_r, which depend solely on the radial distance of the position where F is computed. This formula is that for the flat Earth model with spherical corrections. It is left to readers to simplify it to the case of flat Earth model (Hints: $r^2 d\omega = dS$, and set $r \to \infty$ while restricting to a finite region, i.e., keeping l_r finite). We mention that this formula can also be developed to a form as (6.96) by approximating the kernel using a Taylor series of H', which is also left to readers.

In the kernel of (6.104), i.e., the integrant excluding ρ, it is evident that the first term is dominant when l_r is not significantly larger than H'; however, the second term is dominant when $l_r \gg H'$, which can be proven by expressing the kernel in terms of a power series of H'/l_r. The first term is

$$\left[\frac{1}{(l_r^2 + H'^2)^{1/2}} \right]_{H_{Br}}^{H_{Tr}} = \left[\frac{1}{l_r(1 + H'^2/l_r^2)^{1/2}} \right]_{H_{Br}}^{H_{Tr}}$$

$$= \frac{1}{l_r} \left[1 - \left(\frac{H'^2}{2l_r^2} \right) \right]_{H_{Br}}^{H_{Tr}} = -\frac{1}{2l_r} \left(\frac{H'^2}{l_r^2} \right)_{H_{Br}}^{H_{Tr}}, \tag{6.105}$$

where we have taken into account of the fact that l_r and r are independent of H' and are not to be substituted by the limits H_{Br} and H_{Tr}. The second term is

$$\left[\frac{H'}{2r(l_r^2 + H'^2)^{1/2}} \right]_{H_{Br}}^{H_{Tr}} = \left[\frac{H'}{2r} \frac{1}{l_r} \left(1 - \frac{H'^2}{2l_r^2} \right) \right]_{H_{Br}}^{H_{Tr}} = \frac{1}{2r} \left(\frac{H'}{l_r} - \frac{H'^3}{2l_r^3} \right)_{H_{Br}}^{H_{Tr}}. \tag{6.106}$$

With $l_r \leq 2r$ and $l_r \gg H'$, we understand that $1/r$ may attain the order of magnitude of $1/l_r$ in these two formulae. A part from the factor $1/(2l_r)$ or $1/(2r)$, the first term is on the order $(H'/l_r)^2$, while that of the second term is on the order H'/l_r.

The third and fourth terms in the kernel of (6.104) can be written together as

$$\left[\frac{1}{(l_r^2 + H'^2)^{1/2}} \left(-\frac{3l_r^2}{2r^2} + \frac{l_r^2}{4r^2} \frac{l_r^2}{l_r^2 + H'^2} \right) \right]_{H_{Br}}^{H_{Tr}}. \tag{6.107}$$

When $l_r \ll r$, theses terms are of smaller order of magnitude than the first term by a factor on the order of $(l_r/r)^2$. When $l_r \sim r$, we have $l_r \gg H'$ and $r \gg H'$ and can approximate these terms as, similar to the derivation of (6.105),

$$\left\{ \frac{1}{l_r} \left[1 - \left(\frac{H'^2}{2l_r^2} \right) \right] \left[-\frac{3l_r^2}{2r^2} + \frac{l_r^2}{4r^2} \left(1 - \frac{H'^2}{l_r^2} \right) \right] \right\}_{H_{Br}}^{H_{Tr}}. \tag{6.108}$$

This is on the same order of magnitude as the first term, thus being of smaller order of magnitude than the second term.

6.3.2 A Homogeneous Cap-Shaped Shell Around the Earth's Surface

Exact Formulae

We start by formulating the gravitational potential of a material surface in the form of a spherical cap as shown in Fig. 6.11, where the spherical cap is drawn in thick dashed line. We denote the radius of the sphere, where the cap is located, as R', the density of the cap as μ, and the angular radius of the cap as Θ. We derive the formulae of the gravitational potential and attraction of the cap at a point P on its symmetry axis, whose distance to the center of the sphere is r. The formulation is quite straightforward based on (1.61) and (1.62) by comparing Figs. 1.9 and 6.11. The formulae for $r \geq R'$ and $r \leq R'$ are different. We first consider $r \geq R'$, in which case, the only modifications of (1.61) required are to replace R by R', and the upper limit of the integral by the value of l when $\psi = \Theta$, i.e., $(r^2 + R'^2 - 2rR' \cos \Theta)^{1/2}$. The gravitational potential is

$$V_{r \geq R'} = 2\pi G \mu \int_{r-R'}^{(r^2+R'^2-2rR'\cos\Theta)^{1/2}} \frac{R'}{r} dl$$

$$= 2\pi G \mu \frac{R'}{r} \left[\left(r^2 + R'^2 - 2rR' \cos \Theta \right)^{1/2} - (r - R') \right]. \tag{6.109}$$

The gravitational attraction defined positive downward, $F = -dV/dr$, is

$$F_{r \geq R'} = 2\pi G \mu \frac{R'}{r^2} \left[\left(r^2 + R'^2 - 2rR' \cos \Theta \right)^{1/2} - (r - R') \right]$$

$$- 2\pi G \mu \frac{R'}{r} \left[\frac{r - R' \cos \Theta}{\left(r^2 + R'^2 - 2rR' \cos \Theta \right)^{1/2}} - 1 \right]$$

$$= 2\pi G \mu \frac{R'^2}{r^2} \left[\frac{R' - r \cos \Theta}{\left(r^2 + R'^2 - 2rR' \cos \Theta \right)^{1/2}} + 1 \right]. \tag{6.110}$$

Fig. 6.11 A cap-shaped shell as a plate on sphere

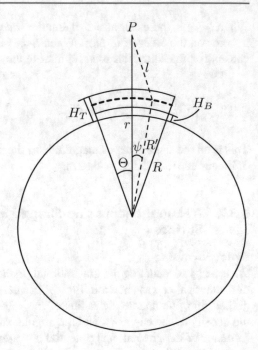

As shown in Fig. 6.11, the gravitational potential and attraction of a cap-shaped shell of density ρ, height from H_B to H_T, and radius Θ with respect to a sphere of radius R can be obtained using (6.109) and (6.110) by replacing μ by $\rho\, dR'$, and then integrating from $R_B = R + H_B$ to $R_T = R + H_T$. As (6.109) and (6.110) are valid only when $r \geq R'$, we can only obtain the result for $r \geq R_T$. Together with the variable substitution $x = R'/r$, we have

$$
V_{r \geq R_T} = 2\pi G\rho \int_{R_B}^{R_T} \frac{R'}{r} \left[\left(r^2 + R'^2 - 2rR'\cos\Theta \right)^{1/2} - (r - R') \right] dR'
$$

$$
= 2\pi G\rho r^2 \int_{R_B/r}^{R_T/r} \left[x \left(1 + x^2 - 2x\cos\Theta \right)^{1/2} - x + x^2 \right] dx , \qquad (6.111)
$$

$$
F_{r \geq R_T} = 2\pi G\rho \int_{R_B}^{R_T} \frac{R'^2}{r^2} \left[\frac{R' - r\cos\Theta}{\left(r^2 + R'^2 - 2rR'\cos\Theta \right)^{1/2}} + 1 \right] dR'
$$

$$
= 2\pi G\rho r \int_{R_B/r}^{R_T/r} \left[\frac{x^3 - x^2\cos\Theta}{\left(1 + x^2 - 2x\cos\Theta \right)^{1/2}} + x^2 \right] dx . \qquad (6.112)
$$

We note that F is the attraction on the symmetry axis of the cap-shaped shell positive downward.

By consulting a table of integrals, we can find the following indefinite integrals:

$$
\int x \left(1 + x^2 - 2x \cos \Theta\right)^{1/2} dx = \frac{1}{3} \left(1 + x^2 - 2x \cos \Theta\right)^{3/2}
$$

$$
+ \frac{1}{2}(x - \cos \Theta) \cos \Theta \left(1 + x^2 - 2x \cos \Theta\right)^{1/2}
$$

$$
+ \frac{1}{2} \sin^2 \Theta \cos \Theta \ln \left[\left(1 + x^2 - 2x \cos \Theta\right)^{1/2} + x - \cos \Theta\right] + C, \qquad (6.113)
$$

$$
\int \frac{x^3 dx}{\left(1 + x^2 - 2x \cos \Theta\right)^{1/2}}
$$

$$
= \left(\frac{x^2}{3} + \frac{5x \cos \Theta}{6} + \frac{5 \cos^2 \Theta}{3} - \frac{2}{3}\right) \left(1 + x^2 - 2x \cos \Theta\right)^{1/2}
$$

$$
+ \left(\frac{5 \cos^3 \Theta}{2} - \frac{3 \cos \Theta}{2}\right) \ln \left[\left(1 + x^2 - 2x \cos \Theta\right)^{1/2} + x - \cos \Theta\right] + C,
$$

$$
(6.114)
$$

$$
\int \frac{x^2 dx}{\left(1 + x^2 - 2x \cos \Theta\right)^{1/2}} = \left(\frac{x}{2} + \frac{3 \cos \Theta}{2}\right) \left(1 + x^2 - 2x \cos \Theta\right)^{1/2}
$$

$$
+ \left(\frac{3 \cos^2 \Theta}{2} - \frac{1}{2}\right) \ln \left[\left(1 + x^2 - 2x \cos \Theta\right)^{1/2} + x - \cos \Theta\right] + C.
$$

$$
(6.115)
$$

Substitute these integral formulae into (6.111) and (6.112). Together with the fundamental integral formula $\int x^n dx = [1/(n + 1)]x^{n+1} + C$, we obtain

$$
V_{r \geq R_T} = 2\pi G \rho r^2 \left\{ \frac{1}{3} \left(1 + x^2 - 2x \cos \Theta\right)^{3/2} \right.
$$

$$
+ \frac{1}{2}(x - \cos \Theta) \cos \Theta \left(1 + x^2 - 2x \cos \Theta\right)^{1/2} - \frac{1}{2}x^2 + \frac{1}{3}x^3
$$

$$
\left. + \frac{1}{2} \sin^2 \Theta \cos \Theta \ln \left[\left(1 + x^2 - 2x \cos \Theta\right)^{1/2} + x - \cos \Theta\right] \right\}_{R_B/r}^{R_T/r},
$$

$$
(6.116)
$$

$$F_{r \geq R_T} = \frac{2\pi G\rho r}{3}\Bigg\{ x^3 + (x^2 + x\cos\Theta + 1 - 3\sin^2\Theta)\left(1 + x^2 - 2x\cos\Theta\right)^{1/2}$$

$$- 3\sin^2\Theta\cos\Theta \ln\left[\left(1 + x^2 - 2x\cos\Theta\right)^{1/2} + x - \cos\Theta\right]\Bigg\}_{R_B/r}^{R_T/r}.$$

$$(6.117)$$

We will also need the result for the spherical cap-shaped shell for $r \leq R_B$, which can be formulated similar to the case of $r \geq R_T$. For the spherical cap drawn in thick dashed line in Fig. 6.11, the gravitational potential for $r \leq R'$ can be obtained based on (1.62) in the same way as deriving (6.109) from (1.61). This is equivalent to replace the lower limit of the integral in (6.109) by $R' - r$

$$V_{r \leq R'} = 2\pi G\mu \int_{R'-r}^{(r^2 + R'^2 - 2rR'\cos\Theta)^{1/2}} \frac{R'}{r} dl$$

$$= 2\pi G\mu \frac{R'}{r}\left[\left(r^2 + R'^2 - 2rR'\cos\Theta\right)^{1/2} - (R' - r)\right]. \qquad (6.118)$$

Similar to (6.110), the gravitational attraction can be obtained to be

$$F_{r \leq R'} = 2\pi G\mu \frac{R'^2}{r^2}\left[\frac{R' - r\cos\Theta}{\left(r^2 + R'^2 - 2rR'\cos\Theta\right)^{1/2}} - 1\right]. \qquad (6.119)$$

The result for the spherical cap-shaped shell for $r \leq R_B$ can be derived similar to (6.116) and (6.117)

$$V_{r \leq R_B} = 2\pi G\rho r^2\Bigg\{\frac{1}{3}\left(1 + x^2 - 2x\cos\Theta\right)^{3/2}$$

$$+ \frac{1}{2}(x - \cos\Theta)\cos\Theta\left(1 + x^2 - 2x\cos\Theta\right)^{1/2} - \frac{1}{3}x^3 + \frac{1}{2}x^2$$

$$+ \frac{1}{2}\sin^2\Theta\cos\Theta \ln\left[\left(1 + x^2 - 2x\cos\Theta\right)^{1/2} + x - \cos\Theta\right]\Bigg\}_{R_B/r}^{R_T/r},$$

$$(6.120)$$

$$F_{r \leq R_B} = \frac{2\pi G\rho r}{3}\Bigg\{ -x^3 + (x^2 + x\cos\Theta + 1 - 3\sin^2\Theta)\left(1 + x^2 - 2x\cos\Theta\right)^{1/2}$$

$$- 3\sin^2\Theta\cos\Theta \ln\left[\left(1 + x^2 - 2x\cos\Theta\right)^{1/2} + x - \cos\Theta\right]\Bigg\}_{R_B/r}^{R_T/r}.$$

$$(6.121)$$

Approximate Formulae

Denote the quantities in the braces of the expressions of $V_{r \geq R_T}$ and $F_{r \geq R_T}$ as Q_V and Q_F, respectively. Simpler approximate formulae of V and F can be obtained by applying the Taylor series expansion (6.63) to Q_V and Q_F. We have, with $R_{Br} = R_B - r$ and $R_{Tr} = R_T - r$, and assuming $| R_{Br}/r | \ll 1$ and $| R_{Tr}/r | \ll 1$,

$$
V_{r \geq R_T} = 2\pi G \rho r^2 \left[\left(\frac{dQ_V}{dx} \right)_{x=1} \frac{H_{Tr} - H_{Br}}{r} \right.
$$
$$
\left. + \left(\frac{d^2 Q_V}{dx^2} \right)_{x=1} \frac{H_{Tr}^2 - H_{Br}^2}{2r^2} + \left(\frac{d^3 Q_F}{dx^3} \right)_{x=1} \frac{H_{Tr}^3 - H_{Br}^3}{6r^3} \right].
$$

$$(6.122)$$

$$
F_{r \geq R_T} = \frac{2\pi G \rho r}{3} \left[\left(\frac{dQ_F}{dx} \right)_{x=1} \frac{H_{Tr} - H_{Br}}{r} \right.
$$
$$
\left. + \left(\frac{d^2 Q_F}{dx^2} \right)_{x=1} \frac{H_{Tr}^2 - H_{Br}^2}{2r^2} + \left(\frac{d^3 Q_F}{dx^3} \right)_{x=1} \frac{H_{Tr}^3 - H_{Br}^3}{6r^3} \right].
$$

$$(6.123)$$

The derivation of the formulae for $dQ_{V,F}/dx$, $d^2 Q_{V,F}/dx^2$, and $d^3 Q_{V,F}/dx^3$ requires some tedious algebraic manipulations. Here we write out the result directly

$$
\frac{dQ_V}{dx} = x \left(1 + x^2 - 2x \cos \Theta \right)^{1/2} - x + x^2,
$$

$$(6.124)$$

$$
\frac{d^2 Q_V}{dx^2} = \left(1 + x^2 - 2x \cos \Theta \right)^{1/2} + \frac{x(x - \cos \Theta)}{\left(1 + x^2 - 2x \cos \Theta \right)^{1/2}} - 1 + 2x,
$$

$$(6.125)$$

$$
\frac{d^3 Q_V}{dx^3} = \frac{x + 2(x - \cos \Theta)}{\left(1 + x^2 - 2x \cos \Theta \right)^{1/2}} - \frac{x(x - \cos \Theta)^2}{\left(1 + x^2 - 2x \cos \Theta \right)^{3/2}} + 2;
$$

$$(6.126)$$

$$
\frac{dQ_F}{dx} = 3x^2 + 3x \left(1 + x^2 - 2x \cos \Theta \right)^{1/2} - \frac{3x(1 - x \cos \Theta)}{\left(1 + x^2 - 2x \cos \Theta \right)^{1/2}},
$$

$$(6.127)$$

$$
\frac{d^2 Q_F}{dx^2} = 6x + \frac{3x(2x - \cos \Theta)}{\left(1 + x^2 - 2x \cos \Theta \right)^{1/2}} + \frac{3x(1 - x \cos \Theta)(x - \cos \Theta)}{\left(1 + x^2 - 2x \cos \Theta \right)^{3/2}},
$$

$$(6.128)$$

$$\frac{d^3 Q_F}{dx^3} = 6 + \frac{3(x - \cos\Theta) + 9x}{\left(1 + x^2 - 2x\cos\Theta\right)^{1/2}}$$

$$+ \frac{3(1 - x\cos\Theta)(x - \cos\Theta) + 3x(1 - x\cos\Theta) - 6x^2(x - \cos\Theta)}{\left(1 + x^2 - 2x\cos\Theta\right)^{3/2}}$$

$$- \frac{9x(1 - x\cos\Theta)(x - \cos\Theta)^2}{\left(1 + x^2 - 2x\cos\Theta\right)^{5/2}} . \tag{6.129}$$

With $1 - \cos\Theta = 2\sin^2(\Theta/2)$ and

$$a_r = r(2 - 2\cos\Theta)^{1/2} = 2r\sin\frac{\Theta}{2} , \tag{6.130}$$

we obtain

$$\left(\frac{dQ_V}{dx}\right)_{x=1} = 2\sin\frac{\Theta}{2} = \frac{a_r}{r} , \tag{6.131}$$

$$\left(\frac{d^2 Q_V}{dx^2}\right)_{x=1} = 1 + 3\sin\frac{\Theta}{2} = 1 + \frac{3a_r}{2r} , \tag{6.132}$$

$$\left(\frac{d^3 Q_V}{dx^3}\right)_{x=1} = 2 + \frac{1}{2\sin(\Theta/2)} + \frac{3}{2}\sin\frac{\Theta}{2} = 2 + \frac{r}{a_r} + \frac{3a_r}{4r}$$

$$= \frac{r}{a_r}\left(1 + \frac{2a_r}{r} + \frac{3a_r^2}{4r^2}\right) ; \tag{6.133}$$

$$\left(\frac{dQ_F}{dx}\right)_{x=1} = 3 + 3\sin\frac{\Theta}{2} = 3\left(1 + \frac{a_r}{2r}\right) , \tag{6.134}$$

$$\left(\frac{d^2 Q_F}{dx^2}\right)_{x=1} = 6 + \frac{3}{2\sin(\Theta/2)} + \frac{9}{2}\sin\frac{\Theta}{2} = 6 + \frac{3r}{a_r} + \frac{9a_r}{4r}$$

$$= \frac{3r}{a_r}\left(1 + \frac{2a_r}{r} + \frac{3a_r^2}{4r^2}\right) , \tag{6.135}$$

$$\left(\frac{d^3 Q_F}{dx^3}\right)_{x=1} = 6 + \frac{15}{4\sin(\Theta/2)} + \frac{9}{4}\sin\frac{\Theta}{2} = 6 + \frac{15r}{2a_r} + \frac{9a_r}{8r}$$

$$= \frac{15r}{2a_r}\left(1 + \frac{4a_r}{5r} + \frac{3a_r^2}{20r^2}\right) . \tag{6.136}$$

The substitution of (6.131) to (6.136) into (6.122) and (6.123) yields the results

$$
\begin{aligned}
V_{r \geq R_T} = 2\pi G \rho a_r \Bigg[& (H_{Tr} - H_{Br}) + \frac{H_{Tr}^2 - H_{Br}^2}{2a_r} \left(1 + \frac{3a_r}{2r} \right) \\
& + \frac{H_{Tr}^3 - H_{Br}^3}{6a_r^2} \left(1 + \frac{2a_r}{r} + \frac{3a_r^2}{4r^2} \right) \Bigg],
\end{aligned}
\tag{6.137}
$$

$$
\begin{aligned}
F_{r \geq R_T} = 2\pi G \rho \Bigg[& (H_{Tr} - H_{Br}) \left(1 + \frac{a_r}{2r} \right) + \frac{H_{Tr}^2 - H_{Br}^2}{2a_r} \left(1 + \frac{2a_r}{r} + \frac{3a_r^2}{4r^2} \right) \\
& + \frac{5 \left(H_{Tr}^3 - H_{Br}^3 \right)}{4 r a_r} \left(1 + \frac{4a_r}{5r} + \frac{3a_r^2}{20r^2} \right) \Bigg].
\end{aligned}
\tag{6.138}
$$

We see that, if $\mid H_{Tr} \mid \sim \mid H_{Br} \mid \ll a_r \ll r$, we have $V_{r \geq R_T} \approx 2\pi G \rho a_r (H_{Tr} - H_{Br})$ and $F_{r \geq R_T} \approx 2\pi G \rho (H_{Tr} - H_{Br})$. Readers are suggested to compare these with the formulae of a homogeneous cylinder formulated in Sect. 1.4.4.

The formulae of $V_{r \leq R_B}$ and $F_{r \leq R_B}$ can be similarly obtained to be

$$
\begin{aligned}
V_{r \leq R_B} = 2\pi G \rho a_r \Bigg[& (H_{Tr} - H_{Br}) + \frac{H_{Tr}^2 - H_{Br}^2}{2a_r} \left(-1 + \frac{3a_r}{2r} \right) \\
& + \frac{H_{Tr}^3 - H_{Br}^3}{6a_r^2} \left(1 - \frac{2a_r}{r} + \frac{3a_r^2}{4r^2} \right) \Bigg],
\end{aligned}
\tag{6.139}
$$

$$
\begin{aligned}
F_{r \leq R_B} = 2\pi G \rho \Bigg[& (H_{Tr} - H_{Br}) \left(-1 + \frac{a_r}{2r} \right) + \frac{H_{Tr}^2 - H_{Br}^2}{2a_r} \left(1 - \frac{2a_r}{r} + \frac{3a_r^2}{4r^2} \right) \\
& + \frac{5 \left(H_{Tr}^3 - H_{Br}^3 \right)}{4 r a_r} \left(1 - \frac{4a_r}{5r} + \frac{3a_r^2}{20r^2} \right) \Bigg].
\end{aligned}
\tag{6.140}
$$

It is left to readers to study the case when $R_B \leq r \leq R_T$.

As will be seen in the next two sections, the formula of a cap-shaped shell, which is indeed a finite spherical plate, can be used to alleviate the singularity of integrals in the computation of gravity anomaly and indirect effect.

6.4   Gravity Reduction for the Spherical Earth Model

6.4.1   Some Relations Between the Formulations for the Flat and Spherical Earth Models

We have assumed the Earth to be flat in formulating gravity reduction, which is appropriate if the influence of only a relatively small region around the location of interest is considered. However, if the influence of faraway regions is to be considered, the curvature of the Earth must be taken into account, which requires to treat the Earth as a sphere.

The Bouguer Plate Corrections

Just by inspecting the formulae of the gravitational attraction of a homogeneous cylinder and a homogeneous sphere, we can already realize that there may be a conflict between the flat and spherical Earth models for the Bouguer plate correction. For the flat Earth model, the correction is $-2\pi G\rho H$ for a cylindrical plate of thickness H and infinite radius. For the spherical Earth model, the correction is $-4\pi G\rho H$ for a spherical plate of thickness H covering the whole Earth's surface. The latter is twice as large as the former. This conflict is true if the plate covers the whole Earth's surface. However, this cannot be true in the real world, where the plate extends only to within a limited distance from the location of interest. For example, H vanishes over the oceans that cover the majority of the Earth's surface. Hence, as to be demonstrated below, $-2\pi G\rho H$ is an approximate formula for practical applications.

The assertion that $-2\pi G\rho H$ is an approximate formula can be verified by using the formula of gravitational attraction of a homogeneous cap-shaped shell lying on a sphere formulated in the previous section (6.117). We first derive an exact formula. In the application here, P is on top of the shell, $r = R_T = R + H$. Furthermore, the bottom of the shell lies on the surface of the sphere of radius R, i.e., $R_B = R$. What we need is the Bouguer plate correction, which has an opposite sign as the attraction, and we denote it as $\Delta_2 g_\Theta$ to indicate that the plate has a limited angular radius of Θ. We also denote $t = R/R_T$ for brevity. The result can be written as (Hagiwara, 1975)

$$
\Delta_2 g_\Theta = -\frac{2\pi G\rho R_T}{3}\left\{1 - t^3 + 2(2 + \cos\Theta - 3\sin^2\Theta)\sin\frac{\Theta}{2}\right.
$$

$$
- (t^2 + t\cos\Theta + 1 - 3\sin^2\Theta)\left(1 + t^2 - 2t\cos\Theta\right)^{1/2}
$$

$$
\left.- 3\sin^2\Theta\cos\Theta\ln\left[\frac{2\sin(\Theta/2)[1 + \sin(\Theta/2)]}{\left(1 + t^2 - 2t\cos\Theta\right)^{1/2} + t - \cos\Theta}\right]\right\}. \quad (6.141)
$$

However, it is not evident to verify analytically the assertion by comparing the exact formula (6.141) with the approximate formula for the flat Earth model given by (6.10), $-2\pi G\rho H$. Hence, we use (6.138) to derive an approximate formula for the spherical Earth model, which is straightforward with $r = R_T$, $H_{Br} = -H$, and $H_{Tr} = 0$. Denoting a_r defined in (6.130) as a_T when $r = R_T$, i.e., $a_T = 2R_T \sin(\Theta/2)$, the result is (keeping only two terms)

$$
\begin{aligned}
\Delta_2 g_\Theta &= -2\pi G\rho H \left[\left(1 + \frac{a_T}{2R_T}\right) - \frac{H}{2a_T}\left(1 + \frac{2a_T}{R_T} + \frac{3a_T^2}{4R_T^2}\right)\right] \\
&= -2\pi G\rho H \left[\left(1 + \sin\frac{\Theta}{2}\right) - \frac{H}{2a_T}\left(1 + 4\sin\frac{\Theta}{2} + 3\sin^2\frac{\Theta}{2}\right)\right].
\end{aligned}
\tag{6.142}
$$

This formula corresponds to the more accurate approximate formula for the flat Earth model, (6.9), with spherical corrections. In the less accurate approximate formula for the flat Earth model, $-2\pi G\rho H$, the term corresponding to the second term in the brackets of this formula is neglected. The spherical correction to the leading term is $\sin(\Theta/2)$, which increases with Θ practically linearly when Θ is small. For example, it attains $0.026, 0.052, 0.078$ and 0.105 when $\Theta = 3°, 6°, 9°$ and $12°$, corresponding to plate radii of about $333, 666, 999$ and 1332 km in metric length of arc on the spherical surface.

In Fig. 6.12, we show a comparison among the results computed using both the exact and approximate formulae for both the flat and spherical Earth models, i.e., (6.8), (6.10) and (6.141), (6.142). Results are shown for $H = 0.5, 1, 2$, and 3 km. The values of the approximate formulae are closer to those of the exact formulae when H is smaller. Our subsequent discussions in this paragraph will be based on $H = 3$ km, which is a reasonable extreme circumstance, while keeping in mind that the spherical Earth model represents better the Earth. A first observance is that the most known and simplest approximate formula $-2\pi G\rho H$ given in (6.10) for the flat Earth model is still a good approximation for a spherical plate of radius within around $0.25° \leq \Theta \leq 5°$, or $28 \leq a_T \leq 555$ km. Beyond $\Theta > 5°$, the difference between the values of the two exact formulae becomes larger and larger with the increase of Θ, but the value of each approximate formula is practically indifferent from that of the corresponding exact formula. The two exact formulae lead to practically the same result when Θ does not exceed around $5°$, or a_T does not exceed around 555 km. As already mentioned, the approximate formula (6.142) for the spherical Earth model has one more term than $-2\pi G\rho H$ given in (6.10) for the flat Earth model; therefore, the former approximates better the exact formula than the later when Θ is small, which becomes more and more apparent when Θ decreases below around $0.25°$. However, even (6.142) becomes inaccurate when the radius of the plate is not larger enough than its height (this is already apparent in the figure for $H = 2$ km, and the far left dot is missing in the figure for $H = 3$ km, as it is too far away). Finally, we note that the more accurate approximate formula (6.9) for the flat Earth model would yield practically the same result as the approximate

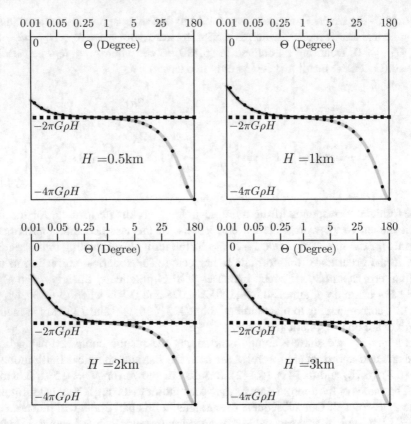

Fig. 6.12 A comparison among the results of the Bouguer plate correction for both the flat and spherical Earth models computed using both the exact and approximate formulae. Thin black line: Exact formula (6.8) for the flat Earth model; Squares: Approximate formula (6.10) for the flat Earth model; Thick gray line: Exact formula (6.141) for the spherical Earth model; Dots (the far left dot is missing for $H = 3$ km, as it is too far away): Approximate formula (6.142) for the spherical Earth model

formula (6.142) for the spherical Earth model when Θ is small, particularly when $\Theta \leq 0.25°$. This can be inferred straightforwardly by inspecting the two formulae.

Density of the Helmert Condensation Layer

For the flat Earth model, the condensed surface density is $\mu = \rho H$. In the case of a sphere, this does not conserve mass. For the mass to be conserved, the surface density should be determined according to

$$(R^2 \sin\theta d\theta d\lambda)\mu = \int_R^{R+H} (r^2 \sin\theta d\theta d\lambda)\rho dr = \frac{\rho}{3}[(R+H)^3 - R^3]\sin\theta d\theta d\lambda,$$

$$(6.143)$$

resulting in

$$\mu = \rho H \left(1 + \frac{H}{R} + \frac{H^2}{3R^2} \right).$$ (6.144)

We mention that, even in the formulation of gravity anomaly and indirect effect for the spherical Earth model, the density for the flat Earth model, $\mu = \rho H$, is still often chosen. In this book, we choose to use the spherical condensation density provided in (6.144), so that the mass of the whole Earth is not altered. In applications, the choice should be consistent throughout, since the effects based on different choices are different.

Isostatic Compensation Depth

We consider only the Airy–Heiskanen model, which is most used. If the formula for the flat Earth model, $H' = (\rho/\Delta\rho)H$, is applied to the spherical case, the mass is no longer conserved. For the mass to be conserved, the compensation is governed by

$$\int_{R-(T+H')}^{R-T} (r^2 \sin\theta d\theta d\lambda)\Delta\rho dr = \int_{R}^{R+H} (r^2 \sin\theta d\theta d\lambda)\rho dr.$$ (6.145)

After integration analytically, we obtain an algebraic equation for H'

$$\frac{\Delta\rho}{3} \left\{ (R-T)^3 - (R-(T+H')]^3 \right\} = \frac{\rho}{3}[(R+H)^3 - R^3].$$ (6.146)

The exact formula for H' can only be written in an iterative form

$$H' = \frac{R^2\rho}{(R-T)^2\Delta\rho} \left(H + \frac{H^2}{R} + \frac{H^3}{3R^2} \right) + \frac{H'^2}{R-T} - \frac{H'^3}{3(R-T)^2}.$$ (6.147)

A formula accurate to the relative order of magnitude of H^2/R^2 can be derived iteratively

$$H' = \frac{R^2\rho H}{(R-T)^2\Delta\rho} \left\{ 1 + \frac{H}{R} \left[1 + \frac{R^3\rho}{(R-T)^3\Delta\rho} \right] \right.$$

$$\left. + \frac{H^2}{3R^2} \left[1 + \frac{6R^3\rho}{(R-T)^3\Delta\rho} + \frac{5R^6\rho^2}{(R-T)^6\Delta\rho^2} \right] \right\}.$$ (6.148)

The spherical correction is on the order of $\sim1\%$ for $T \sim 30$ km. This is the formulation for land. The formulation for ocean is left to readers.

Similar to the case of Helmert anomaly, even in the formulation of gravity anomaly and indirect effect for the spherical Earth model, the compensation depth for the flat Earth model $H' = (\rho/\Delta\rho)H$ is still often chosen. In this book, we choose to use the spherical compensation depth provided in (6.147) or (6.148), so that the mass of the whole Earth is not altered. In applications, the choice should be consistent throughout, since the effects based on different choices are different.

6.4.2   The Complete Bouguer Anomaly

The complete Bouguer anomaly is obtained by removing the effect of topographic mass, i.e., the mass above the geoid, and then projecting the gravity to the geoid by applying the free air correction. We consider only the topographic correction, as applying free air correction is straightforward.

For the flat Earth model, the topographic correction is divided into a plate correction $\Delta_2 g$ and a terrain correction $\Delta_3 g$. Here we consider the topographic correction as a whole and denote it as $\Delta_{2+3} g = \Delta_2 g + \Delta_3 g$, which is the gravitational attraction of the topographic mass positive downward with sign reversed.

We consider the topographic mass as one layer with density varying laterally but not vertically. The formulae derived in the previous section can be used. The point P is on the ground; thus $r = R + H_P$, where H_P is the orthometric height of the point P. The bottom of the layer is on the geoid, which is approximated as a sphere of radius R, i.e., $R_B = R$. We denote $R_T = R + H$, where H is the orthometric height of the topography of the running point of integration, i.e., the location of $d\omega$. According to (6.81) and (6.86), the exact formula is

$$\Delta_{2+3} g = G \int_\omega \rho f'_{2+3}(H_P, H, \psi) d\omega, \tag{6.149}$$

where

$$
\begin{aligned}
f'_{2+3}(H_P, H, \psi) = -r \Bigg\{ & \frac{x^3}{(1 + x^2 - 2x \cos \psi)^{1/2}} \\
& - (x + 3 \cos \psi)(1 + x^2 - 2x \cos \psi)^{1/2} - (3 \cos^2 \psi - 1) \\
& \times \ln \left[(1 + x^2 - 2x \cos \psi)^{1/2} + x - \cos \psi \right] \Bigg\}_{R/r}^{(R+H)/r} .
\end{aligned}
\tag{6.150}
$$

When $\psi = 0$, we have $H = H_P$ and $(R + H)/r = 1$. Hence, this is a singular point where $f'_{2+3}(H_P, H, \psi)$ tends to infinity and is in need of a special treatment. The simplest approach is to treat a small region around P separately. For the flat Earth model, we understand this small region to be a disk of radius a. In the more accurate spherical Earth model, we understand this small region as a cap-shaped shell of angular radius $\Theta = a/R$. As such, we write (6.149) and (6.150) as

$$\Delta_{2+3}g = G \int_{\omega_{\psi \geq \Theta}} \rho f'_{2+3}(H_P, H, \psi)d\omega + \Delta_2 g_\Theta + \Delta_3 g_\Theta , \qquad (6.151)$$

where $\omega_{\psi \geq \Theta}$ is the $\psi \geq \Theta$ portion of the unit sphere ω, and $\Delta_2 g_\Theta$ and $\Delta_3 g_\Theta$ are the Bouguer plate and terrain corrections of the mass within an angular distance of Θ.

For the flat Earth model, $\Delta_2 g_\Theta$ and $\Delta_3 g_\Theta$ degenerate to the Bouguer plate and terrain corrections within a distance of a as formulated in Sect. 6.1 (using the exact formula without making $a \to \infty$). For the spherical Earth model, the formula of $\Delta_2 g_\Theta$ is provided in the previous subsection, and two expressions of $\Delta_3 g_\Theta$ are to be derived in the rest of this subsection.

The exact expression of $\Delta_3 g_\Theta$ for the spherical Earth model can be derived according to (6.81) and (6.86). Here, $r = R + H_P$, $R_B = R + H_P$, and $R_T = R + H$, and hence,

$$\Delta_3 g_\Theta = G \int_{\omega_{\psi \leq \Theta}} \rho f'_3(H_P, H, \psi)d\omega , \qquad (6.152)$$

where $\omega_{\psi \leq \Theta}$ is the $\psi \leq \Theta$ portion of the unit sphere ω, and

$$f'_3(H_P, H, \psi) = -r \left\{ \frac{x^3}{(1 + x^2 - 2x \cos \psi)^{1/2}} \right.$$
$$- (x + 3 \cos \psi)(1 + x^2 - 2x \cos \psi)^{1/2} - (3 \cos^2 \psi - 1)$$
$$\left. \times \ln \left[(1 + x^2 - 2x \cos \psi)^{1/2} + x - \cos \psi \right] \right\}_1^{(R+H)/r} . \qquad (6.153)$$

We see that, when $H = H_P$, we have $(R + H)/r = 1$, and thus $f'_3(H_P, H, \psi) = 0$ when $\psi \neq 0$. If we set $\omega_{\psi \leq \Theta}$ to be an extremely small region around P where the topography is approximated by a plane, we then have $H = H_P$, and thus $f'_3(H_P, H, \psi) = 0$ and $\Delta_3 g_\Theta = 0$. Physically, this is to approximate the topographic mass within $\psi \leq \Theta$ by a cylinder of height H_P. Without this approximation, it is not evident what $f'_3(H_P, H, \psi)$ is when $\psi \to 0$.

The computation of $\Delta_3 g_\Theta$ is more straightforward using an approximate formula. A first approximate formula can be derived according to (6.96). Here, as $r = R + H_P$, $R_B = R + H_P$, and $R_T = R + H$, we have $H_{Tr} = R_T - r = H - H_P$ and $H_{Br} = R_B - r = 0$. We obtain, by reversing the sign per definition,

$$\Delta_3 g_\Theta = -\frac{1}{2}Gr \int_{\omega_{\psi \leq \Theta}} \rho \frac{H - H_P}{l_r} d\omega + \frac{1}{2}Gr^2 \int_{\omega_{\psi \leq \Theta}} \rho \frac{(H - H_P)^2}{l_r^3} \left(1 - \frac{3l_r^2}{4r^2}\right) d\omega$$
$$+ \frac{1}{12}Gr \int_{\omega_{\psi \leq \Theta}} \rho \frac{(H - H_P)^3}{l_r^3} \left(5 - \frac{3l_r^2}{4r^2}\right) d\omega , \qquad (6.154)$$

where l_r is defined in (6.71) and is copied to here for convenience of reference

$$l_r = r(2 - 2\cos\psi)^{1/2} = 2r\sin\frac{\psi}{2}. \tag{6.155}$$

In (6.154), the expression $(3l_r^2)/(4r^2)$ in the parentheses of the second and third terms represents quantities on the relative order of H/r and H^2/r^2 in magnitude as compared to the first term, and the last term is on the relative order of $(H - H_P)/r$ as compared to the second term. Hence, they could be neglected as further approximations. Actually, the second term corresponds to the formula for the flat Earth model, and is the leading term, as to be made clear below.

Another approximate formula can be obtained according to (6.104). This is an extension of the formula for the flat Earth model formulated in Sect. 6.1. We have, denoting $\Delta H = H - H_P$ for brevity,

$$\begin{aligned}
\Delta_3 g_\Theta &= -Gr^2 \int_{\omega_{\psi \le \Theta}} \rho \left[\frac{1}{(l_r^2 + H'^2)^{1/2}} + \frac{H'}{2r(l_r^2 + H'^2)^{1/2}} \right. \\
&\qquad\qquad \left. - \frac{3l_r^2}{2r^2(l_r^2 + H'^2)^{1/2}} + \frac{l_r^4}{4r^2(l_r^2 + H'^2)^{3/2}} \right]_0^{\Delta H} d\omega \\
&= -Gr^2 \int_{\omega_{\psi \le \Theta}} \rho \left\{ \left[\frac{1}{(l_r^2 + \Delta H^2)^{1/2}} - \frac{1}{l_r} \right] + \left[\frac{\Delta H}{2r(l_r^2 + \Delta H^2)^{1/2}} \right] \right. \\
&\qquad\qquad \left. - \left[\frac{3l_r^2}{2r^2(l_r^2 + \Delta H^2)^{1/2}} - \frac{3l_r}{2r^2} \right] + \left[\frac{l_r^4}{4r^2(l_r^2 + \Delta H^2)^{3/2}} - \frac{l_r}{4r^2} \right] \right\} d\omega.
\end{aligned} \tag{6.156}$$

It can be readily seen that the first term corresponds to the formula for the flat Earth model, (6.14), which is dominant when l_r is small, and the rest terms are the spherical correction.

When (6.156) is developed into the form of (6.154), we see differences in the smallest terms, which is due to the fact that different approximations were made in deriving the two formulae. Approximation is made after integration in deriving (6.154), but it is made before integration in deriving (6.156). We mention that the first term in (6.156) corresponds to the second term in (6.154), and the second to the first.

6.4.3 The Helmert Anomaly

The Helmert anomaly is computed by removing the topographic mass, restoring the mass back to the geoid as a material surface, and then projecting the gravity from the station onto the geoid using the free air correction. The removal of topographic mass and projection onto the geoid have been formulated. Here we focus on the Helmert condensation layer.

The gravitational attraction of the Helmert condensation layer, called Helmert condensation correction, can be formulated straightforwardly as a surface integral. By inspecting the derivation of (6.52), keeping in mind that the mass here is condensed on the sphere of radius R, we can readily infer that, at the point P,

$$\Delta_4 g = G \int_\omega \mu \frac{R^2(r - R\cos\psi)}{l^3} d\omega, \tag{6.157}$$

where l is defined as

$$l = (r^2 + R^2 - 2rR\cos\psi)^{1/2}. \tag{6.158}$$

According to (6.158), we can write (6.157) in the form of

$$\Delta_4 g = G \int_\omega \mu \frac{R^2(r^2 - R^2 + l^2)}{2rl^3} d\omega = G \int_\omega \mu \left[\frac{R^2}{2rl} + \frac{R^2(r^2 - R^2)}{2rl^3} \right] d\omega. \tag{6.159}$$

Substitute $r = R + H_P$ and (6.144) into this formula. We obtain

$$\Delta_4 g = G \int_\omega \rho H \left(1 + \frac{H}{R} + \frac{H^2}{3R^2}\right) \left[\frac{R^2}{2rl} + \frac{R^3 H_P}{rl^3} \left(1 + \frac{H_P}{2R}\right) \right] d\omega. \tag{6.160}$$

There is no singularity in this integral. The quantities H/R or H_P/R in the parentheses are of smaller orders of magnitude and can be neglected as approximation.

Similar to the case of complete Bouguer correction, we consider a small region $\omega_{\psi \le \Theta}$. We separate this region into a homogeneous spherical cap of density $\mu_{H_P} = \mu(H_P)$ and a fluctuating density $\mu - \mu_{H_P}$ and denote their gravitational attractions as $\Delta_4 g_\Theta^{H_P}$ and $\Delta_4 g_\Theta^{H - H_P}$, respectively. The term $\Delta_4 g_\Theta^{H_P}$ can be computed according to (6.110) by replacing R' by R. We obtain, by also taking reference to (6.158) with $\psi = \Theta$,

$$\Delta_4 g_\Theta^{H_P} = 2\pi G \mu_{H_P} \frac{R^2}{r^2} \left(1 + \frac{R - r\cos\Theta}{l_\Theta}\right)$$

$$= 2\pi G \mu_{H_P} \frac{R^2}{r^2} \left(1 + \frac{l_\Theta}{2R} - \frac{r^2 - R^2}{2Rl_\Theta}\right), \tag{6.161}$$

where l_Θ is the value of l computed using (6.158) with $\psi = \Theta$. According to (6.144), we further obtain, remembering $r = R + H_P$,

$$\Delta_4 g_\Theta^{H_P} = 2\pi G\rho H_P \frac{R^2}{r^2}\left(1 + \frac{H_P}{R} + \frac{H_P^2}{3R^2}\right)\left[1 + \frac{l_\Theta}{2R} - \frac{H_P}{l_\Theta}\left(1 + \frac{H_P}{2R}\right)\right].$$
(6.162)

We see that it can be approximated as a Bouguer plate correction $\Delta_4 g_\Theta = 2\pi G\rho H_P$ when $H_P \ll l_\Theta \ll R$. The formula for $\Delta_4 g_\Theta^{H-H_P}$ can be derived using (6.160) by replacing ω by $\omega_{\psi \leq \Theta}$ and H by $H - H_P$

$$\Delta_4 g_\Theta^{H-H_P} = G\int_{\omega_{\psi \leq \Theta}} \rho\left(H - H_P + \frac{H^2 - H_P^2}{R} + \frac{H^3 - H_P^3}{3R^2}\right)$$
$$\times\left[\frac{R^2}{2rl} + \frac{R^3 H_P}{rl^3}\left(1 + \frac{H_P}{2R}\right)\right] d\omega.$$
(6.163)

Finally, the gravitational attraction of the spherical cap is $\Delta_4 g_\Theta = \Delta_4 g_\Theta^{H_P} + \Delta_4 g_\Theta^{H-H_P}$. In the above two formulae, the expressions of smaller order of magnitude in the parentheses can be neglected as approximations.

In practical computations, it would be beneficial to compute $\Delta_{2+3+4}g = \Delta_2 g + \Delta_3 g + \Delta_4 g$ altogether by evaluating the integral

$$\Delta_{2+3+4}g = G\int_\omega \rho\left[f'_{2+3}(H_P, H, \psi) + f'_4(H_P, H, \psi)\right] d\omega,$$
(6.164)

where $f'_4(H_P, H, \psi)$ is the kernel in (6.160), i.e., the integrand excluding ρ. The reason is that $f'_{2+3}(H_P, H, \psi)$ and $f'_4(H_P, H, \psi)$ are, by their nature, of similar order of magnitude with opposite signs, and better numerical accuracy could be assured by analytically removing the leading quantities canceling each other.

The Helmert condensation is envisaged purely for gravity reduction in the determination of the gravity field. It is irrelevant in places where no mass exists above the geoid such as over the oceans. In our formulation, the density of the condensed surface layer is computed by strictly conserving mass in the spherical geometry. It is left to readers to examine the case when the simpler formula of surface density for the flat Earth model is used.

6.4.4   The Isostatic Anomaly

As in the case of Helmert anomaly, we focus on the isostatic correction, which is also denoted as $\Delta_4 g$. We consider again the Airy–Heiskanen model only. The isostatic correction is the contribution of the mass, which is within the compensation region of depth T to $T + H'$ with a density $\Delta\rho = \rho_1 - \rho$, to the gravity. We assume that

$\Delta\rho$, T, and H' are all known. We implicitly assume that T is position dependent, which does not render the formulation more complicated.

As in the case of $\Delta_{2+3}g$, the formula of $\Delta_4 g$ can be inferred from (6.81) and (6.86) straightforwardly. Here we have $r = R + H_P$, $R_B = R - (T + H')$, and $R_T = R - T$. Hence,

$$\Delta_4 g = G \int_\omega \Delta\rho f_4'(T, H_P, H', \psi) d\omega, \tag{6.165}$$

where

$$f_4'(T, H_P, H', \psi)$$

$$= r \left\{ \frac{x^3}{(1 + x^2 - 2x \cos\psi)^{1/2}} - (x + 3\cos\psi)(1 + x^2 - 2x \cos\psi)^{1/2} \right.$$

$$\left. - (3\cos^2\psi - 1) \ln \left[(1 + x^2 - 2x \cos\psi)^{1/2} + x - \cos\psi \right] \right\}_{[R-(T+H')]/r}^{(R-T)/r} . \tag{6.166}$$

There is no singularity in the integral in (6.165).

An approximate form can be obtained according to (6.104)

$$\Delta_4 g = G r^2 \int_\omega \Delta\rho \left[\frac{1}{(l_r^2 + H'^2)^{1/2}} + \frac{H'}{2r(l_r^2 + H'^2)^{1/2}} \right.$$

$$\left. - \frac{3 l_r^2}{2r^2 (l_r^2 + H'^2)^{1/2}} + \frac{l_r^4}{4r^2 (l_r^2 + H'^2)^{3/2}} \right]_{R-(T+H')}^{R-T} d\omega, \tag{6.167}$$

which is an extension of the formula for the flat Earth model, (6.44), with spherical correction terms. Readers are suggested to simplify this formula to (6.44) by neglecting the spherical correction terms.

It is to be noted that the Vening Meinesz model is different from the Airy–Heiskanen model only in the values of the parameters T and H'. Therefore, the above formulae may apply to the Vening Meinesz model by simply replacing these parameters.

The same as in the case of Helmert anomaly, in practical computation, it would be beneficial to compute $\Delta_{2+3+4}g = \Delta_2 g + \Delta_3 g + \Delta_4 g$ altogether by evaluating the integral

$$\Delta_{2+3+4}g = G \int_\omega \rho \left[f_{2+3}'(H_P, H, \psi) + \frac{\Delta\rho}{\rho} f_4'(T, H_P, H', \psi) \right] d\omega. \tag{6.168}$$

The reason is also the same: $f_{2+3}'(H_P, H, \psi)$ and $(\Delta\rho/\rho) f_4'(T, H_P, H', \psi)$ are, by their nature, of similar order of magnitude with opposite signs, and better numerical accuracy could be assured by analytically removing the leading quantities canceling each other, particularly for the case of Airy–Heiskanen model, where the crustal mass removed is identical to the mass added right below. Readers are suggested to formulate the problem by applying the approximate formula (6.96) to both the topographic and isostatic corrections.

In gravity reduction, the isostacy over the oceans could be either considered or not, as there is no mass above the geoid to be removed there. However, it should be considered in the development of topographic–isostatic models, which is understood as the gravitational potential of the mass of the topography and isostatic compensation.

6.5   Indirect Effect of Gravity Reduction on the Gravity Field

6.5.1   Indirect Effect and Computational Procedure of the Earth's Gravity Field

Gravity reduction is the projection of the gravity measured on the Earth's surface onto the geoid for computing gravity anomaly so that Stokes' theory can be applied to determine the Earth's external gravity field. Stokes' theory requires that no mass exists above the geoid; hence, the Earth should be altered by moving away the mass above the geoid. Although the mass removed can be put back as a Helmert condensation layer or into the isostatic compensation region, the mass distribution of the Earth is inevitably adjusted, and what we obtain according to Stokes' theory is the gravity field of an adjusted Earth and is referred to as adjusted gravity field. The adjusted geoid is often referred to as cogeoid in the literature. We use the superscript c to indicate quantities of the adjusted gravity field. We then have

$$ W = W^c + \delta W\,, \qquad N = N^c + \delta N\,, \qquad \xi = \xi^c + \delta\xi\,, \qquad \eta = \eta^c + \delta\eta\,, \qquad (6.169) $$

where δ is used to indicate the correction by restoring the mass distribution from the adjusted Earth to the real Earth and is referred to as indirect effect.

The indirect effect is not negligible. While the formulae will be derived in subsequent subsections, we first provide a conceptual procedure to take into account the indirect effect in the computation based on isostatic anomaly (the Helmert anomaly is understood as an extreme case of isostatic anomaly): (1) choose the isostatic model to be used throughout the procedure; (2) compute the indirect effect of the geoidal height δN using the known orthometric height H; (3) perform gravity reduction to compute the adjusted gravity anomaly using the adjusted orthometric height $H^c = H + \delta N$, which is the height above the adjusted geoid; (4) compute the adjusted gravity field based on Stokes' theory using the adjusted gravity anomaly; (5) compute the indirect effect using the adjusted orthometric height $H^c = H + \delta N$, and then add the result to the adjusted gravity field. An iteration of step 3 through 5 is expected to improve the result. Only the adjusted geoidal height, or cogeoidal

Fig. 6.13 Use of Stokes'
formula on an equipotential
surface of gravity above the
Earth's surface

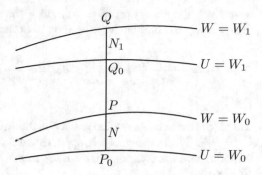

height , is required for the iteration. We mention that, if the Helmert condensation is
used for gravity reduction, this procedure is referred to as Stokes–Helmert method.

The indirect effect could be large. Take for example the case of geoid, which
peaks at a magnitude of ~ 100 m. The indirect effect attains ~ 440 m for the complete
Bouguer anomaly, and ~ 10 m for the isostatic anomaly. These are expected; barely
removing the topographic mass is a much larger mass adjustment than relocating
the topographic mass into the isostatic compensation region.

Even though the free air anomaly does not comply with Stokes' theory that no
mass exists above the geoid, it can be inferred that a skillful use of the free air
anomaly in Stokes' theory can limit the indirect effect[2] to an order of magnitude
below ~ 1 m in the geoidal height. In order to estimate the indirect effect, we
imagine projecting the gravity to an equipotential surface of gravity that encloses
the whole Earth inside (neglecting atmosphere). This surface then complies with
Stokes' theory; it is an equipotential surface of gravity as the geoid, and no mass
exists outside it. As shown in Fig. 6.13, we define $W = W_0$ to be the geoid, $U = W_0$
the reference Earth ellipsoid, $W = W_1$ an equipotential surface of gravity that
encloses the whole Earth inside, and $U = W_1$ an equipotential surface of normal
potential. The relation between $W = W_1$ and $U = W_1$ is identical to that between
$W = W_0$ and $U = W_0$. We draw a straight line perpendicular to the reference Earth
ellipsoid, intersecting with $W = W_0$ and $W = W_1$ at P and Q, respectively, and
with $U = W_0$ and $U = W_1$ at P_0 and Q_0, respectively. We then have $N = P_0 P$, and
we denote $N_1 = Q_0 Q$. We define the gravity anomaly on $W = W_1$ to be

$$\Delta g_1 = g_1 - \gamma_1 , \tag{6.170}$$

---

[2] As no mass allocation is involved, this should be, strictly speaking, the error of using the free air
anomaly without taking care of the mass above the geoid. We use the term *indirect effect* instead
of *error* as we are considering the effect of ignoring the mass above the geoid, but not the error
brought by the approximations in the formulation of the theory.

where g_1 is the gravity on $W = W_1$, and γ_1 the normal gravity on $U = W_1$. We can then use Δg_1 to compute N_1 according to Stokes' formula

$$N_1 = \frac{R + PQ}{4\pi\bar{\gamma}_1} \int_\omega \Delta g_1 S(\psi) d\omega = \frac{R}{4\pi\bar{\gamma}_1} \left(1 + \frac{PQ}{R}\right) \int_\omega \Delta g_1 S(\psi) d\omega, \tag{6.171}$$

where $\bar{\gamma}_1$ is the average normal gravity on the equipotential surface of normal gravity $U = W_1$. While the geoid is approximated as a sphere of radius R in Stokes' formula, here the equipotential surface of gravity $W = W_1$ is approximated as a sphere of radius $R + PQ$.

The rest is to formulate the differences between N_1 and N and between Δg_1 and Δg, so that, together with (6.171), an expression of N computed using Δg could be obtained.

The key to formulate the difference between N_1 and N is to find $P_0 Q_0$ assuming PQ to be known. By applying (1.49) to both the gravity field and normal gravity field, we obtain

$$W_0 - W_1 = \gamma_0 P_0 Q_0 = g_0 PQ, \tag{6.172}$$

where g_0 and γ_0 are still the gravity at P and the normal gravity at P_0, respectively. Hence,

$$P_0 Q_0 = \frac{g_0}{\gamma_0} PQ = \left(1 + \frac{g_0 - \gamma_0}{\gamma_0}\right) PQ = \left(1 + \frac{\Delta g}{\gamma_0}\right) PQ. \tag{6.173}$$

We then have, as $N_1 - N = PQ - P_0 Q_0$,

$$N = N_1 - (PQ - P_0 Q_0) = N_1 + \frac{\Delta g}{\gamma_0} PQ. \tag{6.174}$$

It should be kept in mind that, in this formula, Δg is the gravity anomaly at P, and γ_0 the normal gravity at P_0.

We need the difference between Δg_1 and Δg at any place. We use H_1 to denote the distance between the surfaces $W = W_1$ and $W = W_0$, which is in fact the orthometric height of the surface $W = W_1$, and use H to denote the orthometric height of the ground where the gravity g is assumed to be known. In the computation of the free air anomaly, the gravity on $W = W_0$ is computed by projecting the gravity on the ground downward by adding the free air correction term according to the average normal gravity gradient (6.4)

$$g_0 = g - \left(\frac{\overline{\partial \gamma}}{\partial h}\right)_{U=W_0} H. \tag{6.175}$$

In the same logic, the gravity on $W = W_1$ can also be computed by adding a free air correction term

$$g_1 = g + \left(\frac{\overline{\partial \gamma}}{\partial h}\right)_{U=W_0} (H_1 - H), \tag{6.176}$$

which is a projection of the gravity upward based on the free air correction. From now on, we proceed to seek the relation between Δg_1 and Δg. We denote the distance between the surfaces $U = W_1$ and $U = W_0$ as h_1, which is the geodetic height of the surface $U = W_1$. We have

$$\gamma_1 = \gamma_0 + \left(\frac{\overline{\partial \gamma}}{\partial h}\right)_{U=W_0} h_1. \tag{6.177}$$

It can be readily realized that PQ and P_0Q_0 are in fact H_1 and h_1, respectively, at the location where N is to be computed. The relation (6.173) can be applied to any location. Hence, we have

$$h_1 = \frac{g_0}{\gamma_0} H_1 = \left(1 + \frac{g_0 - \gamma_0}{\gamma_0}\right) H_1 = \left(1 + \frac{\Delta g}{\gamma_0}\right) H_1. \tag{6.178}$$

Finally, we obtain

$$\Delta g_1 = g_1 - \gamma_1 = \Delta g + \left(\frac{\overline{\partial \gamma}}{\partial h}\right)_{U=W_0} (H_1 - h_1) = \Delta g \left[1 - \frac{1}{\gamma_0}\left(\frac{\overline{\partial \gamma}}{\partial h}\right)_{U=W_0} H_1\right]. \tag{6.179}$$

By substituting the approximate formulae (5.24) into this formula, we obtain

$$\Delta g_1 = \Delta g \left(1 + 2\frac{H_1}{R}\right). \tag{6.180}$$

As already mentioned, PQ is the value of H_1 at a particular location. The variation of H_1 from location to location is small; its relative order of magnitude is on the order of the percentage of the variation of g_0. Therefore, replacing H_1 by PQ in (6.180) induces an error of even smaller order of magnitude than H_1/R. Hence, we can approximate (6.180) as

$$\Delta g_1 = \Delta g \left(1 + 2\frac{PQ}{R}\right). \tag{6.181}$$

Now, an expression of N can be obtained according to (6.174), (6.171), and (6.181)

$$N = \frac{R}{4\pi \bar{\gamma}_1} \left(1 + \frac{PQ}{R}\right) \left(1 + 2\frac{PQ}{R}\right) \int_\omega \Delta g S(\psi) d\omega + \frac{\Delta g}{\gamma_0} PQ. \qquad (6.182)$$

According to (6.177) and the approximate formulae (5.24), we have, with $h_1 = P_0 Q_0$ here,

$$\frac{1}{\bar{\gamma}_1} = \frac{1}{\gamma_0} \left[1 - \frac{1}{\gamma_0} \left(\overline{\frac{\partial \gamma}{\partial h}}\right)_{U=W_0} P_0 Q_0\right] = \frac{1}{\gamma_0} \left[1 + 2\frac{PQ}{R}\right], \qquad (6.183)$$

where we approximated $P_0 Q_0$ by PQ in the second identity, which brings only an error of smaller order of magnitude. We obtain finally

$$N = \frac{R}{4\pi \bar{\gamma}_0} \left(1 + 5\frac{PQ}{R}\right) \int_\omega \Delta g S(\psi) d\omega + \frac{\Delta g}{\gamma_0} PQ. \qquad (6.184)$$

The adjusted geoidal height computed using the free air anomaly is

$$N^c = \frac{R}{4\pi \bar{\gamma}_0} \int_\omega \Delta g S(\psi) d\omega. \qquad (6.185)$$

Hence, the indirect effect is

$$\delta N = N - N^c = 5\frac{PQ}{R} N^c + \frac{\Delta g}{\gamma_0} PQ. \qquad (6.186)$$

We now discuss the magnitude of the indirect effect given by (6.186). In the derivation, we assumed that the equipotential surface of gravity $W = W_1$ encloses the whole Earth inside and that the geoid used to define the reference Earth ellipsoid, $W = W_0$, is chosen to be an equipotential surface of gravity that best fits the mean ocean surface according to prescribed criteria. This requires PQ to be on the order of 8 km. With $R \sim 6371$ km and $N \sim 100$ m, the magnitude of the first term in (6.186) is $\sim$60 cm. As Δg rarely exceeds 100 mGal, and $\gamma_0 \sim 10^6$ mGal, the magnitude of the second term is $\sim$80 cm.

However, the requirement of $PQ \sim 8$ km may be loosened by applying the formulation tactically. For example, we can choose the geoid used to define the reference Earth ellipsoid, $W = W_0$, such that the average orthometric height of the Earth's surface of the region of interest is close to zero. Furthermore, we can choose PQ to be large enough to include the topography of only a large enough region nearby within the equipotential surface of gravity $W = W_1$, as a mountain too far away has little influence on the geoidal height in a region. In this regard,

$PQ = 2$ km is more than enough except extreme mountainous regions. In this case, the magnitude of the first term in (6.186) is $\sim$15 cm, and the second term $\sim$20 cm.

We emphasize that these estimates are purely the effect of ignoring the mass above the geoid while assuming the theories of the Stokes formula and free air correction exact. These are also the basis of (6.171) and (6.176), which are straightforward applications of the theories.

6.5.2 Indirect Effect of the Global Gravitational Potential, the Topographic–Isostatic Model

The indirect effect is the correction to restore the mass distribution from the adjusted Earth after gravity reduction to the state before gravity reduction.

The Helmert Anomaly
For the case of Helmert anomaly, the indirect effect is computed by adding back the topographic mass and removing the Helmert condensation layer. The gravitational potential of the topographic mass can be inferred from (6.54) to (6.56). Here, as $R_B = R$ and $R_T = R + H$, we write this potential in the form of

$$V_T = \frac{GM}{r} \sum_{n=0}^{\infty} \left(\frac{R}{r}\right)^n \sum_{k=0}^{n} ({}_T C_n^k \cos k\lambda + {}_T S_n^k \sin k\lambda) P_n^k(\cos\theta). \qquad (6.187)$$

The spherical harmonic coefficients are

$$\begin{cases} {}_T C_n^k = \dfrac{2}{1+\delta_k} \dfrac{(n-k)!}{(n+k)!} \displaystyle\int_\omega {}_T I_n^k P_n^k(\cos\theta') \cos k\lambda' d\omega, \\[4mm] {}_T S_n^k = \dfrac{2}{1+\delta_k} \dfrac{(n-k)!}{(n+k)!} \displaystyle\int_\omega {}_T I_n^k P_n^k(\cos\theta') \sin k\lambda' d\omega, \end{cases} \qquad (6.188)$$

where

$$ {}_T I_n^k = \frac{R^3 \rho}{(n+3)M} \left[\left(\frac{R+H}{R}\right)^{n+3} - 1 \right]. \qquad (6.189)$$

The gravitational potential of the Helmert condensation layer of surface density μ can be readily inferred from the derivation of (6.54) to (6.56), which we write in the form of

$$V_\mu = \frac{GM}{r} \sum_{n=0}^{\infty} \left(\frac{R}{r}\right)^n \sum_{k=0}^{n} ({}_\mu C_n^k \cos k\lambda + {}_\mu S_n^k \sin k\lambda) P_n^k(\cos\theta), \qquad (6.190)$$

where

$$\begin{cases} {}_\mu C_n^k = \dfrac{2}{1+\delta_k} \dfrac{(n-k)!}{(n+k)!} \displaystyle\int_\omega \dfrac{R^2\mu}{M} P_n^k(\cos\theta')\cos k\lambda'\,d\omega, \\[4mm] {}_\mu S_n^k = \dfrac{2}{1+\delta_k} \dfrac{(n-k)!}{(n+k)!} \displaystyle\int_\omega \dfrac{R^2\mu}{M} P_n^k(\cos\theta')\sin k\lambda'\,d\omega. \end{cases} \tag{6.191}$$

The indirect effect is the sum of adding back the topographic mass and removing the Helmert condensation layer. We write it in the form of

$$\delta W = V_T - V_\mu = \frac{GM}{r}\sum_{n=0}^{\infty}\left(\frac{R}{r}\right)^n \sum_{k=0}^{n}(\delta C_n^k\cos k\lambda + \delta S_n^k\sin k\lambda)P_n^k(\cos\theta). \tag{6.192}$$

The spherical harmonic coefficients are

$$\begin{cases} \delta C_n^k = \dfrac{2}{1+\delta_k}\dfrac{(n-k)!}{(n+k)!}\displaystyle\int_\omega \delta I_n^k P_n^k(\cos\theta')\cos k\lambda'\,d\omega, \\[4mm] \delta S_n^k = \dfrac{2}{1+\delta_k}\dfrac{(n-k)!}{(n+k)!}\displaystyle\int_\omega \delta I_n^k P_n^k(\cos\theta')\sin k\lambda'\,d\omega, \end{cases} \tag{6.193}$$

where, with μ given in (6.144),

$$\begin{aligned} \delta I_n^k &= {}_T I_n^k - \frac{R^2\mu}{M} \\ &= \frac{R^3\rho}{M}\left\{\frac{1}{(n+3)}\left[\left(\frac{R+H}{R}\right)^{n+3}-1\right] - \frac{H}{R}\left(1+\frac{H}{R}+\frac{H^2}{3R^2}\right)\right\}. \end{aligned} \tag{6.194}$$

As we used the Helmert condensation in spherical geometry that does conserve mass, the degree 0 term of δW should vanish, which is confirmed by $\delta I_0 = 0$ computed using (6.194). However, as a relocation of mass is involved, the center of mass of the adjusted Earth may be of that of the real Earth, which may bring in a degree 1 term, and this is also the reason we use δW instead of δT to denote the indirect effect. As the degree 1 term can be formulated separately according to (6.192) to (6.194), it can be handled individually to be included or excluded.

The Isostatic Anomaly

For the case of isostatic anomaly, the indirect effect is computed by adding back the topographic mass and removing the mass added in the isostatic correction. For the Airy–Heiskanen or Vening Meinesz model, the gravitational potential of the topographic mass is given in (6.187) to (6.189). The gravitational potential of the mass added in the isostatic correction can also be inferred from (6.54) to (6.56).

Here, we have $R_B = R - (T + H')$ and $R_T = R - T$, and we write the potential in the form of

$$
V_I = \frac{GM}{r} \sum_{n=0}^{\infty} \left(\frac{R}{r}\right)^n \sum_{k=0}^{n} ({}_I C_n^k \cos k\lambda + {}_I S_n^k \sin k\lambda) P_n^k (\cos \theta). \tag{6.195}
$$

The spherical harmonic coefficients are

$$
\begin{cases}
{}_I C_n^k = \dfrac{2}{1 + \delta_k} \dfrac{(n - k)!}{(n + k)!} \displaystyle\int_\omega {}_I I_n^k P_n^k (\cos \theta') \cos k\lambda' d\omega, \\[3mm]
{}_I S_n^k = \dfrac{2}{1 + \delta_k} \dfrac{(n - k)!}{(n + k)!} \displaystyle\int_\omega {}_I I_n^k P_n^k (\cos \theta') \sin k\lambda' d\omega,
\end{cases} \tag{6.196}
$$

where

$$
{}_I I_n^k = \frac{R^3 \Delta\rho}{(n + 3)M} \left\{ \left(\frac{R - T}{R}\right)^{n+3} - \left[\frac{R - (T + H')}{R}\right]^{n+3} \right\}. \tag{6.197}
$$

The indirect effect is the sum of adding back the topographic mass and removing the mass added in the isostatic correction. As in the case of Helmert anomaly, we write it in the form of

$$
\delta W = V_T - V_I = \frac{GM}{r} \sum_{n=0}^{\infty} \left(\frac{R}{r}\right)^n \sum_{k=0}^{n} (\delta C_n^k \cos k\lambda + \delta S_n^k \sin k\lambda) P_n^k (\cos \theta). \tag{6.198}
$$

The spherical harmonic coefficients are

$$
\begin{cases}
\delta C_n^k = \dfrac{2}{1 + \delta_k} \dfrac{(n - k)!}{(n + k)!} \displaystyle\int_\omega \delta I_n^k P_n^k (\cos \theta') \cos k\lambda' d\omega, \\[3mm]
\delta S_n^k = \dfrac{2}{1 + \delta_k} \dfrac{(n - k)!}{(n + k)!} \displaystyle\int_\omega \delta I_n^k P_n^k (\cos \theta') \sin k\lambda' d\omega,
\end{cases} \tag{6.199}
$$

where

$$
\delta I_n^k = {}_T I_n^k - {}_I I_n^k = \frac{R^3 \rho}{(n + 3)M} \left[\left(\frac{R + H}{R}\right)^{n+3} - 1 \right]
$$
$$
- \frac{R^3 \Delta\rho}{(n + 3)M} \left\{ \left(\frac{R - T}{R}\right)^{n+3} - \left[\frac{R - (T + H')}{R}\right]^{n+3} \right\}. \tag{6.200}
$$

If mass conservation is assured in the isostatic model, the degree 0 term of δW vanishes. For the strict Airy–Heiskanen model expressed in (6.146), we do obtain

$\delta I_0 = 0$. Again, as the mass distribution is adjusted in the computation of the isostatic anomaly, the degree 1 term may be involved in δW and should be handled particularly to be rigorous.

As already mentioned, in gravity reduction, the consideration of isostatic compensation in oceanic regions is not really necessary, as the ocean surface can be considered as the geoid.

The Topographic–Isostatic Model

The topographic–isostatic model is conceptually defined as the gravitational potential of the irregularly distributed mass within the isostatically compensated region of the Earth. Over land, it includes the contributions of the surplus of mass in the topography and the deficit of mass in the compensation region. Over oceans, it includes the contributions of the deficit of mass in the ocean (as water density is smaller than that of the crust) and the surplus of mass in the compensation region.

By definition, the topographic–isostatic model is identical to the indirect effect of the gravitational potential for the isostatic anomaly with the isostatic compensation of both the continental and oceanic regions taken into account. For the Airy–Heiskanen or Vening Meinesz model, the contribution from the continental region is given in (6.198) to (6.200). It is left to readers to infer the contribution from the oceanic region. Here we assume that the degree 1 term of δW is handled separately and define the topographic–isostatic model T_{TI} as δW without the degree 1 term. The formulae (6.198) to (6.200), together with those for the contribution from the oceanic region, can be used to compute T_{TI} when the isostatic model is prescribed. However, as the isostatic model is supposed to represent the reality as much as possible, the topographic–isostatic model is then expected to be as close as possible to the disturbing potential. Hence, an appropriate criterion to determine the parameters of the isostatic model based on global gravitational potential is

$$\int_\omega (T_{TI} - T)^2 d\omega = \min. \tag{6.201}$$

In this formula, the notation T is the disturbing potential, which is to be distinguished from the normal crust thickness T in the isostatic model. As the few lowest degree terms of the spherical harmonic series of the disturbing potential T may be mainly due to irregular mass distribution in the deep interior of the Earth, they may be excluded in (6.201).

6.5.3   Indirect Effect of the Geoidal Height

The indirect effect of the geoidal height is, according to Bruns' formula,

$$\delta N = \frac{\delta W}{\bar{\gamma_0}}. \tag{6.202}$$

Different from the previous subsection, here δW is the value on the geoid approximated as a sphere of radius R, and its computation will be the focus in the rest of this subsection. We mention that, theoretically strictly speaking, the degree 1 term of δW, if brought by gravity reduction, should be removed in the result based on the indirect effect of the global gravitational potential.

The Helmert Anomaly

For the case of Helmert anomaly, δW includes the contribution of adding back the topographic mass and removing the Helmert condensation layer. An exact formula of the gravitational potential of the topographic mass on the geoid can be obtained from (6.58) and (6.61). Here, as the geoid is approximated as a sphere of radius R, we have $r = R$, $R_B = R$, and $R_T = R + H$. The gravitational potential of the topographic mass can be written as

$$V_T = G \int_\omega \rho f_T(H, \psi) d\omega, \tag{6.203}$$

where the kernel $f_T(H, \psi)$ is defined as

$$f_T(H, \psi) = \frac{R^2}{2} \Big\{ (x + 3\cos\psi)(1 + x^2 - 2x\cos\psi)^{1/2}$$

$$+ (3\cos^2\psi - 1) \ln\Big[(1 + x^2 - 2x\cos\psi)^{1/2} + x - \cos\psi \Big] \Big\}_1^{1+H/R}. \tag{6.204}$$

The gravitational potential of the Helmert condensation layer can be inferred from the formula for a layer of mass around the Earth's surface (6.57). Here we just need to set $r = R$, $R' = R$, and $\rho dR' = \mu$ and remove the integral over R', where μ is given by (6.144) assuming strict mass conservation in spherical geometry. We obtain

$$V_\mu = G \int_\omega \mu \frac{R^2}{R(2 - 2\cos\psi)^{1/2}} d\omega = G \int_\omega \rho f_\mu(H, \psi) d\omega. \tag{6.205}$$

The kernel $f_\mu(H, \psi)$ is

$$f_\mu(H, \psi) = \frac{R^2}{l_0} \left(H + \frac{H^2}{R} + \frac{H^3}{3R^2} \right), \tag{6.206}$$

where l_0 is defined as

$$l_0 = 2R \sin\frac{\psi}{2}. \tag{6.207}$$

The indirect effect of gravitational potential is thus

$$\delta W = V_T - V_\mu = G \int_\omega \rho \left[f_T(H, \psi) - f_\mu(H, \psi) \right] d\omega. \tag{6.208}$$

The two terms in the kernel, $f_T(H, \psi)$ and $-f_\mu(H, \psi)$, are of about the same magnitude, but with different signs. Hence, it is beneficial to first analytically remove the common part of them for better computation accuracy, which is left to readers who are interested in.

Both $f_T(H, \psi)$ and $f_\mu(H, \psi)$ tend to infinitely large when $\psi \to 0$. Hence, it is not evident to numerically evaluate the integral (6.208) around the point P where δW is computed, particularly due to the presence of the logarithm function in $f_T(H, \psi)$. As in the case of gravity reduction, we consider separately the contribution from a small region $\psi \le \Theta$ in (6.208), denoted as δW_Θ. We have

$$\delta W = G \int_{\omega_{\psi \ge \Theta}} \rho \left[f_T(H, \psi) - f_\mu(H, \psi) \right] d\omega + \delta W_\Theta. \tag{6.209}$$

In the region $\psi \le \Theta$, we separate the topography into a cap-shaped shell of height H_P and a terrain undulating around the height H_P. Accordingly, we also separate the surface density of the Helmert condensation layer into two terms. Thus, we have

$$\delta W_{\omega_{\psi \le \Theta}} = V_T^{0 \sim \Theta, 0 \sim H_P} - V_\mu^{0 \sim \Theta, 0 \sim H_P} + V_T^{0 \sim \Theta, H_P \sim H} - V_\mu^{0 \sim \Theta, H_P \sim H}, \tag{6.210}$$

where, in the superscript, $0 \sim \Theta$ is the range of ψ, and $0 \sim H_P$ and $H_P \sim H$ are the ranges of height, remembering that H is position dependent. We also neglect the variation of the density ρ within this region. The term $V_T^{0 \sim \Theta, 0 \sim H_P}$ is the gravitational potential of a cap-shaped shell, which can be obtained from (6.120) by setting $r = R$, $R_B = R$, and $R_T = R + H_P$

$$V_T^{0 \sim \Theta, 0 \sim H_P} = 2\pi G\rho R^2 \left\{ \frac{1}{3} \left(1 + x^2 - 2x \cos\Theta \right)^{3/2} \right.$$

$$+ \frac{1}{2}(x - \cos\Theta) \left(1 + x^2 - 2x \cos\Theta \right)^{1/2} - \frac{1}{3}x^3 + \frac{1}{2}x^2$$

$$+ \frac{1}{2}\sin^2\Theta \cos\Theta \ln \left[\left(1 + x^2 - 2x \cos\Theta \right)^{1/2} \right.$$

$$\left. \left. + x - \cos\Theta \right] \right\}_1^{1 + H_P/R}. \tag{6.211}$$

The term $V_\mu^{0\sim\Theta,0\sim H_P}$ is the gravitational potential of a material surface of spherical cap, which can be obtained from (6.118) by setting $r = R$, $R' = R$, and $H = H_P$

$$V_\mu^{0\sim\Theta,0\sim H_P} = 2\pi G\rho a_R \left(H_P + \frac{H_P^2}{R} + \frac{H_P^3}{3R^2}\right), \tag{6.212}$$

where (6.144) is used, and a_R is defined as

$$a_R = 2R \sin \frac{\Theta}{2}, \tag{6.213}$$

which can be understood as the radius of the cap. By definition, the other two terms in (6.210) can be inferred from (6.203) to (6.206) to be

$$V_T^{0\sim\Theta,H_P\sim H} = G\rho \int_\omega [f_T(H,\psi) - f_T(H_P,\psi)]\, d\omega, \tag{6.214}$$

$$f_T(H,\psi) - f_T(H_P,\psi) = \frac{R^2}{2}\left\{(x + 3\cos\psi)(1 + x^2 - 2x\cos\psi)^{1/2}\right.$$
$$\left. + (3\cos^2\psi - 1)\ln\left[(1 + x^2 - 2x\cos\psi)^{1/2} + x - \cos\psi\right]\right\}_{1+H_P/R}^{1+H/R}, \tag{6.215}$$

and

$$V_\mu^{0\sim\Theta,H_P\sim H} = G\rho \int_\omega [f_\mu(H,\psi) - f_\mu(H_P,\psi)]\, d\omega, \tag{6.216}$$

$$f_\mu(H,\psi) - f_\mu(H_P,\psi) = \frac{R^2}{l_0}\left[(H - H_P) + \frac{H^2 - H_P^2}{R} + \frac{H^3 - H_P^3}{3R^2}\right], \tag{6.217}$$

where l_0 is defined in (6.207). In (6.214) and (6.216), as $H = H_P$ when $\psi = 0$, we could consider to set the kernel of the integral to vanish within an extremely small cap around P, where the height H can be reasonably approximated with the constant H_P.

It is of interest to express the contributions from the region $\psi \leq \Theta$ using approximate formulae, which are more appropriate for numerical evaluation. The term $V_T^{0\sim\Theta,0\sim H_P}$ can be obtained from (6.139) by setting $r = R$, $R_B = R$, and

$R_T = R + H_P$, so that $H_{Br} = 0$ and $H_{Tr} = H_P$. We have

$$V_T^{0 \sim \Theta, 0 \sim H_P} = 2\pi G \rho a_R \left[H_P + \frac{H_P^2}{2a_R} \left(-1 + \frac{3a_R}{2R} \right) + \frac{H_P^3}{6a_R^2} \left(1 - \frac{2a_R}{R} + \frac{3a_R^2}{4R^2} \right) \right],$$

$$(6.218)$$

where a_R is defined in (6.213). The term $V_T^{0 \sim \Theta, H_P \sim H}$ can be obtained from (6.72) by setting $r = R$, $R_B = R + H_P$, and $R_T = R + H$, and thus $H_{Br} = H_P$ and $H_{Tr} = H$

$$V_T^{0 \sim \Theta, H_P \sim H} = G R^2 \rho \int_{\omega_\psi \leq \Theta} \frac{H - H_P}{l_0} d\omega + \frac{3}{4} G R \rho \int_{\omega_\psi \leq \Theta} \frac{H^2 - H_P^2}{l_0} d\omega$$

$$- \frac{1}{6} G R^2 \rho \int_{\omega_\psi \leq \Theta} \frac{H^3 - H_P^3}{l_0^3} \left(1 - \frac{3l_0^2}{4R^2} \right) d\omega, \qquad (6.219)$$

where l_0 is defined in (6.207). Substitute (6.212) and (6.216) to (6.219) into (6.210). We obtain, after some algebraic reductions,

$$\delta W_\Theta = \pi G \rho H_P^2 \left[\left(-1 - \frac{a_R}{2R} \right) + \frac{H_P}{3a_R} \left(1 - \frac{2a_R}{R} - \frac{5a_R^2}{4R^2} \right) \right]$$

$$- \frac{1}{4} G R \rho \int_{\omega_\psi \leq \Theta} \frac{H^2 - H_P^2}{l_0} d\omega - \frac{1}{6} G R^2 \rho \int_{\omega_\psi \leq \Theta} \frac{H^3 - H_P^3}{l_0^3} \left(1 + \frac{5l_0^2}{4R^2} \right) d\omega.$$

$$(6.220)$$

A particular case is the limit $\Theta = \pi/2$, meaning that the whole Earth's crust is divided into a shell of constant height H_P and an undulation around the height H_P. In this case, as $a_R = 2R$, we obtain, according to (6.220),

$$\delta W = -2\pi G \rho H_P^2 \left(1 + \frac{4H_P}{3R} \right) - \frac{1}{4} G R \rho \int_\omega \frac{H^2 - H_P^2}{l_0} d\omega$$

$$- \frac{1}{6} G R^2 \rho \int_\omega \frac{H^3 - H_P^3}{l_0^3} \left(1 + \frac{5l_0^2}{4R^2} \right) d\omega, \qquad (6.221)$$

where the whole topographic mass is assumed to be homogeneous. The expression $(5l_0^2)/(4R^2)$ within the parentheses of the last term represents a quantity that is on the relative order of H/R as compared to the second term. Therefore, as a further approximation, it can be neglected, together with the expression $(2H_P)/(3R)$ within the parentheses of the first term.

Another approximate formula, which is an extension of the formula for the flat Earth model, can be derived using (6.79), which is left to readers who are interested in.

In the formulation, strict mass conservation in spherical geometry is assured. If the simpler surface density formula for the flat Earth model, which does not conserve mass, were used, the formulae obtained would be different.

The Isostatic Anomaly

The case of the isostatic anomaly is different from that of the Helmert anomaly by only one component: the Helmert condensation layer on the geoid should be replaced by the mass added in the isostatic correction at around the bottom of the crust. Hence, δW includes the contributions of adding back the topographic mass and removing the mass added around the bottom of the crust in computing the isostatic correction. The contribution of adding back topographic mass has already been formulated in the case of Helmert anomaly. Here we focus on the gravitational potential of the mass added in the isostatic correction for the Airy–Heiskanen or Vening Meinesz model. The same as the case of topographic mass, the fundamental formulae are (6.58) and (6.61). Here, we have $r = R$, $R_B = R - (T + H')$, and $R_T = R - T$. Hence,

$$V_I = G \int_\omega \Delta \rho f_I(T, H', \psi) d\omega, \qquad (6.222)$$

where the kernel $f_I(T, H', \psi)$ is defined as

$$f_T(T, H', \psi) = \frac{R^2}{2} \left\{ (x + 3 \cos \psi)(1 + x^2 - 2x \cos \psi)^{1/2} + (3 \cos^2 \psi - 1) \ln \right.$$
$$\left. \times \left[(1 + x^2 - 2x \cos \psi)^{1/2} + x - \cos \psi \right] \right\}_{1-(T+H')/R}^{1-T/R}.$$
$$(6.223)$$

There is no singularity in this integral. Hence, no further development is necessary.

The indirect effect is the sum of the effects of adding back the topographic mass and removing the mass added in the isostatic correction

$$\delta W = V_T - V_I = G \int_\omega \rho \left[f_T(H, \psi) - \frac{\Delta \rho}{\rho} f_I(T, H', \psi) \right] d\omega. \qquad (6.224)$$

Again, it is more favorable for better accuracy to compute the two components together according to this formula by first analytically removing the common parts of the two terms, which are of about the same magnitude, but with different signs.

As in the case of Helmert anomaly, it is also more practicable to consider a small region $\psi \leq \Theta$ separately, where the contribution to V_T is to be computed as the sum of $V_T^{0\sim\Theta,0\sim H_P}$ and $V_T^{0\sim\Theta, H_P\sim H}$ according to (6.218) and (6.219), respectively.

However, the contribution of this small region to V_I can still be computed using the exact formulae (6.222) and (6.223).

Readers are suggested to apply the approximate formula (6.72) to within $\psi \leq \Theta$ for the contributions of both the topographic mass and isostatic compensation mass based on the Airy–Heiskanen model. An even finer formulation is, analogous to (6.210), to divide both the topographic mass and isostatic compensation mass within $\psi \leq \Theta$ into a cap-shaped shell and an undulation. Such a practice has been applied to the topographic gravity correction, which is separated into a plate correction and a terrain correction. The same can be done for the isostatic correction.

6.5.4   Indirect Effect of the Deflection of the Vertical

Preparatory Formulations

In order to compute $\delta\xi$ and $\delta\eta$, we make use of the indirect effect of the gravity disturbance as an intermediate quantity, with its horizontal components denoted as $\delta_i g_\theta$ and $\delta_i g_\lambda$. Taking reference to the relation between the gravity disturbance and the disturbing potential, (5.3), and that between the gravity disturbance and the deflection of the vertical, (5.13), we have

$$\delta_i g_\theta = \frac{1}{r} \frac{\partial \delta W}{\partial \theta}, \qquad \delta_i g_\lambda = \frac{1}{r \sin\theta} \frac{\partial \delta W}{\partial \lambda}; \tag{6.225}$$

$$\delta\xi = \frac{\delta_i g_{0\theta}}{\bar{\gamma_0}}, \qquad \delta\eta = -\frac{\delta_i g_{0\lambda}}{\bar{\gamma_0}}. \tag{6.226}$$

The subscript "0" implies that the value is taken on the geoid, which is approximated as a sphere of radius R. In the reminder of this subsection, we focus on $\delta_i g_\theta$ and $\delta_i g_\lambda$, with which the computation of $\delta\xi$ and $\delta\eta$ is evident. We mention that, as δW is used to replace the disturbing potential in (6.225), the effect of the degree 1 term of δW on the deflection of the vertical is brought in. Theoretically strictly speaking, this effect should be removed based on the indirect effect of the global gravitational potential.

Evidently, for computing the topographic and isostatic effects, we need to formulate the horizontal components of the gravitational attraction of a layer of mass around the Earth's surface considered in Sect. 6.3, which are related to the gravitational potential in the form of

$$F_\theta = \frac{1}{r} \frac{\partial V}{\partial \theta}, \qquad F_\lambda = \frac{1}{r \sin\theta} \frac{\partial V}{\partial \lambda}. \tag{6.227}$$

The gravitational potential V is given in (6.58), which is copied to here,

$$V = G \int_\omega \rho f(R_B, R_T, r, \psi) d\omega, \tag{6.228}$$

where $f(R_B, R_T, r, \psi)$ is given in (6.61). Taking reference to the derivation of gravity disturbance expressed in (5.58) using the disturbing potential expressed in (5.35), we realize right away that

$$
\begin{cases}
F_\theta = \dfrac{G}{r} \displaystyle\int_\omega \rho \frac{\partial}{\partial \psi} f(R_B, R_T, r, \psi) \cos A \, d\omega, \\[2mm]
F_\lambda = -\dfrac{G}{r} \displaystyle\int_\omega \rho \frac{\partial}{\partial \psi} f(R_B, R_T, r, \psi) \sin A \, d\omega,
\end{cases}
\tag{6.229}
$$

where A is the azimuth of the integral running point, i.e., the location of $d\omega$, with respect to the point P, as shown in Fig. 5.6. The rest is then to derive $(\partial/\partial \psi) f(R_B, R_T, r, \psi)$. This consists of some algebraic manipulations, and we write out the result directly

$$
\frac{\partial}{\partial \psi} f(R_B, R_T, r, \psi)
$$

$$
= \frac{r^2}{2} \left\{ -3 \sin \psi (1 + x^2 - 2x \cos \psi)^{1/2} + \frac{x \sin \psi (x + 3 \cos \psi)}{(1 + x^2 - 2x \cos \psi)^{1/2}} \right.
$$

$$
+ \frac{\sin \psi \left(3 \cos^2 \psi - 1\right) \left[(1 + x^2 - 2x \cos \psi)^{1/2} + x\right]}{(1 + x^2 - 2x \cos \psi)^{1/2} \left[(1 + x^2 - 2x \cos \psi)^{1/2} + x - \cos \psi\right]}
$$

$$
\left. - 6 \sin \psi \cos \psi \ln \left[(1 + x^2 - 2x \cos \psi)^{1/2} + x - \cos \psi\right] \right\}_{R_B/r}^{R_T/r}.
\tag{6.230}
$$

For the Helmert condensation, we also need to derive the formulae for a material surface of density μ located on a sphere of radius R. Its gravitational potential can be inferred from the formula for a layer of mass around the Earth's surface, (6.57), by substituting $\rho \, dR'$ by μ, replacing R' by R, and removing the integral over R'

$$
V = G \int_\omega \mu \frac{R^2}{\left(r^2 + R^2 - 2rR \cos \psi\right)^{1/2}} d\omega = G \int_\omega \mu f(r, \psi) d\omega,
\tag{6.231}
$$

where the kernel $f(r, \psi)$ is defined as

$$
f(r, \psi) = \frac{R^2}{\left(r^2 + R^2 - 2rR \cos \psi\right)^{1/2}}.
\tag{6.232}
$$

Similar to the derivation in the previous paragraph, we obtain

$$
F_\theta = \frac{G}{r} \int_\omega \mu \frac{\partial}{\partial \psi} f(r, \psi) \cos A \, d\omega, \qquad F_\lambda = -\frac{G}{r} \int_\omega \mu \frac{\partial}{\partial \psi} f(r, \psi) \sin A \, d\omega,
\tag{6.233}
$$

where

$$\frac{\partial}{\partial \psi} f(r, \psi) = -\frac{2r R^3 \sin \psi}{\left(r^2 + R^2 - 2r R \cos \psi\right)^{3/2}}. \tag{6.234}$$

The Helmert Anomaly

Similar to the indirect effect of the gravitational potential (6.208), the exact formulae for the indirect effect of the horizontal components of gravity disturbance on the geoid can be written as

$$\begin{cases} \delta_i g_\theta = \dfrac{G}{R} \displaystyle\int_\omega \rho \left[\dfrac{\partial}{\partial \psi} f_T(H, \psi) - \dfrac{\partial}{\partial \psi} f_\mu(H, \psi) \right] \cos A \, d\omega, \\[3mm] \delta_i g_\lambda = -\dfrac{G}{R} \displaystyle\int_\omega \rho \left[\dfrac{\partial}{\partial \psi} f_T(H, \psi) - \dfrac{\partial}{\partial \psi} f_\mu(H, \psi) \right] \sin A \, d\omega, \end{cases} \tag{6.235}$$

where $(\partial/\partial \psi) f_T(H, \psi)$ and $(\partial/\partial \psi) f_\mu(H, \psi)$ represent the contributions from the topographic mass and the Helmert condensation layer, respectively. The kernel $(\partial/\partial \psi) f_T(H, \psi)$ is obtained by setting $r = R$, $R_B = R$, and $R_T = R + H$ in (6.229) and (6.230)

$$\frac{\partial}{\partial \psi} f_T(H, \psi) = \frac{R^2}{2} \left\{ -3 \sin \psi (1 + x^2 - 2x \cos \psi)^{1/2} + \frac{x \sin \psi (x + 3 \cos \psi)}{(1 + x^2 - 2x \cos \psi)^{1/2}} \right.$$
$$+ \frac{\sin \psi \left(3 \cos^2 \psi - 1\right) \left[(1 + x^2 - 2x \cos \psi)^{1/2} + x\right]}{(1 + x^2 - 2x \cos \psi)^{1/2} \left[(1 + x^2 - 2x \cos \psi)^{1/2} + x - \cos \psi\right]}$$
$$\left. - 6 \sin \psi \cos \psi \ln \left[(1 + x^2 - 2x \cos \psi)^{1/2} + x - \cos \psi\right] \right\}_1^{1+H/R}. \tag{6.236}$$

The kernel $(\partial/\partial \psi) f_\mu(H, \psi)$ is obtained by setting $r = R$ in (6.233) and (6.234) and replacing μ by its expression (6.144)

$$\frac{\partial}{\partial \psi} f_\mu(H, \psi) = -\frac{2R \sin \psi}{(2 - 2 \cos \psi)^{3/2}} H \left(1 + \frac{H}{R} + \frac{H^2}{3R^2}\right)$$
$$= -\frac{R \cos(\psi/2)}{2 \sin^2(\psi/2)} H \left(1 + \frac{H}{R} + \frac{H^2}{3R^2}\right). \tag{6.237}$$

Again, the kernel in (6.235) tends to infinitely large when $\psi \to 0$, particularly the logarithm function is not evident to handle. Hence, we also provide the formula for treating a region $\psi \le \Theta$ separately to make numerical evaluation easier. The most straightforward way is to use the expression of δW_Θ given by (6.220). The first term is the contribution of a mass in circular symmetry around the point P_0; therefore, it does not contribute to the horizontal components of the gravity. For the other two terms in integral form, the relations between them and their contributions to $\delta_i g_\theta$ and $\delta_i g_\lambda$ are similar to that between (6.228) and (6.229), here with $r = R$. We have

$$
\begin{cases}
\delta_i g_{\Theta\theta} = -\dfrac{1}{4}G\rho \displaystyle\int_{\omega_{\psi\le\Theta}} \left(H^2 - H_P^2\right) \dfrac{d}{d\psi}\left(\dfrac{1}{l_0}\right)\cos A\,d\omega \\[2ex]
\qquad\qquad -\dfrac{1}{6}GR\rho \displaystyle\int_{\omega_{\psi\le\Theta}} \left(H^3 - H_P^3\right) \dfrac{d}{d\psi}\left[\dfrac{1}{l_0^3}\left(1 + \dfrac{5l_0^2}{4R^2}\right)\right]\cos A\,d\omega, \\[2ex]
\delta_i g_{\Theta\lambda} = \dfrac{1}{4}G\rho \displaystyle\int_{\omega_{\psi\le\Theta}} \left(H^2 - H_P^2\right) \dfrac{d}{d\psi}\left(\dfrac{1}{l_0}\right)\sin A\,d\omega \\[2ex]
\qquad\qquad +\dfrac{1}{6}GR\rho \displaystyle\int_{\omega_{\psi\le\Theta}} \left(H^3 - H_P^3\right) \dfrac{d}{d\psi}\left[\dfrac{1}{l_0^3}\left(1 + \dfrac{5l_0^2}{4R^2}\right)\right]\sin A\,d\omega,
\end{cases}
$$
$$(6.238)$$

where $l_0 = 2R\sin(\psi/2)$ is defined in (6.207). After evaluating the derivatives analytically, we obtain

$$
\begin{cases}
\delta_i g_{\Theta\theta} = \dfrac{G\rho}{16R} \displaystyle\int_{\omega_{\psi\le\Theta}} \left(H^2 - H_P^2\right) \dfrac{\cos(\psi/2)}{\sin^2(\psi/2)}\cos A\,d\omega \\[2ex]
\qquad\quad +\dfrac{G\rho}{32R^2} \displaystyle\int_{\omega_{\psi\le\Theta}} \left(H^3 - H_P^3\right) \dfrac{\cos(\psi/2)}{\sin^4(\psi/2)}\left(1 + \dfrac{5}{3}\sin^2\dfrac{\psi}{2}\right)\cos A\,d\omega, \\[2ex]
\delta_i g_{\Theta\lambda} = -\dfrac{G\rho}{16R} \displaystyle\int_{\omega_{\psi\le\Theta}} \left(H^2 - H_P^2\right) \dfrac{\cos(\psi/2)}{\sin^2(\psi/2)}\sin A\,d\omega \\[2ex]
\qquad\quad -\dfrac{G\rho}{32R^2} \displaystyle\int_{\omega_{\psi\le\Theta}} \left(H^3 - H_P^3\right) \dfrac{\cos(\psi/2)}{\sin^4(\psi/2)}\left(1 + \dfrac{5}{3}\sin^2\dfrac{\psi}{2}\right)\sin A\,d\omega.
\end{cases}
$$
$$(6.239)$$

It is left to readers to analyze the relative order of magnitude of the various terms in these formulae. The contribution from $\omega_{\psi\ge\Theta}$ could be computed by replacing ω by $\omega_{\psi\ge\Theta}$ in (6.235).

The Isostatic Anomaly

Analogously, for the Airy–Heiskanen or Vening Meinesz model, we have

$$
\begin{cases}
\delta_i g_\theta = \dfrac{G}{R} \displaystyle\int_\omega \rho\left[\dfrac{\partial}{\partial\psi}f_T(H,\psi) - \dfrac{\Delta\rho}{\rho}\dfrac{\partial}{\partial\psi}f_I(T,H',\psi)\right]\cos A\,d\omega, \\[2ex]
\delta_i g_\lambda = -\dfrac{G}{R} \displaystyle\int_\omega \rho\left[\dfrac{\partial}{\partial\psi}f_T(H,\psi) - \dfrac{\Delta\rho}{\rho}\dfrac{\partial}{\partial\psi}f_I(T,H',\psi)\right]\sin A\,d\omega.
\end{cases}
$$
$$(6.240)$$

As in the case of Helmert anomaly, the term $(\partial/\partial\psi)f_T(H,\psi)$ represents the contribution from the topographic mass and is given in (6.236). The term

$(\partial/\partial\psi) f_I(T, H', \psi)$ represents the contribution from the mass added in the isostatic correction, which can be inferred from (6.229) and (6.230). Here we have $r = R$, $R_B = R - (T + H)$, and $R_T = R - T$, and ρ should be replaced by $\Delta\rho$. Hence, we obtain

$$
\frac{\partial}{\partial\psi} f_I(T, H', \psi)
$$

$$
= \frac{R^2}{2} \Bigg\{ -3 \sin\psi (1 + x^2 - 2x \cos\psi)^{1/2} + \frac{x \sin\psi (x + 3 \cos\psi)}{(1 + x^2 - 2x \cos\psi)^{1/2}}
$$

$$
+ \frac{\sin\psi \, (3\cos^2\psi - 1) \left[(1 + x^2 - 2x \cos\psi)^{1/2} + x \right]}{(1 + x^2 - 2x \cos\psi)^{1/2} \left[(1 + x^2 - 2x \cos\psi)^{1/2} + x - \cos\psi \right]}
$$

$$
- 6 \sin\psi \cos\psi \ln \left[(1 + x^2 - 2x \cos\psi)^{1/2} + x - \cos\psi \right] \Bigg\}_{1-(T+H')/R}^{1-T/R}.
$$

$$(6.241)$$

In (6.240), the topographic term $(\partial/\partial\psi) f_T(H, \psi)$ tends to infinity when $\psi \to 0$, but the isostatic term $(\partial/\partial\psi) f_I(T, H', \psi)$ does not. Hence, we only need to formulate the topographic term for the region $\psi \leq \Theta$. As already mentioned, the plate of height H_P does not contribute to $\delta_i g_\theta$ and $\delta_i g_\lambda$; therefore, we have to consider only the undulating part of topographic mass around the height H_P, whose gravitational potential is given in (6.219). The derivation is similar to that of (6.239). Here we directly write out the result

$$
\begin{cases}
\delta_i g_{\Theta\theta}^T = -\dfrac{G\rho}{4} \displaystyle\int_{\omega_{\psi \leq \Theta}} (H - H_P) \frac{\cos(\psi/2)}{\sin^2(\psi/2)} \cos A \, d\omega \\[2mm]
\qquad - \dfrac{3G\rho}{16R} \displaystyle\int_{\omega_{\psi \leq \Theta}} \left(H^2 - H_P^2 \right) \frac{\cos(\psi/2)}{\sin^2(\psi/2)} \cos A \, d\omega \\[2mm]
\qquad + \dfrac{G\rho}{32R^2} \displaystyle\int_{\omega_{\psi \leq \Theta}} \left(H^3 - H_P^3 \right) \frac{\cos(\psi/2)}{\sin^4(\psi/2)} \left(1 + \sin^2 \frac{\psi}{2} \right) \cos A \, d\omega, \\[2mm]
\delta_i g_{\Theta\lambda}^T = \dfrac{G\rho}{4} \displaystyle\int_{\omega_{\psi \leq \Theta}} (H - H_P) \frac{\cos(\psi/2)}{\sin^2(\psi/2)} \sin A \, d\omega \\[2mm]
\qquad + \dfrac{3G\rho}{16R} \displaystyle\int_{\omega_{\psi \leq \Theta}} \left(H^2 - H_P^2 \right) \frac{\cos(\psi/2)}{\sin^2(\psi/2)} \sin A \, d\omega \\[2mm]
\qquad - \dfrac{G\rho}{32R^2} \displaystyle\int_{\omega_{\psi \leq \Theta}} \left(H^3 - H_P^3 \right) \frac{\cos(\psi/2)}{\sin^4(\psi/2)} \left(1 + \sin^2 \frac{\psi}{2} \right) \sin A \, d\omega,
\end{cases}
$$

$$(6.242)$$

where the superscript T implies that these are the contribution from topographic mass alone. The contribution from the isostatic compensation mass can also be formulated similarly. Again, it is left to readers to analyze the relative order of magnitude of the various terms in these formulae.

Practical computation could be done by combining (6.240) and (6.242), which is quite straightforward.

References

Hagiwara, Y. (1975). Conventional and spherical Bouguer corrections (In Japanese). *Journal of the Geodetic Society of Japan, 21*(1), 16–18.

Laske, G., Masters., G., Ma, Z., & Pasyanos, M. (2013). Update on CRUST1.0—A 1-degree global model of Earth's crust. *Geophysical Research Abstracts, 15*, Abstract EGU2013-2658.

Pasyanos, M. E., Masters, T. G., Laske, G., & Ma, Z. (2014). LITHO1.0: An updated crust and lithospheric model of the Earth. *Journal of Geophysical Research, 119*, 2153–2173. https://doi.org/10.1002/2013JB010626

Sjöberg, L.E. and Nahavandchi, H. (1999). On the indirect effect in the Stokes-Helmert method of geoid determination. *Journal of Geodesy, 73*, 87–93. https://doi.org/10.1007/s001900050222

Further Reading

Forsberg, R. (1984). *A study of terrain reductions, density anomalies and geophysical inversion methods in gravity field modelling*. Report 355, Department of Geodetic Science and Surveying, The Ohio State University.

Göttl, F., & Rummel, R. (2009). A geodetic view on isostatic models. *Pure and Applied Geophysics, 166*, 1247–1260. https://doi.org/10.1007/s00024-004-0489-x

Moritz, H. (1990). *The figure of the Earth: Theoretical geodesy and the Earth's interior* (Chapter 8). Wichmann, Karlsruhe, Wichmann, Germany.

Odera, P. A., & Fukuda, Y. (2015). Comparison of Helmert and rigorous orthometric heights over Japan. *Earth, Planets and Space, 67*, 27. https://doi.org/10.1186/s40623-015-0194-2

Santos, M. C., Venícek, P., Featherstone, W. E., Kingdon, R., Ellmann, A., Martin, B. A., Kuhn, M., & Tenzer, R. (2006). The relation between rigorous and Helmert's definitions of orthometric heights. *Journal of Geodesy, 80*, 691–704. https://doi.org/10.1007/s00190-006-0086-0

Sunkel, H. (1986). Global topographic-isostatic models. *Lecture Note in Earth Science, Vol. 17: Mathematical and Numerical Techniques in Physical Geodesy* (pp. 417–462). Springer.

Watts, A. B. (2001). *Isostasy and flexure of the lithosphere*. Cambridge University Press.

Molodensky's Theory and Beyond

7

Abstract

This chapter formulates the solution of the disturbing gravity field by solving Laplace's equation with the surface of the Earth, or an approximate surface of the Earth called the telluroid, as boundary. The most fundamental quantities in Molodensky's theory are the normal height and gravity anomaly defined by the gravity on the ground; both can be obtained from data of gravimetry and spirit levelling without relocating topographic mass. The disturbing gravity field is also represented by the height anomaly (also called quasigeoidal height) and deflection of vertical defined by the direction of the gravity on the ground. Two surfaces, the telluroid and quasigeoid, are defined using the normal height and height anomaly. This is an oblique boundary value problem, and several methods of solution—Molodensky's shrinking, analytical downward continuation, and harmonic downward continuation—are presented. In particular, a formulation in combination with a gravity reduction is included, which is expected to circumvent a problem of convergence of the spherical harmonic series of the gravitational potential. The solution of the GNSS-gravimetric boundary value problem, where the gravity disturbance and geodetic height on the ground are known, is also formulated. Finally, the combination of different kinds of data is addressed.

7.1 Molodensky's Boundary Value Problem

Stokes' theory has some imperfections. First, an approximation of the crustal density is required in gravity reduction, which would inevitably include errors; even the crustal models determined using seismic tomography include assumptions of geological structures. Hence, the mass above the geoid cannot be correctly removed in reality. Second, the gradient of the normal gravity has to be used to approximate that of the real gravity in the computation of free air correction. In the computation of both Helmert and isostatic anomalies, this approximation of gravity gradient is applied after the mass adjustment, i.e., to the adjusted Earth. Even for the isostatic anomaly, where the lateral heterogeneity in the crust is largely removed, approximating the gradient of the gravity using that of the normal gravity can still not be considered exact; the error should be location dependent and is mostly not quantified. These approximations also make it impossible to exactly determine the orthometric height, since the average of gravity along a plumb line in between the ground and the geoid depends on crustal density and gravity gradient.

To overcome these imperfections of Stokes' theory, Molodensky introduced a new height—the normal height, and subsequently the height anomaly, leading to the definition of a new gravity anomaly, resulting in a new fundamental gravimetric equation and the associated boundary value problem.

7.1.1   Normal Height, Height Anomaly, Telluroid, and Quasigeoid

The normal height H_γ is defined by modifying the definition of the orthometric height, (1.53) to (1.55) or (6.17) to (6.19). Let A be a reference point where H_γ is defined to be zero, and spirit leveling has been performed from A to P, as shown in Fig. 1.7. The normal height at P is defined as

$$H_\gamma = \frac{C}{\gamma_m}, \tag{7.1}$$

$$C = W_A - W_P = \int_A^P g \, dH_L, \qquad \gamma_m = \frac{1}{H_\gamma} \int_0^{H_\gamma} \gamma(H') dH', \tag{7.2}$$

where γ is the normal gravity, and H' is the height above the reference Earth ellipsoid. To better understand the meaning of this definition, we draw a straight line PP_0 perpendicular to the reference Earth ellipsoid with P_0 on it. The height H' is then the height of the line element dH' above P_0 along this line, and the normal height H_γ is the height of a point N on this line above P_0. The property of the point N is to be discussed later. At present, we just recognize the facts that C can be determined using measurements of spirit leveling and gravity as in the case of orthometric height, and γ_m can be computed exactly using parameters of the reference Earth ellipsoid. Therefore, approximations of crustal density and gravity gradient are not required in the determination of H_γ according to (7.1) and (7.2).

In (7.2), H' is the height above the reference Earth ellipsoid, which is equivalent to h in the formula of normal gravity, (4.174), from which we obtain

$$\gamma_m = \frac{1}{H_\gamma} \int_0^{H_\gamma} \left\{ \gamma_0 - 2\gamma_e \left[1 + \alpha + m + \left(-3\alpha + \frac{5}{2}m \right) \cos^2\theta \right] \frac{H'}{a} + 3\gamma_e \frac{H'^2}{a^2} \right\} dH'$$

$$= \gamma_0 - \gamma_e \left[1 + \alpha + m + \left(-3\alpha + \frac{5}{2}m \right) \cos^2\theta \right] \frac{H_\gamma}{a} + \gamma_e \frac{H_\gamma^2}{a^2}. \tag{7.3}$$

Fig. 7.1 Relation between height anomaly and disturbing potential

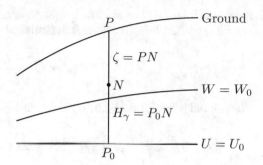

Neglect the second order term, and approximate the first term with its global average. We obtain an approximate formula

$$
\gamma_m = \gamma_0 - \left\{ \frac{1}{4\pi} \int_\omega \gamma_e \left[1 + \alpha + m + \left(-3\alpha + \frac{5}{2}m \right) \cos^2 \theta \right] d\omega \right\} \frac{H_\gamma}{a}
$$

$$
= \gamma_0 - \frac{\gamma_e}{a} \left(1 + \frac{11}{6}m \right) H_\gamma
$$

$$
= \gamma_0 - 0.1543 H_\gamma . \tag{7.4}
$$

In the numerical result, the unit of H_γ is m, and that of γ_m is mGal. In the computation of γ_m, the height obtained by spirit leveling can be used to approximate H_γ or as an initial value for an iterative computation.

It is to be noted that the normal height is not defined with respect to a predefined reference surface; instead, it is defined by modifying the definition of orthometric height in favor of computability without making approximations of crustal density and gravity gradient.

The background postulation in Molodensky's theory is that the geometrical figure of the Earth is to be determined using spirit leveling and gravity measurements, which is equivalent to determining the geodetic height (horizontal positioning is assumed to be known in this book). In the case of Stokes' theory, the geodetic height is very naturally expressed as the sum of the orthometric height and the geoidal height, which are apparent both physically and geometrically. In the case of Molodensky's theory, a height anomaly is defined such that the geodetic height is the sum of the normal height and the height anomaly, which are not yet made physically or geometrically apparent at present, and thus should be temporarily understood only mathematically based on the definition of the normal height given in (7.1) and (7.2).

Analogous to the case of Stokes' theory where the geoidal height is related to the disturbing potential, a relation between the height anomaly and the disturbing potential can be derived. As shown in Fig. 7.1, P is a point on the ground, and P_0 its projection on the reference Earth ellipsoid, where $P P_0$ is perpendicular to the reference Earth ellipsoid. We define a point N on $P P_0$ with $P_0 N$ equal to the normal height at P, i.e., $P_0 N = H_\gamma$. The height anomaly ζ is then defined to be PN, which can be expressed using the normal potential as

$$\zeta_P = \frac{U_N - U_P}{\gamma_N} .$$ (7.5)

The normal potential at N, U_N, is related to U_0 in the form of

$$U_N = U_0 - \int_0^{H_\gamma} \gamma(H')dH' = U_0 - C = -(W_A - U_0) + W_P ,$$ (7.6)

where the second and third identities are obtained according to (7.1) and (7.2). The height anomaly at P is then

$$\zeta_P = \frac{-(W_A - U_0) + W_P - U_P}{\gamma_N} = -\frac{W_A - U_0}{\gamma_N} + \frac{T_P}{\gamma_N} .$$ (7.7)

We note that γ_N can be replaced by γ_P without altering the accuracy of (7.5) and this formula. We used γ_N as it can be computed when the normal height is known.

We further assume that A is on the geoid, thus $W_A = W_0$, and that the reference Earth ellipsoid is chosen such that $U_0 = W_0$. These simplifies the above equation to

$$\zeta = \frac{T}{\gamma} ,$$ (7.8)

where we have removed the subscripts P and N, as this formula holds everywhere on the Earth's surface. By approximating γ with its average on the reference Earth ellipsoid, we obtain

$$\zeta = \frac{T}{\bar{\gamma}_0} ,$$ (7.9)

where it is to be kept in mind that T is the value on the Earth's surface. These last two formulae are Molodensky's version of Bruns' formula.

A physical characteristic of the normal height and height anomaly can be inferred from (7.6). With the zero normal height point A chosen to be on the geoid and $U_0 = W_0$, we have $U_N = W_P$, i.e., the normal potential at N is equal to the gravity potential at P. This permits us to give a slightly more apparent definition of the normal height and height anomaly: For any point P on the ground, find a point N right below it, such that the normal potential at N is equal to the gravity potential at P. The height of N above the reference Earth ellipsoid is then defined to be the normal height, and the height of P above N, the height anomaly.

A relation between the height anomaly and the geoidal height can be readily derived. Based on definition, we have two formulae for the geodetic height

$$h = N + H \quad \text{and} \quad h = \zeta + H_\gamma . \tag{7.10}$$

With A on the geoid, a relation between orthometric height and normal height can be readily obtained

$$C = \gamma_m H_\gamma = g_m H . \tag{7.11}$$

Hence,

$$N = \zeta + H_\gamma - H = \zeta + \frac{\gamma_m - g_m}{\gamma_m} H_\gamma , \quad \zeta = N + H - H_\gamma = N + \frac{g_m - \gamma_m}{g_m} H . \tag{7.12}$$

These transformations involve the computation of g_m and cannot be exact. The difference between N and ζ can reach the order of a meter. It should be mentioned that some countries choose to use the orthometric height, and some other countries choose to use the normal height. Therefore, maps of different countries may refer to either the orthometric height or the normal height; geodetic height is rarely used for mapping. In contemporary surveying, the geodetic height is measured using GNSS. In order to convert it into the height for mapping, either the geoidal height or the height anomaly should be subtracted based on the type of height system adopted.

Although the normal height H_γ is not defined with respect to a reference surface, two surfaces are defined based on the normal height: the telluroid and the quasigeoid. The former is H_γ above the reference Earth ellipsoid (i.e., the surface formed by all the Ns defined in Fig. 7.1), which is required for formulating Molodensky's gravity anomaly and gravimetric boundary value problem; the latter is H_γ below the Earth's surface, which could be understood as a surface assigned to be the reference surface of normal height—the normal height is the height above the quasigeoid. With these definitions, the height anomaly can be understood as the height of the Earth's surface above the telluroid, or the height of the quasigeoid above the reference Earth ellipsoid. For this reason, the height anomaly is also referred to as quasigeoidal height.

Apparently, the vertical datum of a normal height system is a quasigeoid, which can be provided as a levelling network or as a model of the quasigeoidal height; the quasigeoidal height ζ can be used to convert the GNSS measured geodetic height h into the normal height H_γ according to (7.10).

7.1.2   Molodensky's Gravity Anomaly and Gravimetric Boundary Value Problem

The relation between the Earth's surface and the telluroid is similar to that between the geoid and the reference Earth ellipsoid; in both cases, the gravity potential on the former is equal to the normal potential on the later. Analogously, Molodensky's gravity anomaly is defined to be the difference between the absolute value of gravity on the Earth's surface and that of normal gravity on the telluroid. Taking reference to Fig. 7.1, we write it in the following form for more evident further derivation

$$\Delta G_P = g_P - \gamma_N . \tag{7.13}$$

The physical nature of ΔG is completely different from that of Δg. Obviously, the formula for practical computation of ΔG using the measurement g_P is similar to that of the free air anomaly with the orthometric height replaced by the normal height; the free air correction here is used to convert γ_0 into γ_N. As no assumption like crustal density or gravity gradient is involved, the conversion from γ_0 to γ_N is exact. Hence, if the observations of gravity and spirit leveling are exact, ΔG can be computed exactly. We mention that ΔG is also referred to as free air anomaly, but its different physical meaning from Δg should be kept in mind.

A relation between ΔG and T has to be derived in order to determine T using data of ΔG. We use H to denote the height defined along the direction of gravity at P and h the height defined along the direction of normal gravity at N. We then have

$$\Delta G_P = -\left(\frac{\partial W}{\partial H}\right)_P + \left(\frac{\partial U}{\partial h}\right)_N = -\frac{1}{\cos(\hat{e}_H, \hat{e}_h)}\left(\frac{\partial W}{\partial h}\right)_P + \left(\frac{\partial U}{\partial h}\right)_N , \tag{7.14}$$

where $(\hat{e}_H, \hat{e}_h)$ is the angle between the direction of gravity at P and that of normal gravity at N, which is the deflection of the vertical in Molodensky's approach, and is very small, just as in the case of deflection of the vertical in Stokes' approach. Hence, we make the approximation $\cos(\hat{e}_H, \hat{e}_h) = 1$, which is equivalent to neglect the difference between the derivatives with respect to H and h,

$$\Delta G_P = -\left(\frac{\partial W}{\partial h}\right)_P + \left(\frac{\partial U}{\partial h}\right)_N . \tag{7.15}$$

The error introduced by this approximation is of second order of the angle $(\hat{e}_H, \hat{e}_h)$. According to the Taylor series expansion, we have

$$\left(\frac{\partial U}{\partial h}\right)_N = \left(\frac{\partial U}{\partial h}\right)_P - \zeta\left(\frac{\partial^2 U}{\partial h^2}\right)_P . \tag{7.16}$$

Hence,

$$\Delta G_P = -\zeta \left(\frac{\partial^2 U}{\partial h^2} \right)_P - \left[\frac{\partial (W - U)}{\partial h} \right]_P. \tag{7.17}$$

Drop off the subscript P while remembering that all quantities in the formula are the values on the Earth's surface, and then substitute (7.8) into it. We obtain the fundamental gravimetric equation of Molodensky's approach

$$\Delta G = \frac{1}{\gamma} \frac{\partial \gamma}{\partial h} T - \frac{\partial T}{\partial h}, \tag{7.18}$$

where we took into account of $\gamma = \partial U/\partial h$ and $T = W - U$. In the derivation, we assumed that h is the height measured along the direction of the normal gravity at N and made the approximation of not distinguishing this direction from that of the gravity at P. Now we make the further approximation of not distinguishing it from the direction of the normal gravity at P_0, meaning that we also ignore the influence of the small angle between the directions of the normal gravity at N and P_0. We mention that this small angle can be evaluated by applying the formula of curvature of the plumb line derived in Sect. 1.6.3 to the normal gravity field, which will be the topic in Sect. 8.4.1. It is left to readers to quantify the level of error introduced by this approximation. The desired simplification brought by this approximation is that h can be understood as the geodetic height.

Due to the approximation made for h, the formula (7.18) refers to two surfaces: One is the reference Earth ellipsoid based on which h is defined; and the other is the Earth's surface where the values of all other quantities are defined. Following this definition of h, we refer (7.18) to as ellipsoidal approximation, where, as T is a small quantity, we also approximate the coefficient $(1/\gamma)\partial \gamma/\partial h$ by its value on the reference Earth ellipsoid.

As compared to Stokes' approach, gravity reduction is avoided in Molodensky's approach by taking the Earth's surface as the boundary where the fundamental gravimetric equation is established. Therefore, approximating the Earth's surface using a simpler figure such as an ellipsoid or a sphere is irrelevant. Hence, as the reference Earth ellipsoid deviates from a sphere by only about 1/300, we defined the spherical approximation of (7.18) as an approximation of the reference Earth ellipsoid by a sphere of radius R, which would introduce the same level of error. Under spherical approximation, derivatives with respect to h and r are identical; thus the fundamental gravimetric equation can be simplified to

$$\Delta G = \left(-2\frac{T}{r} - \frac{\partial T}{\partial r} \right)_{r=R+h}, \tag{7.19}$$

where we have also replaced the coefficient $(1/\gamma)\partial \gamma/\partial h$ by the spherical approximations given in (5.24).

Readers may have already realized that the geodetic height has appeared in the fundamental gravimetric equation (7.18), or its spherical approximation (7.19), which is a natural consequence of the approximation by understanding h as the height in the direction perpendicular to the reference Earth ellipsoid, but the height that we are supposed to know is the normal height. Purely mathematically, the same as T, $\zeta = h - H_\gamma$ is a first order small term, and replacing h by H_γ in (7.18) or (7.19) introduces only an error on the second order of ζ, while second order terms have been neglected in our formulation. This means that it is legitimate to understand h in (7.18) or (7.19) as H_γ within the accuracy of the formulation, as if the boundary is the telluroid rather than the Earth's surface. Alternatively, we can understand h in (7.18) or (7.19) as the normal height by approximating it as the height measured along the direction perpendicular to the quasigeoid, and meanwhile approximating the quasigeoid as an ellipsoid or sphere, just as approximating the geoid as an ellipsoid or sphere in Stokes' theory.

With the fundamental gravimetric equation (7.18) or (7.19), Molodensky's boundary value problem can be stated as follows: Find the solution of Laplace's equation $\Delta T = 0$ outside the Earth's surface (or the telluroid) that satisfies the boundary condition (7.18) or (7.19) on the Earth's surface (or the telluroid). As the disturbing potential is the unknown, which does not include the degrees 0 and 1 terms if expressed as a spherical harmonic series, an additional condition $\lim_{r \to \infty} r^2 T = 0$ should also be satisfied.

In the literature, the telluroid is customarily understood as the boundary of Molodensky's boundary value problem, which is the most straightforward, as the telluroid is known following the fact that the reference Earth ellipsoid and the normal height are known. This understanding implies that the not yet known Earth's surface is approximated by the known telluroid. However, rigorously speaking, the disturbing potential T in the fundamental gravimetric equation should be understood as a function defined on the Earth's surface rather than that on the telluroid. If the Earth's surface is taken as the boundary, the not yet known quasigeoid should be approximated as an ellipsoid or a sphere. Both understandings introduce the same level of approximation, but physically, mass may exist outside the telluroid, but not outside the Earth's surface.

Like Stokes' boundary value problem, Molodensky's boundary value problem is also a free boundary value problem, where the not yet known boundary—the Earth's surface—has to be approximated by a known boundary. Molodensky's boundary value problem is more complicated than Stokes' in two aspects. First, the boundary can no longer be approximately considered as an ellipsoid or a sphere. Second, the derivative in the boundary condition is no longer along the normal to the boundary; such kinds of boundary value problems are referred to as oblique derivative boundary value problems.

7.1.3 A Generalized Definition of Gravity Anomaly in Space

So far, two kinds of gravity anomaly have been defined as a function on a surface: one is on the geoid and is denoted as Δg, and the other is on the Earth's surface and is denoted as ΔG. However, they do have a common characteristic: in both of them, the gravity anomaly at a point is the difference between the gravity at this point and the normal gravity at another point on the same vertical, where the normal potential at the later point equals the gravity potential at the former point. As this common characteristic does not rely on any surface, we do use it as the definition of gravity anomaly at any point in the space, which is indeed an extension of the definitions on the surfaces. Meanwhile, we extend the definition of height anomaly as the distance between the two points. For any point P, we write the definition of gravity anomaly as

$$\Delta G_P = g_P - \gamma_{U=W_P} \,. \tag{7.20}$$

It can be readily shown that the height anomaly and disturbing potential satisfy the same equations as (7.8) and (7.18),

$$\zeta_P = \frac{T_P}{\gamma_{U=W_P}} \,, \tag{7.21}$$

$$\Delta G_P = \frac{1}{\gamma_{U=W_P}} \left(\frac{\partial \gamma}{\partial h}\right)_{U=W_P} T_P - \left(\frac{\partial T}{\partial h}\right)_P \,. \tag{7.22}$$

Here $\partial \gamma/\partial h$ and $\partial T/\partial h$ should be understood as directional derivatives, where h represents the direction of normal gravity at the point $U = W_P$, i.e., the direction of $\gamma_{U_-W_P}$. Under spherical approximation, these equations become

$$\zeta = \frac{T}{\gamma} \,, \tag{7.23}$$

$$\Delta G = -\frac{2}{r} T - \frac{\partial T}{\partial r} \,, \tag{7.24}$$

where all quantities are understood as functions of r, θ, and λ. Evidently, Stokes' boundary value problem is a special case with the gravity anomaly defined on the geoid, where the geoidal height is equivalent to the height anomaly. Molodensky's boundary value problem is a special case with the gravity anomaly defined on the Earth's surface.

Unless otherwise explicitly specified, the notations ζ and ΔG are always those defined on the Earth's surface in the subsequent text.

7.2   Molodensky's Solution

7.2.1   Molodensky's Integral Equation

Molodensky provided a solution of T on the Earth's surface by converting his boundary value problem into the solution of an integral equation for the case of spherical approximation. According to (2.55), we express the harmonic function T outside the Earth as the gravitational potential of a material surface of density μ/G located on the Earth's surface S

$$T = \int_S \frac{\mu}{l} dS,   \tag{7.25}$$

where l is the distance between the fixed point where T is to be computed and the running point of integral where dS is located. Evidently, μ has to be position dependent. Conceptually, as T expressed in the form of (7.25) already satisfies Laplace's equation outside the Earth, what remains is to determine μ through the boundary condition, and then use the result of μ to compute T. However, it is T on the Earth's surface that will be directly obtained through the formulation.

In (7.25), two sets of coordinates are implicitly defined. One set is r, θ, and λ, which are the coordinates of the point outside or on the Earth's surface where T is to be computed. The other set is r', θ', and λ', which are the coordinates of the point where the surface element dS is located, with $r' = R + h$, where h is a function of θ' and λ'. The distance between the two points is

$$l = \left(r^2 + r'^2 - 2rr' \cos \psi \right)^{1/2},   \tag{7.26}$$

where ψ is the geocentric angle between the two points. The surface density μ is a function of θ' and λ'. As shown in Fig. 7.2, we consider the boundary condition (7.19) at the point P with T substituted by its integral form given in (7.25). In the boundary condition (7.19), all terms should be understood as the limit obtained by approaching the Earth's surface from the outside. Therefore, the discontinuity of the derivative of the gravitational potential of a material surface expressed in (1.128) should be taken into account in the term $(\partial T/\partial r)_{r=R+h}$. At P, we have

$$\left(\frac{\partial T}{\partial r} \right)_{r=R+h_P} = -2\pi \mu_P \cos \beta_P + \int_S \mu \frac{\partial}{\partial r} \left(\frac{1}{l} \right)_{r=R+h_P} dS,   \tag{7.27}$$

where β_P is the angle between the normal of the Earth's surface and the radial direction at P, $\beta_P = (\hat{e}_{nS}, \hat{e}_r)_P$. By substituting (7.25) and (7.27) into (7.19), we

Fig. 7.2 Derivation of
Molodensky's integral
equation

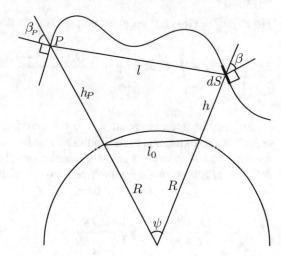

obtain

$$2\pi \mu_P \cos \beta_P - \int_S \mu \left[\frac{\partial}{\partial r} \left(\frac{1}{l} \right)_{r=R+h_P} + \frac{2}{(R+h_P)l_P} \right] dS = \Delta G_P, \qquad (7.28)$$

where l_P is the value of l in (7.26) after replacing r by $R+h_P$. According to (7.26),

$$\frac{\partial}{\partial r} \left(\frac{1}{l} \right)_{r=R+h_P} = - \left(\frac{r - r' \cos \psi}{l^3} \right)_{r=R+h_P} = \left(\frac{r'^2 - r^2 - l^2}{2rl^3} \right)_{r=R+h_P}$$

$$= \frac{r'^2 - (R+h_P)^2 - l_P^2}{2(R+h_P)l_P^3}. \qquad (7.29)$$

We can thus write (7.28) as, after substituting $r' = R+h$,

$$2\pi \mu_P \cos \beta_P - \int_S \mu \left[\frac{3}{2(R+h_P)l_P} + \frac{(R+h)^2 - (R+h_P)^2}{2(R+h_P)l_P^3} \right] dS = \Delta G_P, \qquad (7.30)$$

where we emphasize that h is the height of the surface element dS.

Equation (7.30) includes a surface integral over the Earth's surface that is quite irregular. The first step to simplify this equation is to reduce the integral onto a unit sphere. The radial coordinate of the surface element dS is known to be $R+h$. The projection of dS on the sphere of radius $R+h$ is $\cos \beta \, dS$, where β is the angle between the normal of the Earth's surface and the radial direction, $\beta = (\hat{e}_{ns}, \hat{e}_r)$. Furthermore, the relation between the surface element on the sphere of radius $R+h$, $\cos \beta \, dS$, and the surface element on a unit sphere, $d\omega$, is $(R+h)^2 d\omega = \cos \beta \, dS$.

Hence, $dS = (R + h)^2 \sec \beta d\omega$, and the Eq. (7.30) can be written as

$$2\pi \mu_P \cos \beta_P - \int_\omega \mu \sec \beta \left[\frac{3}{2l_P} + \frac{(R + h)^2 - (R + h_P)^2}{2l_P^3} \right] \frac{(R + h)^2}{R + h_P} d\omega = \Delta G_P .$$

(7.31)

No further approximation has been introduced into Eq. (7.31) besides the spherical approximation of the boundary condition (7.19). The next step to simplify the equation is to neglect small quantities of relative magnitude on the order of $h/R \sim 1/700$ or smaller, which consists of introducing further approximations. With $R + h = R(1 + h/R)$, we have

$$\frac{(R + h)^2}{R + h_P} = R \frac{(1 + h/R)^2}{1 + h_P/R} = R \left(1 + 2\frac{h}{R} - \frac{h_P}{R} \right) ,$$

(7.32)

$$(R + h)^2 - (R + h_P)^2 = 2R \left[(h - h_P) + \frac{h^2 + h_P^2}{2R} \right] = 2R(h - h_P) ,$$

(7.33)

$$
\begin{aligned}
l_P &= \left[2R^2(1 - \cos \psi) + 2R(h + h_P)(1 - \cos \psi) + h^2 + h_P^2 - 2hh_P \cos \psi \right]^{1/2} \\
&= \left[4R^2 \sin^2 \frac{\psi}{2} \left(1 + \frac{h + h_P}{R} + \frac{hh_P}{R^2} \right) + (h - h_P)^2 \right]^{1/2} \\
&= \left[l_0^2 + (h - h_P)^2 \right]^{1/2} \\
&= l_0 \left[1 + \left(\frac{h - h_P}{l_0} \right)^2 \right]^{1/2} ,
\end{aligned}
$$

(7.34)

where l_0 is the distance between the projections of the two points, P and the point where dS is located, on the sphere of radius R, as shown in Fig. 7.2,

$$l_0 = 2R \sin \frac{\psi}{2} .$$

(7.35)

The substitution of (7.32) to (7.34) into (7.31) yields a simplified equation to be solved

$$2\pi \mu_P \cos \beta_P - \int_\omega \mu \sec \beta \left\{ \frac{3R}{2l_0 \left[1 + \left(\frac{h - h_P}{l_0} \right)^2 \right]^{1/2}} + \frac{R^2(h - h_P)}{l_0^3 \left[1 + \left(\frac{h - h_P}{l_0} \right)^2 \right]^{3/2}} \right\} d\omega$$

$$= \Delta G_P .$$

(7.36)

The integral equation (7.30), its equivalent form (7.31), or its approximate form (7.36) are all referred to as Molodensky's integral equation.

7.2.2 Molodensky's Shrinking

Introduce an auxiliary function

$$\chi - \mu \sec \beta \tag{7.37}$$

to substitute μ. The integral equation (7.36) becomes

$$\frac{2\pi \chi_P}{1+\tan^2 \beta_P} - \int_\omega \chi \left\{ \frac{3R}{2l_0 \left[1+\left(\dfrac{h-h_P}{l_0}\right)^2 \right]^{1/2}} + \frac{R^2(h-h_P)}{l_0^3 \left[1+\left(\dfrac{h-h_P}{l_0}\right)^2 \right]^{3/2}} \right\} d\omega$$
$$= \Delta G_P . \tag{7.38}$$

In this equation, replace h by kh, and $\tan \beta$ by $k \tan \beta$, where k is a number in the range $0 \le k \le 1$. We obtain

$$\frac{2\pi \chi_P}{1+k^2 \tan^2 \beta_P} - \frac{3R}{2} \int_\omega \frac{\chi}{l_0 \left[1+k^2 \left(\dfrac{h-h_P}{l_0}\right)^2 \right]^{1/2}} d\omega - R^2 \int_\omega \frac{k(h-h_P)\chi}{l_0^3 \left[1+k^2 \left(\dfrac{h-h_P}{l_0}\right)^2 \right]^{3/2}} d\omega$$
$$= \Delta G_P . \tag{7.39}$$

The next step is to develop this equation into a power series of k. Here we keep terms only up to k^2. We have

$$\frac{1}{1+k^2 \tan^2 \beta_P} = 1 - k^2 \tan^2 \beta_P + \cdots , \tag{7.40}$$

$$\left[1+k^2 \left(\frac{h-h_P}{l_0}\right)^2 \right]^{-1/2} = 1 - \frac{1}{2}k^2 \left(\frac{h-h_P}{l_0}\right)^2 + \cdots , \tag{7.41}$$

$$\left[1+k^2 \left(\frac{h-h_P}{l_0}\right)^2 \right]^{-3/2} = 1 - \frac{3}{2}k^2 \left(\frac{h-h_P}{l_0}\right)^2 + \cdots . \tag{7.42}$$

We also express χ as

$$\chi = \chi_0 + k\chi_1 + k^2\chi_2 + \cdots . \tag{7.43}$$

Equation (7.39) becomes

$$2\pi(\chi_{0P} + k\chi_{1P} + k^2\chi_{2P} + \cdots)(1 - k^2\tan\beta_P + \cdots)$$

$$- \frac{3R}{2}\int_\omega \frac{1}{l_0}(\chi_0 + k\chi_1 + k^2\chi_2 + \cdots)\left[1 - \frac{1}{2}k^2\left(\frac{h - h_P}{l_0}\right)^2 + \cdots\right]d\omega$$

$$- R^2\int_\omega \frac{k(h - h_P)}{l_0^3}(\chi_0 + k\chi_1 + k^2\chi_2 + \cdots)\left[1 - \frac{3}{2}k^2\left(\frac{h - h_P}{l_0}\right)^2 + \cdots\right]d\omega$$

$$- \Delta G_P = 0. \tag{7.44}$$

Group terms in the same power of k. We obtain

$$\left\{2\pi\chi_{0P} - \frac{3R}{2}\int_\omega \frac{\chi_0}{l_0}d\omega - \Delta G_P\right\}$$

$$+ k\left\{2\pi\chi_{1P} - \frac{3R}{2}\int_w \frac{\chi_1}{l_0}d\omega - R^2\int_\omega \chi_0\frac{h - h_P}{l_0^2}d\omega\right\}$$

$$+ k^2\left\{2\pi\chi_{2P} - \frac{3R}{2}\int_\omega \frac{\chi_2}{l_0}d\omega - 2\pi\chi_{0A}\tan^2\beta_P - R^2\int_\omega \chi_1\frac{h - h_P}{l_0^3}d\omega\right.$$

$$\left. + \frac{3R}{4}\int_\omega \chi_0\frac{(h - h_P)^2}{l_0^3}d\omega\right\}$$

$$+ \cdots = 0. \tag{7.45}$$

The left-hand side of this equation is a power series of k. In order for this equation to hold for any value of k, all coefficients of the same power of k must vanish. Hence, we obtain a chain of similar equations:

$$\begin{cases} 2\pi\chi_{0P} - \dfrac{3R}{2}\displaystyle\int_\omega \frac{\chi_0}{l_0}d\omega = \Delta G_P, \\[2mm] 2\pi\chi_{1P} - \dfrac{3R}{2}\displaystyle\int_\omega \frac{\chi_1}{l_0}d\omega = \Delta G_{1P}, \\[2mm] 2\pi\chi_{2P} - \dfrac{3R}{2}\displaystyle\int_\omega \frac{\chi_2}{l_0}d\omega = \Delta G_{2P}, \\[2mm] \cdots, \end{cases} \tag{7.46}$$

where $\Delta G_{1P}, \Delta G_{2P}, \cdots$ are defined as

$$
\begin{cases}
\Delta G_{1P} = R^2 \int_\omega \chi_0 \dfrac{h - h_P}{l_0^3} d\omega, \\[2ex]
\Delta G_{2P} = 2\pi \chi_{0P} \tan^2 \beta_P + R^2 \int_\omega \chi_1 \dfrac{h - h_P}{l_0^3} d\omega - \dfrac{3R}{4} \int_\omega \chi_0 \dfrac{(h - h_P)^2}{l_0^3} d\omega, \\[2ex]
\cdots.
\end{cases}
$$

$$(7.47)$$

The functions to be obtained by solving (7.46) and (7.47) are $\chi_0, \chi_1, \chi_2, \cdots$. Therefore, we need to express the disturbing potential T in terms of these functions. By substituting $dS = (R+h)^2 \sec \beta d\omega$, $l = l_P$, and (7.37) into (7.25), we can write the disturbing potential at P as

$$
T_P = \int_\omega \frac{\chi}{l_P} (R + h)^2 d\omega.
$$

$$(7.48)$$

As T_P itself is a small quantity, we can neglect relative smaller terms in its expression. By also using (7.34), we obtain

$$
T_P = R^2 \int_\omega \frac{\chi}{l_P} \left(1 + 2\frac{h}{R} \right) d\omega = R^2 \int_\omega \frac{\chi}{l_P} d\omega = R^2 \int_\omega \frac{\chi}{l_0 \left[1 + \left(\dfrac{h - h_P}{l_0} \right)^2 \right]^{1/2}} d\omega.
$$

$$(7.49)$$

Similar to the treatment of the integral equation, we replace h by kh, substitute χ into (7.43), and then develop it into a power series of k. This yields

$$
T_P = R^2 \int_\omega \frac{1}{l_0} (\chi_0 + k\chi_1 + k^2\chi_2 + \cdots) \left[1 - \frac{1}{2} k^2 \left(\frac{h - h_P}{l_0} \right)^2 + \cdots \right] d\omega
$$

$$
= R^2 \int_\omega \frac{\chi_0}{l_0} d\omega + k R^2 \int_\omega \frac{\chi_1}{l_0} d\omega + k^2 \left[R^2 \int_\omega \frac{\chi_2}{l_0} d\omega - \frac{1}{2} R^2 \int_\omega \frac{(h - h_P)^2}{l_0^3} \chi_0 d\omega \right]
$$

$$
+ \cdots.
$$

$$(7.50)$$

The value of T_P is obtained by setting $k \to 1$. We write it in the form of

$$
T_P = T_{0P} + T_{1P} + T_{2P} + \cdots,
$$

$$(7.51)$$

where

$$
\begin{cases}
T_{0P} = R^2 \displaystyle\int_\omega \frac{\chi_0}{l_0} d\omega \,, \\[2mm]
T_{1P} = R^2 \displaystyle\int_\omega \frac{\chi_1}{l_0} d\omega \,, \\[2mm]
T_{2P} = R^2 \displaystyle\int_\omega \frac{\chi_2}{l_0} d\omega - \frac{1}{2} R^2 \displaystyle\int_\omega \frac{(h - h_P)^2}{l_0^3} \chi_0 d\omega \,, \\[2mm]
\cdots
\end{cases}
\tag{7.52}
$$

The approach of introducing the parameter k into the equations and then developing the equations into a power series of k to obtain a chain of simpler equations is referred to as Molodensky's shrinking. Conceptually, the functions χ_0, χ_1, χ_2, $\cdots$ can be obtained by solving (7.46) and (7.47), and then the disturbing potential can be computed using (7.51) and (7.52).

7.2.3 Solution of the Height Anomaly

The subject of this subsection is to solve the chain of integral equations developed in the previous subsection. The solution is inferred from that of Stokes' boundary value problem based on an analogy among the equations.

Evidently, if the Earth's surface degenerates to a sphere, the equation to be solved in Molodensky's boundary value problem degenerates to the same form as that in Stokes' boundary value problem. Under this circumstance, β, h, and h_A vanish, and Molodensky's integral equation degenerates to

$$
2\pi \mu_P - \frac{3R}{2} \int_\omega \frac{\mu}{l_0} d\omega = \Delta G_P \,.
\tag{7.53}
$$

The formula of T_P degenerates to, according to (7.25),

$$
T_P = R^2 \int_\omega \frac{\mu}{l_0} d\omega \,.
\tag{7.54}
$$

As these equations are the same as those of Stokes' boundary value problem, their solution is also the same as that of Stokes' boundary value problem, i.e.,

$$
T_P = \frac{R}{4\pi} \int_\omega \Delta G S(\psi) d\omega \,.
\tag{7.55}
$$

Purely mathematically, this implies that (7.55) is the solution of (7.53) and (7.54), which is the basis for constructing the solutions to the chain of similar integral equations of Molodensky's boundary value problem.

Now we consider the chain of integral equations expressed in (7.46) and (7.52). The first equations in (7.46) and (7.52) are analogous to (7.53) and (7.54), yielding a solution of T_0 analogous to (7.55),

$$T_{0P} = \frac{R}{4\pi} \int_\omega \Delta G S(\psi) d\omega. \tag{7.56}$$

For the same reason, the solution of T_1 can be obtained from the second equations in (7.46) and (7.52)

$$T_{1P} = \frac{R}{4\pi} \int_\omega \Delta G_1 S(\psi) d\omega. \tag{7.57}$$

The solution of T_2 is not evident, as the third equation in (7.52) has an extra term as compared to (7.54), although the third equation in (7.46) is in the same form as (7.53). Here we separate T_2 into two terms, $T_2 = T_2' + T_2''$, where T_2' and T_2'' are the first and second terms, respectively, of the third equation in (7.52). The solution of T_2' can then be inferred by analogy of its equations to (7.53) and (7.54). We thus obtain

$$T_{2P} = \frac{R}{4\pi} \int_\omega \Delta G_2 S(\psi) d\omega - \frac{1}{2} R^2 \int_\omega \frac{(h - h_P)^2}{l_0^3} \chi_0 d\omega, \tag{7.58}$$

where the first term is T_2', and the second, T_2''. The solutions of χ_0 and χ_1 can be obtained by substituting (7.52) into (7.46)

$$\chi_{0P} = \frac{1}{2\pi} \left(\Delta G_P + \frac{3}{2} \frac{T_{0P}}{R} \right), \tag{7.59}$$

$$\chi_{1P} = \frac{1}{2\pi} \left(\Delta G_{1P} + \frac{3}{2} \frac{T_{1P}}{R} \right). \tag{7.60}$$

These are required in order to compute ΔG_1 and ΔG_2 according to (7.47). Also, the computation of the second term in T_{2P} requires χ_0.

Up to now all necessary formulae are derived for the disturbing potential on the Earth's surface. Here we group them in the sequential order of computations. In formulae where the subscript P does not appear under the integral signs, it is omitted outside the integral signs, which does not bring confusion:

1. Compute T_0. This is done according to (7.56), which is copied to here,

$$T_0 = \frac{R}{4\pi} \int_\omega \Delta G S(\psi) d\omega. \tag{7.61}$$

2. Compute ΔG_1. The formula is obtained by substituting (7.59) into the first formula of (7.47)

$$\Delta G_1 = \frac{R^2}{2\pi} \int_\omega \left(\Delta G + \frac{3}{2} \frac{T_0}{R} \right) \frac{h - h_P}{l_0^3} d\omega . \tag{7.62}$$

3. Compute T_1. This is done according to (7.57), which is copied to here,

$$T_{1P} = \frac{R}{4\pi} \int_\omega \Delta G_1 S(\psi) d\omega . \tag{7.63}$$

4. Compute ΔG_2. The formula is obtained by substituting (7.59) and (7.60) into the second formula of (7.47)

$$\Delta G_{2P} = \left(\Delta G_P + \frac{3}{2} \frac{T_{0P}}{R} \right) \tan^2 \beta_P + \frac{R^2}{2\pi} \int_\omega \left(\Delta G_1 + \frac{3}{2} \frac{\pi}{R} \right) \frac{h - h_P}{l_0^3} d\omega$$

$$- \frac{3R}{8\pi} \int_\omega \left(\Delta G + \frac{3}{2} \frac{T_0}{R} \right) \frac{(h - h_P)^2}{l_0^3} d\omega . \tag{7.64}$$

5. Compute T_2. The formula is obtained by substituting (7.59) into (7.58)

$$T_{2P} = \frac{R}{4\pi} \int_\omega \Delta G_2 S(\psi) d\omega - \frac{R^2}{4\pi} \int_\omega \left(\Delta G + \frac{3}{2} \frac{T_0}{R} \right) \frac{(h - h_P)^2}{l_0^3} d\omega . \tag{7.65}$$

Higher order terms are not included in the solution here.

The solution obtained, as summarized in the previous paragraph, is for computing the disturbing potential on the Earth's surface. The evident reason to derive this solution is for computing the height anomaly, whose formulae can be written out quite straightforwardly according to (7.9),

$$\zeta = \zeta_0 + \zeta_1 + \cdots , \tag{7.66}$$

$$\zeta_0 = \frac{R}{4\pi \bar{\gamma}_0} \int_\omega \Delta G S(\psi) d\omega , \tag{7.67}$$

$$\Delta G_{1P} = \frac{R^2}{2\pi} \int_\omega \left(\Delta G + \frac{3\bar{\gamma}_0}{2R} \zeta_0 \right) \frac{h - h_P}{l_0^3} d\omega , \tag{7.68}$$

$$\zeta_1 = \frac{R}{4\pi \bar{\gamma}_0} \int_\omega \Delta G_1 S(\psi) d\omega . \tag{7.69}$$

Terms of smaller order of magnitude are not to be further studied.

7.2.4 Solution of the Deflection of the Vertical

Since the Earth's surface is taken as the boundary in Molodensky's theory, the deflection of the vertical is also defined on the Earth's surface. The deflection of the vertical at a point is defined as the angle between the direction of gravity at this point and the direction of normal gravity at the corresponding point on the telluroid. Figure 7.3 shows a cross section, where P is a point on the Earth's surface, N the corresponding point on the telluroid, $W = W_P$ the equipotential surface of gravity through P, and $U = W_P$ the equipotential surface of normal gravity through N. By definition, the normal gravity $\vec{\gamma}_N$ is within the cross section. The deflection of the vertical at P, ϵ, is the angle between the normals of $W = W_P$ and $U = W_P$, the latter being in the same direction as the normal gravity $\vec{\gamma}_N$. Let Q be another point on the Earth's surface very close to P. Draw a straight line perpendicular to $U = W_P$ from Q that intersects $W = W_P$ at Q', and $U = W_P$ at N_Q. We use ζ' to denote the distance between the two surfaces $W = W_P$ and $U = W_P$. With Δl and $\Delta\zeta' = \zeta'_{Q'} - \zeta'_P$ defined in Fig. 7.3, the projection of the deflection of the vertical at P within the cross section can be expressed as

$$\epsilon = \lim_{\Delta l \to 0} \frac{\Delta\zeta'}{\Delta l} = \frac{\partial\zeta'}{\partial l}. \tag{7.70}$$

The rest is to express ζ' in terms of ζ. First, by definition, $\zeta'_P = \zeta_P$; thus

$$\Delta\zeta' = \zeta'_{Q'} - \zeta_P. \tag{7.71}$$

Second, based on the extended definition of height anomaly in space, the surface $U = W_P$ can be understood as the telluroid corresponding to the surface $W = W_P$,

Fig. 7.3 Definition of the deflection of the vertical

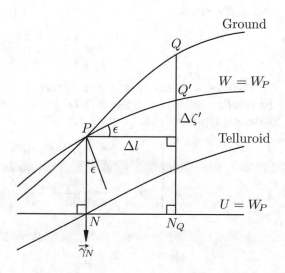

and hence, $\zeta'_{Q'} = Q'N_Q = \zeta_{Q'}$ is the height anomaly at Q'. We have

$$\Delta\zeta' = \zeta_{Q'} - \zeta_P. \tag{7.72}$$

Third, based again on the extended definition of height anomaly in space, we realize that $\zeta_{Q'}$ is related to ζ_Q according to

$$\zeta_{Q'} = \zeta_Q - \frac{\partial\zeta}{\partial h}QQ' = \zeta_Q - \frac{\partial\zeta}{\partial h}(h_Q - h_P), \tag{7.73}$$

where the second identity is obtained by taking into account of the fact that the difference between QQ' and $h_Q - h_P$, $\Delta\zeta'$, is a quantity of smaller order of magnitude. Together with the more obvious relations $\zeta_Q = \zeta_P + (\partial\zeta/\partial l)\Delta l$ and $h_Q = h_P + (\partial h/\partial l)\Delta l$, we can write (7.72) as

$$\Delta\zeta' = \frac{\partial\zeta}{\partial l}\Delta l - \frac{\partial\zeta}{\partial h}\frac{\partial h}{\partial l}\Delta l. \tag{7.74}$$

Hence, we obtain

$$\epsilon = \frac{\partial\zeta}{\partial l} - \frac{\partial\zeta}{\partial h}\frac{\partial h}{\partial l}. \tag{7.75}$$

The function $\partial\zeta/\partial h$ can be further developed according to Bruns' formula (7.8) and the fundamental gravimetric equation (7.18), yielding

$$\frac{\partial\zeta}{\partial h} = \frac{\partial}{\partial h}\left(\frac{T}{\gamma}\right) = -\frac{1}{\gamma^2}\frac{\partial\gamma}{\partial h}T + \frac{1}{\gamma}\frac{\partial T}{\partial h} = -\frac{\Delta G}{\gamma}. \tag{7.76}$$

We finally obtain

$$\epsilon = \frac{\partial\zeta}{\partial l} + \frac{\Delta G}{\gamma}\frac{\partial h}{\partial l}. \tag{7.77}$$

The southward and westward components of the deflection of the vertical can be readily obtained by applying (7.77) in the meridian and prime vertical. Under spherical approximation, we have

$$\begin{cases} \xi = \dfrac{1}{R}\dfrac{\partial\zeta}{\partial\theta} + \dfrac{\Delta G}{\bar{\gamma}_0}\left(\dfrac{1}{R}\dfrac{\partial h}{\partial\theta}\right), \\[2ex] \eta = -\dfrac{1}{R\sin\theta}\dfrac{\partial\zeta}{\partial\lambda} - \dfrac{\Delta G}{\bar{\gamma}_0}\left(\dfrac{1}{R\sin\theta}\dfrac{\partial h}{\partial\lambda}\right). \end{cases} \tag{7.78}$$

Similar to the case of ζ, we express ξ and η as

$$\xi = \xi_0 + \xi_1 + \cdots, \quad \eta = \eta_0 + \eta_1 + \cdots, \tag{7.79}$$

where

$$\xi_0 = \frac{1}{R}\frac{\partial \zeta_0}{\partial \theta}, \quad \eta_0 = -\frac{1}{R \sin \theta}\frac{\partial \zeta_0}{\partial \lambda}, \tag{7.80}$$

$$\begin{cases} \xi_1 = \dfrac{1}{R}\dfrac{\partial \zeta_1}{\partial \theta} + \dfrac{\Delta G}{\bar{\gamma}_0}\left(\dfrac{1}{R}\dfrac{\partial h}{\partial \theta}\right), \\[3mm] \eta_1 = -\dfrac{1}{R \sin \theta}\dfrac{\partial \zeta_1}{\partial \lambda} - \dfrac{\Delta G}{\bar{\gamma}_0}\left(\dfrac{1}{R \sin \theta}\dfrac{\partial h}{\partial \lambda}\right). \end{cases} \tag{7.81}$$

These formulae can be written in more concrete forms using the expressions of ζ_0 and ζ_1, i.e., (7.67) and (7.69), and taking reference to the derivation of the deflection of the vertical in Stokes' theory,

$$\begin{cases} \xi_0 = \dfrac{1}{4\pi \bar{\gamma}_0}\displaystyle\int_{\omega}\Delta G\dfrac{d}{d\psi}S(\psi)\cos A\,d\omega, \\[4mm] \eta_0 = \dfrac{1}{4\pi \bar{\gamma}_0}\displaystyle\int_{\omega}\Delta G\dfrac{d}{d\psi}S(\psi)\sin A\,d\omega. \end{cases} \tag{7.82}$$

$$\begin{cases} \xi_1 = \dfrac{1}{4\pi \bar{\gamma}_0}\displaystyle\int_{\omega}\Delta G_1\dfrac{d}{d\psi}S(\psi)\cos A\,d\omega + \dfrac{\Delta G}{\bar{\gamma}_0}\left(\dfrac{1}{R}\dfrac{\partial h}{\partial \theta}\right), \\[4mm] \eta_1 = \dfrac{1}{4\pi \bar{\gamma}_0}\displaystyle\int_{\omega}\Delta G_1\dfrac{d}{d\psi}S(\psi)\sin A\,d\omega - \dfrac{\Delta G}{\bar{\gamma}_0}\left(\dfrac{1}{R \sin \theta}\dfrac{\partial h}{\partial \lambda}\right). \end{cases} \tag{7.83}$$

7.2.5 Reduction of the Solution to a Practical Form

In the result developed in the previous subsections, three integrals have to be evaluated for every point on the Earth's surface successively to compute the height anomaly according to (7.66) to (7.69), and the computation of the deflection of the vertical is practically impossible according to (7.79), (7.82), and (7.83) due to the fact that the factors in the second term of (7.83), $(1/R)\partial h/\partial \theta$ and $[1/(R \sin \theta)]\partial h/\partial \lambda$, are slopes of the Earth's surface, which, as well as their errors, could be very large. The subject of this subsection is to further develop the result to a form more suitable for practical computation.

We start from the formulae for computing the disturbing potential. The leading term is T_0 given in (7.61), which is rewritten here for convenience of reference in subsequent formulations

$$T_0 = \frac{R}{4\pi} \int_\omega \Delta G S(\psi) d\omega. \tag{7.84}$$

This is a relation between T_0 and ΔG, which is identical to Stokes' formula. Hence, the inverse Stokes' formula (5.134) also holds, which is rewritten using the notations here as

$$\Delta G_P = -\frac{1}{R} T_{0P} - \frac{R^2}{2\pi} \int_\omega \frac{T_0 - T_{0P}}{l_0^3} d\omega. \tag{7.85}$$

This pair of equations will serve as the basis in subsequent formulation.

We decompose the formula of ΔG_1, (7.62), into two terms

$$\Delta G_1 = \Delta G_{11} + \Delta G_{12}, \tag{7.86}$$

where

$$\Delta G_{11P} = -h_P \frac{R^2}{2\pi} \int_\omega \frac{1}{l_0^3} \left[\left(\Delta G + \frac{3}{2} \frac{T_0}{R} \right) - \left(\Delta G + \frac{3}{2} \frac{T_0}{R} \right)_P \right] d\omega$$
$$- \frac{h_P}{R} \left(\Delta G + \frac{3}{2} \frac{T_0}{R} \right)_P, \tag{7.87}$$

$$\Delta G_{12P} = \frac{R^2}{2\pi} \int_\omega \frac{1}{l_0^3} \left[h \left(\Delta G + \frac{3}{2} \frac{T_0}{R} \right) - h_P \left(\Delta G + \frac{3}{2} \frac{T_0}{R} \right)_P \right] d\omega$$
$$+ \frac{h_P}{R} \left(\Delta G + \frac{3}{2} \frac{T_0}{R} \right)_P. \tag{7.88}$$

The formula of T is correspondingly decomposed as

$$T = T_0 + T_{11} + T_{12}, \tag{7.89}$$

where

$$T_0 + T_{11} = \frac{R}{4\pi} \int_\omega (\Delta G + \Delta G_{11}) S(\psi) d\omega, \tag{7.90}$$

$$T_{12} = \frac{R}{4\pi} \int_\omega \Delta G_{12} S(\psi) d\omega. \tag{7.91}$$

The same decomposition is made for the formulae of height anomaly and deflection of the vertical. For the height anomaly, according to Bruns' formula,

$$\zeta = \zeta_0 + \zeta_{11} + \zeta_{12}, \tag{7.92}$$

where

$$\zeta_0 + \zeta_{11} = \frac{R}{4\pi \bar{\gamma_0}} \int_\omega (\Delta G + \Delta G_{11}) S(\psi) d\omega, \qquad (7.93)$$

$$\zeta_{12} = \frac{T_{12}}{\bar{\gamma_0}}. \qquad (7.94)$$

For the deflection of the vertical, according to (7.49) to (7.83),

$$\xi = \xi_0 + \xi_{11} + \xi_{12}, \qquad \eta = \eta_0 + \eta_{11} + \eta_{12}, \qquad (7.95)$$

where, in a form in favor of convenience in subsequent derivations,

$$\begin{cases} \xi_0 + \xi_{11} = \dfrac{1}{R} \dfrac{\partial}{\partial \theta} (\zeta_0 + \zeta_{11}) = \dfrac{1}{4\pi \bar{\gamma_0}} \displaystyle\int_\omega (\Delta G + \Delta G_{11}) \dfrac{d}{d\psi} S(\psi) \cos A d\omega, \\[4mm] \eta_0 + \eta_{11} = -\dfrac{1}{R \sin \theta} \dfrac{\partial}{\partial \lambda} (\zeta_0 + \zeta_{11}) = \dfrac{1}{4\pi \bar{\gamma_0}} \displaystyle\int_\omega (\Delta G + \Delta G_{11}) \dfrac{d}{d\psi} S(\psi) \sin A d\omega; \end{cases} \qquad (7.96)$$

$$\begin{cases} \xi_{12} = \dfrac{1}{R} \dfrac{\partial \zeta_{12}}{\partial \theta} + \dfrac{\Delta G}{\bar{\gamma_0}} \left(\dfrac{1}{R} \dfrac{\partial h}{\partial \theta} \right), \\[4mm] \eta_{12} = -\dfrac{1}{R \sin \theta} \dfrac{\partial \zeta_{12}}{\partial \lambda} - \dfrac{\Delta G}{\bar{\gamma_0}} \left(\dfrac{1}{R \sin \theta} \dfrac{\partial h}{\partial l} \right). \end{cases} \qquad (7.97)$$

First, we further develop G_{11} by writing (7.87) as

$$\Delta G_{11P} = -h_P \frac{R^2}{2\pi} \int_\omega \frac{\Delta G - \Delta G_P}{l_0^3} d\omega - \frac{h_P}{R} \Delta G_P$$

$$+ \frac{3}{2} \frac{h_P}{R} \left(-\frac{R^2}{2\pi} \int_\omega \frac{T_0 - T_{0P}}{l_0^3} d\omega - \frac{1}{R} T_{0P} \right). \qquad (7.98)$$

According to (7.85), the quantity within the parentheses in the second term is ΔG_P. Hence,

$$\Delta G_{11P} = -h_P \frac{R^2}{2\pi} \int_\omega \frac{\Delta G - \Delta G_P}{l_0^3} d\omega + \frac{1}{2} \frac{h_P}{R} \Delta G_P. \qquad (7.99)$$

This formula does not depend on T_0; therefore, it can be evaluated before evaluating T_0. In other words, $T_0 + T_{11}$ can be evaluated by computing two integrals instead of three. We thus have

$$\Delta G_P + \Delta G_{11P} = \Delta G_P \left(1 + \frac{1}{2}\frac{h_P}{R} \right) - h_P \frac{R^2}{2\pi} \int_\omega \frac{\Delta G - \Delta G_P}{l_0^3} d\omega$$

$$= \Delta G_P - h_P \Delta G_P' \,, \qquad (7.100)$$

where we have neglected the term of relative order of magnitude of h_P/R, and

$$\Delta G_P' = \frac{R^2}{2\pi} \int_\omega \frac{\Delta G - \Delta G_P}{l_0^3} d\omega \,. \qquad (7.101)$$

Second, by comparing (7.88) with (7.85) and (7.91) with (7.84), we immediately realize that these two pairs of equations would have the same exact form if we set

$$T_{12} = -h \left(\Delta G + \frac{3}{2}\frac{T_0}{R} \right) \,. \qquad (7.102)$$

Hence, this must be the solution of T_{12}, which can be further approximated as

$$T_{12} = -h\Delta G - \frac{3}{2}\frac{h}{R}T_0 = -h\Delta G \,, \qquad (7.103)$$

where the term of order of magnitude $(h/R)T_0$ has been neglected based on the facts that T_{12} is to be added to T_0 and that all terms of relative order of magnitude of (h/R) have been neglected throughout the formulation.

At this stage, a new formula of T can be readily obtained by substituting (7.100) and (7.103) into (7.89) and (7.90)

$$T_P = \frac{R}{4\pi} \int_\omega (\Delta G_P - h_P \Delta G_P') S(\psi) d\omega - h_P \Delta G_P \,. \qquad (7.104)$$

For the height anomaly and the deflection of the vertical, we first derive the formulae for ζ_{12}, ξ_{12}, and η_{12}. According to (7.103), we have, without introducing spherical approximation,

$$\zeta_{12} = -h\frac{\Delta G}{\gamma} \,. \qquad (7.105)$$

For the deflection of the vertical, we first consider the first term of ξ_{12} given in (7.97). We have, using the above formula,

$$\frac{1}{R}\frac{\partial \zeta_{12}}{\partial \theta} = -\frac{\Delta G}{\gamma} \left(\frac{1}{R}\frac{\partial h}{\partial \theta} \right) - \frac{h}{\gamma} \left(\frac{1}{R}\frac{\partial \Delta G}{\partial \theta} \right) + \frac{h\Delta G}{\gamma^2} \left(\frac{1}{R}\frac{\partial \gamma}{\partial \theta} \right) \,. \qquad (7.106)$$

As γ is practically a function of h alone, we have

$$\frac{\partial \gamma}{\partial \theta} = \frac{\partial \gamma}{\partial h}\frac{\partial h}{\partial \theta} = \frac{\partial \gamma}{\partial r}\frac{\partial h}{\partial \theta}. \tag{7.107}$$

Substituting the last formula into the preceding one, we obtain

$$\frac{1}{R}\frac{\partial \zeta_{12}}{\partial \theta} = -\frac{\Delta G}{\gamma}\left(\frac{1}{R}\frac{\partial h}{\partial \theta}\right) - \frac{h}{\gamma}\left(\frac{1}{R}\frac{\partial \Delta G}{\partial \theta}\right) + \frac{h\Delta G}{\gamma^2}\frac{\partial \gamma}{\partial r}\left(\frac{1}{R}\frac{\partial h}{\partial \theta}\right). \tag{7.108}$$

Under spherical approximation, it becomes

$$\frac{1}{R}\frac{\partial \zeta_{12}}{\partial \theta} = -\frac{\Delta G}{\bar{\gamma}_0}\left(\frac{1}{R}\frac{\partial h}{\partial \theta}\right) - \frac{h}{\bar{\gamma}_0}\left(\frac{1}{R}\frac{\partial \Delta G}{\partial \theta}\right) - 2\frac{h}{R}\frac{\Delta G}{\bar{\gamma}_0}\left(\frac{1}{R}\frac{\partial h}{\partial \theta}\right), \tag{7.109}$$

where the spherical approximations given in (5.24) are used to substitute $(1/\gamma)\partial\gamma/\partial h$. The formula of ξ_{12} can be obtained by substituting the last formula into (7.97). The formula of η_{12} can also be similarly derived. Here we write them together

$$\begin{cases} \xi_{12} = -\dfrac{h}{\bar{\gamma}_0}\left(\dfrac{1}{R}\dfrac{\partial \Delta G}{\partial \theta}\right) - 2\dfrac{h}{R}\dfrac{\Delta G}{\bar{\gamma}_0}\left(\dfrac{1}{R}\dfrac{\partial h}{\partial \theta}\right), \\[3mm] \eta_{12} = -\dfrac{h}{\bar{\gamma}_0}\left(\dfrac{1}{R\sin\theta}\dfrac{\partial \Delta G}{\partial \lambda}\right) - 2\dfrac{h}{R}\dfrac{\Delta G}{\bar{\gamma}_0}\left(\dfrac{1}{R\sin\theta}\dfrac{\partial h}{\partial \lambda}\right). \end{cases} \tag{7.110}$$

In the formulae of ξ_{12}, the second term in (7.97) and the first term in (7.109) are identical in magnitude, which is extremely large, but they have opposite signs and cancel each other out. The case of η_{12} is similar.

Finally, the result for height anomaly is obtained by substituting (7.100) and (7.105) into (7.92) and (7.93)

$$\zeta_P = \frac{R}{4\pi\bar{\gamma}_0}\int_\omega (\Delta G - h\Delta G')\, S(\psi)d\omega - h_P\frac{\Delta G_P}{\bar{\gamma}_0}, \tag{7.111}$$

where the spherical approximation has been made to (7.105). The result of the deflection of the vertical is obtained by substituting (7.100) and (7.110) into (7.95) and (7.96)

$$\begin{cases} \xi_P = \dfrac{1}{4\pi\bar{\gamma}_0}\int_\omega (\Delta G - h\Delta G')\dfrac{d}{d\psi}S(\psi)\cos A\,d\omega - \dfrac{h_P}{\bar{\gamma}_0}\left(\dfrac{1}{R}\dfrac{\partial \Delta G}{\partial \theta}\right)_P, \\[3mm] \eta_P = \dfrac{1}{4\pi\bar{\gamma}_0}\int_\omega (\Delta G - h\Delta G')\dfrac{d}{d\psi}S(\psi)\sin A\,d\omega + \dfrac{h_P}{\bar{\gamma}_0}\left(\dfrac{1}{R\sin\theta}\dfrac{\partial \Delta G}{\partial \lambda}\right)_P. \end{cases} \tag{7.112}$$

We mention that the second term in (7.110) has been neglected, which can be justified to be small enough. In the parentheses of this term is the slope of the terrain, which could be large. However, the coefficient of it, $2(h/R)(\Delta G/\bar{\gamma}_0)$, is extremely small. The ratios h/R and $\Delta G/\bar{\gamma}_0$ can barely exceed 10^{-3} and 10^{-4}, respectively. Even if the slope is $45°$, the term neglected attains only $0.03''$, which is indeed a quite extreme estimate. Readers are suggested to estimate the magnitude of the second term in (7.97) or the first term in (7.109), which canceled each other.

In (7.111) and (7.112), the height h is measured from a sphere of radius R, which has been used as an approximation of the reference surface of height. Now we redefine the reference surface of height to be h_P higher, i.e., the sphere of radius $R + h_P$. The consequence is that h should be replaced by $h - h_P$ and R by $R + h_P$. We see immediately that the second terms of these formulae disappear, since the symbol h_P in them is the height at P. Hence, the formulae for the height anomaly and deflection of the vertical are simplified to the following forms, which are most referenced in practical computations,

$$\zeta_P = \frac{R}{4\pi\bar{\gamma}_0} \int_\omega [\Delta G - (h - h_P)\Delta G']S(\psi)d\omega \qquad (7.113)$$

and

$$\begin{cases} \xi_P = \dfrac{1}{4\pi\bar{\gamma}_0} \displaystyle\int_\omega \left[\Delta G - (h - h_P)\Delta G'\right] \dfrac{d}{d\psi}S(\psi)\cos A\,\omega, \\[3mm] \eta_P = \dfrac{1}{4\pi\bar{\gamma}_0} \displaystyle\int_\omega \left[\Delta G - (h - h_P)\Delta G'\right] \dfrac{d}{d\psi}S(\psi)\sin A\,d\omega. \end{cases} \qquad (7.114)$$

In these formulae, the alteration brought by replacing R by $R + h_P$ is of relative order of magnitude h_P/R, thus being ignored. For convenience of further reference, we also copy the formula for $\Delta G'$, i.e., (7.100), to here

$$\Delta G'_P = \frac{R^2}{2\pi} \int_\omega \frac{\Delta G - \Delta G_P}{l_0^3}d\omega. \qquad (7.115)$$

Finally, we spend a few words on the uniqueness of the solution, while the existence of solution has already been proven by the process of derivation. Like in the case of Stokes' boundary value problem, it can be seen using spherical harmonic series that ΔG does not include any information of the degree 1 term of T. Hence, the solution of the original Molodensky's boundary value problem is indeed not unique, and it can be made unique by setting the degree 1 term of T to vanish, just like in the case of Stokes' boundary value problem. As we used Stokes' formula in the derivation of the solution, the degree 1 term is excluded in the solution by default. As a matter of fact, the degree 0 term is excluded too, which should be and is in agreement with the physical assumptions of the problem.

7.3 Solution by Downward Continuation

7.3.1 A Property of Molodensky's Solution

Molodensky's solutions (7.113) and (7.114) have the same exact forms as those of Stokes' boundary value problem with Δg replaced by $\Delta G - (h - h_P)\Delta G'$. This suggests that $\Delta G - (h - h_P)\Delta G'$ may be understood as the gravity anomaly on the equipotential surface of gravity through P, $W = W_P$, considering that the boundary in Stokes' boundary value problem, the geoid, is an equipotential surface of gravity. Furthermore, as no mass should be present above the geoid in Stokes' boundary value problem, this also suggests that Molodensky's solution behaves as if no mass were present outside the equipotential surface of gravity through P.

An immediate realization is that, in order to understand $\Delta G - (h - h_P)\Delta G'$ as the gravity anomaly on the surface $W = W_P$, $\Delta G'$ should be understood as the vertical gradient of gravity anomaly. This requires to understand the gravity anomaly ΔG in the generalized sense defined in (7.24).

Before justifying how $\Delta G'$ could be understood as the vertical gradient of gravity anomaly, we first formulate $\partial \Delta G / \partial h$ on a spherical surface of radius R when ΔG is known on this surface, and T is assumed harmonic outside this surface. It can be readily verified using (7.24) that "T is harmonic" implies "$r\Delta G$ is harmonic." Hence, the formula for harmonic functions, (5.132), can be used. We obtain, by replacing V in (5.132) by $r\Delta G$,

$$\frac{\partial}{\partial r}\Delta G(r, \theta, \lambda)_R = -\frac{2}{R}\Delta G(R, \theta, \lambda) + \frac{R^2}{2\pi}\int_\omega \frac{\Delta G(R, \theta', \lambda') - \Delta G(R, \theta, \lambda)}{l_0^3}d\omega,$$

(7.116)

where θ' and λ' are the variables of integration. This equation is written following the notations of (5.132) to make it more evident that the values of ΔG under the integral sign are taking on the spherical surface. Using the more brief notations of this chapter, it is

$$\left(\frac{\partial \Delta G}{\partial h}\right)_P = -\frac{2}{R}\Delta G_P + \frac{R^2}{2\pi}\int_\omega \frac{\Delta G - \Delta G_P}{l_0^3}d\omega.$$

(7.117)

Within the accuracy of formulation in the previous section, the first term at the right-hand side can be neglected; the same level of approximation has been made in (7.100). We obtain

$$\left(\frac{\partial \Delta G}{\partial h}\right)_P = \frac{R^2}{2\pi}\int_\omega \frac{\Delta G - \Delta G_P}{l_0^3}d\omega.$$

(7.118)

We emphasize that the values of ΔG under the integral sign should be taken on the spherical surface.

In order to compare with $\Delta G'$, we make the assumption that no mass exists outside the surface $W = W_P$. We can then apply (7.118) while approximating this surface as a sphere of radius $R + h_P$. Neglecting terms of relative order of magnitude h_P / R, we obtain

$$\left(\frac{\partial \Delta G}{\partial h} \right)_P = \frac{R^2}{2\pi} \int_\omega \frac{\Delta G_{W_P} - \Delta G_P}{l_0^3} d\omega, \qquad (7.119)$$

where the subscript W_P is used to indicate that the value should be taken on the surface $W = W_P$. With $\Delta G = \Delta G_{W_P} + (h - h_P)(\partial \Delta G / \partial h)$, we obtain, using also the expression of $\Delta G'$,

$$\left(\frac{\partial \Delta G}{\partial h} \right)_P = \frac{R^2}{2\pi} \int_\omega \frac{\Delta G - \Delta G_P}{l_0^3} d\omega - \frac{R^2}{2\pi} \int_\omega \frac{h - h_P}{l_0^3} \frac{\partial \Delta G}{\partial h} d\omega$$

$$= \Delta G_P' - \frac{R^2}{2\pi} \int_\omega \frac{h - h_P}{l_0^3} \frac{\partial \Delta G}{\partial h} d\omega, \qquad (7.120)$$

which shows how $\Delta G'$ and $\partial \Delta G / \partial h$ are related. It is expected that the second term be of smaller magnitude than the first, so that the formula can be used to compute $\partial \Delta G / \partial h$ iteratively, such that $\Delta G'$ can be regarded as a first approximation.

At this point, we see that $\Delta G'$ is not exactly the vertical gradient of ΔG, which has been demonstrated for the degenerate case where no mass exists outside the surface $W = W_P$. Nevertheless, in Molodensky's solution (7.113) and (7.114), $\Delta G'$ does behave like a vertical gradient to project gravity anomaly from the Earth's surface to the surface $W = W_P$, such that Stokes' theory can be applied to compute the height anomaly and deflection of the vertical at P with $W = W_P$ as the boundary. Stated differently, it projects the gravity anomaly from the Earth's surface to the surface $W = W_P$ while maintaining T to be regarded as harmonic as far as only the height anomaly and deflection of the vertical at P are computed. We should keep in mind that this is only how the formulae manifest mathematically.

7.3.2   Analytical Downward Continuation to a Surface Through the Point of Interest

The analytical downward continuation can be regarded as a generalization of the projection of the gravity anomaly from the Earth's surface to the surface $W = W_P$ (which is more rigorously the surface h_P above the quasigeoid, as the normal height is known) in Molodensky's solution using $\Delta G'$ as the vertical gradient of the gravity anomaly on $W = W_P$. As a gradient is a kind of partial derivative, here we introduce a gradient operator $\tilde{\partial}_h$, such that

$$\left(\tilde{\partial}_h \Delta G \right)_{W_P} = \Delta G', \qquad (7.121)$$

which is indeed different from $\partial \Delta G / \partial h$, as discussed in the previous subsection. The projection of the gravity anomaly from the Earth's surface to the surface $W = W_P$ in Molodensky's solution can be understood as the relation

$$\Delta G = \Delta G_{W_P} + z \left(\tilde{\partial}_h \Delta G \right)_{W_P}, \tag{7.122}$$

where z is a brief notation for $h - h_P$, i.e., $z = h - h_P$. The idea of analytical downward continuation is to understand (7.122) as a Taylor series with only the first order term kept and generalize it to a complete Taylor series

$$\Delta G = \Delta G_{W_P} + z \left(\tilde{\partial}_h \Delta G \right)_{W_P} + \frac{z^2}{2!} \left(\tilde{\partial}_h^2 \Delta G \right)_{W_P} + \frac{z^3}{3!} \left(\tilde{\partial}_h^3 \Delta G \right)_{W_P} + \cdots$$

$$= \Delta G_{W_P} + \sum_{n=1}^{\infty} \frac{z^n}{n!} \left(\tilde{\partial}_h^n \Delta G \right)_{W_P}, \tag{7.123}$$

where $\tilde{\partial}_h^n \Delta G$ is defined such that

$$\tilde{\partial}_h^2 \Delta G = \tilde{\partial}_h \left(\tilde{\partial}_h \Delta G \right), \cdots, \tilde{\partial}_h^n \Delta G = \tilde{\partial}_h \left(\tilde{\partial}_h^{n-1} \Delta G \right), \cdots. \tag{7.124}$$

This is to assume that (7.122) is an approximation of (7.123) and that Molodensky's solutions (7.113) and (7.114) are an approximation of the exact solution

$$\zeta_P = \frac{R}{4\pi \overline{\gamma}_0} \int_\omega \Delta G_{W_P} S(\psi) d\omega, \tag{7.125}$$

$$\begin{cases} \xi_P = \dfrac{1}{4\pi \overline{\gamma}_0} \displaystyle\int_\omega \Delta G_{W_P} \dfrac{d}{d\psi} S(\psi) \cos A \omega, \\[3mm] \eta_P = \dfrac{1}{4\pi \overline{\gamma}_0} \displaystyle\int_\omega \Delta G_{W_P} \dfrac{d}{d\psi} S(\psi) \sin A \omega. \end{cases} \tag{7.126}$$

In summary, the basic assumptions of the analytical downward continuation are (7.123), (7.125), and (7.126), where $\tilde{\partial}_h^n \Delta G$ is defined in (7.121) and (7.124).

Before deriving the formulae for computation, we make a few remarks. First, this solution is based on an inference rather than a rigorous derivation obtained by solving the differential or integral equation of the problem; hence, its validity is not proven by the derivation. Nevertheless, there exist proofs that this solution is equivalent to Molodensky's solution extended to include more, and even an infinite number of terms (Moritz, 1980, Section 46). Second, we are presenting this theory differently from former literature where the strict derivative $\partial^n / \partial h^n$ was used, and $\Delta G'$ was understood as $\partial \Delta G / \partial h$. Here we replaced the strict derivative with the operator $\tilde{\partial}_h^n$ based on the discussion in the previous subsection that $\Delta G'$ is not exactly $\partial \Delta G / \partial h$. This will not alter the result. Third, as an extension of

Molodensky's solution, (7.123) is expected to project the gravity anomaly from the Earth's surface to the surface $W = W_P$ while maintaining T to be regarded as harmonic, which is a property of Molodensky's solution discussed in the previous subsection and is the logic leading to the conjecture of (7.125) and (7.126).

In order to derive the formulae for computation, we write (7.123) symbolically in the form of

$$\Delta G = \left(1 + \sum_{n=1}^{\infty} \frac{z^n}{n!} \tilde{\partial}_h^n\right) \Delta G_{W_P} = \left(I + \sum_{n=1}^{\infty} z^n L_n\right) \Delta G_{W_P} = U\left(\Delta G_{W_P}\right),$$

(7.127)

where I is the identity operator,

$$I(\Delta G) = \Delta G,$$

(7.128)

L_n is a vertical differential operator defined as

$$L_n = \frac{1}{n!} \tilde{\partial}_h^n,$$

(7.129)

and U is referred to as upward continuation operator defined as

$$U = I + \sum_{n=1}^{\infty} z^n L_n.$$

(7.130)

Our purpose is to formulate the inverse operation

$$\Delta G_{W_P} = U^{-1}(\Delta G) = D(\Delta G),$$

(7.131)

where

$$D = U^{-1}$$

(7.132)

is the inverse of U and is referred to as downward continuation operator. Hence, what we seek is D.

We adopt again the approach of Molodensky's shrinking to replace z by kz, with k a parameter in the range $0 \leq k \leq 1$. We have

$$U = I + \sum_{n=1}^{\infty} k^n z^n L_n = \sum_{n=0}^{\infty} k^n U_n,$$

(7.133)

where

$$U_0 = I, \quad U_n = z^n L_n.$$ (7.134)

We assume D to be in the form of

$$D = \sum_{n=0}^{\infty} k^n D_n.$$ (7.135)

What remains is to determine D_n according to

$$UD = I.$$ (7.136)

Substitute (7.133) and (7.135) into (7.136). We obtain

$$\sum_{p=0}^{\infty} k^p U_p \sum_{q=0}^{\infty} k^q D_q = I,$$ (7.137)

which can be further written as

$$\sum_{p=0}^{\infty} \sum_{q=0}^{\infty} k^{p+q} U_p D_q = I.$$ (7.138)

Substitute q by $n = p + q$. As the lower and higher limits of both p and q are 0 and ∞, respectively, the lower and higher limits of n should also be 0 and ∞, respectively. Hence, we obtain

$$\sum_{n=0}^{\infty} \sum_{p=0}^{n} k^n U_p D_{n-q} = I,$$ (7.139)

where the upper limit of p becomes n due to the fact that $q = n - p \geq 0$ must hold. Moving the identity operator to the left-hand side, we obtain

$$\sum_{n=0}^{\infty} \sum_{p=0}^{n} k^n U_p D_{n-q} - I = 0.$$ (7.140)

As this equation must hold for any $0 \leq k \leq 1$, the coefficient of k^n must vanish for any value of n. When $n = 0$, we have

$$D_0 = I.$$ (7.141)

When $n > 0$, we have

$$\sum_{p=0}^{n} U_p D_{n-q} = 0,$$ (7.142)

which can be written as

$$D_n + \sum_{p=1}^{n} U_p D_{n-q} = 0,$$ (7.143)

and consequently,

$$D_n = -\sum_{p=1}^{n} U_p D_{n-q}.$$ (7.144)

This is a recurrence formula to compute $D_1, D_2, \cdots$ successively starting from $D_0 = I$ and using U_p.

To find the formula to compute ΔG_{W_P}, we define

$$\Delta G_n = D_n(\Delta G).$$ (7.145)

We then have, by setting $k = 1$ in (7.135) and substituting it into (7.131),

$$\Delta G_{W_P} = D(\Delta G) = \sum_{n=0}^{\infty} D_n(\Delta G) = \sum_{n=0}^{\infty} \Delta G_n.$$ (7.146)

What remains is to find a formula for computing ΔG_n, which is straightforward according to (7.144),

$$\Delta G_n = -\sum_{p=1}^{n} U_p D_{n-q}(\Delta G) = -\sum_{p=1}^{n} U_p(\Delta G_{n-p}) = -\sum_{p=1}^{n} z^p L_P(\Delta G_{n-p}),$$
(7.147)

where (7.134) is used to derive the last identity, and by definition,

$$\Delta G_0 = \Delta G.$$ (7.148)

The last formula needed is the one for L_n. According to its definition (7.129), we have

$$L_n = \frac{1}{n!}\tilde{\partial}_h^n = \frac{1}{n}\tilde{\partial}_h\left[\frac{1}{(n-1)!}\tilde{\partial}_h^{n-1}\right] = \frac{1}{n}L_1 L_{n-1}.$$ (7.149)

This is a formula for successively computing $L_2, L_3, \cdots$ starting from the formula of L_1. What we compute is the values on $W = W_P$. Hence, by definition, L_1 is

$$L_1(\Delta G) = (\tilde{\partial}_h \Delta G)_{W_P} = \Delta G' = \frac{R^2}{2\pi} \int_\omega \frac{\Delta G - \Delta G_P}{l_0^3} d\omega, \qquad (7.150)$$

and $L_2, L_3, \cdots, L_n$ are successively

$$\begin{cases} L_2(\Delta G) = \dfrac{1}{2} L_1[L_1(\Delta G)], \\[2mm] L_3(\Delta G) = \dfrac{1}{3} L_1[L_2(\Delta G)], \\[2mm] \cdots \\[2mm] L_n(\Delta G) = \dfrac{1}{n} L_1[L_{n-1}(\Delta G)]. \end{cases} \qquad (7.151)$$

The formulation is now complete. The downward continued gravity anomaly ΔG_{W_P} should be first computed using (7.146), where ΔG_n should be computed using (7.148), (7.147) and (7.150), (7.151). The height anomaly ζ and the deflection of the vertical are then computed using (7.125) and (7.126).

Before concluding this subsection, we mention again that this solution is not obtained by solving the differential or integral equation of the problem, and nor is the solution substituted into the equation to verify its validity. Its validity is ensured by a proof that it is equivalent to Molodensky's solution (Moritz, 1980, Section 46).

7.3.3   A Discussion on the Possible Analytical Downward Continuation to the Quasigeoid

If the analytical downward continuation to the surface $W = W_P$ formulated in the previous subsection were plausible, it may be conjectured then that the analytical downward continuation to the quasigeoid be also plausible. In fact, as will be discussed below, Molodensky's solution (7.111) and (7.112) could be appealingly interpreted physically assuming the term $\Delta G - h\Delta G'$ to be the gravity anomaly analytically downward continued to the quasigeoid, but it should be prudent in extending the implication of this interpretation.

As an extension of the inference in the previous subsections, we further understand $\tilde{\partial}_h = \Delta W'$ as a general gradient operator that could be used to analytically project gravity anomaly from the Earth's surface down to the quasigeoid. The term $\Delta G - h\Delta G'$ in (7.111) and (7.112) could then be interpreted as the gravity anomaly on the quasigeoid as if no mass were present above the quasigeoid. Neglecting the difference between the geoid and quasigeoid, the height anomaly and the deflection of the vertical on the quasigeoid could then be computed using the solution of Stokes' boundary value problem, which are the first terms in (7.111) and (7.112).

Consequently, it could be inferred that the second terms in (7.111) and (7.112) are the correction that projects the value on the quasigeoid upward to the Earth's surface. According to (7.76), the second term in (7.111) is

$$-h\frac{\Delta G}{\bar{\gamma_0}} = h\frac{\partial \zeta}{\partial h},$$ (7.152)

which confirms our inference. For (7.112), we take as example the component ξ. According to (1.172) with the x-axis assumed to point toward the north, the curvatures of the plumb line and normal plumb line toward north are

$$\kappa_g = -\frac{1}{g}\left(\frac{1}{R}\frac{\partial g}{\partial \theta}\right) \quad \text{and} \quad \kappa_\gamma = -\frac{1}{\gamma}\left(\frac{1}{R}\frac{\partial \gamma}{\partial \theta}\right),$$ (7.153)

respectively. While projecting upward from the quasigeoid to the Earth's surface, the directions of the plumb line and normal plumb line shift southward by $h\kappa_g$ and $h\kappa_\gamma$, respectively. However, $h\kappa_g$ is the change of direction of the plumb line from the quasigeoid to the Earth's surface, while $h\kappa_\gamma$ is the change of direction of the normal plumb line from the reference Earth ellipsoid to the telluroid. Per definition, the change of the deflection of the vertical is, with the approximations $\gamma = \bar{\gamma_0}$ and $g = \bar{\gamma_0}$,

$$h\kappa_g - h\kappa_\gamma = -\frac{h}{\bar{\gamma_0}}\left[\frac{1}{R}\left(\frac{\partial g}{\partial \theta} - \frac{\partial \gamma}{\partial \theta}\right)\right] = -\frac{h}{\bar{\gamma_0}}\left(\frac{1}{R}\frac{\partial \Delta G}{\partial \theta}\right),$$ (7.154)

which also confirms our inference. An additional remark, which should have been evident, is that, in (7.154), $\partial g/\partial \theta$ is evaluated at one point, while $\partial \gamma/\partial \theta$ is evaluated at another point, where the normal potential at the latter point equals to the gravity potential at the former point.

Based on the interpretation that the term $\Delta G - h\Delta G'$ is the gravity anomaly on the quasigeoid as if no mass were present above the quasigeoid, it may be further inferred that Molodensky's boundary value problem may be replaced by Stokes' boundary value problem with $\Delta G - h\Delta G'$ as gravity anomaly on the quasigeoid as boundary, and the solution of the latter problem outside the boundary could be interpreted as the solution of the former on and outside the Earth's surface, but the solution is meaningless within the topographic mass where the disturbing potential does not satisfy Laplace's equation. This inference, if realizable, would be very meaningful for the determination of global models of the Earth's gravitational potential in terms of spherical harmonic series as the sum of normal and disturbing gravitational potentials. Based on this inference, it is expected that the gravitational potential on and outside the Earth's surface be expressed using only one spherical harmonic series, which is valid not only beyond the Brillouin sphere, but also between this sphere and the Earth's surface. The validity of this inference should meanwhile guaranty that the gravitational potential and attraction down to the Earth's surface be correctly computed using the spherical harmonic series in

Fig. 7.4 A simple
hypothetical Earth model

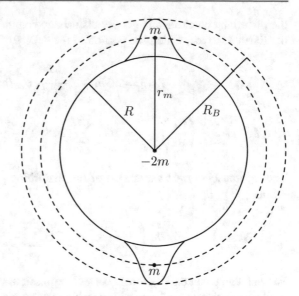

the same way as that outside the Brillouin sphere. However, based on a simple
hypothetical model of the Earth, this inference is not guaranteed to be correct.

As shown in Fig. 7.4, we consider a simple hypothetical Earth model, which is
a homogeneous sphere of radius R and mass M superimposed with three particles
producing the disturbing potential. We assume that a mountain is situated at each
pole, which is hollow except that a particle of mass m is within, and is situated right
on the poles with a radial distance r_m. We also assume that a particle of mass $-2m$
is situated at the center of the sphere. Hence, for this model, both the degrees 0 and
1 spherical harmonic terms of the disturbing potential vanish. The surfaces of the
mountains and the sphere of radius R define the surface of the hypothetical Earth
model. The Brillouin sphere of radius R_B is also drown in the figure.

The default judgment is that, if the disturbing potential on and outside the Earth's
surface can be expressed by only one spherical harmonic series, the only achievable
choice should be that it is identical to the one outside the Brillouin sphere, which
is also the one obtained by satellite gravimetry missions. Therefore, for the simple
hypothetical Earth model, the global model of the disturbing potential should be
identical to the spherical harmonic series of the gravitational potential of the three
particles outside the Brillouin sphere, which can be readily written out to be

$$T(r, \theta, \lambda) = 2Gm \sum_{n=1}^{\infty} \frac{r_m^{2n}}{r^{2n+1}} P_{2n}(\cos\theta) \,, \tag{7.155}$$

where the odd degree terms of the two masses m cancel each other, and their degree
0 terms cancel with the gravitational potential of the mass $-2m$. Following the
inference, if the same expression of T is to be used down to the Earth's surface,

the disturbing potential and gravity disturbance on the non-mountainous portion of
the Earth's surface, which is a portion of the sphere of radius R, should be

$$T_{r=R} = 2Gm \sum_{n=1}^{\infty} \frac{r_m^{2n}}{R^{2n+1}} P_{2n}(\cos\theta),\qquad(7.156)$$

$$\delta g_{r=R} = -\left(\frac{\partial T}{\partial r}\right)_{r=R} = 2Gm \sum_{n=1}^{\infty} (2n+1)\frac{r_m^{2n}}{R^{2n+2}} P_{2n}(\cos\theta).\qquad(7.157)$$

The rigorously correct expression of the disturbing potential within the sphere of
radius $r = r_m$ is

$$T'(r,\theta,\lambda) = -2Gm\left(\frac{1}{r} - \frac{1}{r_m}\right) + 2Gm \sum_{n=1}^{\infty} \frac{r^{2n}}{r_m^{2n+1}} P_{2n}(\cos\theta).\qquad(7.158)$$

We also know that the rigorously correct expressions of the disturbing potential and
gravity disturbance on the non-mountainous portion of the Earth's surface are

$$T'_{r=R} = -2Gm\left(\frac{1}{R} - \frac{1}{r_m}\right) + 2Gm \sum_{n=1}^{\infty} \frac{R^{2n}}{r_m^{2n+1}} P_{2n}(\cos\theta),\qquad(7.159)$$

$$\delta g'_{r=R} = -\left(\frac{\partial T'}{\partial r}\right)_{r=R} = -2Gm\frac{1}{R^2} - 2Gm \sum_{n=1}^{\infty} (2n)\frac{R^{2n-1}}{r_m^{2n+1}} P_{2n}(\cos\theta).$$
$$(7.160)$$

In order that (7.155) be downward continued to the Earth's surface validly, the
identity $T = T'$ must hold everywhere in between the sphere of radius r_m and
the Earth's surface. In particular, $T_{r=R} = T'_{r=R}$ and $\delta g_{r=R} = \delta g'_{r=R}$ must hold
on the Earth's surface, so that surface data be correctly interpreted. We are not
going into the detail of the comparison. Readers interested in are encouraged to
implement the formulae to compare numerically for the whole domain of θ. Here
we show the contradiction with an extreme case. The spherical harmonic series
must be truncated, say, at the degree N, which has a resolution of half wavelength
$180°/N$. We now assume that the mountain at the north pole is extremely narrow
as compared to the resolution, so that $P_{2n}(\cos\theta) > 0$ holds from the pole to the
foot of the mountain for any value of $n \leq N$. This is envisagable based on the facts
$P_{2n}(\cos\theta) = 1$ and $(d/d\theta)P_{2n}(\cos\theta) = 0$ at $\theta = 0$. The most evident difference
is then that $\delta g_{r=R}$ given in (7.157) and $\delta g'_{r=R}$ given in (7.160) are of opposite signs
at the foot of the mountain! This is physically evident. At a point very close to
the pole and on the sphere, the gravitational attraction of the mass above the pole
points upward, which is correctly formulated as $\delta g'_{r=R}$. However, $\delta g_{r=R}$ is obtained
by extrapolating the gravitational attraction at points above the mass, which points
downward. In reality, this contradiction should be common at places right at the foot

of high mounts, where the contribution of the masses above the horizontal plane to the gravity points upward.

This contradiction implies that it is impossible in the strict theoretical sense to use a spherical harmonic series outside a sphere fully or partly embedded within the Earth's surface to represent the disturbing potential on and outside the Earth's surface, and meanwhile, its gradient represents the gravity disturbance on and outside the Earth's surface, thus correctly interpreting the gravity anomaly.

Hence, in the context of Molodensky's boundary value problem, it is impossible in the strict theoretical sense to represent the disturbing potential on and outside the Earth's surface using a spherical harmonic series outside the quasigeoid, whose gradient also represents the gravity disturbance on and outside the Earth's surface. Therefore, we suppose that the interpretation of analytical downward continuation in Molodensky's solution (7.111) and (7.112) be limited to within its own context for computing the solution on the Earth's surface—the analytical downward continuation to the surface $W = W_P$ discussed in the previous subsection is of this kind—but it should not be used as a firm evidence that a solution of Stokes' boundary value problem outside the quasigeoid can be found to represent the disturbing potential on and outside the Earth's surface, and meanwhile to represent the gravity disturbance by its gradient on and outside the Earth's surface. The possible approximate use of a spherical harmonic series outside a sphere embedded within the Earth for representing the gravitational potential on and outside the Earth's surface will be discussed in the appendix of this chapter.

7.3.4 An Iterative Approach of Harmonic Downward Continuation

Assuming that the disturbing potential on and outside the Earth's surface can be exactly expressed using a harmonic function on and outside the quasigeoid, whose gradient is also exactly the gravity disturbance, a rigorous formula can be derived to project the gravity anomaly from the Earth's surface to the quasigeoid. Although such an exact expression of the disturbing potential has been demonstrated to be not likely exist for the real Earth in the previous subsection, this approach can be made plausible in combination of a gravity reduction that will be discussed in the next section.

Approximating the quasigeoid as a sphere of radius R, the formulation is nothing else other than the reverse of the Poisson integral. As shown in Fig. 7.5, P is a point on the Earth's surface, and P_0 is its projection on the quasigeoid. We use ΔG_0 to denote the gravity anomaly on the quasigeoid and ΔG its extension in space. We obtain, by applying the Poisson integral to the harmonic function $r\Delta G$,

$$(R + h_P)\,\Delta G_P = \frac{R\left[(R + h_P)^2 - R^2\right]}{4\pi} \int_\omega \frac{R\Delta G_0}{l^3}d\omega, \qquad (7.161)$$

Fig. 7.5 Projection of
gravity anomaly from the
Earth's surface to quasigeoid

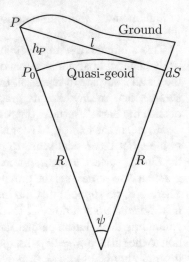

where l is the distance between P and the surface element dS on the sphere of radius R corresponding to $d\omega$ on the unit sphere, with $dS = R^2 d\omega$. The formula for ΔG_P is

$$\Delta G_P = \frac{R^2 \left[(R + h_P)^2 - R^2\right]}{4\pi \, (R + h_P)} \int_\omega \frac{\Delta G_0}{l^3} d\omega \,. \tag{7.162}$$

This is an integral equation of ΔG_0 assuming ΔG to be known.

A rigorous iterative approach can be formulated to compute ΔG_0. By applying the Poisson integral to the harmonic function $1/r$, we obtain

$$\frac{1}{(R + h_P)} = \frac{R \left[(R + h_P)^2 - R^2\right]}{4\pi} \int_\omega \frac{1}{R l^3} d\omega \,, \tag{7.163}$$

which can be written as

$$\frac{R^2}{(R + h_P)^2} = \frac{R^2 \left[(R + h_P)^2 - R^2\right]}{4\pi \, (R + h_P)} \int_\omega \frac{1}{l^3} d\omega \,. \tag{7.164}$$

Multiply (7.164) by the value of ΔG_0 at P_0, which is denoted as $\Delta G_0^{P_0}$, and then subtract (7.162) from the result. We obtain

$$\frac{R^2}{(R + h_P)^2} \Delta G_0^{P_0} - \Delta G_P = -\frac{R^2 \left[(R + h_P)^2 - R^2\right]}{4\pi \, (R + h_P)} \int_\omega \frac{\Delta G_0 - \Delta G_0^{P_0}}{l^3} d\omega \,. \tag{7.165}$$

Finally, an iterative formula for ΔG_0 is obtained

$$\Delta G_0^{P_0} = \frac{(R+h_P)^2}{R^2}\Delta G_P - \frac{(R+h_P)\left[(R+h_P)^2 - R^2\right]}{4\pi}\int_\omega \frac{\Delta G_0 - \Delta G_0^{P_0}}{l^3}d\omega.$$

(7.166)

It is beneficial to use a simplified form of the iterative formula (7.166). Define

$$x = \frac{R}{R+h_P},$$

(7.167)

$$d = \left(1 + x^2 - 2x\cos\psi\right)^{1/2},$$

(7.168)

where ψ is shown in Fig. 7.5. We can write (7.166) as

$$\Delta G_0^{P_0} = \frac{\Delta G_P}{x^2} - \frac{1-x^2}{4\pi}\int_\omega \frac{\Delta G_0 - \Delta G_0^{P_0}}{d^3}d\omega.$$

(7.169)

In computations, the first iteration is

$$\Delta G_{0,1}^{P_0} = \frac{\Delta G_P}{x^2},$$

(7.170)

the second

$$\Delta G_{0,2}^{P_0} = \frac{\Delta G_P}{x^2} - \frac{1-x^2}{4\pi}\int_\omega \frac{\Delta G_{0,1} - \Delta G_{0,1}^{P_0}}{d^3}d\omega,$$

(7.171)

the third

$$\Delta G_{0,3}^{P_0} = \frac{\Delta G_P}{x^2} - \frac{1-x^2}{4\pi}\int_\omega \frac{\Delta G_{0,2} - \Delta G_{0,2}^{P_0}}{d^3}d\omega,$$

(7.172)

and so forth, till a prescribed accuracy is achieved. It may be even more beneficial to compute the increment of ΔG_0 between successive iterations, $\delta\Delta G_{0,i} = \Delta G_{0,i} - \Delta G_{0,i-1}$, whose formulation is straightforward and is left to readers.

As the physical logic of this downward continuation method and that discussed in the previous two subsections are essentially the same, it is certainly interesting to compare them. It can be readily inferred that the result of iteration accurate to the first order of h_P is

$$\Delta G_0^{P_0} = \Delta G_P - h_P\left(\frac{R^2}{2\pi}\int_\omega \frac{\Delta G - \Delta G_P}{l_0^3}d\omega\right) = \Delta G_P - h_P\Delta G_P',$$

(7.173)

where terms of relative order of magnitude h_P/R or smaller are neglected, and $\Delta G'$ is given in (7.115). This is equivalent to the first order approximation of the analytical downward continuation discussed in the previous two subsections.

The integral equation (7.166) can also be solved using Molodensky's shrinking method, which is quite straightforward. Here we directly give the result. The gravity anomaly on the quasigeoid is written as

$$\Delta G_0 = \Delta G_{00} + \Delta G_{01} + \Delta G_{02} + \Delta G_{03} + \cdots . \tag{7.174}$$

Neglecting terms of relative order of magnitude h_P/R or smaller, we have

$$\Delta G_{00}^{P_0} = \Delta G_P , \tag{7.175}$$

$$\Delta G_{01}^{P_0} = -h_P \left(\frac{R^2}{2\pi} \int_\omega \frac{\Delta G_{00} - \Delta G_{00}^{P_0}}{l_0^3} d\omega \right) , \tag{7.176}$$

$$\Delta G_{02}^{P_0} = -h_P \left(\frac{R^2}{2\pi} \int_\omega \frac{\Delta G_{01} - \Delta G_{01}^{P_0}}{l_0^3} d\omega \right) , \tag{7.177}$$

$$\Delta G_{03}^{P_0} = -h_P \left(\frac{R^2}{2\pi} \int_\omega \frac{\Delta G_{02} - \Delta G_{02}^{P_0}}{l_0^3} d\omega \right) + h_P^3 \left(\frac{3R^2}{4\pi} \int_\omega \frac{\Delta G_{00} - \Delta G_{00}^{P_0}}{l_0^5} d\omega \right) . \tag{7.178}$$

We see that $\Delta G_{00} + \Delta G_{01}$ is the same as (7.173). However, the first order derivative inferred from the above formulae includes more terms, which is the coefficient of $-h_p$,

$$\left(\frac{\partial \Delta G_0}{\partial h} \right)_{P_0} = \frac{R^2}{2\pi} \int_\omega \frac{\Delta G_{00} - \Delta G_{00}^{P_0}}{l_0^3} d\omega + \frac{R^2}{2\pi} \int_\omega \frac{\Delta G_{01} - \Delta G_{01}^{P_0}}{l_0^3} d\omega$$

$$+ \frac{R^2}{2\pi} \int_\omega \frac{\Delta G_{02} - \Delta G_{02}^{P_0}}{l_0^3} d\omega + \cdots$$

$$= \frac{R^2}{2\pi} \int_\omega \frac{\left[\begin{array}{c} (\Delta G_{00} + \Delta G_{01} + \Delta G_{02} + \cdots) \\ - (\Delta G_{00} + \Delta G_{01} + \Delta G_{02} + \cdots)^{P_0} \end{array} \right]}{l_0^3} d\omega . \tag{7.179}$$

This result, with (7.174) taken into account, agrees with the rigorous formula (7.118), while the first order term $\tilde{\partial}_h \Delta G$ in the analytical downward continuation formula (7.123) or (7.127) does not. A remarkable fact is that no term of the order h_P^2 is present in this solution, meaning that it is of relative order of magnitude h_P/R or smaller, thus being neglected.

With ΔG_0, the disturbing potential and other quantities derived from it can be determined straightforwardly according to Stokes' theory. We mention again that the assumption of this theory, i.e., the disturbing potential is harmonic outside the quasigeoid, is not realistic for the Earth. To make it physically plausible, a combination with some kind of gravity reduction is necessary, which will be further discussed in the next section.

This method was originally developed by Bjerhammar who projected from the Earth's surface to a sphere that can even be embedded within the Earth, called Bjerhammar sphere. As a harmonic function is downward continued, we refer the method to as harmonic downward continuation. The theory behind is in fact an iterative use of upward continuation of a harmonic function, which is $r \Delta G$ here. Therefore, this method can be modified for applications to other quantities. Readers are suggested to apply the method to the downward continuation of the disturbing potential.

7.4 Gravity Reduction

7.4.1 General Description

In Molodensky's theory, the computation of the gravity anomaly requires only to compute the normal gravity on the telluroid. As the telluroid is supposed to be known from the known normal height determined using spirit leveling and gravity measurements, this computation can be done exactly. Hence, there is no need to introduce physical approximations as in the gravity reduction of Stokes' theory, namely, the approximation of crustal density and the approximate use of normal gravity gradient as gravity gradient.

Even though theoretically not necessary, a gravity reduction can still be combined with Molodensky's theory, which does bring advantages.

Any gravity reduction to be used in combination with Molodensky's theory should be based on the normal height, which is supposed to be known; therefore, throughout this section, the notation h is understood as the normal height, $h = H_\gamma$. For reasons justified during our formulation of Stokes' theory, the Helmert and isostatic anomalies are to be discussed. It is evident that the computation of the gravity anomaly is identical to that in Stokes' theory; the only difference is that the orthometric height is replaced by the normal height. Hence, the detail is not repeated here. We only emphasize that the free air correction is a projection of the normal gravity from the reference Earth ellipsoid to the telluroid and can be considered error free, but an approximation of crustal density is still necessary for computing topographic and isostatic corrections, which introduces errors. What we obtain, ΔG^c, is the adjusted gravity anomaly in Molodensky's theory, which is defined on the Earth's surface, and the Earth is adjusted so that it is hollow between the its surface and the quasigeoid. It is to be recognized that the hollow is corrupted by the error of the crustal density adopted.

Geographical coverage of gravity data is quite inhomogeneous, and there are regions where gravity data are sparse. Hence, interpolation of gravity from insufficient or even sparse data is necessary, for which the small and smooth Helmert or isostatic anomaly is more suitable. In regions of very complex topography, such as mountainous regions, not only gravity data are most likely sparse, but also the computation of the integrals in Molodensky's solution is more difficult due to large and abrupt variations of the topographic height. If the Helmert or isostatic anomaly is adopted, as the topographic mass is relocated on or below the adjusted quasigeoid, a somewhat smoothed topography could be used to alleviate the complexity of computation. A compromise, which is not to be discussed further in this book, is to perform gravity reduction by applying a terrain correction with respect to a smoothed topography used to replace the Earth's surface in the approximate evaluation of Molodensky's solution.

Another advantage to adopt the Helmert or isostatic anomaly is that the adjusted disturbing potential T^c is harmonic outside the adjusted quasigeoid. Therefore, harmonic downward continuations, such as that formulated in Sect. 7.3.4, are consistent with the physical model. Not only the expression of the adjusted disturbing potential on and outside the Earth's surface using a harmonic function on and outside the adjusted quasigeoid is rigorous theoretically, but also the gradient of the adjusted disturbing potential is guaranteed to be the adjusted gravity disturbance. Without removing the topographic mass, this kind of representation of the disturbing gravity field should be at least supplemented with a formulation of the influence of the topographic mass.

Different from the case in Stokes' theory, the height anomaly and deflection of the vertical in Molodensky's theory are defined on the Earth's surface, thus requiring to formulate the indirect effect of the gravitational potential and the horizontal components of the gravitational attraction on the Earth's surface. The same as in the case of Stokes' theory, we still use the superscript c to indicate quantities of the adjusted gravity field, as in ΔG^c and T^c defined in the previous paragraphs. We have

$$W = W^c + \delta W, \quad \zeta = \zeta^c + \delta\zeta, \quad \xi = \xi^c + \delta\xi, \qquad \eta = \eta^c + \delta\eta, \qquad (7.180)$$

where δ is used to indicate the indirect effect, i.e., the correction by restoring mass distribution from the adjusted Earth to the real Earth.

The detailed formulation of the indirect effect will be the subject of the next two subsections. Here we first provide a procedure to take into account of the indirect effect in the result, which is similar to that in Stokes' theory: (1) choose the isostatic model to be used throughout the procedure (we still understand the Helmert anomaly as an extreme case of isostatic anomaly); (2) compute the indirect effect of the height anomaly $\delta\zeta$ using the known normal height h; (3) perform gravity reduction to compute the adjusted gravity anomaly using the adjusted normal height $h^c = h + \delta\zeta$, which is the height of the Earth's surface above the adjusted quasigeoid, or the height of the adjusted telluroid above the reference Earth ellipsoid; (4) compute the adjusted height anomaly and deflection of the vertical based on Molodensky's theory using the adjusted gravity anomaly; (5) compute the indirect effect using the adjusted normal height $h^c = h + \delta\zeta$, and then add the result to the adjusted height anomaly and deflection of the vertical. An iteration of steps 3 to 5 is expected to improve the result. Only the adjusted height anomaly is required for the iteration.

7.4.2 Indirect Effect of the Height Anomaly

The indirect effect of the height anomaly is related to that of the gravitational potential according to Bruns' formula

$$\delta \zeta = \frac{\delta W}{\bar{\gamma}_0}. \tag{7.181}$$

Hence, we will only give the formula of δW. What differs from that in Stokes' theory is that, here, δW is the value on the Earth's surface.

The Helmert Anomaly

If the Helmert anomaly is used, δW includes the contribution of adding back topographic mass and removing the Helmert condensation layer. An exact formula of the gravitational potential of the topographic mass can be obtained from (6.58) and (6.61). Here, we have $r = R + h_P$, $R_B = R$, and $R_T = R + h$. The gravitational potential of the topographic mass can be written as

$$V_T = G \int_\omega \rho f_T(h_P, h, \psi) d\omega, \tag{7.182}$$

where $f_T(h_P, h, \psi)$ is defined as

$$f_T(h_P, h, \psi) = \frac{r^2}{2} \left\{ (x + 3 \cos \psi)(1 + x^2 - 2x \cos \psi)^{1/2} \right.$$

$$\left. + (3 \cos^2 \psi - 1) \ln \left[(1 + x^2 - 2x \cos \psi)^{1/2} + x - \cos \psi \right] \right\}_{R/r}^{(R+h)/r}. \tag{7.183}$$

In this formula, the factor r^2 can be approximated as R^2 if terms of relative order of magnitude h_P/R or smaller are to be neglected. However, we will always keep it in the original form throughout this section, so that readers can immediately realize that the formulae are for a point on the Earth's surface of height h_P, with $r = R + h_P$.

The gravitational potential of the Helmert condensation layer can be inferred from the formula for a layer of mass around the Earth's surface (6.57). With $R' = R$, $\rho dR' = \mu$, and omitting the integral over height, we have

$$V_\mu = G \int_\omega \mu \frac{R^2}{\left(R^2 + r^2 - 2Rr \cos \psi\right)^{1/2}} d\omega = G \int_\omega \rho f_\mu(h_P, h, \psi) d\omega .$$

(7.184)

The kernel $f_\mu(h_P, h, \psi)$ is, with $r = R + h_P$ and μ given by (6.144), which assures strict mass conservation in spherical geometry,

$$f_\mu(h_P, h, \psi) = \frac{R^2}{l_R^r} \left(h + \frac{h^2}{R} + \frac{h^3}{3R^2}\right) ,$$

(7.185)

where l_R^r is defined as

$$l_R^r = \left(R^2 + r^2 - 2Rr \cos \psi\right)^{1/2} .$$

(7.186)

The indirect effect of the gravitational potential is thus

$$\delta W = V_T - V_\mu = G \int_\omega \rho \left[f_T(h_P, h, \psi) - f_\mu(h_P, h, \psi)\right] d\omega .$$

(7.187)

The two terms of the kernel, $f_T(h_P, h, \psi)$ and $-f_\mu(h_P, h, \psi)$, are of about the same magnitude, but with opposite signs. Hence, it is beneficial to first analytically remove the common part of them for better computation accuracy, which is left to readers who are interested in.

The kernel $f_T(h_P, h, \psi)$ tends to infinitely large when $\psi \to 0$. Hence, it is not evident to numerically evaluate the integral (7.187) around the point P where δW is computed, particularly due to the presence of the logarithm function. As in the case of Stokes' theory, we consider separately a region $\psi \le \Theta$ in (7.187)

$$\delta W = G \int_{\omega_{\psi \ge \Theta}} \rho \left[f_T(h, \psi) - f_\mu(h, \psi)\right] d\omega + \delta W_\Theta .$$

(7.188)

In the region $\psi \le \Theta$, we also separate the topography into a shell of height h_P and a terrain undulating around the height h_P. Accordingly, we also separate the surface density of the Helmert condensation layer into two terms. Thus, we have

$$\delta W_{\omega_{\psi \le \Theta}} = V_T^{0\sim\Theta, 0\sim h_P} - V_\mu^{0\sim\Theta, 0\sim h_P} + V_T^{0\sim\Theta, h_P\sim h} - V_\mu^{0\sim\Theta, h_P\sim h} ,$$

(7.189)

where, in the superscript, $0 \sim \Theta$ is the range of ψ, and $0 \sim h_P$ and $h_P \sim h$ are the ranges of height, remembering that h is position dependent. We neglect the variation of the density ρ within the region $\psi \leq \Theta$. The term $V_T^{0\sim\Theta,0\sim h_P}$ is the gravitational potential of a cap-shaped shell, which can be obtained from (6.116) by setting $r = R + h_P$, $R_B = R$ and $R_T = R + h_P$,

$$
\begin{aligned}
V_T^{0\sim\Theta,0\sim h_P} =2\pi G\rho r^2 \bigg\{ &\frac{1}{3}\left(1 + x^2 - 2x\cos\Theta\right)^{3/2} \\
&+ \frac{1}{2}(x - \cos\Theta)\left(1 + x^2 - 2x\cos\Theta\right)^{1/2} + \frac{1}{3}x^3 - \frac{1}{2}x^2 \\
&+ \frac{1}{2}\sin^2\Theta\cos\Theta\ln\left[\left(1 + x^2 - 2x\cos\Theta\right)^{1/2} + x - \cos\Theta\right]\bigg\}\bigg|_{R/r}^{1}.
\end{aligned}
$$
(7.190)

The term $V_\mu^{0\sim\Theta,0\sim h_P}$ is the gravitational potential of a spherical cap, which can be obtained from (6.109) by setting $r = R + h_P$, $R' = R$, and $h = h_P$,

$$
V_\mu^{0\sim\Theta,0\sim h_P} = 2\pi G\rho \frac{R}{r}\left(a_R^r - h_P\right)\left(h_P + \frac{h_P^2}{R} + \frac{h_P^3}{3R^2}\right),
$$
(7.191)

where (6.144) is used, and a_R^r is defined as

$$
a_R^r = \left[R^2 + r^2 - 2Rr\cos\Theta\right]^{1/2}.
$$
(7.192)

By definition, the other two terms in (7.189) can be inferred from (7.182) to (7.185) to be

$$
V_T^{0\sim\Theta,h_P\sim h} = G\rho \int_\omega \left[f_T(h_P, h, \psi) - f_T(h_P, h_P, \psi)\right]d\omega,
$$
(7.193)

$$
\begin{aligned}
f_T(h_P, h, \psi) - f_T(h_P, h_P, \psi) = \frac{r^2}{2}\bigg\{ &(x + 3\cos\psi)(1 + x^2 - 2x\cos\psi)^{1/2} \\
&+ (3\cos^2\psi - 1)\ln\left[(1 + x^2 - 2x\cos\psi)^{1/2} + x - \cos\psi\right]\bigg\}\bigg|_1^{(R+h)/r};
\end{aligned}
$$
(7.194)

and

$$
V_\mu^{0\sim\Theta,h_P\sim h} = G\rho \int_\omega \left[f_\mu(h_P, h, \psi) - f_\mu(h_P, h_P, \psi)\right]d\omega,
$$
(7.195)

$$
f_\mu(h_P, h, \psi) - f_\mu(h_P, h_P, \psi) = \frac{R^2}{l_R^r}\left[(h - h_P) + \frac{h^2 - h_P^2}{R} + \frac{h^3 - h_P^3}{3R^2}\right],
$$
(7.196)

where l_R^r is defined in (7.186). In (7.193) and (7.195), as $h = h_P$ when $\psi = 0$, we could consider setting the kernel of the integral to vanish within an extremely small cap around P, where the height h can be reasonably approximated with the constant h_P.

It is of interest to express the contribution of the region $\psi \leq \Theta$ using approximate formulae, which are more appropriate for numerical evaluation. The term $V_T^{0 \sim \Theta, 0 \sim h_P}$ can be obtained from (6.137) by setting $r = R + h_P$, $R_B = R$, and $R_T = R + h_P$, so that $H_{Br} = -h_P$ and $H_{Tr} = 0$. We have

$$V_T^{0 \sim \Theta, 0 \sim h_P} = 2\pi G \rho a_r \left[h_P - \frac{h_P^2}{2a_r} \left(1 + \frac{3a_r}{2r} \right) + \frac{h_P^3}{6a_r^2} \left(1 + \frac{2a_r}{r} + \frac{3a_r^2}{4r^2} \right) \right],$$

$$(7.197)$$

where a_r is defined as

$$a_r = 2r \sin \frac{\Theta}{2}. \tag{7.198}$$

The term $V_T^{0 \sim \Theta, h_P \sim h}$ can be obtained from (6.72) by setting $r = R + h_P$, $R_B = R + h_P$, and $R_T = R + h$, and thus $H_{Br} = 0$ and $H_{Tr} = h - h_P$. We have

$$V_T^{0 \sim \Theta, h_P \sim h} = Gr^2 \rho \int_{\omega_{\psi \leq \Theta}} \frac{h - h_P}{l_r} d\omega + \frac{3}{4} Gr\rho \int_{\omega_{\psi \leq \Theta}} \frac{(h - h_P)^2}{l_r} d\omega$$

$$- \frac{1}{6} Gr^2 \rho \int_{\omega_{\psi \leq \Theta}} \frac{(h - h_P)^3}{l_r^3} \left(1 - \frac{3l_r^2}{4R^2} \right) d\omega, \quad (7.199)$$

where l_r is defined as

$$l_r = 2r \sin \frac{\psi}{2}. \tag{7.200}$$

In (7.199), the expression $(3l_r^2)/(4R^2)$ in the parentheses of the last term represents a quantity on the relative order of h^2/R^2 as compared to the first term ($h - h_P$ is on the same order as h), and the second term is on the relative order of h/R. They could be neglected as further approximations.

In summary, δW_Θ could be computed using (7.191), (7.195), (7.197), and (7.199).

The Isostatic Anomaly

In the case of isostatic anomaly, instead of the Helmert condensation layer, it is the mass added in the isostatic correction that should be removed. We still consider the Airy–Heiskanen or Vening Meinesz model, which are both represented by the values of T and H'. The same as in the case of topographic mass, the fundamental formulae of the gravitational potential of this mass are (6.58) and (6.61). Here, we have $r = R + h_P$, $R_B = R - (T + h')$, and $R_T = R - T$. Hence,

$$V_I = G \int_\omega \Delta \rho f_I(T, h_P, h', \psi) d\omega, \qquad (7.201)$$

where the kernel $f_I(T, h_P, h', \psi)$ is defined as

$$f_T(T, h_P, h', \psi) = \frac{r^2}{2} \left\{ (x + 3\cos\psi)(1 + x^2 - 2x\cos\psi)^{1/2} \right.$$

$$\left. + (3\cos^2\psi - 1) \ln \left[(1 + x^2 - 2x\cos\psi)^{1/2} + x - \cos\psi \right] \right\}_{[R-(T+h')]/r}^{(R-T)/r}.$$
$$(7.202)$$

As there is no singularity in this integral, no further development is necessary. The indirect effect is

$$\delta W = V_T - V_I = G \int_\omega \rho \left[f_T(h_P, h, \psi) - \frac{\Delta\rho}{\rho} f_I(T, h_P, h', \psi) \right] d\omega. \qquad (7.203)$$

Again, it is more favorable for better accuracy to compute the two components together according to (7.203) by first analytically removing the common parts of the two terms, which are of about the same magnitude, but with opposite signs.

It is also more practicable to consider a small region $\psi \leq \Theta$ separately, where the contribution to V_T is computed as the sum of $V_T^{0\sim\Theta, 0\sim H_P}$ and $V_T^{0\sim\Theta, H_P\sim H}$ according to (7.197) and (7.199), respectively. Although the contribution of this small region to V_I can still be computed using the exact formulae (7.201) and (7.202), the isostatic compensation mass can also be divided into a cap-shaped shell and an undulation.

Readers are suggested to further develop (7.203) by applying the approximate formula (6.72) to both the topographic and isostatic corrections based on the Airy–Heiskanen model.

7.4.3 Indirect Effect of the Deflection of the Vertical

Taking reference to Fig. 7.3, we realize that the definition of the deflection of the vertical at a point P in Molodensky's theory would be identical to that of Stokes' theory with the geoid replaced by the equipotential surface $W = W_P$ and the reference Earth ellipsoid by the surface $U = W_P$. Therefore, the preparatory formulation in Sect. 6.5.4, i.e., (6.225) to (6.234), which are general enough, is still valid by understanding the geoid as the surface $W = W_P$ and the reference Earth ellipsoid as the surface $U = W_P$. Hence, here we provide only the formulae for gravity disturbance.

The Helmert Anomaly

The exact formulae for the indirect effect of the horizontal components of gravity disturbance at a point P on the Earth's surface can be written as, with $r = R + h_P$,

$$
\begin{cases}
\delta_i g_\theta = \dfrac{G}{r} \displaystyle\int_\omega \rho \left[\dfrac{\partial}{\partial \psi} f_T(h_P, h, \psi) - \dfrac{\partial}{\partial \psi} f_\mu(h_P, h, \psi) \right] \cos A \, d\omega, \\
\delta_i g_\lambda = -\dfrac{G}{r} \displaystyle\int_\omega \rho \left[\dfrac{\partial}{\partial \psi} f_T(h_P, h, \psi) - \dfrac{\partial}{\partial \psi} f_\mu(h_P, h, \psi) \right] \sin A \, d\omega,
\end{cases}
\tag{7.204}
$$

where $(\partial/\partial\psi) f_T(h_P, h, \psi)$ and $-(\partial/\partial\psi) f_\mu(h_P, h, \psi)$ represent the contributions from adding back the topographic mass and removing the Helmert condensation layer, respectively. The kernel $(\partial/\partial\psi) f_T(h_P, h, \psi)$ is obtained by setting $r = R + h_P$, $R_B = R$, and $R_T = R + h$ in (6.229) and (6.230)

$$
\frac{\partial}{\partial \psi} f_T(h_P, h, \psi) = \frac{r^2}{2} \left\{ -3 \sin \psi (1 + x^2 - 2x \cos \psi)^{1/2} + \frac{x \sin \psi (x + 3 \cos \psi)}{(1 + x^2 - 2x \cos \psi)^{1/2}} \right.
$$

$$
+ \frac{\sin \psi \, (3 \cos^2 \psi - 1) \left[(1 + x^2 - 2x \cos \psi)^{1/2} + x \right]}{(1 + x^2 - 2x \cos \psi)^{1/2} \left[(1 + x^2 - 2x \cos \psi)^{1/2} + x - \cos \psi \right]}
$$

$$
\left. - 6 \sin \psi \cos \psi \ln \left[(1 + x^2 - 2x \cos \psi)^{1/2} + x - \cos \psi \right] \right\} \Big|_{R/r}^{(R+h)/r} .
\tag{7.205}
$$

The kernel $(\partial/\partial\psi) f_\mu(h_P, h, \psi)$ is obtained by setting $r = R + h_P$ and replacing μ by its expression (6.144) in (6.233) and (6.234)

$$
\frac{\partial}{\partial \psi} f_\mu(h_P, h, \psi) = -\frac{2R^3 r \sin \psi}{\left(l_R^r \right)^{3/2}} h \left(1 + \frac{h}{R} + \frac{h^2}{3R^2} \right),
\tag{7.206}
$$

where l_R^r is defined in (7.186).

The topographic part of the kernel in (7.204) tends to infinitely large when $\psi \to 0$, especially, there is no evident way to handle the logarithm function. Hence, we provide an alternative formulation for the topographic term for the region $\psi \le \Theta$. As the plate of height h_P does not contribute to $\delta_i g_\theta$ and $\delta_i g_\lambda$, we have to consider only the undulating part of the topographic mass around the height h_P, whose gravitational potential is given in (7.199). The derivation is similar to those of (6.239) and (6.242) . Here we directly write out the result:

$$
\begin{cases}
\delta_i g_{\Theta\theta}^T = -\dfrac{G\rho}{4} \displaystyle\int_{\omega_{\psi}\leq\Theta} (h-h_P)\, \dfrac{\cos(\psi/2)}{\sin^2(\psi/2)}\cos A\, d\omega \\[2ex]
\qquad -\dfrac{3G\rho}{16R}\displaystyle\int_{\omega_{\psi}\leq\Theta}(h-h_P)^2\,\dfrac{\cos(\psi/2)}{\sin^2(\psi/2)}\cos A\, d\omega \\[2ex]
\qquad +\dfrac{G\rho}{32R^2}\displaystyle\int_{\omega_{\psi}\leq\Theta}(h-h_P)^3\,\dfrac{\cos(\psi/2)}{\sin^4(\psi/2)}\left(1+\sin^2\dfrac{\psi}{2}\right)\cos A\, d\omega, \\[2ex]
\delta_i g_{\Theta\lambda}^T = \dfrac{G\rho}{4}\displaystyle\int_{\omega_{\psi}\leq\Theta}(h-h_P)\,\dfrac{\cos(\psi/2)}{\sin^2(\psi/2)}\sin A\, d\omega \\[2ex]
\qquad +\dfrac{3G\rho}{16R}\displaystyle\int_{\omega_{\psi}\leq\Theta}(h-h_P)^2\,\dfrac{\cos(\psi/2)}{\sin^2(\psi/2)}\sin A\, d\omega \\[2ex]
\qquad -\dfrac{G\rho}{32R^2}\displaystyle\int_{\omega_{\psi}\leq\Theta}(h-h_P)^3\,\dfrac{\cos(\psi/2)}{\sin^4(\psi/2)}\left(1+\sin^2\dfrac{\psi}{2}\right)\sin A\, d\omega,
\end{cases}
$$

$$(7.207)$$

where the superscript T implies that these are the contribution from topographic mass alone.

The Isostatic Anomaly

Analogously, for the case of isostatic anomaly, we have

$$
\begin{cases}
\delta_i g_\theta = \dfrac{G}{r}\displaystyle\int_\omega \rho\left[\dfrac{\partial}{\partial\psi}f_T(h_P,h,\psi)-\dfrac{\Delta\rho}{\rho}\dfrac{\partial}{\partial\psi}f_I(T,h_P,h',\psi)\right]\cos A\, d\omega, \\[2ex]
\delta_i g_\lambda = -\dfrac{G}{r}\displaystyle\int_\omega \rho\left[\dfrac{\partial}{\partial\psi}f_T(h_P,h,\psi)-\dfrac{\Delta\rho}{\rho}\dfrac{\partial}{\partial\psi}f_I(T,h_P,h',\psi)\right]\sin A\, d\omega.
\end{cases}
$$

$$(7.208)$$

As in the case of Helmert anomaly, the term $(\partial/\partial\psi)f_T(h_P,h,\psi)$ represents the contribution from adding back the topographic mass and is given in (7.205). The term $-(\partial/\partial\psi)f_I(T,h_P,h',\psi)$ represents the contribution from removing the mass added in the isostatic correction. It can be inferred from (6.229) and (6.230) for the Airy–Heiskanen or Vening Meinesz model. Here we have $r = R + h_P$, $R_B = R - (T+h')$, and $R_T = R - T$, and ρ should be replaced by $\Delta\rho$. Hence, we obtain

$$
\frac{\partial}{\partial\psi}f_I(T,h_P,h',\psi) = \frac{r^2}{2}\Bigg\{-3\sin\psi\,(1+x^2-2x\cos\psi)^{1/2}
$$

$$
+\frac{x\sin\psi\,(x+3\cos\psi)}{(1+x^2-2x\cos\psi)^{1/2}}
$$

$$+ \frac{\sin\psi \left(3\cos^2\psi - 1\right)\left[(1 + x^2 - 2x\cos\psi)^{1/2} + x\right]}{(1 + x^2 - 2x\cos\psi)^{1/2}\left[(1 + x^2 - 2x\cos\psi)^{1/2} + x - \cos\psi\right]}$$

$$\left. -6\sin\psi\cos\psi\ln\left[(1+x^2-2x\cos\psi)^{1/2} + x - \cos\psi\right]\right\}_{[R-(T+h')]/r}^{(R-T)/r}.$$

$$(7.209)$$

Again, the kernel of the topographic term, $(\partial/\partial\psi) f_T (h_P, h, \psi)$, tends to infinitely large when $\psi \to 0$, and it has been particularly treated in the formula for the region $\psi \le \Theta$ in (7.207). Computation could be done by combining (7.207) and (7.208), which is quite straightforward and is left to readers.

7.4.4 Downward Continuation After Gravity Reduction: A Conceptual Procedure

An iterative downward continuation procedure consistent with the physical model can be devised based on the solution of the adjusted disturbing potential. We continue to use T^c to denote the adjusted disturbing potential, and ΔG^c, the adjusted Helmert or isostatic anomaly. Both of them are understood in the generalized sense defined in Sect. 7.1.3. We also use the subscript h to denote a value on the Earth's surface of adjusted normal height h and the subscript 0 for a value on the adjusted quasigeoid. We still discuss only the simple case of spherical approximation, where

$$\Delta G^c = -\frac{2}{r}T^c - \frac{\partial T^c}{\partial r}.$$

$$(7.210)$$

Our basis is the truncated Taylor series

$$\Delta G_0^c = \Delta G_h^c - \sum_{k=1}^{k_{max}} \frac{h^k}{k!} \left(\frac{\partial^k \Delta G^c}{\partial r^k}\right)_0,$$

$$(7.211)$$

with

$$\left(\frac{\partial^k \Delta G^c}{\partial r^k}\right)_0 = -\left[2\frac{\partial^k}{\partial r^k}\left(\frac{T^c}{r}\right) + \frac{\partial^{k+1} T^c}{\partial r^{k+1}}\right]_{r=R}$$

$$(7.212)$$

derived according to (7.210).

The procedure is to compute the adjusted disturbing potential T^c from the known Helmert or isostatic anomaly ΔG_h^c iteratively using (7.210) to (7.212). The basic principle is that T^c can be computed based on Stokes' theory from ΔG_0^c, and ΔG_0^c can be updated using the solution of T^c and the known gravity anomaly ΔG_h^c. The initial value of ΔG_0^c could be assumed to be ΔG_h^c or could be computed based on a preliminary downward continuation such as (7.173). The computation can be done using the solution of Stokes' theory either in integral form or in spherical harmonic series.

The computation of T^c based on Stokes' theory in integral form includes two distinct steps. The first step is to compute ΔG_0^c iteratively using (7.211) and (7.212), where, as will be formulated in the next paragraph, $(\partial^k T^c / \partial r^k)_{r=R}$ is to be evaluated based on the solution of T_0^c computed using Stokes' formula (5.48) with the kernel, i.e., the Stokes' function, given by (5.49). The second step is to evaluate T^c on the Earth's surface using the generalized Stokes' formula (5.35) with the kernel, i.e., the generalized Stokes' function, given by (5.46).

As indicated in the previous paragraph, we now complete the formulation of the problem by deriving the formula for computing $(\partial^k T^c / \partial r^k)_{r=R}$ using T_0^c. As T^c is harmonic on and outside the adjusted quasigeoid, which is here approximated as a sphere of radius R, the formula (5.132) is readily applicable, using the notations of this section. Let P be a point on the Earth's surface of height h, and P_0 the corresponding point on the adjusted quasigeoid. The formula of $(\partial T^c / \partial r)_{r=R}$ at P_0 is obtained by replacing V by T^c in (5.132)

$$\left(\frac{\partial T^c}{\partial r} \right)_{P_0} = -\frac{1}{R} T_{P_0}^c + \frac{R^2}{2\pi} \int_\omega \frac{T_0^c - T_{P_0}^c}{l_0^3} d\omega, \tag{7.213}$$

where the subscript P_0 indicates that the value is taken at P_0. This formula holds when T^c is replaced by any function harmonic on and outside the adjusted quasigeoid, and it can be easily shown that $r^k(\partial^k T^c / \partial r^k)$ is the case. Hence, we have

$$\frac{\partial}{\partial r} \left(r^k \frac{\partial^k T^c}{\partial r^k} \right)_{P_0} = -\frac{R^k}{R} \left(\frac{\partial^k T^c}{\partial r^k} \right)_{P_0}$$

$$+ \frac{R^2}{2\pi} \int_\omega \frac{R^k \left[(\partial^k T^c / \partial r^k)_0 - (\partial^k T^c / \partial r^k)_{P_0} \right]}{l_0^3} d\omega, \tag{7.214}$$

which can be readily reduced to

$$\left(\frac{\partial^{k+1} T^c}{\partial r^{k+1}} \right)_{P_0} = -\frac{k+1}{R} \left(\frac{\partial^k T^c}{\partial r^k} \right)_{P_0}$$

$$+ \frac{R^2}{2\pi} \int_\omega \frac{\left[(\partial^k T^c / \partial r^k)_0 - (\partial^k T^c / \partial r^k)_{P_0} \right]}{l_0^3} d\omega. \tag{7.215}$$

This is a recurrence formula to compute $(\partial^k T^c / \partial r^k)_0$ for $k = 1, 2, \cdots$ successively. The formula for the first order derivative, (7.213), is the degenerate case of this formula with $k = 0$.

A conceptual procedure of computation is as follows: (1) Set $\Delta G_0^c = \Delta G_h^c$, or compute it using a primitive downward continuation formula such as (7.173). (2) Compute T_0^c using ΔG_0^c according to Stokes' formula (5.48) with the Stokes' function given by (5.49). (3) Compute $(\partial^k T^c / \partial r^k)_0$ for $k = 1, 2, \cdots, k_{\max} + 1$ successively using (7.215). (4) Evaluate $(\partial^k \Delta G^c / \partial r^k)_0$ for $k = 1, 2, \cdots, k_{\max}$ according to (7.212), which we further develop as

$$\left(\frac{\partial \Delta G^c}{\partial r} \right)_0 = \left(-\frac{2}{r^2} T^c + \frac{2}{r} \frac{\partial T^c}{\partial r} + \frac{\partial^2 T^c}{\partial r^2} \right)_{r=R}, \tag{7.216}$$

$$\left(\frac{\partial^2 \Delta G^c}{\partial r^2} \right)_0 = \left(\frac{4}{r^3} T^c - \frac{4}{r^2} \frac{\partial T^c}{\partial r} + \frac{2}{r} \frac{\partial^2 T^c}{\partial r^2} + \frac{\partial^3 T^c}{\partial r^3} \right)_{r=R}, \tag{7.217}$$

and so forth. A recurrence formulae can be derived, which is left to readers. (5) Compute ΔG_0^c according to (7.211) using data of ΔG_h^c. (6) Repeat (2) to (5) till the result converges to within a prescribed accuracy. (7) Compute T^c on the Earth's surface using ΔG_0^c according to the generalized Stokes' formula (5.35) with the generalized Stokes' function given by (5.46), which can then be used to compute the height anomaly. Lastly, the deflection of the vertical on the Earth's surface can be computed using ΔG_0^c. Taking reference to Fig. 7.3 and (7.70), we see that ζ' behaves like the geoidal height N in Stokes' theory, and we can thus infer that the relation between the deflection of the vertical and the gravity disturbance is

$$\delta \xi^c = \frac{\delta g_{h\theta}^c}{\bar{\gamma}_0}, \qquad \delta \eta^c = -\frac{\delta g_{h\lambda}^c}{\bar{\gamma}_0}. \tag{7.218}$$

Hence, we can compute $\delta g_{h\theta}^c$ and $\delta g_{h\lambda}^c$ using ΔG_0^c according to (5.58) where the kernel $\partial S(r, \psi) / \partial \psi$ is given in (5.62) and then compute the deflection of the vertical according to (7.218). A more rigorous form is to replace $\bar{\gamma}_0$ by the normal gravity on the telluroid, γ_h.

The iterative computation can also be done using spherical harmonic series of the quantities involved. This requires expressing T^c in (7.210) and (7.212) as a spherical harmonic series. The procedure of the computation is similar to the case using integral formulae.

However, the convergence of the spherical harmonic series involved may be critical. By expressing T^c as a spherical harmonic series $T^c = \sum_{n=2}^{n_{\max}} T_n^c$, where T_n^c is the degree n term of T^c, we obtain, according to (7.212),

$$\frac{h^k}{k!} \left(\frac{\partial^k \Delta G^c}{\partial r^k} \right)_0 = -\frac{h^k}{k!} \sum_{n=2}^{n_{\max}} \left[2 \frac{\partial^k}{\partial r^k} \left(\frac{T_n^c}{r} \right) + \frac{\partial^{k+1} T_n^c}{\partial r^{k+1}} \right]_{r=R}, \tag{7.219}$$

which should be evaluated in order to evaluate $\left(\partial^k \Delta G^c / \partial r^k\right)_0$ using (7.211). For degrees higher than 4, the spherical harmonic coefficients of the normal gravitational potential are negligibly small as compared to those of the disturbing potential. Hence, Kaula's rule (5.176) or (5.177) applies to the disturbing potential T. Here we take the general expression (5.177) and write it as, taking also (5.173) into account,

$$(T_n)_{r=R} \sim \frac{GM}{R}\alpha^{1/2}n^{-\beta/2}, \tag{7.220}$$

meaning that the degree n term of T is on the order of $\alpha^{1/2}(GM/R)n^{-\beta/2}$ in magnitude when $r = R$. According to (5.181), we have

$$\left(\frac{\partial T_n}{\partial r}\right)_{r=R} \sim \frac{GM}{R}\frac{\alpha^{1/2}n^{1-\beta/2}}{R}, \tag{7.221}$$

which is on the order of n/R times that of $(T_n)_{r=R}$. Higher order derivatives can also be successively derived. In general, $(\partial^k T_n/\partial r^k)_{r=R}$ is on the order of $(n/R)^k$ times that of $(T_n)_{r=R}$. We have

$$\left(\frac{\partial^k T_n}{\partial r^k}\right)_{r=R} \sim \frac{GM}{R}\frac{\alpha^{1/2}n^{k-\beta/2}}{R^k}. \tag{7.222}$$

Assuming $T_n^c \sim T_n$, as the second term in the brackets at the right-hand side of (7.219) is dominant, we would have

$$\frac{h^k}{k!}\left(\frac{\partial^k \Delta G^c}{\partial r^k}\right)_0 \sim \frac{h^k}{k!}\left(\frac{\partial^{k+1} T_n^c}{\partial r^{k+1}}\right)_{r=R}$$

$$\sim \frac{GM}{R}\frac{1}{k!}\left(\frac{h}{R}\right)^k \frac{\alpha^{1/2}n^{k+1-\beta/2}}{R} \sim \frac{GM}{R^2}\frac{\alpha^{1/2}n^{1-\beta/2}}{k!}\left(\frac{nh}{R}\right)^k. \tag{7.223}$$

For any value of h, when n approaches or exceeds the order of magnitude of R/h, this term may decrease too slowly or even increase with n, rendering the spherical harmonic series in (7.219) poorly convergent or divergent.

The estimate (7.223) is to be understood qualitatively. According to (5.176), $\alpha \sim 2 \times 10^{10}$ and $\beta = 3$. In reality, T_n may decrease faster, as Kaula's rule may be different when n is extremely large. Furthermore, T_n^c may decrease faster than T_n with n, as the ultra-high degree part of T is mainly due to topographic mass that could have been removed in gravity reduction. Despite these possible alleviative factors, convergence of the spherical harmonic series in (7.219) should still be carefully verified in its evaluation—whatever the values of α and β are, for any $h \neq 0$, the spherical harmonic series is doomed to diverge when the truncation degree $n_{\max}$ exceeds a certain value.

A hybrid iterative process is also possible. For example, ΔG_0^c may be computed iteratively using the integral formulae, and T^c may be finally solved in the form of a spherical harmonic series for a global gravitational model. This combination may have the potential to avoid the divergence problem in the pure spherical harmonic approach. However, the flaw of divergence shown to exist in the spherical harmonic approach puts a doubt on whether the iteration in the integral approach converges, as they are the formulations of the same physical problem, and may behave the same way. Hence, the convergence must also be verified in the computation of ΔG_0^c based on the iterative integral approach. We also mention that similar convergence issue may even exist in the iterative approach of harmonic downward continuation discussed in Sect. 7.3.4, thus also being in need of verification in numerical computation.

Finally, we mention that, although the indirect effect has not been mentioned in the discussion, it should be handled consistently throughout the procedure.

7.5   The GNSS-Gravimetric Boundary Value Theory

7.5.1   The Gravity Disturbance and the GNSS-Gravimetric Boundary Value Problem

We put this topic in this chapter as a supplement to Molodensky's theory, since this is also an oblique derivative boundary value problem with the Earth's surface as the boundary, and the formulation of the solution is also similar.

In both Stokes' and Molodensky's theories, the height is determined using spirit leveling and gravity measurements. The difference is that the former adopts the orthometric height, and the latter the normal height.

At present, the GNSS technique is more and more widely used for positioning, providing the geodetic height h with unprecedented accuracy, including gravity measurement stations. In this circumstance, a normal gravity must also be subtracted from the gravity in order to obtain a small quantity related to the disturbing gravity field. As only the normal gravity at the gravity measurement station can be computed when geodetic height is known, what can be defined is the gravity disturbance on the Earth's surface,

$$\delta g = g - \gamma . \tag{7.224}$$

We emphasize that γ is the normal gravity at the gravity measurement station, and δg is the difference between the values of g and γ at the same point. It has been made evident that

$$\gamma = \gamma_0 + \left(\frac{\partial \gamma}{\partial h}\right)_0 h , \tag{7.225}$$

where the subscript "0" indicates the value on the reference Earth ellipsoid. An approximation of it is

$$\gamma = \gamma_0 - 0.3086h \,, \tag{7.226}$$

where the unit of normal gravity is mGal, and that of h is m. Numerically, we have

$$\delta g = g + 0.3086h - \gamma_0 \,. \tag{7.227}$$

Hence, loosely speaking, $0.3086h$ is a free air correction using the geodetic height. Readers are suggested to take reference to Sect. 6.1.1 for a more accurate formula of γ.

As in the derivation of the fundamental gravimetric equations in both Stokes' and Molodensky's boundary value problems, we neglect the differences among the directions of the normal to the reference Earth ellipsoid, the normal gravity, and the gravity to obtain

$$\delta g = -\frac{\partial T}{\partial h} \,. \tag{7.228}$$

Under spherical approximation, i.e., by approximating the reference Earth ellipsoid as a sphere of radius R, we have

$$\delta g = -\left(\frac{\partial T}{\partial r}\right)_{r=R+h} \,. \tag{7.229}$$

We refer the above two equations as the fundamental GNSS-gravimetric equation. The GNSS-gravimetric boundary value problem is then to solve Laplace's equation $\Delta T = 0$ on and outside the Earth's surface with the boundary condition (7.228) or (7.229), which is, like Molodensky's theory, an oblique derivative boundary value problem. However, the boundary is known in the present case, and it is a fixed boundary value problem.

Like ΔG, the domain of definition of δg can also be extended to the space with the expression

$$\delta g = -\frac{\partial T}{\partial r} \,. \tag{7.230}$$

Furthermore, like $r\Delta G$, $r\delta g$ is also a harmonic function if T is a harmonic function.

In the context of this book, we still formulate the solution only for the case of spherical approximation.

7.5.2 Solution of Disturbing Potential by Molodensky's Shrinking

The formulation entirely follows that of Molodensky's theory with only a little difference. Here we follow Stock (1983) and give only some critical steps to follow the logic. Readers are suggested to read this subsection in parallel with the formulation of Molodensky's theory.

As in (7.25), we express T outside the Earth as the gravitational potential of a material surface of density μ/G located on the Earth's surface S,

$$T = \int_S \frac{\mu}{l} dS, \tag{7.231}$$

where l is the distance between the point where T is computed and the position of the surface element dS. The substitution of (7.231) into the boundary condition (7.229) yields

$$2\pi \mu_P \cos \beta_P - \int_S \mu \frac{\partial}{\partial r} \left(\frac{1}{l} \right)_{r=R+h_P} dS = \delta g_P, \tag{7.232}$$

where all symbols have the same meaning as those in (7.28). See also Fig. 7.2. Similar to (7.30), this equation can be further developed to

$$2\pi \mu_P \cos \beta_P - \int_S \mu \left[-\frac{1}{2(R+h_P)l_P} + \frac{(R+h)^2 - (R+h_P)^2}{2(R+h_P)l_P^3} \right] dS = \delta g_P, \tag{7.233}$$

which can be written as an integral equation over a unit sphere like (7.31),

$$2\pi \mu_P \cos \beta_P - \int_\omega \mu \sec \beta \left[-\frac{1}{2l_P} + \frac{(R+h)^2 - (R+h_P)^2}{2l_P^3} \right] \frac{(R+h)^2}{R+h_P} d\omega = \delta g_P. \tag{7.234}$$

Finally, the approximate equation to be solved like (7.36) is

$$2\pi \mu_P \cos \beta_P - \int_\omega \mu \sec \beta \left\{ -\frac{R}{2l_0 \left[1 + \left(\frac{h-h_P}{l_0} \right)^2 \right]^{1/2}} + \frac{R^2(h-h_P)}{l_0^3 \left[1 + \left(\frac{h-h_P}{l_0} \right)^2 \right]^{3/2}} \right\} d\omega$$

$$= \delta g_P. \tag{7.235}$$

Introduce an auxiliary function

$$\chi = \mu \sec \beta \tag{7.236}$$

to substitute μ, and then replace h by kh, and $\tan \beta$ by $k \tan \beta$. We obtain

$$\frac{2\pi \chi_P}{1 + k^2 \tan^2 \beta_P} + \frac{R}{2} \int_\omega \frac{\chi}{l_0 \left[1 + k^2 \left(\dfrac{h - h_P}{l_0}\right)^2\right]^{1/2}} d\omega$$

$$- R^2 \int_\omega \frac{k(h - h_P)\chi}{l_0^3 \left[1 + \left(\dfrac{h - h_P}{l_0}\right)^2\right]^{3/2}} d\omega = \delta g_P . \tag{7.237}$$

Note that the coefficient $-3R/2$ of the second term in (7.39) becomes $R/2$ here, which is the only difference between Molodensky's theory and the present one, and is to be persistent in forthcoming derivations. Actually, we can already guess that the solution of the disturbing potential would have the same form as in Molodensky's theory with the Stokes' function replaced by the Hotine function. We will anyway outline some major steps for readers to go through the whole logic.

We express χ as

$$\chi = \chi_0 + k\chi_1 + k^2\chi_2 + \cdots , \tag{7.238}$$

develop (7.237) into a power series of k, and then set the coefficients of all powers of k to vanish, to obtain a set of equations like (7.46) and (7.47),

$$\begin{cases} 2\pi \chi_{0P} + \dfrac{R}{2} \displaystyle\int_\omega \dfrac{\chi_0}{l_0} d\omega = \delta g_P , \\[2mm] 2\pi \chi_{1P} + \dfrac{R}{2} \displaystyle\int_\omega \dfrac{\chi_1}{l_0} d\omega = \delta g_{1P} , \\[2mm] 2\pi \chi_{2P} + \dfrac{R}{2} \displaystyle\int_\omega \dfrac{\chi_2}{l_0} d\omega = \delta g_{2P} , \\[2mm] \cdots \end{cases} \tag{7.239}$$

where $\delta g_{1P}, \delta g_{2P}, \cdots$ are defined as

$$\begin{cases} \delta g_{1P} = R^2 \displaystyle\int_\omega \chi_0 \dfrac{h - h_P}{l_0^3} d\omega , \\[2mm] \delta g_{2P} = 2\pi \chi_{0P} \tan^2 \beta_P + R^2 \displaystyle\int_\omega \chi_1 \dfrac{h - h_P}{l_0^3} d\omega + \dfrac{R}{4} \displaystyle\int_\omega \chi_0 \dfrac{(h - h_P)^2}{l_0^3} d\omega . \\[2mm] \cdots \end{cases}$$

$$\tag{7.240}$$

As T is expressed in the same exact way as in Molodensky's theory, (7.51) and (7.52) are also valid here, which are

$$T_A = T_{0P} + T_{1P} + T_{2P} + \cdots ; \tag{7.241}$$

$$
\begin{cases}
T_{0P} = R^2 \int_\omega \frac{\chi_0}{l_0} d\omega , \\[2mm]
T_{1P} = R^2 \int_\omega \frac{\chi_1}{l_0} d\omega , \\[2mm]
T_{2P} = R^2 \int_\omega \frac{\chi_2}{l_0} d\omega - \frac{1}{2} R^2 \int_\omega \frac{(h - h_P)^2}{l_0^3} \chi_0 d\omega , \\[2mm]
\cdots
\end{cases}
\tag{7.242}
$$

Equations (7.239) to (7.242) obtained by Molodensky's shrinking are to be solved. The procedure of deriving the solution is also almost identical to that of Molodensky's theory. In the degenerate case of $h = 0$, the integration equation to be solved becomes

$$2\pi \mu_P + \frac{R}{2} \int_\omega \frac{\mu}{l_0} d\omega = \delta g_P \tag{7.243}$$

with

$$T_P = R^2 \int_\omega \frac{\mu}{l_0} d\omega . \tag{7.244}$$

These equations are the same as for the problem when the gravity disturbance is known on a spherical boundary formulated in Sect. 5.3.1. The solution is given in (5.87), which is copied here for convenience of reference,

$$T_P = \frac{R}{4\pi} \int_\omega \delta g K(\psi) d\omega . \tag{7.245}$$

Based on the logic that (7.245) is the solution of the pair of Eqs. (7.243) and (7.244), we can readily obtain, similar to the case of Molodensky's theory, the solutions of the three pairs of equations given in (7.239) and (7.242),

$$T_{0P} = \frac{R}{4\pi} \int_\omega \delta g S(\psi) d\omega, \tag{7.246}$$

$$T_{1P} = \frac{R}{4\pi} \int_\omega \delta g_1 S(\psi) d\omega, \tag{7.247}$$

$$T_{2P} = \frac{R}{4\pi} \int_\omega \delta g_2 S(\psi) d\omega - \frac{1}{2} R^2 \int_\omega \frac{(h - h_P)^2}{l_0^3} \chi_0 d\omega. \tag{7.248}$$

In the solution of T_{2P}, the first term is obtained from (7.239) and the first term in (7.242), and the second term is just the second term in (7.242). The solutions of χ_0 and χ_1 can be obtained by substituting (7.242) into (7.239)

$$\chi_{0P} = \frac{1}{2\pi} \left(\delta g_P - \frac{1}{2} \frac{T_{0P}}{R} \right), \tag{7.249}$$

$$\chi_{1P} = \frac{1}{2\pi} \left(\delta g_{1P} - \frac{1}{2} \frac{T_{1P}}{R} \right). \tag{7.250}$$

These are required in order to compute δg_1 and δg_2 according to (7.47). Also, the computation of the second term in T_{2P} requires χ_0.

The solution of disturbing potential is now complete. The solution of this problem is unique. If the degrees 0 and 1 terms are present in data of δg, their effect on the result is included in the solution. However, if the reference Earth ellipsoid is defined such that degrees 0 and 1 terms are missing in the disturbing potential, these terms would be neither in the data nor in the result. A conceptual procedure of computation is similar to (7.61) to (7.65) in Molodensky's theory.

7.5.3 Simplification of the Solution of the Disturbing Potential

The Stokes' formula (7.84) and its inverse (7.85) in Molodensky's theory should be replaced by Hotine formula

$$T_0 = \frac{R}{4\pi} \int_\omega \delta g K(\psi) d\omega, \tag{7.251}$$

and its inverse (5.133), written as, with N substituted by T, and using the notations here,

$$\delta g_P = \frac{1}{R} T_{0P} - \frac{R^2}{2\pi} \int_\omega \frac{T - T_{0P}}{l_0^3} d\omega. \tag{7.252}$$

This equation could also be derived from (7.85) with the relation $\Delta G = -2T/r + \delta g$, which is the fundamental gravimetric equation in Molodensky's theory. We substitute (7.249) into (7.240) and decompose δg_1 into two parts

$$\delta g_1 = \delta g_{11} + \Delta g_{12} \,, \tag{7.253}$$

where

$$\delta g_{11P} = -h_P \frac{R^2}{2\pi} \int_\omega \frac{1}{l_0^3} \left[\left(\delta g - \frac{1}{2} \frac{T_0}{R} \right) - \left(\delta g - \frac{1}{2} \frac{T_0}{R} \right)_P \right] d\omega + \frac{h_P}{R} \left(\delta g - \frac{1}{2} \frac{T_0}{R} \right)_P \,, \tag{7.254}$$

$$\delta g_{12P} = \frac{R^2}{2\pi} \int_\omega \frac{1}{l_0^3} \left[h \left(\delta g - \frac{1}{2} \frac{T_0}{R} \right) - h_P \left(\delta g - \frac{1}{2} \frac{T_0}{R} \right)_P \right] d\omega - \frac{h_P}{R} \left(\delta g - \frac{1}{2} \frac{T_0}{R} \right)_P \,. \tag{7.255}$$

We also decompose T accordingly as

$$T = T_0 + T_{11} + T_{12} \,, \tag{7.256}$$

where

$$T_0 + T_{11} = \frac{R}{4\pi} \int_\omega (\delta g + \delta g_{11}) k(\psi) d\omega \,, \tag{7.257}$$

$$T_{12} = \frac{R}{4\pi} \int_\omega \delta g_{12} K(\psi) d\omega \,. \tag{7.258}$$

The expression of the term δg_{11}, (7.254), can be written as

$$\delta g_{11P} = -h_P \frac{R^2}{2\pi} \int_\omega \frac{\delta g - \delta g_P}{l_0^3} d\omega + \frac{h_P}{R} \delta g_P - \frac{1}{2} \frac{h_P}{R} \left(-\frac{R^2}{2\pi} \int_\omega \frac{T_0 - T_{0P}}{l_0^3} d\omega + \frac{1}{R} T_{0P} \right) . \tag{7.259}$$

According to (7.252), the quantity within the parentheses in the second term is δg_P. Hence,

$$\delta g_{11P} = -h_P \frac{R^2}{2\pi} \int_\omega \frac{\delta g - \delta g_P}{l_0^3} d\omega + \frac{1}{2} \frac{h_P}{R} \delta g_P \,, \tag{7.260}$$

and we have

$$\delta g_P + \delta g_{11P} = \delta g_P \left(1 + \frac{1}{2} \frac{h_P}{R} \right) - h_P \frac{R^2}{2\pi} \int_\omega \frac{\delta g - \delta g_P}{l_0^3} d\omega = \delta g_P - h_P \delta g_P' \,, \tag{7.261}$$

where terms on the relative order of h_P/R in magnitude are neglected, and $\delta g'_P$ is defined as

$$\delta g'_P = \frac{R^2}{2\pi} \int_\omega \frac{\delta g - \delta g_P}{l_0^3} d\omega. \tag{7.262}$$

For the same reason in obtaining (7.102), by comparing (7.255) with (7.252) and (7.258) with (7.251), we obtain

$$T_{12} = -h\left(\delta g - \frac{1}{2}\frac{T_0}{R}\right) = -h\delta g + \frac{1}{2}\frac{h}{R}T_0 = -h\delta g, \tag{7.263}$$

where terms on the relative order of $(h/R)T_0$ in magnitude have also been neglected. Finally, we obtain

$$T_P = \frac{R}{4\pi} \int_\omega (\delta g_P - h\delta g'_P)K(\psi)d\omega - h_P\delta g_P, \tag{7.264}$$

noting that the last term is $h(\partial T/\partial h)$ at P.

7.5.4 Solutions of the Height Anomaly and the Deflection of the Vertical

In Stokes' or Molodensky's theory, the geoidal height or height anomaly is indispensable, thus being an inherent quantity, but this is not the case in the GNSS-gravimetric boundary value theory. However, determination of geoidal height or height anomaly is one of the major purposes in geodesy to determine the Earth's gravity field, which is the basis of vertical datum, and is particularly necessary in the space geodesy era to convert the GNSS measured geodetic height into the orthometric or normal height for mapping.

In the GNSS-gravimetric boundary value problem, the geoidal height or height anomaly is a derived quantity, and the deflection of the vertical can be understood as a derived quantity in all the boundary value theories discussed so far.

As it is the disturbing potential on the Earth's surface that is determined, computation of the height anomaly is more straightforward than the geoidal height. Therefore, we formulate the height anomaly here, while assuming the geoidal height is to be computed using the relation between the height anomaly and the geoidal height.

In Bruns' formula (7.8), the normal gravity γ is assumed to be its value on the telluroid. However, for evident reason, replacing it by its value on the Earth's surface results in the same accuracy. Here we use the approximate formula (7.9), which

Fig. 7.6 The deflection of the vertical defined as the angle between the gravity and normal gravity vectors at the same point

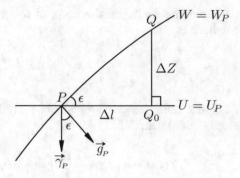

remains indifferent for both Molodensky's theory and the GNSS-gravimetric one. We readily obtain

$$\zeta_P = \frac{R}{4\pi \bar{\gamma_0}} \int_\omega \left(\delta g - h \delta g'\right) K(\psi) d\omega - h_P \frac{\delta g_P}{\bar{\gamma_0}} . \tag{7.265}$$

This is the result corresponding to (7.111) in Molodensky's theory.

For this problem, the natural way is to define the deflection of the vertical as the angle between the plumb line and the normal plumb line at the same point on the Earth's surface. Very likely, this definition will progressively become the standard, as GNSS is taking over the positioning in gravimetry. Airborne gravimetry, for which GNSS positioning is most appropriate, will likely become the main technique of gravity measurement for the purpose of higher resolution static gravity field determination, while the lower resolution part is best measured using satellite gravimetry techniques.

We start to derive the formula of the deflection of the vertical based on the definition. In Fig. 7.6, we show an cross section plane through a point P. This plane is vertical according to the normal gravity, i.e., the normal gravity vector $\vec{\gamma}_P$ is within this plane. The vector $\vec{g}_P$ is the projection of the gravity in this plane. We also draw the equipotential surface of gravity $W = W_P$ and the equipotential surface of normal gravity $U = U_P$. The projection of the deflection of the vertical in this plane, ϵ, is the angle between $\vec{g}_P$ and $\vec{\gamma}_P$, which equals the angle between the surfaces $W = W_P$ and $U = U_P$. We also draw a point Q very close to P on the surface $W = W_P$, and its projection on the surface $U = U_P$ is Q_0. Keep in mind that ϵ is a very small angle, which is very much exaggerated in the figure. With Δl and ΔZ defined in the figure, we have $\epsilon = \Delta Z / \Delta l$ assuming $\Delta l \to 0$. The rest of the formulation is to express ΔZ using Δl. By definition, we have $W_Q = W_P$ and $U_{Q_0} = U_P$. By applying (1.49) to the normal gravity field, we obtain

$$\Delta Z = \frac{U_{Q_0} - U_Q}{\gamma_{Q_0}} = \frac{U_{Q_0} - W_Q + T_Q}{\gamma_{Q_0}} = \frac{U_P - W_P + T_Q}{\gamma_{Q_0}} = \frac{T_Q - T_P}{\gamma_{Q_0}} . \tag{7.266}$$

We also have

$$T_Q - T_P = (T_{Q_0} - T_P) + (T_Q - T_{Q_0}) = \left(\frac{\partial T}{\partial l}\right)_P \Delta l + \left(\frac{\partial T}{\partial Z}\right)_{Q_0} \Delta Z,$$

$$(7.267)$$

where the directional derivatives are along the directions of Δl and ΔZ defined in Fig. 7.6. We finally obtain

$$\epsilon = \lim_{\Delta l \to 0} \frac{\Delta Z}{\Delta l} = \frac{1}{\gamma_P} \left(\frac{\partial T}{\partial l}\right)_P + \frac{1}{\gamma_P} \left(\frac{\partial T}{\partial Z}\right)_P \epsilon,$$

$$(7.268)$$

where we used the fact $Q_0 \to P$ when $\Delta l \to 0$. This formula holds at any point. Hence, the subscript "P" can be dropped. We have

$$\epsilon = \frac{1}{\gamma}\frac{\partial T}{\partial l} + \frac{1}{\gamma}\frac{\partial T}{\partial Z}\epsilon = \frac{\delta g_l}{\gamma} + \frac{\delta g_z}{\gamma}\epsilon = \frac{\delta g_l}{\gamma},$$

$$(7.269)$$

where $\delta g_l = \partial T/\partial l$ and $\delta g_z = \partial T/\partial Z$ are the projections of the gravity disturbance δg along the directions of Δl and ΔZ, respectively. A term of smaller order of magnitude has been neglected based on the fact that, as ΔZ is a length measured upward, $-\delta g_z = \delta g$, and $\delta g/\gamma \ll 1$. The two components of the deflection of the vertical are

$$\xi = \frac{\delta g_\theta}{\gamma}, \qquad \eta = -\frac{\delta g_\lambda}{\gamma}.$$

$$(7.270)$$

This result can be inferred directly based on the definition. Take the component ξ for example. The directions of the gravity and normal gravity in the meridional plane are represented by g_θ/g and γ_θ/γ, respectively. The component ξ is defined to be their difference, which is $\delta g_\theta/\gamma$ neglecting a term of relative order $\delta g/\gamma$ of magnitude.

We now derive the formulae of the deflection of the vertical using the solution of T given by (7.264). An immediate observation is that the function T in (7.264) is a function defined on the Earth's surface, which is two dimensional. Take reference to Fig. 7.7, which is a vertical cross section plane through the point P on the Earth's

Fig. 7.7 Derivative along the Earth's surface and that along the horizontal direction

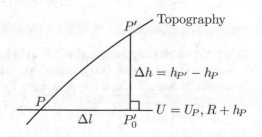

surface. Another point P' on the Earth's surface is very close to P. We approximate the equipotential surface of normal gravity, $U = U_P$, as a sphere of radius $R + h_P$. The projection of P' on this surface is denoted as P_0'. The solution of T in (7.264) has the property

$$T_{P'} - T_P = \left(\frac{\partial T}{\partial l}\right)_P \Delta l . \tag{7.271}$$

By applying (1.49) to the disturbing gravity field, we obtain

$$T_{P_0'} - T_{P'} = (\delta g)_{P_0'} \Delta h . \tag{7.272}$$

By definition, the component of the gravity disturbance defined in (7.269), δg_l, is

$$\delta g_l = \lim_{\Delta l \to 0} \frac{T_{P_0'} - T_P}{\Delta l} = \lim_{\Delta l \to 0} \frac{(T_{P'} - T_P) + (T_{P_0'} - T_{P'})}{\Delta l} = \frac{\partial T}{\partial l} + \frac{\partial h}{\partial l} \delta g , \tag{7.273}$$

where we have taken into account of the fact $P_0' \to P$ when $\Delta l \to 0$, and omitted the subscript "P" in the final result. Under spherical approximation, the two components of the deflection of the vertical are, according to (7.270),

$$\begin{cases} \xi = \dfrac{1}{\bar{\gamma}_0}\left(\dfrac{1}{R}\dfrac{\partial T}{\partial \theta}\right) + \dfrac{1}{\bar{\gamma}_0}\left(\dfrac{1}{R}\dfrac{\partial h}{\partial \theta}\right)\delta g , \\[2mm] \eta = -\dfrac{1}{\bar{\gamma}_0}\left(\dfrac{1}{R\sin\theta}\dfrac{\partial T}{\partial \lambda}\right) - \dfrac{1}{\bar{\gamma}_0}\left(\dfrac{1}{R\sin\theta}\dfrac{\partial h}{\partial \lambda}\right)\delta g , \end{cases} \tag{7.274}$$

where the disturbing potential T is assumed to be computed according to (7.264), and terms of relative order of magnitude h/R are neglected, leading to the approximation of $1/(R+h)$ by $1/R$ in the formulae.

The substitution of (7.264) into (7.274) yields, taking reference to the derivation of (5.90),

$$\begin{cases} \xi_P = \dfrac{1}{4\pi\bar{\gamma}_0}\displaystyle\int_\omega (\delta g - h\delta g')\dfrac{d}{d\psi}K(\psi)\cos A\, d\omega - \dfrac{h_P}{\bar{\gamma}_0}\left(\dfrac{1}{R}\dfrac{\partial \delta g}{\partial \theta}\right)_P , \\[3mm] \eta_P = \dfrac{1}{4\pi\bar{\gamma}_0}\displaystyle\int_\omega (\delta g - h\delta g')\dfrac{d}{d\psi}K(\psi)\sin A\, d\omega + \dfrac{h_P}{\bar{\gamma}_0}\left(\dfrac{1}{R\sin\theta}\dfrac{\partial \delta g}{\partial \lambda}\right)_P . \end{cases} \tag{7.275}$$

This is the result corresponding to (7.112) in Molodensky's theory.

For comparison with the geometrical interpretation of Molodensky's theory discussed in Sect. 7.3.3, we approximate the second term in (7.265) as, taking reference to (7.76) and within the accuracy of our formulation,

$$
- h \frac{\delta g}{\bar{\gamma}_0} = -h \frac{\Delta G}{\bar{\gamma}_0} - 2 \frac{h}{R} \frac{T}{\bar{\gamma}_0} = -h \frac{\Delta G}{\bar{\gamma}_0} - 2 \frac{h}{R} \zeta = -h \frac{\Delta G}{\bar{\gamma}_0} = h \frac{\partial \zeta}{\partial h} . \tag{7.276}
$$

The is the formula corresponding to (7.152) for the height anomaly. The formula corresponding to (7.154) for the deflection of the vertical is

$$
h \kappa_g - h \kappa_\gamma = - \frac{h}{\bar{\gamma}_0} \left[\frac{1}{R} \left(\frac{\partial g}{\partial \theta} - \frac{\partial \gamma}{\partial \theta} \right) \right] = - \frac{h}{\bar{\gamma}_0} \left(\frac{1}{R} \frac{\partial \delta g}{\partial \theta} \right) . \tag{7.277}
$$

This is the formula for the component ξ, and that for the component η is not written explicitly. The difference from the case of Molodensky's theory is that the change of direction of both the gravity and normal gravity is from the reference Earth ellipsoid to the Earth's surface, and hence, the height involved, h, is the geodetic height.

The same as in Molodensky's theory, if we use a surface h_P above the reference Earth ellipsoid as boundary, within the accuracy of the formulation, (7.265) and (7.275) can be written as

$$
\zeta_P = \frac{R}{4 \pi \bar{\gamma}_0} \int_\omega \left[\delta g - (h - h_P) \delta g' \right] K(\psi) d\omega , \tag{7.278}
$$

$$
\begin{cases}
\xi_P = \dfrac{1}{4 \pi \bar{\gamma}_0} \displaystyle\int_\omega \left[\delta g - (h - h_P) \delta g' \right] \dfrac{d}{d\psi} K(\psi) \cos A d\omega , \\[2mm]
\eta_P = \dfrac{1}{4 \pi \bar{\gamma}_0} \displaystyle\int_\omega \left[\delta g - (h - h_P) \delta g' \right] \dfrac{d}{d\psi} K(\psi) \sin A d\omega ,
\end{cases} \tag{7.279}
$$

where $\delta g'$ is defined in (7.262), which is copied to here for convenience of reference

$$
\delta g'_P = \frac{R^2}{2\pi} \int_\omega \frac{\delta g - \delta g_P}{l_0^3} d\omega . \tag{7.280}
$$

These correspond to the result of Molodensky's theory given in (7.113) to (7.115).

7.5.5 Analytical Downward Continuation and Gravity Reduction

We are putting analytical downward continuation and gravity reduction together here, as most formulations of these subjects in Molodensky's theory are still valid for the GNSS-gravimetric boundary value theory. We are not going into the detail of the formulations. Instead, we simply modify those parts that do not hold for the GNSS-gravimetric boundary value theory. First, the following replacements of quantities or formulae should be made:

- The gravity anomaly ΔG is replaced by the gravity disturbance δg.
- The quasigeoid is replaced by the reference Earth ellipsoid.
- The height h is understood as geodetic height.
- The surface $W = W_P$ is replaced by the surface h_P above the reference Earth ellipsoid.
- Molodensky's solutions (7.113) and (7.114) are replaced by the solutions here, (7.278) and (7.278), together with $\Delta G'$ defined in (7.115) replaced by $\delta g'$ defined in (7.280).

After the replacements of quantities and formulae, the following is a list of validity or validity after modification:

- The discussions in Sect. 7.3.1 are all valid, considering that, like $r\Delta G$, $r\delta g$ is also a harmonic function.
- The discussions in Sect. 7.3.2 are all valid with $S(\psi)$ in (7.125) and (7.126) replaced by $K(\psi)$.
- The discussions in Sect. 7.3.3 are all valid, taking reference to (7.276) and (7.277) for the geometrical interpretation of the height anomaly given in (7.265) and the deflection of the vertical given in (7.275), respectively.
- The discussions in Sect. 7.3.4 are all valid.
- All discussions in Sect. 7.4 are valid with two exceptions. First, in the last paragraph of Sect. 7.4.1, all $h^c = h + \delta\zeta$ should be replaced by h, as the indirect effect does not influence geodetic height, and iteration is no longer required. Second, in Sect. 7.4.4, the Eq. (7.210) should be replaced by

$$\delta g^c = -\frac{\partial T^c}{\partial r},\qquad(7.281)$$

and the subsequent formulae should also be modified accordingly. This is a straightforward task and is left to readers.

7.6   Combination of Different Types of Data

Before the era of space geodesy, height had been always measured using spirit leveling, and either the orthometric or normal height is adopted. Now, a lot of historical data are based on the orthometric or normal height, while a lot of recent data are obtained with GNSS as the mean of positioning, thus being based on geodetic height. Satellite altimetry provides data of mean ocean surface height, which is the height above the reference Earth ellipsoid, thus being geodetic height. The properly corrected mean ocean surface height is either treated as geoidal height, or its slope is inferred, and is treated as the deflection of the vertical. Various corrections are necessary, including particularly the mean dynamic ocean surface topography, which is the height above the geoid, and is mostly provided by ocean circulation models. Recent satellite gravimetry missions permitted the direct determination of the long wavelength portion of the Earth's gravity field, e.g., the spherical harmonic coefficients of the gravitational potential up to a certain degree and order, which can be used to compute the oceanic geoid. Nevertheless, in order to interpret the short wavelength portion of mean ocean surface height determined using satellite altimetry as geoidal height, e.g., beyond the degree and order of the spherical harmonic coefficients determined using satellite gravimetry, the result of mean dynamic ocean surface topography in the same short wavelength band as output of some kind of ocean circulation model must still be subtracted.

We digress a bit to mention that, when the oceanic geoidal height determined using satellite gravimetry is subtracted from the mean ocean surface height determined using satellite altimetry in the same long wavelength band, the mean dynamic ocean surface topography in this wavelength band is obtained, which can be used as an observation for ocean circulation modeling.

In summary, we now may have four kinds of surface data: gravity anomaly and/or gravity disturbance over land, and geoidal height or deflection of the vertical over oceans. We assume that the data of gravity anomaly and gravity disturbance are provided on the Earth's surfaces. As the mean dynamic ocean surface topography is very small, we can understand the data of geoidal height or deflection of the vertical as provided on either the geoid or the Earth's surface. None of the boundary value theories we formulated so far is able to treat all or even a subset of these kinds of data together.

A boundary value problem with the Earth's surface divided into up to four regions could be defined. One type of data, either the gravity anomaly, gravity disturbance, geoidal height, or deflection of the vertical, is provided in each region. This kind of boundary value problem is referred to as mixed boundary value problem.

The mixed boundary value problem is still an idealized situation. In reality, different types of data may exist over the same region. If multiple types of data over the same region are to be included altogether in a boundary value problem, there are then more data than necessary, and the boundary value problem is over determined.

Both mixed and over determined boundary value problems are of advanced theoretical research topics, and no commonly accepted theory for practical data processing is yet available. Here we are not going to deal with these advanced topics. Instead, we proved ideas to convert all types of data into one type, either the gravity anomaly or gravity disturbance, to use the solution of Molodensky's or the GNSS-gravimetric boundary value problem. These are anyway ideas rather than practical computational procedures, as this is also an advanced research topic for best methodologies and data results. However, the most fundamental relations among all these types of data are made clear in our formulations so far, and how these relations may possibly be applied is our focus in this section. With the understanding of these fundamental relations and the possible ways they may be applied, a big picture of the determination of the Earth's gravity field using all available data on the Earth's surface could be depicted.

We assume that all types of data are compatible, i.e., in all data, the same reference Earth ellipsoid is used with both degrees 0 and 1 terms in the disturbing potential vanishing by definition, and use of the same geoid defining the reference Earth ellipsoid to define the orthometric height or normal height. Consequently, the geodetic height can be interpreted as the sum of the geoidal height and orthometric height, or as the sum of the quasigeoidal height and normal height.

7.6.1   Conversion of Different Types of Data into Gravity Anomaly

The main barrier in data conversion is that the computation of one type of data from another type of data always requires the evaluation of a global integral, which needs global coverage of the known data.

We start with the conversion of geoidal height into gravity anomaly. Over the oceans, the Earth's surface is very well approximated by the geoid, which practically coincides with the quasigeoid. We thus have $N = \zeta$ and $\Delta g = \Delta G$. The basic formula is the inverse Stokes' formula (5.134), which is written here using notations of this chapter as

$$\Delta G_P = -\frac{\bar{\gamma}_0}{R} N_P - \frac{R^2 \bar{\gamma}_0}{2\pi} \int_\omega \frac{N - N_P}{l_0^3} d\omega . \tag{7.282}$$

Since the kernel $1/l_0^3$ decreases quickly with the distance to P, we realize that data of N within a domain surrounding P contribute much more than those in the remote region. Hence, a compromise to overcome data shortage in the remote region is to use the result of N of a known model of the Earth's gravity field, so that data of N are patched to have global coverage. As the data in the "patches" are considered less accurate, the result of ΔG computed according to (7.282) in the proximity of the "patches" must be carefully accessed. A more assured, but also more expensive, remedy is to have an overlapped region where both ΔG and N are measured, e.g., to perform gravity measurement in coastal oceans, so that data of ΔG computed according to (7.282) have to be used only far away enough from the coasts.

We mention that (7.282) is derived assuming no mass exists above the geoid. Hence, in order to be strictly physically correct, gravity reduction of relocating the topographic mass of the land has to be performed according to Stokes' theory. The computation here is the reverse of determining N from Δg, where $\Delta g = \Delta G$ over oceans. The procedure is to first compute N^c from N, then compute ΔG^c from N^c, and finally compute ΔG from ΔG^c.

The formula to convert deflection of the vertical into gravity anomaly is the inverse Vening Meinesz formula (5.124), written using notations of this chapter as

$$\Delta G_P = \frac{\bar{\gamma}_0}{4\pi} \int_\omega H'(\psi) \left(\xi \cos A' + \eta \sin A' \right) d\omega , \tag{7.283}$$

where A' is defined in Fig. 5.8, and $H'(\psi)$ is provided in (5.125). All discussions on (7.282) apply equally to (7.283) with N replaced by ξ and η and are not to be repeated.

In order to convert gravity disturbance into gravity anomaly, we make use of the fundamental gravimetric equation, which we write here as

$$\Delta G = -\frac{2}{r}T + \delta g . \qquad (7.284)$$

What we do is to compute T using δg according to the solution of the GNSS-gravimetric boundary value theory and then compute ΔG according to (7.284). The formula for computing T using δg is, by substituting ζ by T in (7.278) according to Bruns' formula,

$$T_P = \frac{R}{4\pi} \int_\omega \left[\delta g - (h - h_P)\delta g' \right] K(\psi) d\omega . \qquad (7.285)$$

The discussion on (7.282) concerning the patching of data to global coverage using an existing model of the gravity field still applies here, with the coastal oceans replaced by the neighboring regions between the regions where δg and ΔG are provided. However, as these formulae are based on the Earth's surface, gravity reduction is not mandatory to make the approach physically strictly correct.

A less strict conversion from δg to ΔG is to directly use the values of T in (7.284) from an existing model of the Earth's gravity field, as the contribution of T is of smaller order of magnitude. This could be best understood in the spherical harmonic domain; the order of magnitude of the degree n term of $(2/r)T$ is about $2/n$ of those of ΔG and δg. For example, the degree 200 term of $(2/r)T$ is only about 1% of the degree 200 term of ΔG.

A last remark is that the accuracy of conversion of different types of data into gravity anomaly may be improved iteratively in the determination of a global model of the gravitational potential.

7.6.2 Conversion of Different Types of Data into Gravity Disturbance

The logic of this subsection is practically the same as that of the previous one. The only difference is the formulae used. Hence, we just list the formulae.

The formula for converting geoidal height into gravity disturbance is provided in (5.133) and is written here as

$$\delta g_P = \frac{\bar{\gamma}_0}{R} N_P - \frac{R^2 \bar{\gamma}_0}{2\pi} \int_\omega \frac{N - N_P}{l_0^3} d\omega . \qquad (7.286)$$

The formula for converting deflection of the vertical into gravity disturbance is provided in (5.122) and is written here as

$$\delta g_P = \frac{\bar{\gamma}_0}{4\pi} \int_\omega \left[-R \frac{\partial}{\partial r} C'(r, \psi)_R \right] \left(\xi \cos A' + \eta \sin A' \right) d\omega, \tag{7.287}$$

where A' is defined in Fig. 5.8, and the kernel $-R(\partial/\partial r)C'(r, \psi)_R$ is provided in (5.123).

The formulae for converting gravity anomaly into gravity disturbance are

$$\delta g = \frac{2}{r} T + \Delta G. \tag{7.288}$$

$$T_P = \frac{R}{4\pi} \int_\omega \left[\Delta G - (h - h_P) \Delta G' \right] S(\psi) d\omega, \tag{7.289}$$

where the first is obtained from (7.284), and the second is Molodensky's solution.

Appendix: Some Remarks on the Use of Runge's Theorem

This section is considered as an appendix for the reason that it is the only content of the book not built starting from first principles; the proof of this theorem is far beyond the reachable scope of this book.

Runge's theorem is an existence theorem that may potentially be used to infer an approximate solution of the disturbing potential, that is not equal, but infinitesimally close to the exact solution, by simplifying a boundary value problem with the Earth's surface as boundary to another one with a sphere embedded within the Earth as boundary. It is challenging to prove whether an approximate solution satisfies Runge's theorem. Here we focus on some concerns that must be carefully taken into consideration in using the theorem.

We quote a derived form of Runge's theorem from Moritz (1980), called Keldysh–Lavrentiev theorem, which is the most relevant to physical geodesy:

Any harmonic function ϕ, harmonic outside the Earth's surface and continuous outside and on it, may be uniformly approximated by a harmonic function ψ regular outside an arbitrarily given sphere inside the Earth, in the sense that for any given $\epsilon > 0$, the relation

$$|\phi - \psi| < \epsilon \tag{7.290}$$

holds everywhere outside and on the Earth's surface.

Here we understand the term *Earth's surface* as a surface above but infinitesimally close to the real surface of the Earth, so that this theorem agrees with the Runge–Krarup theorem, which states that (7.290) holds only above the Earth's surface. We also understand the term *regular* as single-valued, finite and continuous, and continuously differentiable up to any order, which pertain to the disturbing

Fig. 7.8 Runge's theorem

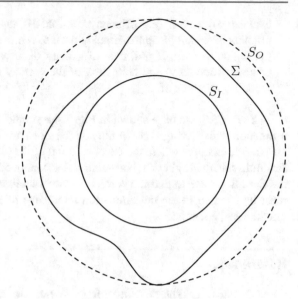

potential outside the Earth's surface. The sphere inside the Earth is referred to as Bjerhammar sphere.

To connect this theorem with our knowledge built thus far in this book, we take reference to Fig. 7.8, where Σ is the Earth's surface, S_I the Bjerhammar sphere, and S_O the Brillouin sphere. As an application of the Keldysh–Lavrentiev theorem to the disturbing gravity field, we understand ϕ as the disturbing potential, and ψ the harmonic function used to represent ϕ. We have two major concerns in the application of the Keldysh–Lavrentiev theorem:

- The Keldysh–Lavrentiev theorem does not tell how to find a function ψ. The most apparent way is to (1) find a function ψ such that $\psi = \phi$ outside S_O expressed as a spherical harmonic series; (2) extend the domain to within S_O using the same expression; and (3) assume (7.290) to be automatically satisfied in between Σ and S_O. Our concern is that the satisfaction of (7.290) by ψ is not guaranteed. We also mention that the theorem does not concern the convergence of the spherical harmonic series of ψ between S_O and Σ.

- Based on the Keldysh–Lavrentiev theorem, it is indeed possible to design a method to find a ψ that satisfies (7.290) for any ϕ known on and outside Σ. However, a function ψ satisfying (7.290) does not guaranty the proximity

$$|\nabla \phi - \nabla \psi| < \sigma , \tag{7.291}$$

to hold, where $\sigma > 0$ is assumed to be of similar order of magnitude as ϵ. Even the statement is mathematically loose, its meaning is clear: "ϕ is approximated by ψ" does not guaranty "$\nabla \phi$ is meanwhile approximated by $\nabla \psi$." This means that even if ψ can be used to approximately represent the disturbing potential between S_O and Σ (e.g., Bucha et al., 2019), it has not been proven that the gradient of ψ can be approximately used to represent the gravity disturbance in the region as well. If not, data of gravity disturbance (and consequently, data of gravity anomaly) and deflection of the vertical may be misinterpreted in the modeling or usage of ψ.

In Sect. 7.3.3, we have shown in Fig. 7.4 a simple synthetic Earth model, where a function ψ satisfying $\psi = \phi$ outside S_O does not satisfy (7.291) on Σ. That is an extreme example where the exact solution of the disturbing potential outside the Brillouin sphere in spherical harmonic series cannot be downward continued as is to the Earth's surface to find the correct gravity disturbance. This kind of controversy must be handled in the determination and usage of global models of the Earth's gravitational potential.

References

Bucha, B., Hirt, C., & Kuhn, M. (2019). Divergence-free spherical harmonic gravity field modeling based on the Runge-Krarup theorem: a case study for the Moon. *Journal of Geodesy, 93*, 489–513. https://doi.org/10.1007/s00190-018-1177-4.

Moritz, H. (1980). *Advanced physical geodesy*. Abacus Press.

Stock, B. (1983). A Molodensky-type solution of the geodetic boundary value problem using the known surface of the Earth. *Manuscripta Geodaetica, 8*, 273–288.

Further Reading

Flury, J., & Rummel, R. (2009). On the geoid–quasigeoid separation in mountain areas. *Journal of Geodesy, 83*, 829–847. https://doi.org/10.1007/s00190-009-0302-9.

Hirt, C., & Kuhn, M. (2017). Convergence and divergence in spherical harmonic series of the gravitational field generated by high-resolution planetary topography—A case study for the Moon. *Journal of Geophysical Research, 122*, 1727–1746. http://dx.doi.org/10.1002/2017JE005298.

Moritz, H. (2000). Molodensky's theory and GPS. In: H. Moritz, & M. A. Yurkina (Eds.), *M.S. Molodensky in Memoriam*. Mitteilungen der geodätischen Institute der Technischen Universität Graz, 88, Graz, Austria (pp. 69–85).

Ogle, C., Costin, O., & Bevis, M. (2021). Non-convergence of the spherical harmonic expansion of gravitational potential below the Brillouin sphere: The continuous case. *Journal of Mathematical Physics, 62*, 102901. https://doi.org/10.1063/5.0044930.

Sansò, F., Barzaghi, R., & Carion, D. (2012). The Geoid today: Still a problem of theory and practice. In: N. Sneeuw et al. (eds.) *VII Hotine-Marussi Symposium on Mathematical Geodesy, International Association of Geodesy Symposia* (vol. 137). Springer-Verlag. https://doi.org/10.1007/978-3-642-22078-4.

Sjöberg, L. E. (1977). *On the Errors of Spherical Harmonic Developments of Gravity at the Surface of the Earth*. Department of Geodetic Science, Report No 257, The Ohio State University.

Sjöberg, L. E. (1979). On the existence of solutions for the method of Bjerhammar in the continuous case. *Bulltin Géodésique, 53*, 227–230.

Sjöberg, L. E. (2010). A strict formula for geoid-to-quasigeoid separation. *Journal of Geodesy, 84*, 699–702. https://doi.org/10.1007/s00190-010-0407-1.

Sjöberg, L. E., & Bgherbandi, M. (2011). A numerical study of the analytical downward continuation error in geoid computation by EGM2008. *Journal of Geodetic Science, 1*, 2–8. https://doi.org/10.2478/v10156-010-0001-8.

Fundamentals of Computation and Determination

8

Abstract

This chapter illustrates the fundamental aspects of practical computations and determinations of the Earth's gravity field. The topics covered are of constant concern in state-of-the-art research. Therefore, the focus is on the cautions that must be taken in the computations. It involves the preparation and assimilation of gravity data, computation of the Earth Gravitational Model, i.e., the spherical harmonic coefficients of the gravitational potential, the geoidal height, the height anomaly, and the deflection of the vertical in both Stokes' and Molodensky's theories. The mathematics concerned—the evaluation of singular integrals, the computation of integrals of the associated Legendre function, the Pellinen smoothing (the Gaussian smoothing is formulated incidentally), and the ideal filter—is also formulated. The determination of the geoidal height and height anomaly when data of astro-geodetic deflection of the vertical are used, i.e., astronomical leveling and astro-gravimetric leveling, is formulated. Finally, the principle of the determination of geoidal height and height anomaly using data of GNSS positioning and spirit leveling, called GNSS-leveling, is presented.

8.1 Preparation of Gravity Data

8.1.1 Transformation Between Different Geodetic Reference Frames

In order to use data of different types or from different sources together, the data must be made compatible in definition. This has already been made clear in Sect. 7.6: *we assume that all types of data are compatible, i.e., in all data, the same reference Earth ellipsoid is used with both degrees 0 and 1 terms in the disturbing potential vanishing by definition and use of the same geoid defining*

the reference Earth ellipsoid to define the orthometric height or normal height.
Consequently, the geodetic height can be interpreted as the sum of the geoidal height
and orthometric height or as the sum of the quasigeoidal height and normal height.
In this subsection, we provide the logic of how to assimilate data for use together.

Transformation of Geodetic Coordinates and Gravity Disturbance

Although any reference Earth ellipsoid is expected to satisfy the same requirements
outlined in Chap. 4, its different realizations adopted in practical usages are different. This difference may be attributed to several reasons. First, with the progress
of observation technologies, more and more data are being collected with higher
and higher accuracy. Hence, the parameters for the reference Earth ellipsoid are
being updated from time to time with more and more accurate estimates. Second,
a reference Earth ellipsoid is required as the basis of mapping for any country or a
group of countries. Once the reference Earth ellipsoid is updated, everything based
on it should be updated too, and confusion may persist in applications for a period
of time. Hence, the official reference Earth ellipsoid for a country or a group of
countries used as basis for mapping is required to stay stable, and the need for update
should be carefully assessed. As a result, different geodetic coordinate systems
are used by different countries, groups of countries, or international scientific
organizations.

The differences among different reference Earth ellipsoids include not only the
differences in the geometrical and physical parameters a, GM, J_2, and Ω but also
those of their origins and orientations. Besides, even the scales of the unit length
may be different.

We first consider the transformation of coordinates of a point from one geodetic
coordinate system to another. Each geodetic coordinate system is associated with
a reference Earth ellipsoid and a Cartesian coordinate system. The geometrical
parameters a and b of the two reference Earth ellipsoids are assumed to be different
and are known. The parameters representing the relation between the two Cartesian
coordinate systems as defined in Appendix A.5—the offset between the origins, the
Euler angles between the axes, and the scale factor between the unit lengths—are
also assumed to be known. Based on the formulae derived in this book, a conceptual
procedure to transform the geodetic coordinates from system 1 to system 2 is as
follows:

- Step 1: Compute the Cartesian coordinates x_1, y_1, and z_1 from the geodetic
  coordinates B_1, λ_1, and h_1 according to the formulae derived in Appendix A.4.
- Step 2: Compute the Cartesian coordinates x_2, y_2, and z_2 from x_1, y_1, and z_1
  according to the formulae derived in Appendix A.5, which requires the offset
  between the origins, the Euler angles between the axes, and the scale factor
  between the unit lengths.
- Step 3: Compute the geodetic coordinates B_2, λ_2, and h_2 from the Cartesian
  coordinates x_2, y_2, and z_2 according to the formulae derived in Appendix A.4.

The parameters representing the relation between the two coordinate systems, i.e., the offset between the origins, the Euler angles between the axes, and the scale factor between the unit lengths, can be estimated by the values of x_1, y_1, and z_1 as well as x_2, y_2, and z_2 of a set of stations based on least squares according to their relation (A.81). This requires both x_1, y_1, z_1 and x_2, y_2, z_2 to be known at the stations involved.

With the transformation of coordinates and the parameters of the two reference Earth ellipsoids, the transformation of gravity disturbance defined on the Earth's surface can be straightforwardly performed: add back γ_1 and then remove γ_2, each relying on the parameters of the reference Earth ellipsoid and the geodetic height in the respective system. This is based on the fact that the gravity at a point on the Earth's surface is coordinate system independent.[1]

Transformation of Orthometric or Normal Height and Gravity Anomaly

The transformation of orthometric or normal height and gravity anomaly is more complicated, as the relation between two different vertical datums over two continents can hardly be determined.

The horizontal coordinates B and λ can be transformed according to the procedure outlined above. However, the geodetic height is not known here, and it has to be computed approximately using a known model of geoidal or quasigeoidal height, based on whether the orthometric or normal height is adopted. This should provide accurate enough B and λ for our purpose, but the transformation of height must be considered in a more rigorous way, as gravity anomaly is very sensitive to the orthometric or normal height, which could have been measured based on different datums.

Spirit leveling is not considered as a standalone measurement to determine height. The results of height of a point P purely measured by spirit leveling through different paths are different. Gravity measurement is necessary, and the quantity that uniquely defines height is the potential number, which is then converted to orthometric or normal height by dividing by a kind of mean gravity or mean normal gravity. Hence, the coordinate transformation of orthometric or normal height is the coordinate transformation of the potential number, which is the difference of gravity potential between the geoid and the point P,

$$C_P = W_A - W_P = \int_A^P g \, dh_L, \tag{8.1}$$

where A is a point on the geoid assumed to be the origin of the vertical datum, and the integral is independent of the path from A to P.

---

[1] Strictly speaking, there is a small difference of centrifugal acceleration due to the difference of the coordinates of the point in different reference systems.

Evidently, C_P is independent of the choice of coordinate system; the gravity potential of a point does not rely on the coordinate system where the position of the point is expressed.[2] To define the point A where the height is 0, the common choice is the mean position of sea surface at a tide gauge, or some kind of average of a set of tide gauges, during a certain period of time. The variation of sea surface height is very complicated and includes components unlikely to be modeled. Hence, different realizations of vertical datum are not expected to refer to the same equipotential surface of gravity, meaning that the origin A may be chosen on different equipotential surfaces of gravity in different vertical datums.

Hence, for a point P, the potential number C_P in different vertical datums may be different by an offset, which is the difference of gravity potential between the origins. Therefore, the key to transform orthometric or normal height is to find this offset, and the rest of the computation is straightforward. For two different vertical datums over the same continent, this offset can be determined by spirit leveling to a point from both origins; it is in fact the difference between the two potential numbers obtained. However, connecting the vertical datums over two different continents is challenging.

Here we suggest a conceptual strategy to relate a vertical datum over one continent to a vertical datum over another continent. We assume that we have a very accurate global model of the gravitational potential. We measure the position of a point on each continent using GNSS so that we can compute the gravity potential at each point using the model of the gravitational potential together with the Earth's angular velocity. The difference between the gravity potentials obtained is then equal to the difference between the potential numbers if the origins of the vertical datums over the two continents are on the same equipotential surface of gravity. If not, the difference between the differences—one is the difference between the gravity potentials obtained using a global gravitational model through GNSS positioning and the other is the difference between potential numbers obtained through spirit leveling and gravity measurement on each continent—is then the offset. To alleviate the influence of the limited resolution and precision of the global model of gravitational potential adopted, a network over each continent could be used based on least squares fitting.

Ground data are also used to build high degree and order global models of gravitational potential, while a lower degree and order part is determined using satellite gravimetry. We realize that the offsets among vertical datums, if not properly corrected, would bring errors into the global model obtained. Hence, a more sophisticated approach could be to estimate the offsets among different vertical datums while determining the high degree and order global model by combining ground data with a lower degree and order model determined by satellite gravimetry. The satellite gravimetry determined model is the long wavelength portion of the gravitational potential independent of vertical datums. Hence, it can

---

[2] Similar to the centrifugal acceleration, there is a small difference of centrifugal potential due to the difference of the coordinates of the point in different reference systems.

serve as a control of vertical datums through geoidal height or height anomaly in a regional averaged sense. In this way, the result of offsets among vertical datums can be considered as a byproduct of the high degree and order global model of gravitational potential.

With the offsets of gravity potential among the origins of vertical datums determined, the transformation of orthometric or normal height among different vertical datums is straightforward. Transformation between orthometric and normal heights has already been discussed in Sect. 7.1.1, which is in fact identical to the transformation between geoidal and quasigeoidal heights.

With the transformation of orthometric or normal height between different geodetic coordinate systems established, the transformation of Stokes- or Molodensky-type gravity anomaly is straightforward. We just have to keep in mind that the gravity can be recovered by reversing the gravity reduction in the original system, and a gravity reduction can be performed in the new system to obtain the gravity anomaly. The transformation between the gravity anomalies of Stokes- and Molodensky-types can also be performed this way.

8.1.2 Evaluation of Surface Integrals in Gravity Reduction

Gravity reduction is necessary for computing gravity over a grid of points homogeneously distributed over a region using a limited amount of available gravity measurements. Assuming that we know the topography to the accuracy required, the idea is to (1) reduce the measured gravity to Helmert or isostatic anomaly, which is small and spatially smooth, thus being most suitable for interpolation, (2) interpolate the Helmert or isostatic anomaly onto a regularly spaced grid of points, and (3) reverse the gravity reduction over the grid of points to obtain the gravity.

In this subsection, we explain conceptually how the integrals in the formulae for gravity reduction can be evaluated numerically, which concerns specifically what cares should be taken in the computation. Contemporary computational approach is based on the fast Fourier transform (FFT) (e.g., Jekeli, 2017), which includes heavy further formulations beyond the scope of this book.

All the integrals have the form

$$I(\theta, \lambda) = \int_{\omega} f(\theta', \lambda') K(\theta, \lambda, \theta', \lambda') d\omega, \tag{8.2}$$

where θ' and λ' are the variables of integration. The function $f(\theta', \lambda')$ is provided as data, and the kernel $K(\theta, \lambda, \theta', \lambda')$ is provided analytically, mostly in the form of $K(\psi)$, with ψ being the geocentric angle between (θ', λ') and (θ, λ). Since $K(\theta, \lambda, \theta', \lambda')$ may tend to infinitely large when (θ', λ') approaches (θ, λ), i.e., when $\psi \to 0$, we divide the Earth's surface into an inner region $\omega_{\psi \leq \Theta}$ and an outer region $\omega_{\psi \geq \Theta}$ and separate the integral into two terms

$$I(\theta, \lambda) = I_{\psi \leq \Theta}(\theta, \lambda) + I_{\psi \geq \Theta}(\theta, \lambda), \tag{8.3}$$

Fig. 8.1 A demonstrative
division into cells for
numerical evaluation of
integrals

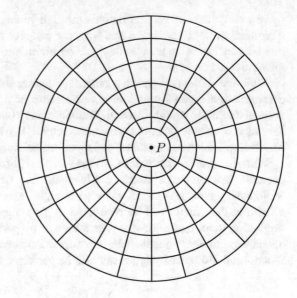

with

$$I_{\psi \leq \Theta}(\theta, \lambda) = \int_{\omega_{\psi \leq \Theta}} f(\theta', \lambda') K(\theta, \lambda, \theta', \lambda') d\omega, \qquad (8.4)$$

$$I_{\psi \geq \Theta}(\theta, \lambda) = \int_{\omega_{\psi \geq \Theta}} f(\theta', \lambda') K(\theta, \lambda, \theta', \lambda') d\omega. \qquad (8.5)$$

The singularity that the kernel $K(\theta, \lambda, \theta', \lambda')$ may tends to infinitely large when
(θ', λ') approaches (θ, λ) is to be handled in the evaluation of the inner term
$I_{\psi \leq \Theta}(\theta, \lambda)$.

As shown in Fig. 8.1, where the point (θ, λ) is denoted as P, the disk around P is
the inner region $\omega_{\psi \leq \Theta}$, and outside the disk is the outer region $\omega_{\psi \geq \Theta}$. We divide the
outer region into quasi-square cells, the closer to P the smaller the cells are, such
that over each cell, both $f(\theta', \lambda')$ and $K(\theta, \lambda, \theta', \lambda')$ can be approximated by the
value at the center of the cell. In this way, the integral over the outer region can be
computed as a sum

$$I_{\psi \geq \Theta}(\theta, \lambda) = \sum_i f(\theta'_i, \lambda'_i) K(\theta, \lambda, \theta'_i, \lambda'_i) \Delta\omega_i, \qquad (8.6)$$

where the subscript i indicates the ith cell, (θ'_i, λ'_i) its center, and $\Delta\omega_i$ its area.
Better accuracy can be achieved using more sophisticated numerical methods. For
example, Gaussian quadrature could be applied in each cell to compute the integral
using the values of $f(\theta', \lambda')$ and $K(\theta, \lambda, \theta', \lambda')$ at a set of predefined quadrature
points.

If $K(\theta, \lambda, \theta', \lambda')$ does not tend to infinitely large when (θ', λ') approaches (θ, λ), the integral $I_{\psi \leq \Theta}(\theta, \lambda)$ can be computed in the same way as $I_{\psi \geq \Theta}(\theta, \lambda)$, which is the case of isostatic gravity correction. Otherwise, $K(\theta, \lambda, \theta', \lambda')$ is mostly inverse proportional to $\sin^\alpha(\psi/2)$, and the integral $I_{\psi \leq \Theta}(\theta, \lambda)$ has to be particularly formulated for each α. The terrain gravity correction given in (6.154) is of this kind, which is written here as, accurate to only the two leading terms and with $l_r = 2r \sin(\psi/2)$

$$
\Delta_3 g_\Theta = -\frac{1}{2} Gr \int_{\omega_{\psi \leq \Theta}} \rho \frac{H - H_P}{l_r} d\omega + \frac{1}{2} Gr^2 \int_{\omega_{\psi \leq \Theta}} \rho \frac{(H - H_P)^2}{l_r^3} d\omega
$$

$$
= -\frac{1}{4} G\rho \int_{\omega_{\psi \leq \Theta}} \frac{H - H_P}{\sin(\psi/2)} d\omega + \frac{1}{16r} G\rho \int_{\omega_{\psi \leq \Theta}} \frac{(H - H_P)^2}{\sin^3(\psi/2)} d\omega, \qquad (8.7)
$$

where ρ is assumed to be constant. It has been explained that terms neglected are of relative order of H/R or $(H/R)^2$ in magnitude as compared to the terms kept.

We have two integrals to evaluate, which we write in general forms as

$$
I_1(\theta, \lambda) = \int_{\omega_{\psi \leq \Theta}} \frac{F_1(\theta', \lambda')}{\sin(\psi/2)} d\omega, \qquad I_2(\theta, \lambda) = \int_{\omega_{\psi \leq \Theta}} \frac{F_2(\theta', \lambda')}{\sin^3(\psi/2)} d\omega. \qquad (8.8)
$$

The same types of integrals are to be evaluated in the computations of the geoidal height and height anomaly, and hence, the more general forms are treated here for later reference. For the present case of gravity reduction, we have $F_1 = H - H_P$ and $F_2 = (H - H_P)^2$ in (8.8). We will have $F_1 = \Delta g$ for the geoidal height, $F_1 = \Delta G - (H - H_P)\Delta G'$ for the height anomaly, and $F_2 = \Delta G - \Delta G_P$ for $\Delta G'$.

We assume Θ to be very small, such that the approximations $\sin(\psi/2) = \psi/2$ and $\sin \psi = \psi$ are accurate enough. We substitute the coordinates θ' and λ' by ψ and the azimuth A, which can be understood as the polar angle and the negative of longitude taking P as the pole of a spherical coordinate system. We can then approximate the surface element on the unit sphere as $d\omega = \sin \psi d\psi dA = \psi d\psi dA$ and consequently approximate the integrals I_1 and I_2 as

$$
I_1(\theta, \lambda) = 2 \int_0^{2\pi} \int_0^\Theta F_1(\psi, A) d\psi dA, \qquad I_2(\theta, \lambda) = 8 \int_0^{2\pi} \int_0^\Theta \frac{F_2(\psi, A)}{\psi^2} d\psi dA.
$$
$$(8.9)$$

The first one is no longer singular, while the second one is still seriously singular.

To proceed with the derivation, analytic expressions of F_1 and F_2 are required. Here we make use of the two-dimensional Cartesian coordinates

$$
X_0 = \psi \cos A, \qquad Y_0 = \psi \sin A. \qquad (8.10)
$$

These constitute a coordinate system $P_0 X_0 Y_0$ defined on a unit sphere, with the origin P_0 corresponding to P, X_0 points to the north, and Y_0 to the east. As P and P_0 are both defined by $\psi = 0$ and the values of F_1 and F_2 at P_0 are in fact those at P, we use interchangeably the position of P and P_0. We express F_1 and F_2 analytically as Taylor series in two variables truncated to second order

$$
F_i(X_0, Y_0) = F_{iP} + X_0 \left(\frac{\partial F_i}{\partial X_0} \right)_P + Y_0 \left(\frac{\partial F_i}{\partial Y_0} \right)_P
$$
$$
+ \frac{1}{2} \left[X_0^2 \left(\frac{\partial^2 F_i}{\partial X_0^2} \right)_P + 2 X_0 Y_0 \left(\frac{\partial^2 F_i}{\partial X_0 \partial Y_0} \right)_P + Y_0^2 \left(\frac{\partial^2 F_i}{\partial Y_0^2} \right)_P \right].
$$
(8.11)

Substitute (8.10) into it. We obtain

$$
F_i(\psi, A) = F_{iP} + \psi \left[\left(\frac{\partial F_i}{\partial X_0} \right)_P \cos A + \left(\frac{\partial F_i}{\partial Y_0} \right)_P \sin A \right]
$$
$$
+ \frac{\psi^2}{2} \left[\left(\frac{\partial^2 F_i}{\partial X_0^2} \right)_P \cos^2 A + 2 \left(\frac{\partial^2 F_i}{\partial X_0 \partial Y_0} \right)_P \cos A \sin A
$$
$$
+ \left(\frac{\partial^2 F_i}{\partial Y_0^2} \right)_P \sin^2 A \right].
$$
(8.12)

Even though we use this expression to represent F_1 and F_2, we do not mean to compute the derivatives based on definitions. This expression should be understood as an approximation to F_1 and F_2 in analytic form, where the derivatives should be understood as parameters to be determined by fitting to known values of F_1 and F_2 according to a criterion such as least squares.

With (8.12), we can readily evaluate I_1 to obtain

$$
I_1(\theta, \lambda) = 4\pi \Theta F_{1P} + \frac{\pi}{3} \Theta^3 \left[\left(\frac{\partial^2 F_1}{\partial X_0^2} \right)_P + \left(\frac{\partial^2 F_1}{\partial Y_0^2} \right)_P \right],
$$
(8.13)

where the following formulae have been used:

$$
\int_0^{2\pi} \cos A\, dA = \int_0^{2\pi} \sin A\, dA = \int_0^{2\pi} \cos A \sin A\, dA = 0,
$$
(8.14)

$$
\int_0^{2\pi} \cos^2 A\, dA = \int_0^{2\pi} \sin^2 A\, dA = \pi,
$$
(8.15)

$$
\int_0^{\Theta} \psi^2 d\psi = \frac{1}{3} \Theta^3.
$$
(8.16)

The evaluation of I_2 is not as evident, and we provide more detail. With (8.12), we can write I_2 as

$$
I_2(\theta, \lambda) = 8 \int_0^{2\pi} \int_0^{\Theta} \frac{F_{2P}}{\psi^2} d\psi \, dA
$$

$$
+ 8 \int_0^{2\pi} \int_0^{\Theta} \frac{1}{\psi} \left[\left(\frac{\partial F_i}{\partial X_0} \right)_P \cos A + \left(\frac{\partial F_i}{\partial Y_0} \right)_P \sin A \right] d\psi \, dA
$$

$$
+ 4 \int_0^{2\pi} \int_0^{\Theta} \left[\left(\frac{\partial^2 F_i}{\partial X_0^2} \right)_P \cos^2 A + 2 \left(\frac{\partial^2 F_i}{\partial X_0 \partial Y_0} \right)_P \cos A \sin A \right.
$$

$$
\left. + \left(\frac{\partial^2 F_i}{\partial Y_0^2} \right)_P \sin^2 A \right] d\psi \, dA. \tag{8.17}
$$

In order for the first term to be finite or integrable, we must have

$$
F_{2P} = 0, \tag{8.18}
$$

which is indeed the case in the formula of gravity reduction and remains true in the formula of $\Delta G'$ required for the computation of the height anomaly. For the second term, the integrals over A vanish, but the integral over ψ tends to infinitely large, or not integrable. Here we understand the integral in the following way:

$$
\int_0^{2\pi} \int_0^{\Theta} \frac{\cos A}{\psi} d\psi \, dA = \int_0^{\pi} \int_0^{\Theta} \frac{\cos A}{\psi} d\psi \, dA + \int_0^{\pi} \int_0^{\Theta} \frac{\cos(\pi + A)}{\psi} d\psi \, dA = 0,
$$

$$
\tag{8.19}
$$

where $\cos A + \cos(\pi + A) = 0$ has been used. The same is for the integral including $\sin A$. Hence, the second term in (8.17) vanishes. We obtain, taking into account of (8.14) and (8.15),

$$
I_2(\theta, \lambda) = 4\pi \Theta \left[\left(\frac{\partial^2 F_2}{\partial X_0^2} \right)_P + \left(\frac{\partial^2 F_2}{\partial Y_0^2} \right)_P \right]. \tag{8.20}
$$

We emphasize that this formula requires (8.18) to hold.

If a regular latitude–longitude grid is used to evaluate these integrals numerically, the singularity must also be handled appropriately.

We mention again that the formulae derived in this subsection are for evaluating the integrals in the corrections $\Delta_2 g$, $\Delta_3 g$, and $\Delta_4 g$ to compute Helmert or isostatic anomaly from gravity or to compute gravity from Helmert or isostatic anomaly by using the corrections reversely.

8.1.3   Interpolation of Point-Wise Gravity Anomaly

In this subsection, we focus on the interpolation of Helmert or isostatic gravity anomaly to regularly spaced grid points using the values at gravity measurement stations. The computation of the gravity anomaly at gravity measurement stations from the gravity measured can be done by adding the respective corrections, where the integrals in the formulae are supposed to be computed according to the method outlined in the previous subsection. The computation of gravity from gravity anomaly at the regularly spaced grid points can be done by applying the corrections reversely, as mentioned in the last paragraph of the previous subsection.

Even though the variation of the Helmert or isostatic gravity anomaly from place to place is smooth, there is still a need of dense enough known data for the interpolation to be meaningful. Unlike the topography of the Earth, for which interpolation can be done with known values of height at critical positions like top of hills and bottom of ditches, no reliable picture exists to serve as a visual reference in the case of gravity anomaly. As a general criterion, the value at any place between two stations must be obtainable based on linear interpolation with reasonable accuracy, and this can be verified only experimentally with data. If this criterion is not fulfilled, it can be hardly envisaged that any reliable algorithm of interpolation be designed. Hence, in this subsection, we always assume that data of gravity anomaly are dense enough for the result obtained by interpolation to be meaningful.

The difference between Δg and ΔG lies in the use of different heights in gravity reduction, but their interpolations are based on the same algorithm. Here we take Δg as an example.

The most primitive idea is to express the value of Δg at any point as a linear function of coordinates, here chosen to be the polar angle θ and longitude λ,

$$\Delta g = a + b\theta + c\lambda. \tag{8.21}$$

There are three coefficients to be determined. Hence three known values of Δg are required. This implies that linear interpolation is suitable for the interior of a triangle, where the three coefficients are determined using the values of Δg at the three vertices,

$$\Delta g_i = a + b\theta_i + c\lambda_i, \quad i = 1, 2, 3. \tag{8.22}$$

The solution in matrix form is

$$\begin{bmatrix} a \\ b \\ c \end{bmatrix} = \begin{bmatrix} 1 & \theta_1 & \lambda_1 \\ 1 & \theta_2 & \lambda_2 \\ 1 & \theta_3 & \lambda_3 \end{bmatrix}^{-1} \begin{bmatrix} \Delta g_1 \\ \Delta g_2 \\ \Delta g_3 \end{bmatrix}. \tag{8.23}$$

This linear interpolation is continuous across adjacent triangles, as shown in Fig. 8.2.

Fig. 8.2 Linear interpolation within triangles

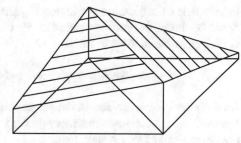

Fig. 8.3 Bi-linear interpolation within quadrilaterals

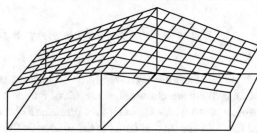

Inspired from linear interpolation within a triangle, if interpolation is to be made within a quadrilateral, at least four coefficients are required

$$\Delta g = a + b\theta + c\lambda + d\,\theta\lambda. \tag{8.24}$$

The last term is linear in both θ and λ. Hence, this expression is referred to as bi-linear. The four coefficients are determined using the values of Δg at the four vertices

$$\Delta g_i = a + b\theta_i + c\lambda_i + d\,\theta_i\lambda_i\,, \qquad i = 1, 2, 3, 4. \tag{8.25}$$

The solution in matrix form is

$$\begin{bmatrix} a \\ b \\ c \\ d \end{bmatrix} = \begin{bmatrix} 1 & \theta_1 & \lambda_1 & \theta_1\lambda_1 \\ 1 & \theta_2 & \lambda_2 & \theta_2\lambda_2 \\ 1 & \theta_3 & \lambda_3 & \theta_3\lambda_3 \\ 1 & \theta_4 & \lambda_4 & \theta_4\lambda_4 \end{bmatrix}^{-1} \begin{bmatrix} \Delta g_1 \\ \Delta g_2 \\ \Delta g_3 \\ \Delta g_4 \end{bmatrix}. \tag{8.26}$$

The bi-linear interpolation is also continuous across adjacent quadrilaterals, as shown in Fig. 8.3.

The above methods of piecewise interpolation ensure the result to be continuous. Methods ensuring the result to be smooth can also be devised. Quite comprehensive lists of various kinds of piecewise interpolation methods can be found in books of finite element method.

Besides interpolation, another way is to approximate Δg within a region of interest by a prescribed function, most conveniently, a polynomial, and use all the known data of Δg within the region to estimate the coefficients. The polynomial could be linear, bi-linear, or include even more terms. Here we take a full second degree polynomial as an example

$$\Delta g = a + b\theta + c\lambda + d\,\theta\lambda + e\theta^2 + f\lambda^2. \tag{8.27}$$

As there are six coefficients in the polynomial, at least six known values of Δg are required. Here we assume that more than six values are known, and the coefficients are determined based on least squares

$$\sum_i W_i \left[(a + b\theta_i + c\lambda_i + d\,\theta_i\lambda_i + e\theta_i^2 + f\lambda_i^2) - \Delta g_i \right]^2 = \min, \tag{8.28}$$

where W_i is a prescribed weight for Δg_i. We realize immediately that if the points of known Δg are clustered, the result of the least squares fitting may not reasonably represent the whole region if all weights are chosen to be equal. Hence, care should be taken to ensure that data are distributed in the region reasonably homogeneously even in the phase of measurement. The weights could be assigned as the areas that the data represent, larger for sparser data and smaller for denser data, to alleviate the bias of the result toward subregions of denser data. Finally, we mention that terms based on other functions, such as sine and cosine, can be included in (8.27) with a coefficient to be determined.

8.1.4   Computation of Grid Mean Gravity Anomaly

Customarily, gravity anomaly is not provided as point values, but as averages within the cells of a regular latitude–longitude grid, and is briefly referred to as grid mean gravity anomaly. Such a grid cell, or simply grid, with (θ_1, λ_1) and (θ_2, λ_2) as upper left and lower right corners, respectively, is shown in Fig. 8.4.

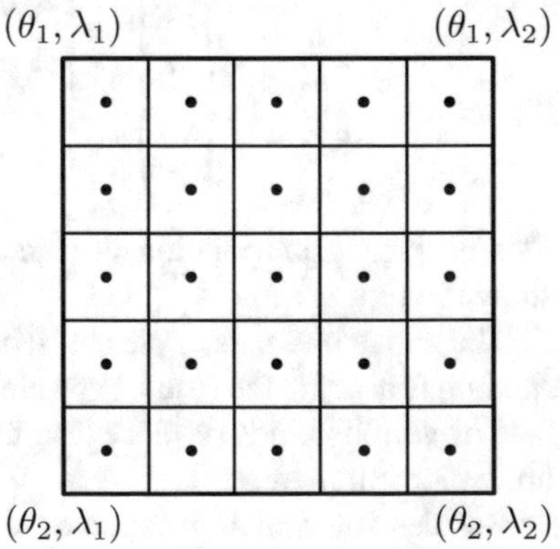

Fig. 8.4 Division of a latitude–longitude grid cell into sub-grid cells

A regular latitude–longitude grid is made by lines of constant θ and constant λ with constant intervals and is mostly referred using grid size, e.g., $5° \times 5°$, $1° \times 1°$, $0.5° \times 0.5°$. The gird intervals in latitude and longitude can be different, e.g., $0.5° \times 1°$.

By definition, the grid mean gravity anomaly as shown in Fig. 8.4 is

$$\Delta \bar{g} = \left(\int_{\lambda_1}^{\lambda_2} \int_{\theta_1}^{\theta_2} d\omega \right)^{-1} \int_{\lambda_1}^{\lambda_2} \int_{\theta_1}^{\theta_2} \Delta g d\omega , \tag{8.29}$$

with $d\omega = \sin \theta d\theta d\lambda$. This average can be computed using the analytic expression of Δg, such as (8.21) or (8.24) used for interpolation, or (8.27) obtained by least squares fitting. The other way is to divide the grid into sub-grids as in Fig. 8.4 and compute the average gravity anomaly according to

$$\Delta \bar{g} = \frac{1}{S} \sum_i \Delta g_i S_i , \tag{8.30}$$

where Δg_i is the gravity anomaly at the center of the ith sub-grid, S_i its area, and $S = \sum_i S_i$ the total area of the grid. In this formula, the gravity anomaly over each sub-grid is approximated by a constant.

8.2 Computation of the Global Model of the Gravitational Potential

8.2.1 Computation by Evaluating Integrals

The determination of the global gravitational potential is equivalent to the determination of the global disturbing potential. Here we demonstrate how the disturbing potential could be computed using data of gravity anomaly. The basic formulae are (5.27), (5.29), (5.30), and (5.31), which we rewrite here using the fully normalized Legendre function defined in (3.313). The spherical harmonic series of the disturbing potential and gravity anomaly truncated up to degree and order $n_{\max}$ are

$$T = \frac{GM}{r} \sum_{n=2}^{n_{\max}} \sum_{k=0}^{n} \left(\frac{R}{r} \right)^n (\bar{A}_n^k \cos k\lambda + \bar{B}_n^k \sin k\lambda) \bar{P}_n^k (\cos \theta) , \tag{8.31}$$

$$\Delta g = \sum_{n=2}^{n_{\max}} \sum_{k=0}^{n} (\bar{c}_n^k \cos k\lambda + \bar{s}_n^k \sin k\lambda) \bar{P}_n^k (\cos \theta) . \tag{8.32}$$

The relations between the coefficients are

$$\bar{A}_n^k = \frac{R^2}{GM} \frac{\bar{c}_n^k}{n-1} , \qquad \bar{B}_n^k = \frac{R^2}{GM} \frac{\bar{s}_n^k}{n-1} . \tag{8.33}$$

We see that in order to determine $\bar{A}_n^k$ and $\bar{B}_n^k$, we have only to determine $\bar{c}_n^k$ and $\bar{s}_n^k$, and the formulae are

$$\begin{cases} \bar{c}_n^k = \dfrac{1}{4\pi} \displaystyle\int_\omega \Delta g(\theta', \lambda') \bar{P}_n^k(\cos\theta') \cos k\lambda' d\omega , \\[3mm] \bar{s}_n^k = \dfrac{1}{4\pi} \displaystyle\int_\omega \Delta g(\theta', \lambda') \bar{P}_n^k(\cos\theta') \sin k\lambda' d\omega . \end{cases} \tag{8.34}$$

In physical geodesy, the spherical harmonic coefficients provided are always fully normalized.

In order to compute $\bar{c}_n^k$ and $\bar{s}_n^k$ according to (8.34), we first divide the sphere into small grids denoted as ω_i. The integral over the sphere is the sum of integrals over all the grids

$$\begin{cases} \bar{c}_n^k = \displaystyle\sum_i \dfrac{1}{4\pi} \int_{\omega_i} \Delta g(\theta', \lambda') \bar{P}_n^k(\cos\theta') \cos k\lambda' d\omega_i , \\[3mm] \bar{s}_n^k = \displaystyle\sum_i \dfrac{1}{4\pi} \int_{\omega_i} \Delta g(\theta', \lambda') \bar{P}_n^k(\cos\theta') \sin k\lambda' d\omega . \end{cases} \tag{8.35}$$

We assume that the grid ω_i is formed by lines of constant θ and λ as shown in Fig. 8.4 with upper left and lower right vertices denoted as $(\theta_1^{(i)}, \lambda_1^{(i)})$ and $(\theta_2^{(i)}, \lambda_2^{(i)})$, respectively, and the value of Δg within the grid is given as an average denoted as $\Delta\bar{g}_i$. By approximating Δg by $\Delta\bar{g}_i$ in ω_i and expressing $d\omega$ as $\sin\theta' d\theta' d\lambda'$, we can write (8.35) as

$$\begin{cases} \bar{c}_n^k = \displaystyle\sum_i \dfrac{\Delta\bar{g}_i}{4\pi} \int_{\lambda_1^{(i)}}^{\lambda_2^{(i)}} d\lambda' \int_{\theta_1^{(i)}}^{\theta_2^{(i)}} \bar{P}_n^k(\cos\theta') \cos k\lambda' \sin\theta' d\theta' , \\[3mm] \bar{s}_n^k = \displaystyle\sum_i \dfrac{\Delta\bar{g}_i}{4\pi} \int_{\lambda_1^{(i)}}^{\lambda_2^{(i)}} d\lambda' \int_{\theta_1^{(i)}}^{\theta_2^{(i)}} \bar{P}_n^k(\cos\theta') \sin k\lambda' \sin\theta' d\theta' , \end{cases} \tag{8.36}$$

where $\Delta\bar{g}_i$ is moved out of the integral sign, as it is a constant in ω_i. After evaluating the integrals over λ' analytically, we obtain

$$\begin{cases} \bar{c}_n^k = \displaystyle\sum_i \dfrac{\Delta\bar{g}_i}{4\pi} \left[\lambda_2^{(i)} - \lambda_1^{(i)} \right] \bar{I}_n^{(i)} , & k = 0 ; \\[3mm] \bar{c}_n^k = \displaystyle\sum_i \dfrac{\Delta\bar{g}_i}{4\pi k} \left[\sin k\lambda_2^{(i)} - \sin k\lambda_1^{(i)} \right] \bar{I}_{nk}^{(i)} , & k \neq 0 ; \\[3mm] \bar{s}_n^k = \displaystyle\sum_i \dfrac{\Delta\bar{g}_i}{4\pi k} \left[-\cos k\lambda_2^{(i)} + \cos k\lambda_1^{(i)} \right] \bar{I}_{nk}^{(i)} , & k \neq 0 . \end{cases} \tag{8.37}$$

The quantity $\bar{I}_{nk}^{(i)}$ is defined as

$$\bar{I}_{nk}^{(i)} = \int_{\theta_1^{(i)}}^{\theta_2^{(i)}} \bar{P}_n^k(\cos\theta') \sin\theta' d\theta' , \qquad (8.38)$$

and $\bar{I}_n^{(i)}$ is its degenerate case with $k = 0$.

In the following, we first discuss (8.37) and the coefficients $\bar{A}_n^k$ and $\bar{B}_n^k$ supposed to be computed according to (8.33) and leave the formulation for computing $\bar{I}_{nk}^{(i)}$ to the next subsection.

In deriving (8.37), we have only assumed that the grids are formed by lines of constant θ and constant λ. Now we further assume that the intervals between the adjacent lines of constant θ and constant λ are identical, and we let them to be $\Delta\theta = \theta_2^{(i)} - \theta_1^{(i)} = 180°/M$ and $\Delta\lambda = \lambda_2^{(i)} - \lambda_1^{(i)} = 180°/M$. We immediately realize that the maximum degree and order of truncation attainable, $n_{\max}$, must be related to M; the larger M is, the larger $n_{\max}$ could be. The first thing we see is that the grid sizes are not equal if measured using metric length on the Earth's surface. Every grid is a slightly trapezoidal rectangle. Both the lengths of the sides along the meridian are $R\Delta\theta$. The lengths of the sides along the parallel are $R\sin\theta_1\Delta\lambda$ and $R\sin\theta_2\Delta\lambda$, which are slightly different. A grid at the equator is practically a square, but a grid when $\theta = 30°$ is practically approximately a rectangle with the side along the parallel to be about half of that along the meridian. This implies that data of Δg are denser at higher latitude regions. Therefore, the attainable $n_{\max}$ should be what the data density in the equatorial region permits, where the grids are square, which is the reason to choose $\Delta\theta = \Delta\lambda$.

We know that according to (3.371), at the equator, the spherical harmonic series (8.31) or (8.32) degenerates to a Fourier series up to order $n_{\max}$. We take (8.32) as an example, with the subscript E indicating the equator,

$$\Delta g_E = \sum_{k=0}^{n_{\max}} (a_k \cos k\lambda + b_k \sin k\lambda) . \qquad (8.39)$$

This means that the resolution is, in maximum, $n_{\max}$ cycles of sine curve from $0°$ to $360°$. The wavelength is then $360°/n_{\max}$ and half wavelength $180°/n_{\max}$. It can be readily realized that a minimum of two values are required to determine one cycle of the sine curve, i.e., one value is required every half cycle. This implies that at least $2n_{\max}$ values of Δg_E are required in the equator to determine the coefficients in (8.39).

By generalizing the inference in the previous paragraph to the case over a sphere, we have $n_{\max} = M$. This means that in order to determine the spherical harmonic coefficients up to degree and order $n_{\max}$, data over a regular latitude–longitude grid of at least $180°/n_{\max} \times 180°/n_{\max}$ are required, which is logical, as the resolution of the spherical harmonic series over any great arc is identical to that over the equator, as discussed in Sect. 5.5.2. This was in fact an established practice in geodesy, and $180°/n_{max} \times 180°/n_{max}$ is referred to as the resolution, which reflects the half wavelength of the smallest sine wave attainable. We must realize the fact that data of Δg are denser at higher latitude. The number of data over a $180°/n_{max} \times 180°/n_{max}$ grid is $2n_{max}^2$, while the total number (including degrees 0 and 1) of spherical harmonic coefficients up to degree and order n_{max} is $n_{max}(n_{max} + 1)$, meaning that the number of grid data is about twice as many as that of the coefficients. In fact, the data at higher latitude include information beyond the resolution attainable by the spherical harmonic series, as the grid sizes there are smaller than those over the equator, and this information is eliminated by truncation up to degree and order n_{max}. Hence, $180°/n_{max}$ should not be misunderstood as a resolution of the spherical harmonic series in longitude, as the grid may suggest, but it is the resolution along any great arc. An alternative way to express resolution is to use length on the Earth's surface, e.g., $(\pi/n_{max})R \times (\pi/n_{max})R$, which avoids the possible misunderstanding mentioned.

Our problem is to compute the global model of the gravitational potential. The above discussions are on the mathematical aspects of the formulation and evaluation of the spherical harmonic coefficients. In the next paragraph, we shortly analyze the physical aspects of the problem: gravity reduction and indirect effect.

The formulation is applicable to both Stokes' and Molodensky's boundary value theories. For the former, a gravity reduction should have been performed, such that the mass above the geoid is removed in the computation of gravity anomaly. For the latter, in addition to gravity reduction that removes the mass above the quasigeoid, the gravity anomaly should have been projected or downward continued to the quasigeoid. Based on the formulations and discussions in the previous chapters, the approach in this subsection should be conceptually understood as for the adjusted Earth. The indirect effect should then be added back to the result. For global models of the gravitational potential, the indirect effect was formulated in Sect. 6.5.2 for Stokes' theory, which can be applied to Molodensky's theory by replacing the orthometric height by the normal height. We mention that this section concerns only the computational approach of the spherical harmonic coefficients. The physical meaning and applicability of the formulae as discussed in the previous chapters must be taken into account.

The readers are encouraged to derive the formulae for the GNSS-gravimetric boundary value problem, which is straightforward.

8.2.2 Computation of Integrals of the Associated Legendre Function

In this subsection, we provide the formulae to compute the integral defined in (8.38). We only derived the un-normalized version of the formulae, from which the normalized version can be easily inferred. The computation of the associated Legendre function as well as its derivative with respect to θ has been discussed in Sect. 3.3.4.

We first consider the un-normalized version of the integral defined in (8.38). For brevity, we omit the superscript "(i)," replace θ' by θ, and perform the variable substitution $x = \cos\theta$. These lead to

$$I_{nk} = \int_{\theta_1}^{\theta_2} P_n^k(\cos\theta)\sin\theta \, d\theta = \int_{\cos\theta_2}^{\cos\theta_1} P_n^k(x)dx \, . \tag{8.40}$$

Several recurrence formulae have been derived for it based on the recurrence formulae of $P_n^k(x)$. Here we provide one that is proven to be numerically stable in the computation.

We start from the following formula that can be readily obtained according the derivative rule of a product

$$\frac{d}{dx}\left[-(2n+1)(1-x^2)P_n^k(x)\right] = (2n+1)\left[2x P_n^k(x) - (1-x^2)\frac{d}{dx}P_n^k(x)\right] . \tag{8.41}$$

Further derivation will be based on the recurrence formulae (3.144) and (3.149) of $P_n^k(x)$, which are copied to here for convenience of reference,

$$(2n+1)x P_n^k(x) = (n+k)P_{n-1}^k(x) + (n-k+1)P_{n+1}^k(x) \, . \tag{8.42}$$

$$(1-x^2)\frac{d}{dx}P_n^k(x) = (n+k)P_{n-1}^k(x) - nx P_n^k(x) \, . \tag{8.43}$$

The successive substitution of (8.43) and (8.42) into (8.41) yields

$$\frac{d}{dx}\left[-(2n+1)(1-x^2)P_n^k(x)\right] = (2n+1)\left[(n+2)x P_n^k(x) - (n+k)P_{n-1}^k(x)\right]$$

$$= (n+2)\left[(2n+1)x P_n^k(x)\right] - (2n+1)(n+k)P_{n-1}^k(x)$$

$$= (n+2)(n-k+1)P_{n+1}^k(x) - (n-1)(n+k)P_{n-1}^k(x) \, . \tag{8.44}$$

By integrating both sides of this formula over x from $\cos\theta_2$ to $\cos\theta_1$, we obtain

$$(2n+1)\left[(1-x^2)P_n^k(x)\right]_{\cos\theta_1}^{\cos\theta_2} = (n+2)(n-k+1)I_{(n+1)k} - (n-1)(n+k)I_{(n-1)k} \, , \tag{8.45}$$

which is the recurrence formula we need. Its practical form for computation is

$$I_{(n+1)k} = \frac{(n-1)(n+k)}{(n+2)(n-k+1)}I_{(n-1)k} + \frac{2n+1}{(n+2)(n-k+1)}\left[(1-x^2)P_n^k(x)\right]_{\cos\theta_1}^{\cos\theta_2} \, . \tag{8.46}$$

Evidently, this formula requires $P_n^k(x)$ to be computed beforehand.

The initial values required by (8.46) can be derived from (3.164), which is written here in terms of θ,

$$P_k^k(\cos\theta) = \frac{(2k)!}{2^k k!} \sin^k\theta \,. \tag{8.47}$$

As $P_0(\cos\theta) = 1$, it is easy to infer

$$I_0 = \int_{\theta_1}^{\theta_2} \sin\theta d\theta = -(\cos\theta)_{\theta_1}^{\theta_2} \,. \tag{8.48}$$

As $P_1^1(\cos\theta) = \sin\theta$, we have

$$I_{1,1} = \int_{\theta_1}^{\theta_2} \sin^2\theta d\theta = \int_{\theta_1}^{\theta_2}\left(\frac{1}{2} - \frac{1}{2}\cos 2\theta\right)d\theta = \left(\frac{1}{2}\theta - \frac{1}{4}\sin 2\theta\right)_{\theta_1}^{\theta_2}$$
$$= \frac{1}{2}(\theta - \sin\theta\cos\theta)_{\theta_1}^{\theta_2} \,. \tag{8.49}$$

Before evaluating I_{kk} for $k > 1$, we derive an auxiliary formula. By integrating by parts, we obtain

$$\int \sin^{k+1}\theta d\theta = -\int \sin^k\theta d\cos\theta = -\cos\theta\sin^k\theta + k\int \sin^{k-1}\theta\cos^2\theta d\theta \,. \tag{8.50}$$

After substituting $\cos^2\theta = 1 - \sin^2\theta$ into it, we can solve the equation to obtain

$$\int \sin^{k+1}\theta d\theta = \frac{1}{k+1}\left[k\int \sin^{k-1}\theta d\theta - \cos\theta\sin^k\theta\right] \,. \tag{8.51}$$

A recurrence formula for I_{kk} can then be derived as

$$I_{kk} = \frac{(2k)!}{2^k k!}\int_{\theta_1}^{\theta_2}\sin^{k+1}\theta d\theta$$
$$= \frac{1}{k+1}\left\{k\frac{(2k-3)(2k-2)(2k-1)(2k)}{4(k-1)k}\int_{\theta_1}^{\theta_2}\frac{[2(k-2)]!}{2^{k-2}(k-2)!}\sin^{k-1}\theta d\theta \right.$$
$$\left. -\left[\cos\theta\frac{(2k)!}{2^k k!}\sin^k\theta\right]_{\theta_1}^{\theta_2}\right\}$$
$$= \frac{1}{k+1}\left\{k(2k-3)(2k-1)I_{(k-2)(k-2)} - \left[\cos\theta P_k^k(\cos\theta)\right]_{\theta_1}^{\theta_2}\right\}$$
$$= \frac{1}{k+1}\left\{k(2k-3)(2k-1)I_{(k-2)(k-2)} - \left[x P_k^k(x)\right]_{\cos\theta_1}^{\cos\theta_2}\right\} \,. \tag{8.52}$$

For any k, I_{kk} can be computed according to (8.48), (8.49), and (8.52), and $I_{(k+1)k}, I_{(k+2)k}, \cdots$ can be computed according to (8.46) with $I_{(k-1)k} = 0$ taken into account. Conversion of the formulae to the fully normalized associated Legendre function is straightforward. Following the procedure of computation, (8.48), (8.49), (8.52), and (8.46) become

$$\bar{I}_0 = -(\cos\theta)_{\theta_1}^{\theta_2} , \qquad (8.53)$$

$$\bar{I}_{1,1} = \frac{\sqrt{3}}{2} (\theta - \sin\theta\cos\theta)_{\theta_1}^{\theta_2} , \qquad (8.54)$$

$$\bar{I}_{kk} = \frac{1}{k+1} \left\{ \frac{1}{2} \left[\frac{(1+\delta_{k-2})k(2k+1)(2k-1)}{k-1} \right]^{1/2} \bar{I}_{(k-2)(k-2)} - \left[x\bar{P}_k^k(x) \right]_{\cos\theta_1}^{\cos\theta_2} \right\} , \qquad (8.55)$$

$$\bar{I}_{(n+1)k} = \frac{n-1}{n+2} \left[\frac{(2n+3)(n+k)(n-k)}{(2n-1)(n+k+1)(n-k+1)} \right]^{1/2} \bar{I}_{(n-1)k}$$
$$+ \frac{1}{n+2} \left[\frac{(2n+3)(2n+1)}{(n+k+1)(n-k+1)} \right]^{1/2} \left[(1-x^2)\bar{P}_n^k(x) \right]_{\cos\theta_1}^{\cos\theta_2} , \qquad (8.56)$$

respectively.

8.2.3 Pellinen and Gaussian Smoothing

We have used grid mean gravity anomaly in the formulation for the determination of global models of the gravitational potential. In this subsection, we discuss how the grid means are related to the point values as computed using a spherical harmonic series and then discuss a possible way to alleviate the effect of using grid means in the result obtained.

The grid mean is defined as a kind of average over the grid cell. The term *average* may be instinctually understood as the simple arithmetic mean. Hence, in this subsection, we use the appearingly more general term *smoothing*, which is also widely used in the literature, to indicate a weighted averaging of the values in the vicinity, leading to a smoother result. Evidently, the grid mean of gravity anomaly is of this category.

In the literature, Pellinen smoothing was approximately used to de-smooth the grid mean gravity anomaly by introducing a parameter known as de-smoothing factor. However, as Gaussian smoothing is currently widely used in the processing of satellite determined gravitational models for studying mass redistribution within the Earth system, we also include it here, as it can be easily and shortly formulated incidentally. We will also mention another one, the ideal filter, which eliminates all terms beyond a prescribed degree in the spherical harmonic series.

General Formulation of Isotropic Smoothing

We start from the general definition of a weighted average of a function defined on a sphere. For convenience, we use θ' and λ' to denote the coordinates for the un-smoothed function, $f(\theta', \lambda')$ and use θ and λ to denote the coordinates for the smoothed function, $\bar{f}(\theta, \lambda)$. We define the weighted average as

$$\bar{f}(\theta, \lambda) = \int_{\omega_{\psi \leq \Theta}} \frac{W(\psi)}{W_s} f(\theta', \lambda') d\omega, \tag{8.57}$$

where ψ is the geocentric angle between the two points (θ, λ) and (θ', λ'), $W(\psi)$ the weight supposed to be prescribed, and W_s the sum of the weight defined by

$$W_s = \int_{\omega_{\psi \leq \Theta}} W(\psi) d\omega. \tag{8.58}$$

We assumed the weight to be a function of ψ instead of a more general form as function of θ, λ and θ', λ'. This is the simplest form that conforms with the most direct intuition that the weight should decrease with the distance to the point (θ, λ) and is still general enough for the content of this book. As this kind of weight behaves the same in all directions as seen from the point (θ, λ), the smoothing is called isotropic. Defining the averaging region to be $\psi \leq \Theta$ is a consequence of the simple form of the weight. What we are going to formulate is the relation between the spherical harmonic series of $f(\theta', \lambda')$ and $\bar{f}(\theta, \lambda)$.

We use the un-normalized associated Legendre function to express $f(\theta', \lambda')$ as

$$f(\theta', \lambda') = \sum_{n=0}^{\infty} \sum_{k=0}^{n} (a_n^k \cos k\lambda' + b_n^k \sin k\lambda') P_n^k(\cos \theta'). \tag{8.59}$$

The spherical harmonic series of $\bar{f}(\theta, \lambda)$ can be obtained by substituting (8.59) into (8.57)

$$\bar{f}(\theta, \lambda) = \sum_{n=0}^{\infty} \sum_{k=0}^{n} \left[a_n^k \int_{\omega_{\psi \leq \Theta}} \frac{W(\psi)}{W_s} P_n^k(\cos \theta') \cos k\lambda' d\omega \right.$$
$$\left. + b_n^k \int_{\omega_{\psi \leq \Theta}} \frac{W(\psi)}{W_s} P_n^k(\cos \theta') \sin k\lambda' d\omega \right]. \tag{8.60}$$

By expressing $d\omega = \sin\psi\, d\psi\, dA$, where A is the azimuth of the point (θ', λ') as observed at the point (θ, λ), we can express (8.60) as

$$\bar{f}(\theta, \lambda) = \sum_{n=0}^{\infty} \sum_{k=0}^{n} \left\{ a_n^k \int_0^\Theta \left[\int_0^{2\pi} P_n^k(\cos\theta') \cos k\lambda'\, dA \right] \frac{W(\psi)}{W_s} \sin\psi\, d\psi \right.$$
$$\left. + b_n^k \int_0^\Theta \left[\int_0^{2\pi} P_n^k(\cos\theta') \sin k\lambda'\, dA \right] \frac{W(\psi)}{W_s} \sin\psi\, d\psi \right\}. \qquad (8.61)$$

The integrals in the brackets can be evaluated according to the formulae of the azimuthal averages of the surface spherical harmonics (3.272). We have

$$\bar{f}(\theta, \lambda) = \sum_{n=0}^{\infty} \sum_{k=0}^{n} \left\{ a_n^k \int_0^\Theta \left[2\pi\, P_n(\cos\psi) P_n^k(\cos\theta) \cos k\lambda \right] \frac{W(\psi)}{W_s} \sin\psi\, d\psi \right.$$
$$\left. + b_n^k \int_0^\Theta \left[2\pi\, P_n(\cos\psi) P_n^k(\cos\theta) \sin k\lambda \right] \frac{W(\psi)}{W_s} \sin\psi\, d\psi \right\}, \qquad (8.62)$$

which can be written in the simple form

$$\bar{f}(\theta, \lambda) = \sum_{n=0}^{\infty} \sum_{k=0}^{n} \beta_n \left(a_n^k \cos k\lambda + b_n^k \sin k\lambda \right) P_n^k(\cos\theta), \qquad (8.63)$$

with β_n defined as

$$\beta_n = 2\pi \int_0^\Theta \frac{W(\psi)}{W_s} P_n(\cos\psi) \sin\psi\, d\psi. \qquad (8.64)$$

Hence, any smoothing in the form defined in (8.57) requires only to multiply the spherical harmonic coefficients with a degree only dependent parameter. We realize that in (8.63), this parameter remains the same even if a normalized associated Legendre function is used.

The formulae (8.63) and (8.64) are the starting formulae for both Pellinen and Gaussian smoothing, which are defined by choosing different $W(\psi)$.

For any weight $W(\psi)$ designed to make the variation of the smoothed function smoother than the original one, β_n decreases with n. Hence, the smoothing is equivalent to a filtering that reduces the contribution of higher degree spherical harmonics. In a more general language, a smoothing in the space domain (the function is considered as depending on spatial position) is a low pass filtering in the spectral domain (the function is considered as depending on the coefficients of a series expansion, such as the Fourier series, with a higher order term varying with shorter wavelength than a lower order term). Therefore, smoothing is also referred to as filtering in the literature, and thus, the terms *Pellinen* and *Gaussian filtering* are also used.

A formula for computing $W(\psi)/W_s$ from β_n can also be derived, which is the inverse of (8.64). Even if $W(\psi)$ is defined within $\psi \leq \Theta$, we can always extend its domain to $0 \leq \psi \leq \pi$ with $W(\psi) = 0$ when $\psi > \Theta$ and write (8.64) as, with the variable substitution $x = \cos\psi$,

$$\beta_n = 2\pi \int_{-1}^{1} \frac{W(x)}{W_s} P_n(x) dx. \tag{8.65}$$

Taking reference to the Legendre series expansion (3.368) and (3.369), we obtain a formula to compute $W(x)/W_s$ from β_n

$$\frac{W(x)}{W_s} = \sum_{n=0}^{\infty} \left[\frac{2n+1}{2} \int_{-1}^{1} \frac{W(x)}{W_s} P_n(x) dx \right] P_n(x) = \sum_{n=0}^{\infty} \frac{2n+1}{4\pi} \beta_n P_n(x). \tag{8.66}$$

This formula permits to prescribe β_n and then compute $W(x)$, or $W(\psi)$ after the variable substitution $x = \cos\psi$, which is the basis of the ideal filter.

Many smoothing methods can be designed by prescribing $W(\psi)$ or β_n. In the space domain, the computation is supposed to be done according to (8.57) and in the spectral domain (8.63). Pellinen and Gaussian smoothing are designed by prescribing $W(\psi)$, and the formulae of the parameter β_n are to be derived in the rest of this subsection. The ideal filter is designed by prescribing β_n, and the computation of $W(\psi)$ is straightforward using (8.66).

Pellinen Smoothing

In Pellinen smoothing, $W(\psi) = 1$ and $\Theta = \psi_0$, i.e., the smoothing is an averaging within a spherical cap. We have

$$W_s = \int_0^{2\pi} \int_0^{\psi_0} \sin\psi\, d\psi\, dA = 2\pi(1 - \cos\psi_0), \tag{8.67}$$

and furthermore,

$$\beta_n = \frac{\int_0^{\psi_0} P_n(\cos\psi) \sin\psi\, d\psi}{1 - \cos\psi_0} = \frac{\int_{\cos\psi_0}^{1} P_n(x) dx}{1 - \cos\psi_0}, \tag{8.68}$$

where a variable substitution $x = \cos\psi$ is made. Substitute $P_n(x)$ according to the recurrence formula (3.97). We obtain

$$\beta_n = \frac{\int_{\cos\psi_0}^{1} \left[\frac{d}{dx} P_{n+1}(x) - \frac{d}{dx} P_{n-1}(x) \right] dx}{(2n+1)(1-\cos\psi_0)} = \frac{P_{n-1}(\cos\psi_0) - P_{n+1}(\cos\psi_0)}{(2n+1)(1-\cos\psi_0)}, \tag{8.69}$$

Fig. 8.5 The factor β_n of Pellinen smoothing for $\psi_0 = 0.5°$, $1°$, $2°$, and $4°$

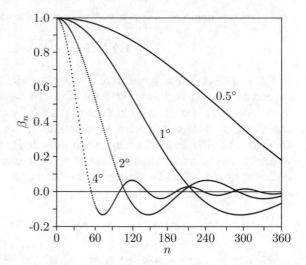

where the fact $P(1) = 1$ is used. This formula permits to compute β_n for $n > 0$ using values of $P_n(\cos \psi_0)$ computed using a recurrence formula. The special case $\beta_0 = 1$ can be obtained from (8.68) with $P_0(x) = 1$. As ψ_0 is usually small, $P_{n-1}(\cos \psi_0)$ and $P_{n+1}(\cos \psi_0)$ are very close, and several significant digits may be lost in the difference $P_{n-1}(\cos \psi_0) - P_{n+1}(\cos \psi_0)$, although the denominator could be computed without losing significant digits with the substitution $1 - \cos \psi_0 = 2\sin^2(\psi_0/2)$. Hence, alternative formulae have been derived for practical computation (Sjöberg, 1980). However, using quadruple precision without modifying the formula would provide enough significant digits, which is not a burden for any contemporary computer.

Graphs of β_n as a function of n for various values of ψ_0 are shown in Fig. 8.5.

Gaussian Smoothing

In Gaussian smoothing, the weight is a generalization of the Gaussian function, or simply Gaussian, which is sometimes referred to as bell curve according to its shape. The Gaussian as a function of x with the bell centered at $x = 0$ is defined as

$$g(x) = \frac{1}{\sqrt{2\pi}\sigma}e^{-x^2/(2\sigma^2)}, \tag{8.70}$$

which is the probability density function of a Gaussian or normal distribution satisfying $\int_{-\infty}^{\infty} g(x)dx = 1$. In the language of statistics, σ is the standard deviation defining the width of the bell, such as $|x| \leq \sigma$ or $|x| \leq 2\sigma$ or $|x| \leq 3\sigma$, with $\int_{-\sigma}^{\sigma} g(x)dx = 68.27\%$, $\int_{-2\sigma}^{2\sigma} g(x)dx = 95.45\%$, and $\int_{-\sigma}^{\sigma} g(x)dx = 99.73\%$. A width particularly used in geodesy is the half width at half maximum (HWHM), $|x_{1/2}|$, which is defined such that $g(x_{1/2}) = g(0)/2$, i.e., the value of $g(x)$ at $x_{1/2}$ is half of its maximum $g(0)$. It can be readily obtained that

$$\text{HWHM} = \sqrt{2 \ln 2}\, \sigma . \tag{8.71}$$

On a sphere, if the probability density function of the normal distribution depending on ψ is written in the form of (8.70), we can see that it is valid when σ is small, with the probability density practically dropping to zero when ψ is small enough. This is equivalent to approximate the sphere by a plane. Here we adopt a probability density function specifically designed for data on a sphere, called Fisher or Langevin distribution, whose degenerate case depending only on ψ alone is $Ce^{\kappa \cos \psi}$, where C and κ are constants (Fisher et al., 1987). For our purpose here, we write the weight function as

$$W(\psi) = e^{-(1-\cos \psi)/\sigma^2} = e^{-2\sin^2(\psi/2)/\sigma^2} , \tag{8.72}$$

which can be easily converted to the form $Ce^{\kappa \cos \psi}$. When ψ is very small, $\sin(\psi/2) = \psi/2$, (8.72) is simplified to

$$W(\psi) = e^{-\psi^2/(2\sigma^2)} , \tag{8.73}$$

which conforms with the probability density function of the normal distribution (8.70). Hence, $Ce^{\kappa \cos \psi}$ can indeed be understood as a simple generalization of the probability density function of the normal distribution onto a sphere. In this case, we set $\Theta = \pi$, and we have

$$W_s = \int_0^{2\pi} \int_0^{\pi} e^{-(1-\cos \psi)/\sigma^2} \sin \psi \, d\psi \, dA = 2\pi \sigma^2 \left(1 - e^{-2/\sigma^2}\right) . \tag{8.74}$$

Hence,

$$
\begin{aligned}
\beta_n &= \frac{1}{\sigma^2 \left(1 - e^{-2/\sigma^2}\right)} \int_0^{\pi} e^{-(1-\cos \psi)/\sigma^2} P_n(\cos \psi) \sin \psi \, d\psi \\
&= \frac{1}{\sigma^2 \left(1 - e^{-2/\sigma^2}\right)} \int_{-1}^{1} e^{-(1-x)/\sigma^2} P_n(x) dx ,
\end{aligned}
\tag{8.75}
$$

where the variable substitution $x = \cos \psi$ has been made to obtain the last expression. Analytically evaluating the integral in (8.75) for any n is not really necessary, as a recurrence formula of β_n can be derived using the recurrence formula of $P_n(x)$, (3.97), here written as

$$\frac{d}{dx} P_n(x) = (2n - 1) P_{n-1}(x) + \frac{d}{dx} P_{n-2}(x) . \tag{8.76}$$

Fig. 8.6 The factor β_n of Gaussian smoothing for $r_s = 0.5°, 1°, 2°$, and $4°$

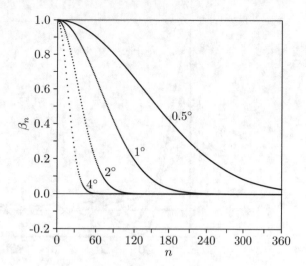

By integrating (8.75) by part and then using (8.76), we obtain

$$\beta_n = \beta_{n-2} - (2n - 1)\sigma^2 \beta_{n-1}. \tag{8.77}$$

The starting valucs for the recurrence, β_0 and β_1, can be obtained by analytically evaluating (8.75)

$$\beta_0 = 1, \qquad \beta_1 = \frac{1 + e^{-2/\sigma^2}}{1 - e^{-2/\sigma^2}} - \sigma^2. \tag{8.78}$$

Integration by part has been required to obtain β_1. We refer (8.77) and (8.78) as Jekeli's formulae to credit the author who first derived them (Jekeli, 1981).

When Gaussian smoothing is applied in physical geodesy, the HWHM is referred to as smoothing radius, which is, according to (8.71),

$$r_s = \sqrt{2 \ln 2}\,\sigma. \tag{8.79}$$

This is the measure of the smoothing radius in angle. The radius of the Earth, R, should be multiplied in order to obtain the smoothing radius in metric length.

Graphs of β_n as a function of n for various values of r_s are shown in Fig. 8.6.

The Ideal Filter
The ideal filter is designed by setting $\beta_n = 1$ when $n <= n_{\max}$ and $\beta_n = 0$ when $n > n_{\max}$. We have, according to (8.66) with $x = \cos\psi$,

$$\frac{W(\psi)}{W_s} = \sum_{n=0}^{n_{\max}} \frac{2n + 1}{4\pi} P_n(\cos\psi). \tag{8.80}$$

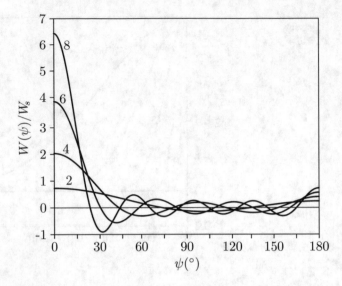

Fig. 8.7 The weight $W(\psi)/W_s$ of the ideal filter for $n_{\max} = 2, 4, 6$, and 8

Graphs of $W(\psi)/W_s$ for various values of $n_{\max}$ are shown in Fig. 8.7.

By definition, the weight $W(\psi)/W_s$ has the property

$$\int_\omega \frac{W(\psi)}{W_s} d\omega = 2\pi \int_0^\pi \frac{W(\psi)}{W_s} \sin \psi d\psi = 1. \tag{8.81}$$

The limit $n_{\max} \to \infty$ corresponds to the unfiltered state. Hence,

$$\lim_{n_{\max} \to \infty} \frac{W(\psi)}{W_s} = \begin{cases} \infty, & \psi = 0; \\ 0, & \psi \neq 0. \end{cases} \tag{8.82}$$

When $n_{\max}$ is not very small, $W(\psi)/W_s$ at $\psi = 0$ would be too large for graphing. Hence, very small values of $n_{\max}$ are chosen in Fig. 8.7.

8.2.4 De-smoothing of the Grid Mean Gravity Anomaly

We look to alleviate the effect of using the grid mean gravity anomaly. We use c_n^k and s_n^k to denote the un-normalized spherical harmonic coefficients of Δg. The average gravity anomaly $\Delta \bar{g}_i$ is computed by averaging according to the definition in (8.29) over the grid domain defined by $\theta_1^{(i)} \leq \theta \leq \theta_2^{(i)}$ and $\lambda_1^{(i)} \leq \lambda \leq \lambda_2^{(i)}$, which is denoted as ω_i. Hence, we have

$$
\Delta \bar{g}_i = \sum_{n=2}^{n_{\max}} \sum_{k=0}^{n} \left[c_n^k \int_{\omega_i} \frac{1}{W_s^{(i)}} P_n^k(\cos \theta') \cos k\lambda' d\omega + s_n^k \int_{\omega_i} \frac{1}{W_s^{(i)}} P_n^k(\cos \theta') \sin k\lambda' d\omega \right],
$$

(8.83)

where

$$
W_s^{(i)} = \int_{\lambda_1^{(i)}}^{\lambda_2^{(i)}} \int_{\theta_1^{(i)}}^{\theta_2^{(i)}} d\omega = \left[\lambda_2^{(i)} - \lambda_1^{(i)} \right] \left[\cos \theta_1^{(i)} - \cos \theta_2^{(i)} \right].
$$

(8.84)

This is an arithmetic mean as in the case of Pellinen smoothing, but the smoothing domain is different.

In order to simplify the formulation, we approximate (8.83) by Pellinen smoothing with $\psi_0 = \psi_0^{(i)}$, such that the areas of their smoothing domains are identical. In this way, $W_s^{(i)}$ also satisfies

$$
W_s^{(i)} = 2\pi \left[1 - \cos \psi_0^{(i)} \right].
$$

(8.85)

We obtain

$$
1 - \cos \psi_0^{(i)} = \frac{1}{2\pi} \left[\lambda_2^{(i)} - \lambda_1^{(i)} \right] \left[\cos \theta_1^{(i)} - \cos \theta_2^{(i)} \right],
$$

(8.86)

which can be written as

$$
\sin^2 \frac{\psi_0^{(i)}}{2} = \frac{1}{4\pi} \left[\lambda_2^{(i)} - \lambda_1^{(i)} \right] \left[\cos \theta_1^{(i)} - \cos \theta_2^{(i)} \right]
$$

(8.87)

for computating $\psi_0^{(i)}$. Consequently, (8.83) is approximated by

$$
\Delta \bar{g}_i = \sum_{n=2}^{n_{\max}} \sum_{k=0}^{n} \beta_n^{(i)} \left[c_n^k \cos k\lambda + s_n^k \sin k\lambda \right] P_n^k(\cos \theta),
$$

(8.88)

where θ and λ are the coordinates of the center of the grid cell, and $\beta_n^{(i)}$ is to be computed according to (8.69) with ψ_0 replaced by $\psi_0^{(i)}$.

In (8.88), $\beta_n^{(i)}$ depends on $\theta_1^{(i)}$ and $\theta_2^{(i)}$, or in other words, it depends on θ. Hence, its right-hand side cannot be understood as a spherical harmonic series with $\beta_n^{(i)}$ considered as a part of the coefficients; the coefficients must be constant. As a further approximation, we replace $\beta_n^{(i)}$ by a constant factor q_n, which can be regarded as a global average of $\beta_n^{(i)}$, or the value of $\beta_n^{(i)}$ at mid-latitude. We can then approximately write (8.88) as

$$\Delta \bar{g}_i = \sum_{n=2}^{n_{max}} \sum_{k=0}^{n} \left[q_n c_n^k \cos k\lambda + q_n s_n^k \sin k\lambda \right] P_n^k (\cos \theta) . \tag{8.89}$$

This formula remains valid for the normalized spherical harmonics

$$\Delta \bar{g}_i = \sum_{n=2}^{n_{max}} \sum_{k=0}^{n} \left[q_n \bar{c}_n^k \cos k\lambda + q_n \bar{s}_n^k \sin k\lambda \right] \bar{P}_n^k (\cos \theta) . \tag{8.90}$$

What we realize is that the coefficients $\bar{c}_n^k$ and $\bar{s}_n^k$ in (8.37) computed using $\Delta \bar{g}_i$ are better interpreted as $q_n \bar{c}_n^k$ and $q_n \bar{s}_n^k$. Therefore, a better approximation than (8.37) is

$$\begin{cases} \bar{c}_n^k = \sum_i \frac{\Delta \bar{g}_i}{4\pi q_n} \left[\lambda_2^{(i)} - \lambda_1^{(i)} \right] \bar{I}_n^{(i)} , & k = 0 ; \\[3mm] \bar{c}_n^k = \sum_i \frac{\Delta \bar{g}_i}{4\pi k q_n} \left[\sin k\lambda_2^{(i)} - \sin k\lambda_1^{(i)} \right] \bar{I}_{nk}^{(i)} , & k \neq 0 ; \\[3mm] \bar{s}_n^k = \sum_i \frac{\Delta \bar{g}_i}{4\pi k q_n} \left[- \cos k\lambda_2^{(i)} + \cos k\lambda_1^{(i)} \right] \bar{I}_{nk}^{(i)} , & k \neq 0 . \end{cases} \tag{8.91}$$

The parameter q_n is referred to as de-smoothing factor. It is understood as an empirical parameter to alleviate not only the influence of smoothing but also the influence of other approximations made in the process. Therefore, it is chosen empirically based on numerical experiments with synthetic data assumed to be error free using $\beta_n^{(i)}$ as a protocol.

8.2.5　Computation of Gravitational Potential Based on Least Squares Fitting

We can also solve for $\bar{c}_n^k$ and $\bar{s}_n^k$ directly from the algebraic equations that relate them to the grid mean gravity anomalies. We assume that the Earth's surface is divided into a grid of N intervals in polar angle and $2N$ intervals in longitude, with $\Delta \theta = \Delta \lambda$. We use i to number the intervals in polar angle, j to number the intervals in longitude, and consequently ω_{ij} to number the grid cell $(\theta_i - \Delta\theta/2 \leq \theta \leq \theta_i + \Delta\theta/2, \lambda_j - \Delta\lambda/2 \leq \lambda \leq \lambda_j + \Delta\lambda/2)$, where (θ_i, λ_j) is the coordinate of the center of the cell. We can then obtain from (8.32), based on the definition of grid mean gravity anomaly (8.29),

$$\Delta \bar{g}_{ij} = \frac{1}{S_{ij}} \int_{\omega_{ij}} \Delta g \, d\omega = \frac{1}{S_{ij}} \int_{\omega_{ij}} \left[\sum_{n=2}^{n_{max}} \sum_{k=0}^{n} (\bar{c}_n^k \cos k\lambda + \bar{s}_n^k \sin k\lambda) \bar{P}_n^k (\cos \theta) \right] d\omega , \tag{8.92}$$

where $S_{ij} = \int_{\omega_{ij}} d\omega$ is the area of the grid cell ω_{ij}, which can be further formulated as

$$
S_{ij} = \int_{\lambda_j - \Delta\lambda/2}^{\lambda_j + \Delta\lambda/2} \int_{\theta_i - \Delta\theta/2}^{\theta_i + \Delta\theta/2} \sin\theta d\theta d\lambda
$$

$$
= \Delta\lambda \left[\cos\left(\theta_i - \frac{\Delta\theta}{2}\right) - \cos\left(\theta_i + \frac{\Delta\theta}{2}\right) \right]
$$

$$
= \sin\theta_i \left(2\sin\frac{\Delta\theta}{2} \right) \Delta\lambda . \tag{8.93}
$$

The equation relating $\bar{c}_n^k$ and $\bar{s}_n^k$ to $\Delta\bar{g}_{ij}$ can be obtained from (8.92) to be

$$
\Delta\bar{g}_{ij} = \sum_{n=2}^{n_{\max}} \sum_{k=0}^{n} (\alpha_{ijnk}\bar{c}_n^k + \beta_{ijnk}\bar{s}_n^k) , \tag{8.94}
$$

where, by definition,

$$
\Delta\bar{g}_{ij} = \frac{1}{S_{ij}} \int_{\lambda_j - \Delta\lambda/2}^{\lambda_j + \Delta\lambda/2} \int_{\theta_i - \Delta\theta/2}^{\theta_i + \Delta\theta/2} \Delta g \sin\theta d\theta d\lambda , \tag{8.95}
$$

$$
\alpha_{ijnk} = \frac{1}{S_{ij}} \int_{\lambda_j - \Delta\lambda/2}^{\lambda_j + \Delta\lambda/2} \int_{\theta_i - \Delta\theta/2}^{\theta_i + \Delta\theta/2} \bar{P}_n^k(\cos\theta) \cos k\lambda \sin\theta d\theta d\lambda , \tag{8.96}
$$

$$
\beta_{ijnk} = \frac{1}{S_{ij}} \int_{\lambda_j - \Delta\lambda/2}^{\lambda_j + \Delta\lambda/2} \int_{\theta_i - \Delta\theta/2}^{\theta_i + \Delta\theta/2} \bar{P}_n^k(\cos\theta) \sin k\lambda \sin\theta d\theta d\lambda . \tag{8.97}
$$

We mention that $\Delta\bar{g}_{ij}$ is provided as data, and α_{ijnk} and β_{ijnk} depend only on the grid, thus being known, and can be easily evaluated according to the formulations of similar quantities in the previous subsections, which is left to the readers.

The coefficients to be determined, $\bar{c}_n^k$ and $\bar{s}_n^k$, are related to the data $\Delta\bar{g}_{ij}$ through the linear equation (8.94). Based on this equation, the least squares problem

$$
\sum_{i=1}^{N} \sum_{j=1}^{2N} W_{ij} \left[\sum_{n=2}^{N-1} \sum_{k=0}^{n} (\alpha_{ijnk}\bar{c}_n^k + \beta_{ijnk}\bar{s}_n^k) - \Delta\bar{g}_{ij} \right] = \min \tag{8.98}
$$

is to be solved for practical computation. Several problems are to be clarified in subsequent paragraphs, including the adoption of the weight W_{ij} and the assertion of $n_{\max} = N - 1$.

The first problem is that the data of gravity anomaly are denser in higher latitude regions. If all data are treated equally in the least squares problem, the solution would be biased toward regions of higher data density. It is for alleviating this problem that we have added a weight W_{ij}. Instead of $W_{ij} = 1$, we may assign

$W_{ij} = \sin \theta_i$, which is proportional to the area of the grid, as shown in (8.93). This implies that data are weighted by the area of the grid it represents, implying that denser data are weighted less to avoid biases.

The second problem is the resolvability. In this approach, the coefficients $\bar{c}_n^k$ and $\bar{s}_n^k$ can be reliably solved only up to degree and order $N - 1$ (Colombo, 1981). This could be demonstrated by the simpler degenerate situation: Δg is a function of λ alone, thus being independent of θ. In this simpler case, according to (3.371), the original equation of the problem (8.32) degenerates to the Fourier series

$$\Delta g(\lambda) = \sum_{k=0}^{n_{\max}} (a_k \cos k\lambda + b_k \sin k\lambda). \tag{8.99}$$

Hence, there are $2n_{\max} + 1$ coefficients to be determined. However, in this case, $\Delta \bar{g}_{ij}$ is independent of i, and hence there are $2N$ independent values. The total number of coefficients to be determined cannot be more than the total number of data. As a result, the coefficients can only be solved up to order $n_{\max} = N - 1$ at most. With $n_{\max} = N - 1$ as chosen in (8.98), the total number of coefficients $\bar{c}_n^k$ and $\bar{s}_n^k$ is $N(N - 1)$, while the total number of data $\Delta \bar{g}_{ij}$ is $2N^2$.

The third problem is that as discussed in Sect. 8.2.1, at higher latitude, the grid sizes are smaller, and data of $\Delta \bar{g}_{ij}$ include information beyond the highest resolvable degree $N - 1$. Besides the corruption of the coefficients to be solved, the residuals obtained from the least squares problem (8.98), which are the misfits to Eq. (8.94), are also largely caused by the information in $\Delta \bar{g}_{ij}$ beyond degree $N-1$, which is not modeled by (8.98). In other words, the information in the data is inconsistent with that modeled by the parameters, and the residuals represent mostly the information not modeled. Therefore, in (8.98), it is inappropriate to understand the residuals as if they were resulted from the random errors of $\Delta \bar{g}_{ij}$ to quantify the errors of the coefficients obtained. It is thus more appropriate to first infer the errors of $\Delta \bar{g}_{ij}$ based on the computations from the original data of observation and then use these errors to estimate the errors of the coefficients obtained through (8.98) or the integral approach in Sect. 8.2.1.

We mention that the conclusion of the previous paragraph is not to be generalized. In the problem formulated there, the data density in the equatorial region is just what is required to resolve the coefficients. In this respect, there is no actual redundant data at all if not for the sphericity of the Earth, which results in denser than necessary data in longitude at higher latitude, but not in latitude. Now we suppose that the spherical harmonic coefficients up to the degree and order $n_{\max}$ are to be solved, with $n_{\max}$ much smaller than $N - 1$, and the information beyond the degree and order $n_{\max}$ in $\Delta \bar{g}_{ij}$ is negligibly small as compared to the errors of the data. In this circumstance, (8.98) with the upper limit of n replaced by $n_{\max}$ would be a legitimate least squares problem, where the information in the data is consistent with that modeled by the coefficients, and there are also enough redundant data so that the residuals do reflect the errors in the data and can be used to access the errors of the coefficients. However, preprocessing the data to match such a supposition is not an evident task to achieve.

As a supplement to the discussions in the four previous paragraphs, we now imagine to filter the data of Δg by applying an ideal filter. Define $\beta_n = 1$ for $n \leq N - 1$ and $\beta_n = 0$ for $n > N - 1$, compute W/W_s according to (8.66), and then filter the data of Δg according to (8.57). This eliminates all spherical harmonic terms beyond degree $N-1$. In this case, the information in the data of $\Delta \bar{g}_{ij}$ computed from the ideal filtered data of Δg is made consistent with that represented by the parameters in (8.98). As a result, when we put the data of $\Delta \bar{g}_{ij}$ into (8.98), if numerical round-off errors are not considered, we would obtain min $= 0$, i.e., there is no misfit at all. This does not mean that there is no error in the data of $\Delta \bar{g}_{ij}$ or in the coefficients obtained, but this just implies that errors in the data within the resolvable degree and order are completely represented by errors in the coefficients, and only when errors of the data are known can errors of the coefficients be evaluated based on error propagation. This fact can also be understood based on the integral formulae of the coefficients given in (8.34): whatever errors are in data of Δg, the spherical harmonic series (8.32) with coefficients computed using (8.34) does represent the data of Δg up to degree and order $n_{\max}$ as is including errors. This implies that in the least squares problem, errors of data within the truncation degree and order of the coefficients to be determined do not contribute to the residuals, and hence, their effects on the errors of the coefficients are not included in the errors estimated using residuals.

Despite the drawbacks discussed above, the least squares problem (8.98) does have the advantage that the effect of smoothing of gravity data is taken into account in the formulation, and the coefficients obtained are not affected.

In summary, if errors in data of $\Delta \bar{g}_{ij}$ are to be taken into account for an optimal estimation of $\bar{c}_n^k$ and $\bar{s}_n^k$ by minimizing errors, particular cares must be taken.

8.2.6 Geoidal Height, Height Anomaly, and Deflection of the Vertical Based on a Global Model of the Gravitational Potential

The geoidal height, height anomaly, and deflection of the vertical can be computed using a model of the disturbing potential, but they are implicitly related to a reference Earth ellipsoid. If the spherical harmonic series of the disturbing potential is provided, a reference Earth ellipsoid is implicitly defined. If the spherical harmonic series of the total gravitational potential is provided, the normal gravitational potential, i.e., the gravitational potential of the reference Earth ellipsoid, should be subtracted to obtain the disturbing potential.

All the formulae needed in the computations to be discussed below have been formulated in previous chapters, and we are not going to repeat here.

Depending on whether Stokes' or Molodensky's theory is adopted, the computation of the geoidal height or height anomaly requires the computation of point values of the disturbing potential on the geoid or the Earth's surface, and the computation of the deflection of the vertical requires the computation of point values of the horizontal components of the gravity disturbance on the geoid or the Earth's surface.

If the adjusted disturbing potential is used in the computation, the indirect effect should be added afterward.

However, if the disturbing potential of the real Earth is used in the computation, the effect of the topographic mass may cause some complication. Take for example the vertical component of the gravity disturbance discussed in Sect. 7.3.3. Its value at a point computed purely from the spherical harmonic series of the disturbing potential may be different from the actual value if there exists nearby topographic mass above the horizon through the point. In fact, there is no theoretical guarantee that the point values of the disturbing potential and gravity disturbance computed purely from the spherical harmonic series of the disturbing potential be identical to the actual values below the Brillouin sphere, although they are exact beyond. In this circumstance, a process assured to be physically stringent could be designed to do the computation based on a remove–compute–restore process: (1) remove the effect of topographic mass from the spherical harmonic series of the disturbing potential, (2) do the computation, and (3) add the effect of the topographic mass back into the result of the previous step. To keep the effect small, removing and restoring of the topographic mass could be combined with filling and removing of a Helmert condensation layer or a mass associated with an isostatic compensation model. The computation could be done according to the formulae of the indirect effect.

8.3   Computation of the Geoidal Height, Height Anomaly, and Deflection of the Vertical Using Gravity Data

In this section we show the most naive way to compute the geoidal height, height anomaly, and deflection of the vertical based on the original Stokes' or Molodensky's theory using gravity anomaly. The computation based on the GNSS-gravimetric boundary value theory using gravity disturbance is similar to that of Molodensky's theory and could be written in parallel with very limited effort, which is left to readers, and hence, it will not be mentioned in the subsequent text of this section.

In practical computations at present, a global model of the gravitational potential is always combined to control the long (to mid-range) wavelength variations based on a remove–compute–restore procedure: (1) remove the gravity anomaly computed from the global model of gravitational potential from the data, and the result is referred to as residual gravity anomaly, (2) compute the geoidal height/height anomaly and deflection of the vertical using the residual gravity anomaly, and (3) compute the geoidal height/height anomaly and deflection of the vertical using the global model of the gravitational potential, and then add the result into that of the previous step. This procedure is for the adjusted Earth; therefore, the effect of gravity reduction should have been removed from the global model of the gravitational potential beforehand.

The formulation in this section corresponds mostly to the step (2) of the remove–compute–restore process outlined in the previous paragraph. The formulations for steps (1) and (3) can be found in previous chapters. However, the most up to date literature must be consulted for practical computation, as modifications of Stokes' and Molodensky's formulae to adapt to the remove–compute–restore procedure, as well as many minor aspects and small details of the problem, are not discussed in this book.

8.3.1   Geoidal Height and Height Anomaly

The geoidal height N is to be computed using Stokes' formula (5.50), where the Stokes function $S(\psi)$ is defined in (5.49), which are rewritten here for easy reference,

$$N(\theta, \lambda) = \frac{R}{4\pi \bar{\gamma}_0} \int_\omega \Delta g(\theta', \lambda') S(\psi) d\omega, \tag{8.100}$$

$$S(\psi) = \frac{1}{\sin(\psi/2)} + 1 - 5\cos\psi - 6\sin\frac{\psi}{2} - 3\cos\psi \ln\left(\sin\frac{\psi}{2} + \sin^2\frac{\psi}{2}\right). \tag{8.101}$$

When ψ is small, a good approximation of ψ is

$$S(\psi) \approx \frac{1}{\sin(\psi/2)} \approx \frac{2}{\psi}, \tag{8.102}$$

as shown in Fig. 5.5.

The integral (8.100) is the same type as that of (8.2) and is hence computed in the same way. We also divide it into two terms

$$N(\theta, \lambda) = N_{\psi \leq \Theta}(\theta, \lambda) + N_{\psi \geq \Theta}(\theta, \lambda). \tag{8.103}$$

The term $N_{\psi \geq \Theta}(\theta, \lambda)$ can be computed according to (8.6)

$$N_{\psi \geq \Theta}(\theta, \lambda) = \frac{R}{4\pi \bar{\gamma}_0} \sum_i \Delta g(\theta'_i, \lambda'_i) S(\psi)_i \Delta\omega_i, \tag{8.104}$$

where (θ'_i, λ'_i) is the coordinate of the center of the ith grid cell, $S(\psi)_i$ the value of $S(\psi)$ with ψ computed using (θ'_i, λ'_i), and $\Delta\omega_i$ the area of the ith grid cell. For the term $N_{\psi \leq \Theta}(\theta, \lambda)$, $S(\psi)$ can be very well approximated by (8.102). Hence, we have

$$N_{\psi \leq \Theta}(\theta, \lambda) = \frac{R}{4\pi \bar{\gamma}_0} \int_{\psi \leq \Theta} \frac{\Delta g(\theta', \lambda')}{\sin(\psi/2)} d\omega. \tag{8.105}$$

This integral is in the same form as I_1 in (8.9), whose result is given in (8.13). Here the result is

$$N_{\psi \leq \Theta}(\theta, \lambda) = \frac{R}{\bar{\gamma_0}} \Theta \Delta G_P + \frac{R}{12 \bar{\gamma_0}} \Theta^3 \left[\left(\frac{\partial^2 \Delta g}{\partial X_0^2} \right)_P + \left(\frac{\partial^2 \Delta g}{\partial Y_0^2} \right)_P \right], \qquad (8.106)$$

where the subscript P is a short representation of the point (θ, λ), noting that X_0 and Y_0 are local Cartesian coordinates on the unit sphere pointing to the north and east, respectively, with the corresponding point of P as the origin.

As Stokes's theory requires gravity reduction to relocate the mass above the geoid, what is obtained is in fact the adjusted geoidal height. The indirect effect must be added back afterward. The computations of the integrals involved are similar to those of gravity reduction.

The height anomaly ζ is to be computed according to (7.113), where $\Delta G'$ is given in (7.115), which are rewritten here for easy reference,

$$\zeta_P = \frac{R}{4\pi \bar{\gamma_0}} \int_\omega [\Delta G - (h - h_P) \Delta G'] S(\psi) d\omega, \qquad (8.107)$$

$$\Delta G'_P = \frac{R^2}{2\pi} \int_\omega \frac{\Delta G - \Delta G_P}{l_0^3} d\omega. \qquad (8.108)$$

We realize immediately that when $\Delta G'$ is known, the computation of the height anomaly is the same as that of the geoidal height with Δg replaced by $\Delta G - (h - h_P) \Delta G'$. Hence, we only have to consider the computation of $\Delta G'$ here.

The integral (8.108) is the same type as that of (8.2). Hence, we also divide it into two terms

$$\Delta G'_P = \Delta G'_{P, \psi \leq \Theta} + \Delta G'_{P, \psi \geq \Theta}, \qquad (8.109)$$

and we have

$$\Delta G'_{P, \psi \geq \Theta} = \frac{R^2}{2\pi} \sum_i \frac{\Delta G_i - \Delta G_P}{l_{0i}^3} \Delta \omega_i. \qquad (8.110)$$

With $l_0 = 2R \sin(\psi/2)$, the term $\Delta G'_{P, \psi \leq \Theta}$ can be written as

$$\Delta G'_{P, \psi \leq \Theta} = \frac{1}{16\pi R} \int_{\psi \leq \Theta} \frac{\Delta G - \Delta G_P}{\sin^3(\psi/2)} d\omega. \qquad (8.111)$$

The integral in this formula is the same type as I_2 in (8.9) with result given in (8.20). We have

$$\Delta G'_{P, \psi \leq \Theta} = \frac{\Theta}{4R} \left[\left(\frac{\partial^2 \Delta G}{\partial X_0^2} \right)_P + \left(\frac{\partial^2 \Delta G}{\partial Y_0^2} \right)_P \right]. \qquad (8.112)$$

We must note that the request (8.18) is satisfied by $\Delta G - \Delta G_P = 0$ at P, and that as ΔG_P is a constant under the integral sign, its derivatives vanish. We also note that X_0 and Y_0 are local Cartesian coordinates on the unit sphere pointing to the north and east, respectively, with the corresponding point of P as the origin.

If the free air anomaly computed using the normal height is used in the computation, there is no indirect effect involved. However, if a smoother gravity anomaly associated with some kind of mass relocation has been used, the indirect effect must be added back. The computation of the integrals involved is similar to those in gravity reduction.

8.3.2 The Deflection of the Vertical

The formula of the deflection of the vertical defined in Stokes' theory is given by Vening Meinesz' formula (5.64) with the kernel, the Vening Meinesz function, $dS(\psi)/d\psi$, given by (5.63). We rewrite them here for easy reference

$$
\begin{cases}
\xi(\theta, \lambda) = \dfrac{1}{4\pi\bar{\gamma}_0} \displaystyle\int_\omega \Delta g(\theta', \lambda') \dfrac{d}{d\psi} S(\psi) \cos A \, d\omega, \\
\eta(\theta, \lambda) = \dfrac{1}{4\pi\bar{\gamma}_0} \displaystyle\int_\omega \Delta g(\theta', \lambda') \dfrac{d}{d\psi} S(\psi) \sin A \, d\omega;
\end{cases}
\tag{8.113}
$$

$$
\frac{d}{d\psi} S(\psi) = -\frac{\cos(\psi/2)}{2\sin^2(\psi/2)} + 8\sin\psi - 6\cos\frac{\psi}{2} - 3\frac{1 - \sin(\psi/2)}{\sin\psi}
$$

$$
+ 3\sin\psi \ln\left(\sin\frac{\psi}{2} + \sin^2\frac{\psi}{2}\right).
\tag{8.114}
$$

We again use P to denote the point (θ, λ) where the deflection of the vertical is computed. The azimuth A is the angle from north to the point (θ', λ') as seen at P.

As (8.113) is similar to (8.2), we separate it into two terms as in (8.3)

$$
\begin{cases}
\xi(\theta, \lambda) = \xi_{\psi \leq \Theta}(\theta, \lambda) + \xi_{\psi \geq \Theta}(\theta, \lambda), \\
\eta(\theta, \lambda) = \eta_{\psi \leq \Theta}(\theta, \lambda) + \eta_{\psi \geq \Theta}(\theta, \lambda);
\end{cases}
\tag{8.115}
$$

with

$$
\begin{cases}
\xi_{\psi \geq \Theta}(\theta, \lambda) = \dfrac{1}{4\pi\bar{\gamma}_0} \displaystyle\sum_i \Delta g(\theta'_i, \lambda'_i) \dfrac{d}{d\psi} S(\psi)_i \cos A_i \, \Delta\omega_i, \\
\eta_{\psi \geq \Theta}(\theta, \lambda) = \dfrac{1}{4\pi\bar{\gamma}_0} \displaystyle\sum_i \Delta g(\theta'_i, \lambda'_i) \dfrac{d}{d\psi} S(\psi)_i \sin A_i \, \Delta\omega_i,
\end{cases}
\tag{8.116}
$$

where (θ'_i, λ'_i) is the coordinate of the center of the ith grid cell, $(d/d\psi)S(\psi)_i$ the value of $(d/d\psi)S(\psi)$ with ψ computed using (θ'_i, λ'_i), A_i the azimuth of (θ'_i, λ'_i) as observed at (θ, λ), and $\Delta\omega_i$ the area of the ith grid cell. What remains is to compute $\xi_{\psi\le\Theta}(\theta, \lambda)$ and $\eta_{\psi\le\Theta}(\theta, \lambda)$.

As shown in Fig. 5.7, when ψ is small, a very good approximation of $(d/d\psi)S(\psi)$ is

$$\frac{d}{d\psi}S(\psi) = -\frac{\cos(\psi/2)}{2\sin^2(\psi/2)} \approx -\frac{2}{\psi^2}. \tag{8.117}$$

With this approximation and the approximate expression $d\omega = \sin\psi\,d\psi\,dA \approx \psi\,d\psi\,dA$, we have

$$\begin{cases} \xi_{\psi\le\Theta}(\theta, \lambda) = -\dfrac{1}{2\pi\bar{\gamma}_0} \displaystyle\int_0^{2\pi}\int_0^{\Theta} \dfrac{\Delta g(\psi, A)}{\psi}\cos A\,d\psi\,dA, \\[4mm] \eta_{\psi\le\Theta}(\theta, \lambda) = -\dfrac{1}{2\pi\bar{\gamma}_0} \displaystyle\int_0^{2\pi}\int_0^{\Theta} \dfrac{\Delta g(\psi, A)}{\psi}\sin A\,d\psi\,dA, \end{cases} \tag{8.118}$$

where we substituted θ' and λ' by ψ and A as variables of integration. For a fixed value of A, we express $\Delta g(\psi, A)$ as

$$\Delta g(\psi, A) = \Delta g_P + \frac{\partial\Delta g}{\partial\psi}\psi, \tag{8.119}$$

where Δg_P is the value of Δg at P of coordinate (θ, λ). The substitution of (8.119) into (8.118) yields

$$\begin{cases} \xi_{\psi\le\Theta}(\theta, \lambda) = -\dfrac{1}{2\pi\bar{\gamma}_0} \displaystyle\int_0^{2\pi}\left[\int_0^{\Theta}\dfrac{\partial\Delta g}{\partial\psi}d\psi\right]\cos A\,dA, \\[4mm] \eta_{\psi\le\Theta}(\theta, \lambda) = -\dfrac{1}{2\pi\bar{\gamma}_0} \displaystyle\int_0^{2\pi}\left[\int_0^{\Theta}\dfrac{\partial\Delta g}{\partial\psi}d\psi\right]\sin A\,dA. \end{cases} \tag{8.120}$$

After evaluating the integral with respect to ψ analytically, we obtain

$$\begin{cases} \xi_{\psi\le\Theta}(\theta, \lambda) = -\dfrac{1}{2\pi\bar{\gamma}_0} \displaystyle\int_0^{2\pi}\left[\Delta g(\Theta, A) - \Delta g_P\right]\cos A\,dA, \\[4mm] \eta_{\psi\le\Theta}(\theta, \lambda) = -\dfrac{1}{2\pi\bar{\gamma}_0} \displaystyle\int_0^{2\pi}\left[\Delta g(\Theta, A) - \Delta g_P\right]\sin A\,dA, \end{cases} \tag{8.121}$$

which can be simplified to

$$\begin{cases} \xi_{\psi\le\Theta}(\theta, \lambda) = -\dfrac{1}{2\pi\bar{\gamma}_0} \displaystyle\int_0^{2\pi}\Delta g(\Theta, A)\cos A\,dA, \\[4mm] \eta_{\psi\le\Theta}(\theta, \lambda) = -\dfrac{1}{2\pi\bar{\gamma}_0} \displaystyle\int_0^{2\pi}\Delta g(\Theta, A)\sin A\,dA. \end{cases} \tag{8.122}$$

The result (8.122) can be discretized by dividing the circle $\psi = \Theta$ into K equal fractions of arc with $\Delta A = 2\pi / K$. The formula for computation is then

$$
\begin{cases}
\xi_{\psi \leq \Theta}(\theta, \lambda) = -\dfrac{1}{K \bar{\gamma_0}} \displaystyle\sum_{k=1}^{K} \Delta g(\Theta, A_k) \cos A_k \,, \\
\eta_{\psi \leq \Theta}(\theta, \lambda) = -\dfrac{1}{K \bar{\gamma_0}} \displaystyle\sum_{k=1}^{K} \Delta g(\Theta, A_k) \sin A_k \,,
\end{cases}
\tag{8.123}
$$

where A_k is the value of A in the middle of the kth fraction of arc.

An alternative way to evaluate $\xi_{\psi \leq \Theta}(\theta, \lambda)$ and $\eta_{\psi \leq \Theta}(\theta, \lambda)$ is to express Δg in the form of (8.12) and then substitute into (8.118). This is left as an excise to the readers.

Finally, we remark that the formulations for the cases of Molodensky's theory are similar and are left to the readers. Furthermore, as in the case of geoidal height and height anomaly, the indirect effect of the deflection of the vertical should also be appropriately considered.

8.4   Astro-geodetic Determination of the Deflection of the Vertical

8.4.1   Astro-geodetic Measurement of the Deflection of the Vertical

The direction of gravity at any point can also be expressed with respect to the reference Earth ellipsoid. The angle between the direction of gravity and the equatorial plane is called astronomical latitude, and that between the local astronomical meridional plane and the Greenwich meridional plane is called astronomical longitude, where the local astronomical meridional plane is the plane perpendicular to the equatorial plane and containing the direction of gravity. More specifically, the local astronomical meridional plane can be understood as the plane defined by two straight lines through the station or point of interest, one being along the direction of gravity and the other being perpendicular to the equatorial plane. The signs of the astronomical coordinates, i.e., the astronomical latitude and longitude, are defined in the same way as respective the geodetic coordinates; see Fig. 8.8. The readers are suggested to compare it with Fig. 4.1, which shows the geodetic latitude and longitude defined by the normal of the reference Earth ellipsoid, $\overrightarrow{n}$.

Assuming the directions of stars with respect to the reference Earth ellipsoid to be known, which can be inferred from a star catalog combined with the Earth orientation parameters (EOPs), the direction of gravity at an observation station on the Earth's surface can then be determined by observing stars using ground based instruments aligned along the direction of gravity. In the scope of this book, we assume that the astronomical latitude and longitude at the stations are known without going into the detail of their determination.

Fig. 8.8 Definition of the astronomical latitude ϕ and longitude Λ, where PP' is perpendicular to the equatorial plane

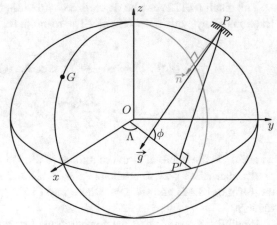

Fig. 8.9 Definition of the astro-geodetic determined deflection of the vertical

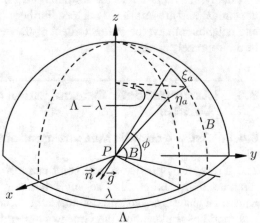

We assume that the geodetic latitude and longitude of the stations are known, which can be measured easily using GNSS nowadays. For a station, the differences between the astronomical and geodetic latitudes and longitudes represent the deviation of the direction of gravity from the normal to the reference Earth ellipsoid, which is referred to as astro-geodetic deflection of the vertical. We denote its southward and westward components using ξ_a and η_a. It will also be simply referred to as deflection of the vertical when no confusion is possible.

The two components of the deflection of the vertical, ξ_a and η_a, the geodetic latitude and longitude, B and λ, and the astronomical latitude and longitude ϕ and Λ, are shown in Fig. 8.9 based on a Cartesian coordinate system $Pxyz$ and the associated spherical coordinate system, where the origin P is on the station, the z-axis is parallel to the minor axis of the reference Earth ellipsoid, thus the xy plane is parallel to the equatorial plane, and the x-axis is parallel to the Greenwich meridional plane. As shown in this figure, we consider a sphere centered at P. The relation among the deflection of the vertical and the astronomical and geodetic latitudes and longitudes can be best seen by looking at the intersections of the opposite directions of gravity and the normal to the reference Earth ellipsoid with this sphere. Recognizing the fact that both ξ_a and η_a are angles with respect to the center of the sphere, P, we immediately realize that

$$\xi_a = \phi - B\,, \qquad \eta_a = (\Lambda - \lambda)\cos B\,. \tag{8.124}$$

By definition, the deflection of the vertical defined in (8.124) is different from that defined in either Stokes' or Molodensky's theory. Hence, in order to use it as data, a correction should be applied to make it commensurate with the definition in the theory adopted.

In Stokes' theory, the deflection of the vertical is defined as if P were on the geoid. In practice, we should define a corresponding point of P on the geoid, as shown in Fig. 8.10, where Σ is the geoid and S the reference Earth ellipsoid. It is the point P_0, which is the intersection point between the normal from P to the reference Earth ellipsoid and the geoid. We take as an example only the component ξ_a. The correction is the difference between ξ_a and ξ_a^S, where the superscript S indicates Stokes' theory or, in other words, the angle between the directions of $\overrightarrow{g}_P$ and $\overrightarrow{g}_{P_0}$. However, what we can compute is only the change of direction of a plumb line using the formula of its curvature. Hence, it is the angle between the directions of $\overrightarrow{g}_P$ and $\overrightarrow{g}_{P_0'}$ that can be computed, where P_0' is the intersection point between the plumb line through P and the geoid. As P_0 and P_0' are really very close, we ignore

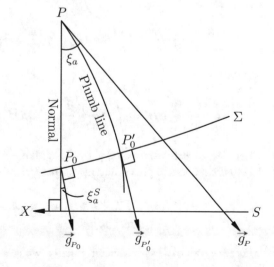

Fig. 8.10 Reduction of the astro-geodetic determined deflection of the vertical to Stokes' deflection of the vertical

the difference between the directions of $\vec{g}_{P_0}$ and $\vec{g}_{P_0'}$. Hence, we have, according to (1.172) and (1.173),

$$
\begin{cases}
\xi_a^S = \xi_a - \displaystyle\int_0^H \frac{1}{g(H')} \frac{\partial g(H')}{\partial x} dH' , \\
\eta_a^S = \eta_a - \displaystyle\int_0^H \frac{1}{g(H')} \frac{\partial g(H')}{\partial y} dH' ,
\end{cases}
\tag{8.125}
$$

where x and y are the Cartesian coordinates with origin on the plumb line H' above P_0', perpendicular to the plumb line, and point toward the north and east, respectively. While there is no room for any ambiguity for the magnitude of the correction, the negative signs in the correction can be understood based on physical inference as follows: according to (1.172), when $[1/g(H')][\partial g(H')/\partial x]$ is positive, the projection of the plumb line within the meridional plane is concave northward, which is the case shown in Fig. 8.10, resulting in a negative correction. The formula in the prime vertical can be explained similarly.

In the practical evaluation of the integrals in (8.125), as the variation of gravity is quite irregular, and the rate of change of its direction can be hardly known, it is normally approximated by the normal gravity. Furthermore, as H is mostly within 2 km and rarely exceeds 8 km, we neglect the dependence of the curvature on H'. Hence, the normal gravity on the reference Earth ellipsoid is used, and the coordinates x and y are replaced by X and Y, which are tangential to the reference Earth ellipsoid and point to the north and south, respectively. According to (4.150), we have

$$
\frac{\partial \gamma_0}{\partial X} = \frac{1}{R} \frac{\partial \gamma_0}{\partial B} = \frac{\gamma_e \beta}{R} \sin 2B , \qquad \frac{\partial \gamma_0}{\partial Y} = \frac{1}{R \cos B} \frac{\partial \gamma_0}{\partial \lambda} = 0 .
\tag{8.126}
$$

Therefore,

$$
\int_0^H \frac{1}{\gamma_0} \frac{\partial \gamma_0}{\partial X} dH' = \int_0^H \frac{1}{\gamma_0} \frac{\gamma_e \beta}{R} \sin 2B \, dH' = \frac{\gamma_e \beta}{\bar{\gamma}_0 R} H \sin 2B ,
\tag{8.127}
$$

where we have approximated γ_0 by $\bar{\gamma}_0$. With the relation $\bar{\gamma}_0 = \gamma_e(1 + \beta/3)$, which is left to the readers to prove, and the numerical values of β and R, we obtain

$$
\int_0^H \frac{1}{\gamma_0} \frac{\partial \gamma_0}{\partial X} dH' = 0.171'' H \sin 2B ,
\tag{8.128}
$$

where the unit of H is kilometer. Finally, we can write (8.125) as

$$
\xi_a^S = \xi_a - 0.171'' H \sin 2B , \qquad \eta_a^S = \eta_a .
\tag{8.129}
$$

Fig. 8.11 Reduction of the astro-geodetic determined deflection of the vertical to Molodensky's deflection of the vertical

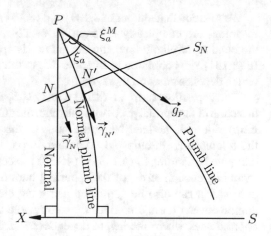

In this formula, B can be replaced by $\pi/2-\theta$ without further degrading the accuracy. It is left to the readers to verify that keeping the term $\gamma_e/\bar{\gamma_0}$ in the approximation makes the numerical result 0.171 more accurate at mid-latitude, where $\sin 2B$ is largest.

In Molodensky's theory, the deflection of the vertical to be determined, ξ_a^M, is defined as the angle between $\vec{g}_P$ and $\vec{\gamma}_N$, as shown in Fig. 8.11, where S is the reference Earth ellipsoid. In order to show the angle ξ_a^M, we draw the equipotential surface of normal gravity through N, which is denoted as S_N in the figure. We draw a straight line perpendicular to S_N from P, which interacts S_N at N'. This line is also in the direction of the normal gravity at N'. As N and N' are extremely close and the equipotential surface of normal gravity is very smooth, we neglect the difference between the directions of the normal gravity at N and N'. In this way, ξ_a^M is the angle between $\vec{g}_P$ and $\vec{\gamma}_{N'}$, as shown in the figure. The correction is the angle between $\vec{\gamma}_N$ and the normal, or between $\vec{\gamma}_{N'}$ and the normal, which is due to the bending of the normal plumb line. Hence, we have

$$\begin{cases} \xi_a^M = \xi_a - \int_0^{H_\gamma} \frac{1}{\gamma(H')} \frac{\partial\gamma(H')}{\partial x} dH' , \\ \eta_a^M = \eta_a - \int_0^{H_\gamma} \frac{1}{\gamma(H')} \frac{\partial\gamma(H')}{\partial y} dH' , \end{cases} \tag{8.130}$$

where x and y are the Cartesian coordinates with origin on the normal plumb line H' above the reference Earth ellipsoid, perpendicular to the normal plumb line, and point toward the north and east, respectively. By neglecting the dependence of γ on H', we obtain the numerical result

$$\xi_a^M = \xi_a - \frac{\gamma_e \beta}{\bar{\gamma_0} R} H_\gamma \sin 2B = \xi_a - 0.171'' H_\gamma \sin 2B , \qquad \eta_a^M = \eta_a . \tag{8.131}$$

We mention that although (8.129) and (8.131) have the same form, their physical meanings are completely different. In (8.129), the correction is an angle due to the bending of plumb line, and the gravity is approximated by the normal gravity. In (8.131), the correction is an angle due to the bending of the normal plumb line, which depends on normal gravity.

The correction term in (8.131), $0.171'' H_\gamma \sin 2B$, is the angle between the direction of the normal gravity on the telluroid and that of the normal to the reference Earth ellipsoid, the later being the same as the direction of the normal gravity on the reference Earth ellipsoid. The direction of the normal gravity can be directly computed according to (4.133) to (4.135) as accurately as the computer permits by including enough terms of the spherical harmonic series. On the reference Earth ellipsoid, it can also be computed using the closed-form expression (4.130). The angle between these directions thus computed is exact, while $0.171'' H_\gamma \sin 2B$ is approximate, where the height dependence of the curvature of the normal plumb line has been neglected in the formulation.

8.4.2 Interpolation Based on the Gravimetric Deflection of the Vertical

Let Σ be the region of interest, where data of the gravity anomaly are available, and the deflection of the vertical is measured by astro-geodetic observations at several points. The two data sets—data of deflection of the vertical and data of gravity anomaly—can then be combined to interpolate the deflection of the vertical.

We take as an example Stokes' theory and write Vening Meinesz' formula as

$$\xi - \xi_\Sigma = \xi_{\omega-\Sigma}, \qquad \eta - \eta_\Sigma = \eta_{\omega-\Sigma}; \tag{8.132}$$

where

$$\begin{cases} \xi_\Sigma = \dfrac{1}{4\pi\bar{\gamma}_0} \displaystyle\int_\Sigma \Delta g \dfrac{d}{d\psi} S(\psi) \cos A \, d\omega, \\[3mm] \eta_\Sigma = \dfrac{1}{4\pi\bar{\gamma}_0} \displaystyle\int_\Sigma \Delta g \dfrac{d}{d\psi} S(\psi) \sin A \, d\omega; \end{cases} \tag{8.133}$$

and

$$\begin{cases} \xi_{\omega-\Sigma} = \dfrac{1}{4\pi\bar{\gamma}_0} \displaystyle\int_{\omega-\Sigma} \Delta g \dfrac{d}{d\psi} S(\psi) \cos A \, d\omega, \\[3mm] \eta_{\omega-\Sigma} = \dfrac{1}{4\pi\bar{\gamma}_0} \displaystyle\int_{\omega-\Sigma} \Delta g \dfrac{d}{d\psi} S(\psi) \sin A \, d\omega. \end{cases} \tag{8.134}$$

What we realize is that $\xi - \xi_\Sigma$ and $\eta - \eta_\Sigma$ should vary much more smoothly than ξ and η, thus being suitable for interpolation. The method of computation is similar to that of the interpolation of gravity anomaly. The procedure is as follows: (1)

compute ξ_Σ and η_Σ at all points where ξ_a^S and η_a^S are measured as well as where ξ and η are to be computed by interpolation. (2) Subtract ξ_Σ and η_Σ from ξ_a^S and η_a^S. As ξ_a^S and η_a^S are actually measurements of ξ and η, what we obtain are $\xi - \xi_\Sigma$ and $\eta - \eta_\Sigma$ at the astro-geodetic stations. (3) Interpolate $\xi - \xi_\Sigma$ and $\eta - \eta_\Sigma$ according to the linear, bi-linear, or polynomial method discussed in Sect. 8.1.3 or some other method. (4) Add ξ_Σ and η_Σ to the values of $\xi - \xi_\Sigma$ and $\eta - \eta_\Sigma$ obtained by interpolation to obtain ξ and η.

In this procedure, the functions interpolated are in fact $\xi_{\omega-\Sigma}$ and $\eta_{\omega-\Sigma}$, which are assumed to be small and smooth. This procedure even does not require that the reference Earth ellipsoid used to compute ξ_a^S and η_a^S to be the same as that used to compute Δg. The result of interpolation is with respect to the reference Earth ellipsoid used to compute ξ_a^S and η_a^S. The reason is simple; the difference between the deflections of the vertical with respect to the two reference Earth ellipsoids, here we denote its two components as κ_ξ and κ_η, is small and varies smoothly within Σ, and the procedure can be understood as an interpolation of $\xi_{\omega-\Sigma}+\kappa_\xi$ and $\eta_{\omega-\Sigma}+\kappa_\eta$.

The readers may have already realized that the region where data of gravity are available should be larger than and enclose the region where the deflection of the vertical is observed and to be interpolated.

We have shown the fundamental principle for Stokes' theory. It is left to the readers to think about the interpolation of deflection of the vertical in Molodensky's theory.

8.4.3 Interpolation Based on Isostatic Deflection of the Vertical

We know that the cause of the deflection of the vertical is the irregularity of mass distribution of the Earth. The excess of topographic mass and the deficit of isostatic compensation around the bottom of the crust contribute the most. In other words, if there is no other irregularity of mass distribution except topographic mass and its isostatic compensation, the observed deflection of the vertical can be completely explained by an isostatic model. From this point of view, if the effect of topographic mass and its isostatic compensation are removed from the deflection of the vertical, the remaining part would be small and vary smoothly. This is just like the case of gravity anomaly, and hence, interpolation can be performed similarly.

We assume that the region of interest is Σ, where the deflection of the vertical is determined by astro-geodetic observations at some points, and we are going to interpolate the deflection of the vertical onto a set of other points within the region. The procedure of computation is practically the same as the interpolation of gravity anomaly: (1) compute the effect of the topography and its isostatic compensation within Σ on the deflection of the vertical, denoted as $\delta\xi$ and $\delta\eta$, at points where ξ_a^S and η_a^S are measured as well as where ξ and η are to be computed by interpolation. (2) Remove $\delta\xi$ and $\delta\eta$ from the data of ξ_a^S and η_a^S. (3) Interpolate for $\xi - \delta\xi$ and $\eta - \delta\eta$ using data of $\eta_a^S - \delta\xi$ and $\eta_a^S - \delta\eta$. (4) Add $\delta\xi$ and $\delta\eta$ into the result of interpolation to obtain the results of ξ and η. This is for the case of Stokes' theory as an example. The case of Molodensky's theory is similar. The formulae for computing the effects of the topographic mass and its isostatic compensation can be found in the previous two chapters.

An alternative way is to make use of the gravity anomaly, compute the contributions of the topographic mass and its isostatic compensation within Σ to the gravity anomaly, and then use the gravity anomaly to interpolate the deflection of the vertical according to the method described in the previous subsection.

The same as in the previous subsection, the region Σ should be larger than and enclose the region where the deflection of the vertical is observed and to be interpolated.

8.5   Astro-geodetic Determination of Geoidal Height and Height Anomaly

8.5.1   Astronomical Leveling

We provide formulations for both the geoidal height and height anomaly, although they can be converted from one to another base on their relation.

Determination of Geoidal Height

The deflection of the vertical represents the slope of the geoid. Therefore, we can compute the geoidal height using the deflection of the vertical. In Fig. 8.12, we show a vertical cross section, where p and q are two closely located points on the geoid, ϵ_a^S the deflection of the vertical within the cross section, and dN_a the increment of the geoidal height from p to q. We have

$$dN_a = \epsilon_a^S dl . \tag{8.135}$$

We use dN_a instead of dN to indicate that the geoidal height is with respect to the same reference ellipsoid as that used in the computation of the deflection of the vertical ξ_a^S and η_a^S, from which ϵ_a^S is computed. We use A to denote the azimuth of q as seen at p, as shown in Fig. 8.13, where X and Y point toward the north and east, respectively. We have

$$\epsilon_a^S = -\xi_a^S \cos A - \eta_a^S \sin A , \tag{8.136}$$

Fig. 8.12 Relation between the deflection of the vertical and the geoidal height

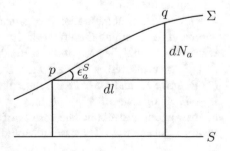

Fig. 8.13 The deflection of
the vertical along the azimuth
A

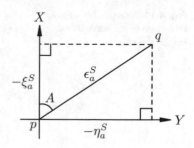

where the negative sign is due to the fact that ξ_a^S and η_a^S are southward and westward components of the deflection of the vertical.

We now assume that P and Q are two points on the geoid. We can obtain from (8.135) and (8.136) that

$$N_a^Q - N_a^P = -\int_P^Q (\xi_a^S \cos A + \eta_a^S \sin A)dl . \tag{8.137}$$

This integral can be computed as a sum

$$N_a^Q - N_a^P = -\sum_{i=0}^{n-1}(\xi_{ai}^S \cos A_i + \eta_{ai}^S \sin A_i)dl_i , \tag{8.138}$$

where the path from P to Q is divided into n segments through the points $P = P_0, P_1, P_2, \cdots, P_n = Q$, ξ_{ai}^S, and η_{ai}^S being the deflection of the vertical at P_i, A_i the azimuth of P_{i+1} as seen at P_i, and dl_i the distance between P_i and P_{i+1}.

If the deflection of the vertical at the points $P = P_0, P_1, P_2, \cdots, P_{n-1}$ is determined using astro-geodetic observations, the difference between the geoidal heights at P and Q can be computed according to (8.138). This method is referred to as astronomical leveling. If the deflection of the vertical at Q is also known, a more accurate formula is

$$N_a^Q - N_a^P = -\sum_{i=0}^{n-1}\left[\frac{\xi_{ai}^S + \xi_{a(i+1)}^S}{2}\cos A_i + \frac{\eta_{ai}^S + \eta_{a(i+1)}^S}{2}\sin A_i\right]dl_i . \tag{8.139}$$

Astronomical leveling can only be used to determine the difference between the geoidal height at two points, but not the geoidal height itself. In order to determine the geoidal height itself, it must already be known at an astro-geodetic station.

Determination of Height Anomaly
The dependence of height anomaly on deflection of the vertical is not as evident as that of geoidal height. In Fig. 8.14, we show a vertical cross section through two closely located points on the Earth's surface, p and q, where S is the reference Earth

Fig. 8.14 Relation between
deflection of the vertical and
height anomaly

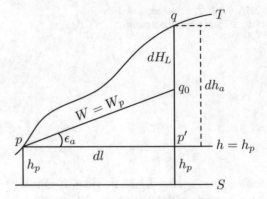

ellipsoid, T the Earth's surface (topography), $W = W_p$ the equipotential surface
of gravity through p, and h_P the geodetic height of p. The line pp' is parallel to
the reference Earth ellipsoid. By definition, ϵ_a is the deflection of the vertical with
respect to the normal to the reference Earth ellipsoid, which should be computed
using ξ_a and η_a as given in (8.124). The segment dH_L is the height of q above the
equipotential surface of gravity, $W = W_p$, thus being indeed the height increment
of spirit leveling. We use dh_a to denote the increment of geodetic height, where
the subscript a indicates also that the reference Earth ellipsoid is the same as the
one used to determine the deflection of vertical using astro-geodetic observations,
as implied by (8.124). We have

$$dh_a = \epsilon_a dl + dH_L \, , \tag{8.140}$$

where dl is the distance between the projections of p and q on the reference Earth
ellipsoid. By dividing the geodetic height into the sum of normal height and height
anomaly

$$dh_a = dH_\gamma + d\zeta_a \, , \tag{8.141}$$

we obtain a formula for height anomaly

$$d\zeta_a = \epsilon_a dl + dH_L - dH_\gamma \, . \tag{8.142}$$

We now reduce (8.142) to a form for practical computation. According to the
definition of normal height, $H_\gamma = C/\gamma_m$, we have

$$dH_\gamma = \frac{dC}{\gamma_m} - \frac{C}{\gamma_m^2} d\gamma_m \, . \tag{8.143}$$

At this point, the readers are suggested to reread the first three paragraphs of
Sect. 7.1.1 for the definitions of H_γ, C, and γ_m, as there is a need of very precise

understanding of these concepts in order to follow the subsequent derivations. As $dC = g\,dH_L$, where g is the gravity at p or q, which are on the Earth's surface, and $H_\gamma = C/\gamma_m$, we have

$$dH_\gamma = \frac{g\,dH_L}{\gamma_m} - \frac{H_\gamma}{\gamma_m}d\gamma_m\,. \tag{8.144}$$

In order to further simplify this formula, we express its first term at the right-hand side as

$$\frac{g\,dH_L}{\gamma_m} = dH_L + \frac{g-\gamma}{\gamma_m}dH_L + \frac{\gamma-\gamma_m}{\gamma_m}dH_L\,, \tag{8.145}$$

where γ is the normal gravity at height H_γ above the reference Earth ellipsoid, and hence, $\Delta G = g - \gamma$ is the gravity anomaly on the Earth's surface defined in Molodensky's theory. Taking the approximations $\gamma = \gamma_0 - 0.3086H_\gamma$ and $\gamma_m = \gamma_0 - 0.3086(H_\gamma/2)$, we have

$$\gamma - \gamma_m = -0.3086\frac{H_\gamma}{2}\,, \qquad d\gamma_m = d\gamma_0 - 0.3086\frac{dH_\gamma}{2}\,. \tag{8.146}$$

According to (8.145) and (8.146) and with $\Delta G = g - \gamma$, we can reduce (8.144) to

$$
\begin{aligned}
dH_\gamma &= dH_L + \frac{\Delta G}{\gamma_m}dH_L - \frac{0.3086}{\gamma_m}\frac{H_\gamma}{2}dH_L - \frac{H_\gamma}{\gamma_m}\left(d\gamma_0 - 0.3086\frac{dH_\gamma}{2}\right) \\
&= dH_L + \frac{\Delta G}{\gamma_m}dH_L - \frac{H_\gamma}{\gamma_m}d\gamma_0 + \frac{0.3086H_\gamma}{\gamma_m}\frac{dH_\gamma - dH_L}{2} \\
&= dH_L + \frac{\Delta G}{\gamma_m}dH_L - \frac{H_\gamma}{\gamma_m}d\gamma_0\,.
\end{aligned} \tag{8.147}
$$

The term neglected in the last step is of smaller order of magnitude. The free correction $0.3086\,H_\gamma$ is on the order of $\sim\!\Delta G$. The difference $dH_\gamma - dH_L$ is the correction to spirit leveling for normal height, which is on the order of $\sim\![(g - \gamma_m)/\gamma_m]dH_L \sim (\Delta G/\gamma_m)dH_L$. Therefore, the term neglected is on the order of $(\Delta G/\gamma_m)^2 dH_L$. By substituting (8.147) into (8.142), we obtain

$$d\zeta_a = \epsilon_a dl + \frac{H_\gamma}{\gamma_m}d\gamma_0 - \frac{\Delta G}{\gamma_m}dH_L\,. \tag{8.148}$$

In this result, $d\gamma_0$ is the increment of normal gravity on the reference Earth ellipsoid, which can be computed using the geodetic coordinates of p and q, while ϵ_a can be computed from ξ_a and η_a

$$\epsilon_a = -\xi_a \cos A - \eta_a \sin A\,, \tag{8.149}$$

where A is the azimuth of q as seen at p.

The result (8.148) can be further simplified. As γ_0 is a function of B alone, the second term at the right-hand side can be developed as

$$\frac{H_\gamma}{\gamma_m} d\gamma_0 = \frac{H_\gamma}{\gamma_m} \frac{\partial \gamma_0}{\partial l} dl = \frac{H_\gamma}{\gamma_m} \frac{1}{R} \frac{\partial \gamma_0}{\partial B} \cos A dl , \qquad (8.150)$$

where A is the azimuth of q as seen at p, as already mentioned in the previous paragraph. According to (4.150), we have

$$\frac{H_\gamma}{\gamma_m} d\gamma_0 = \frac{\gamma_e \beta}{\bar{\gamma_0} R} H_\gamma \sin 2B \cos A dl , \qquad (8.151)$$

where the approximation $\gamma_m = \bar{\gamma_0}$ has been made. We can now write (8.148) as, by making use of (8.149) and (8.151),

$$d\zeta_a = -\left[\left(\xi_a - \frac{\gamma_e \beta}{\bar{\gamma_0} R} H_\gamma \sin 2B\right) \cos A + \eta_a \sin A\right] dl - \frac{\Delta G}{\gamma_m} dH_L$$

$$= -(\xi_a^M \cos A + \eta_a^M \sin A) dl - \frac{\Delta G}{\gamma_m} dH_L , \qquad (8.152)$$

where (8.131) has been used to obtain the second identity. This is the differential of height anomaly. The difference between two points P and Q is

$$\zeta_a^Q - \zeta_a^P = -\int_P^Q (\xi_a^M \cos A + \eta_a^M \sin A) dl - \int_P^Q \frac{\Delta G}{\bar{\gamma_0}} dH_L , \qquad (8.153)$$

where we approximated γ_m by $\bar{\gamma_0}$ in the last term, which brings only a negligible influence.

As in the case of geoidal height, if the deflection of the vertical at the points $P = P_0, P_1, P_2, \cdots , P_{n-1}$ is determined using astro-geodetic observations, the difference between the height anomalies at P and Q can be computed according to

$$\zeta_a^Q - \zeta_a^P = -\sum_{i=0}^{n-1} (\xi_{ai}^M \cos A_i + \eta_{ai}^M \sin A_i) dl_i - \sum_{i=0}^{n-1} \frac{\Delta G_i}{\bar{\gamma_0}} dH_i . \qquad (8.154)$$

The path from P to Q is divided into n segments through the points $P = P_0, P_1, P_2, \cdots , P_n = Q, \xi_{ai}^M$, and η_{ai}^M being the deflection of the vertical at P_i, ΔG_i the gravity anomaly at P_i, A_i the azimuth of P_{i+1} as seen at P_i, dl_i the distance between P_i and P_{i+1}, and dH_i the difference of height between P_i and P_{i+1}. In the derivation, we assumed dH_i to be the difference of height determined by spirit leveling. However, as the term is small, replacing it with the normal height would not influence the accuracy of the result. If the deflection of the vertical is measured using astro-geodetic observations at $P = P_0, P_1, P_2, \cdots, P_{n-1}$, (8.154) could be used to compute the difference of height anomaly between P and Q. Hence, (8.154) is the practical formula for astronomical leveling of height anomaly. It can also be written in a similar form as (8.139).

The last term in (8.154) is small, as $\Delta G / \bar{\gamma}_0$ rarely exceeds 10^{-4}. It may be significant only in mountainous regions where dH_i is large. If the observation stations are carefully chosen to have virtually the same height, this term can be neglected.

We mention that a formula for determining the geoidal height using ϵ_a can be derived by dividing the geodetic height into the sum of orthometric height and geoidal height in (8.141) and (8.142). This is to replace the normal height by the orthometric height and the height anomaly by the geodetic height in these formulae. The subsequent formulation is left to the readers.

8.5.2 Astro-gravimetric Leveling

It is evident that in order to use astronomical leveling to determine geoidal height or height anomaly, the deflection of the vertical has to be known over a dense enough network, which could include astro-geodetic stations, as well as points where deflection of the vertical is obtained by interpolation.

If gravity data are available, there is a method to combine them directly with the deflection of the vertical, rather than interpolating the later. This method is referred to as astro-gravimetric leveling.

We take as an example the case of Stokes' theory. We assume that the astro-geodetic stations are all within the region Σ, where the gravity anomaly is known. Again, we use here the subscript a to indicate that the computation is done with respect to the same reference Earth ellipsoid as the one used in the computation of the deflection of the vertical using astro-geodetic observations. As in Sect. 8.4.2, we separate ξ_a and η_a into two parts:

$$\xi_a^S = \xi_\Sigma + (\xi_a^S - \xi_\Sigma), \qquad \eta_a^S = \eta_\Sigma + (\eta_a^S - \eta_\Sigma), \tag{8.155}$$

where ξ_Σ and η_Σ are defined in (8.133), and $\xi_a^S - \xi_\Sigma$ and $\eta_a^S - \eta_\Sigma$ are assumed to be small and vary smoothly, and linear interpolation can be used to infer their values at unknown points from known values. The difference of geoidal height between two points, P and Q, can then be expressed as

$$N_a^Q - N_a^P = -\int_P^Q (\xi_\Sigma \cos A + \eta_\Sigma \sin A) dl - \int_P^Q [(\xi_a^S - \xi_\Sigma) \cos A + (\eta_a^S - \eta_\Sigma) \sin A] dl. \tag{8.156}$$

The first term represents the contribution of gravity anomaly within Σ to $N_a^Q - N_a^P$, and we denote it as δN_Σ, which can be computed using Stokes' formula. We Assume that the distance between P and Q is not very long, and the region Σ is large enough, so that linear interpolation is applicable to $\xi_a^S - \xi_\Sigma$ and $\eta_a^S - \eta_\Sigma$, and thus the second integral at the right-hand side can be evaluated using the trapezoidal formula. We obtain

$$N_a^Q - N_a^P = \delta N_\Sigma - \left[\frac{(\xi_a^{SQ} - \xi_\Sigma^Q) - (\xi_a^{SP} - \xi_\Sigma^P)}{2} \cos A \right.$$

$$\left. + \frac{(\eta_a^{SQ} - \eta_\Sigma^Q) - (\eta_a^{SP} - \eta_\Sigma^P)}{2} \sin A \right] l_{PQ}, \tag{8.157}$$

where l_{PQ} is the distance between P and Q. This formula can also be written as

$$N_a^Q - N_a^P = -\left(\frac{\xi_a^{SQ} - \xi_a^{SP}}{2} \cos A + \frac{\eta_a^{SQ} - \eta_a^{SP}}{2} \sin A \right) l_{PQ}$$

$$+ \left[\delta N_\Sigma + \left(\frac{\xi_\Sigma^Q - \xi_\Sigma^P}{2} \cos A + \frac{\eta_\Sigma^Q - \eta_\Sigma^P}{2} \sin A \right) l_{PQ} \right], \tag{8.158}$$

where the first term is the result of astronomical leveling and the second the gravimetric correction. This is the basic equation of astro-gravimetric leveling. We mention that the indirect effect associated with the type of gravity anomaly chosen should also be considered in the computations.

In (8.158), ξ_a^{SP}, η_a^{SP} and ξ_a^{SQ}, η_a^{SQ} are assumed to be determined by astro-geodetic observations, the azimuth A of Q as seen at P as well as l_{PQ} are known from the geodetic coordinates of P and Q, δN_Σ is computed by integrating Stokes' formula, and ξ_Σ^P, η_Σ^P and ξ_Σ^Q, η_Σ^Q are computed by integrating Vening Meinesz' formula.

For the height anomaly, if the second term at the right-hand side of (8.153) is neglected and $\xi_a^M - \xi_\Sigma$ and $\eta_a^M - \eta_\Sigma$ are assumed to be interpolated linearly, we can obtain a formula in the same form as (8.158)

$$\zeta_a^Q - \zeta_a^P = -\left(\frac{\xi_a^{MQ} - \xi_a^{MP}}{2} \cos A + \frac{\eta_a^{MQ} - \eta_a^{MP}}{2} \sin A \right) l_{PQ}$$

$$+ \left[\Delta \zeta_\Sigma + \left(\frac{\xi_\Sigma^Q - \xi_\Sigma^P}{2} \cos A + \frac{\eta_\Sigma^Q - \eta_\Sigma^P}{2} \sin A \right) l_{PQ} \right]. \tag{8.159}$$

In this formula, $\Delta \zeta_\Sigma$ as well as ξ_Σ and η_Σ should be computed according to Molodensky's theory.

The removal of the contribution from Σ makes the residual of the deflection of the vertical varying much more smoothly, and the distance between P and Q can be much longer than that in the pure astronomical leveling. The same as in spirit leveling, the error in astronomical or astro-gravimetric leveling accumulates if the geoidal height or height anomaly is to be propagated farther and farther from a starting point.

8.5.3 GNSS and GNSS-Gravimetric Leveling

Geodetic height can be obtained by GNSS measurements. The subtraction of orthometric height or normal height measured by spirit leveling from the geodetic height yields the geoidal height or height anomaly. This method is referred to as GNSS-leveling, which makes direct measurement of the geoidal height or height anomaly much cheaper.

If gravity data are available, they can also be used to interpolate GNSS-leveling, and we refer to it as GNSS-gravimetric leveling. The approach is similar to the case of astro-gravimetric leveling. Take for example the case of geoidal height. We still use the subscript a to indicate the result with respect to the reference Earth ellipsoid used to process GNSS and spirit leveling data. We decompose N_a as

$$N_a = N_\Sigma + (N_a - N_\Sigma). \tag{8.160}$$

We assume that N_a is to be interpolated in a region within Σ, and Σ is large enough so that $N_a - N_\Sigma$ varies smoothly enough for interpolation. The computation is similar to the interpolation of gravity and deflection of the vertical, and it is not reproduced here. If no gravity data are available, interpolation can be performed based on an isostatic model, and again, the process is similar to the interpolation of gravity anomaly and deflection of the vertical.

It also happens that a gravimetric geoid, i.e., a geoid determined using data of gravity anomaly based on Stokes' theory, is available within a region, and GNSS-leveling data are available at some points within the same region. As GNSS-leveling data are supposed to be more precise and are with respect to the reference Earth ellipsoid adopted, a suitable approach of data fusion is to match the gravimetric geoid to the GNSS-leveling data based on least squares by removing a very low degree polynomial from the gravimetric geoid.

The case of height anomaly is similar.

References

Colombo, O. (1981). *Numerical methods for harmonic analysis on the sphere*. Report 310, Deptment of Geodetic Science and Surveying, The Ohio State University.

Fisher, N. I., Lewis, T., & Embleton, B. J. J. (1987). *Statistical analysis of spherical data*. Cambridge University Press.

Jekeli, C. (1981). *Alternative methods to smooth the Earth's gravity field*. Technical Report, 327, Department of Geodetic Science and Surveying, The Ohio State University.

Jekeli, C. (2017). *Spectral methods in geodesy and geophysics*. CRC Press.

Sjöberg L. E. (1980). A recurrence relation for the β_n-function. *Bulltin Géodésique, 54*, 69–72.

Further Reading

Fukushima, T. (2014). Numerical computation of spherical harmonics of arbitrary degree and order by extending exponent of floating pointnumbers: III integral. *Computers and Geosciences, 63*, 17–21. https://doi.org/10.1016/j.cageo.2013.10.010

Gerstl, M. (1980). On the recursive computation of the integrals of the associated Legendre functions. *Manuscripta Geodaetica, 5*, 181–199.

Molodensky, M. S., Eremeev, V. F., & Yurkina, M. I. (1962). *Methods for study of the external gravitational field and figure of the earth*. Transl. from Russian (1960), Israel Program for Scientific Translations, Jerusalem.

Pavlis, N. K., Holmes, S. A., Kenyon, S. C., & Factor, J. K. (2012). The development and evaluation of the Earth Gravitational Model 2008 (EGM2008). *Journal of Geophysical Research, 117*, B04406. https://doi.org/10.1029/2011JB008916

Pavlis, N. K., Holmes, S. A., Kenyon, S. C., & Factor, J. K. (2013). Corrections to "The development and evaluation of the Earth Gravitational Model 2008 (EGM2008)". *Journal of Geophysical Research, 118*, 2633. https://doi.org/10.1002/jgrb.50167

Rapp, R. H. (1986). Global Geopotential solutions. In *Lecture note in Earth science, Vol. 17: Mathematical and numerical techniques in physical geodesy* (pp. 365–415). Springer-Verlag

Sànchez, L., & Sideris, M. G. (2017). Vertical datum unification for the international height reference system (IHRS). *Geophysical Journal International, 209*, 570–586. https://doi.org/10.1093/gji/ggx025

Schwarz, K. P., Sideris, M. G., & Forsberg R. (1990). The use of FFT techniques in physical geodesy. *Geophysical Journal International, 100*, 485–514. https://doi.org/10.1111/j.1365-246X.1990.tb00701.x

Sjöberg L. E. (2009). The terrain correction in gravimetric geoid computation—is it needed? *Geophysical Journal International, 176*, 14–18. https://doi.org/10.1111/j.1365-246X.2008.03851.x

Wang, Y. M., Huang, J. L., Jiang, T., & Sideris, M. G. (2016). Local geoid computation. In: E.W. Grafarend (Ed.), *Encyclopedia of geodesy*. Springer.

Flattening and Gravity Inside the Earth

9

Abstract

This chapter models the flattening and gravity inside the Earth. The Earth is assumed to be in hydrostatic equilibrium, i.e., it behaves like a body of fluid rotating as whole without any relative motion inside. In this state, an equi-density surface is also an equipotential surface of gravity. This coincidence allows the determination of the shape of an equi-density surface or an equipotential surface of gravity inside the Earth, which turns out to be an ellipsoid of revolution accurate to the first order of flattening. A spherically symmetrical model of density inside the Earth determined using seismic data is required. Preparatory formulation includes the gravitational potential of a homogeneous aspherical shell, and then that of a heterogeneous aspherical body, both in spherical harmonic series. The imposition of the coincidence mentioned yields a second order differential equation of the flattening, called Clairaut's equation. The flattening of the Earth's fluid outer core and that of the solid inner core are parts of the solution of this equation. An approximate analytic solution of the equation is presented. Finally, the gravity inside the Earth and the moments of inertia of the body inside an equi-density surface are formulated.

9.1 Hydrostatic Equilibrium and the Spherical Earth Model

The meaning of hydrostatic equilibrium of a body is of twofold: (1) The body behaves like a fluid. (2) The body is in standstill state, i.e., there is no motion of one part with respect to another part within the body in standstill. Being fluid means that, within the body, the force acting by the portion of the body at one side of a surface on the portion at the other side of the surface is perpendicular to the surface, i.e., there is no shear force inside. Being in standstill implies that the total force acting on any portion of the body vanishes.

A simple example of a body in hydrostatic equilibrium is a number of different non-mixable fluids filling a bottle layer by layer in standstill state. The lower the layer, the higher the density.

We now derive the equation of hydrostatic equilibrium for the Earth. Let v_i be any portion within the Earth, and S its surface assumed to be smooth, so that an outward normal unit vector $\hat{e}_n$ could be defined. For the material within v_i to be in equilibrium, the total force acting on it must vanish. We only consider the largest forces here, namely, the gravity acting in the inside of v_i, $\rho \nabla W$, where ρ is the density, and W the gravity potential, and the pressure p exerted on the surface S. The electromagnetic force, which is relatively much smaller and is not really known within the Earth, is not considered. Hence, we have

$$\int_{v_i} \rho \nabla W dv - \int_S p \hat{e}_n dS = 0 \,, \tag{9.1}$$

where the negative sign in the second term is due to the facts that the pressure pushes to the inside of v_i, but $\hat{e}_n$ points to the outside of v_i. The surface integral can be converted into a volume integral according to Gauss' theorem (2.1), which we copy to here for convenience of reference,

$$\int_{v_i} \left(\frac{\partial P}{\partial x} + \frac{\partial Q}{\partial y} + \frac{\partial R}{\partial z} \right) dv = \int_S \left[P \cos(\hat{e}_n, \hat{e}_x) + Q \cos(\hat{e}_n, \hat{e}_y) + R \cos(\hat{e}_n, \hat{e}_z) \right] dS \,. \tag{9.2}$$

Three equations for p can be derived

$$\begin{cases} \displaystyle \int_{v_i} \frac{\partial p}{\partial x} dv = \int_S p \cos(\hat{e}_n, \hat{e}_x) dS \,, \\[2ex] \displaystyle \int_{v_i} \frac{\partial p}{\partial y} dv = \int_S p \cos(\hat{e}_n, \hat{e}_y) dS \,, \\[2ex] \displaystyle \int_{v_i} \frac{\partial p}{\partial z} dv = \int_S p \cos(\hat{e}_n, \hat{e}_z) dS \,, \end{cases} \tag{9.3}$$

where the first one is obtained by setting $P = p$ and $Q = R = 0$, the second by setting $Q = p$ and $P = R = 0$, and the third by setting $R = p$ and $P = Q = 0$. By multiplying $\hat{e}_x$, $\hat{e}_y$, and $\hat{e}_z$ with the three equations in (9.3), respectively, and then adding together, we obtain

$$\int_{v_i} \nabla p \, dv = \int_S p \hat{e}_n dS \,, \tag{9.4}$$

where we used the relations

$$\nabla p = \frac{\partial p}{\partial x} \hat{e}_x + \frac{\partial p}{\partial y} \hat{e}_y + \frac{\partial p}{\partial z} \hat{e}_z \,, \tag{9.5}$$

$$\hat{e}_n = \cos(\hat{e}_n, \hat{e}_x)\hat{e}_x + \cos(\hat{e}_n, \hat{e}_y)\hat{e}_y + \cos(\hat{e}_n, \hat{e}_z)\hat{e}_z \,. \tag{9.6}$$

The substitution of (9.4) into (9.1) yields

$$\int_{v_i} (\rho \nabla W - \nabla p) dv = 0 \,. \tag{9.7}$$

As v_i is arbitrarily chosen, for this equation to hold, we must have

$$\rho \nabla W - \nabla p = 0 \,. \tag{9.8}$$

This is the equation of hydrostatic equilibrium.

In (9.8), p can be eliminated by taking the curl at both sides of it. We first obtain

$$\nabla \times (\rho \nabla W) - \nabla \times (\nabla p) = 0 \,. \tag{9.9}$$

The curl of a vector $\vec{A} = A_x \hat{e}_x + A_y \hat{e}_y + A_z \hat{e}_z$ is defined as, written in the form of the three components,

$$\nabla \times \vec{A} = \begin{bmatrix} \dfrac{\partial A_z}{\partial y} - \dfrac{\partial A_y}{\partial z} \\[2mm] \dfrac{\partial A_x}{\partial z} - \dfrac{\partial A_z}{\partial x} \\[2mm] \dfrac{\partial A_y}{\partial x} - \dfrac{\partial A_x}{\partial y} \end{bmatrix} \,. \tag{9.10}$$

We have

$$\nabla \times (\rho \nabla W) = \begin{bmatrix} \dfrac{\partial}{\partial y}\left(\rho \dfrac{\partial W}{\partial z}\right) - \dfrac{\partial}{\partial z}\left(\rho \dfrac{\partial W}{\partial y}\right) \\[2mm] \dfrac{\partial}{\partial z}\left(\rho \dfrac{\partial W}{\partial x}\right) - \dfrac{\partial}{\partial x}\left(\rho \dfrac{\partial W}{\partial z}\right) \\[2mm] \dfrac{\partial}{\partial x}\left(\rho \dfrac{\partial W}{\partial y}\right) - \dfrac{\partial}{\partial y}\left(\rho \dfrac{\partial W}{\partial x}\right) \end{bmatrix} = \nabla \rho \times \nabla W \,, \tag{9.11}$$

where $\nabla \rho \times \nabla W$ is the usual vector product of the two vectors $\nabla \rho$ and ∇W. We can easily obtain $\nabla \times (\nabla p) = 0$ by setting $\rho = 1$ and $W = p$ in the above formula. Hence, (9.9) can be finally reduced to

$$\nabla \rho \times \nabla W = 0 \,, \tag{9.12}$$

which is another equation of hydrostatic equilibrium.

We obtained the equation of hydrostatic equilibrium in two forms (9.8) and (9.12). The former implies that the gradient of the gravity potential and that of the pressure are in the same direction, and thus the equipotential surfaces of gravity and the isobar surfaces (i.e., equi-pressure surfaces) coincide. The latter implies that the gradient of the density and that of the gravity potential are in the same direction; hence, the equi-density surfaces and equipotential surfaces of gravity coincide. This property is the basis for studying the shape of equi-value surfaces of density, gravity potential, and pressure inside the Earth. Under the long term action of gravity, the materials within the Earth do flow like fluids toward the hydrostatic equilibrium state.

The simplest Earth model in hydrostatic equilibrium is the spherically symmetrical non-rotating Earth model, where the density is a function of the radial distance r alone determined using seismic tomography. For this kind of the Earth model, the gravity potential W degenerates to the gravitational potential V, and (9.8) is simplified to

$$\frac{dp}{dr} = \rho \frac{dV}{dr} = -\rho g \,, \tag{9.13}$$

and (9.12) is trivial. The Poisson's equation satisfied by V, here written in the form $-\Delta V = 4\pi G \rho$, can be simplified to, by taking reference to (3.14),

$$- \Delta V = -\frac{1}{r^2} \frac{d}{dr} \left(r^2 \frac{dV}{dr} \right) = \frac{1}{r^2} \frac{d}{dr} \left(r^2 g \right) = \frac{dg}{dr} + \frac{2}{r} g = 4\pi G \rho \,. \tag{9.14}$$

A first observance by setting $r = 0$ in this formula is that, as both dg/dr and ρ are finite at $r = 0$, g must vanish when $r = 0$. The formula for g can be easily obtained by solving the ordinary differential equation (9.14), which we write in the form

$$\frac{d}{dr}(r^2 g) = 4\pi G r^2 \rho \,. \tag{9.15}$$

Integrate it over r from 0 to r_0, and then divide both sides by $1/r_0^2$. We obtain

$$g(r_0) = 4\pi G \frac{1}{r_0^2} \int_0^{r_0} r^2 \rho dr \,. \tag{9.16}$$

The pressure can be obtained according to (9.13) as, by taking into account of the fact that $p(R) = 0$, i.e., the pressure on the Earth's surface vanishes,

$$p(r_0) = \int_R^{r_0} \frac{dp}{dr} dr = \int_{r_0}^R \rho g dr \,. \tag{9.17}$$

In spherically symmetrical Earth models, e.g., the preliminary reference Earth model (PREM)[1] (Dziewonski & Anderson, 1981), the Earth's interior is composed

---

[1] We mention that $G = 6.67 \times 10^{-11} \text{m}^3 \cdot \text{kg}^{-1} \cdot \text{s}^{-2}$ was adopted in this classical model.

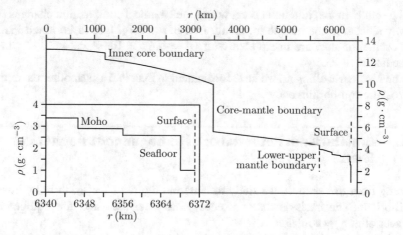

Fig. 9.1 Profile of density of the spherically symmetrical Earth model PREM

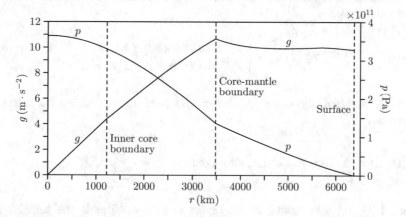

Fig. 9.2 Profiles of gravity and pressure of the PREM

of, from the center to the surface (the values are slightly different in different models), the solid inner core of radius 1221.5 km, the fluid outer core of radius 1221.5 to 3480 km, the solid lower mantle of radius 3480 to 5701 km, the solid upper mantle of radius 5701 to 6346.6 km, the solid crust of radius 6346.6 to 6368 km, and the fluid ocean of radius 6368 to 6371 km. A profile of the density is shown in Fig. 9.1, where the major boundaries of jump, the inner core boundary, core–mantle boundary, lower–upper mantle boundary, Moho, and seafloor are marked. A profile of the gravity, together with that of the pressure, is shown in Fig. 9.2. The values of gravity are 0, 4.400, 10.682, and 9.816 m · s^{-2} at the Earth's center, the inner core boundary, the core–mantle boundary, and the Earth's surface, respectively. It remains almost constant from the Earth's surface down to the core–mantle boundary and then almost linearly drops to zero at the Earth's center. This is due to the large jump of density at the core–mantle boundary from 5.56 g · cm^{-3} in the mantle to

$9.90 \, \text{g} \cdot \text{cm}^{-3}$ in the fluid outer core, where the chemical composition changes from mainly silicate in the mantle to mainly iron in the core. It is to be noted that the free oscillation data are interpreted with $G = 6.67 \times 10^{-11} \, \text{m}^3 \cdot \text{kg}^{-1} \cdot \text{s}^{-2}$ in the computation of PREM.

The Maclaurin ellipsoid we have formulated in Sect. 4.5.1 is another example of hydrostatic equilibrium body.

9.2   Gravitational Potential of a Homogeneous Aspherical Shell

Starting from this section, we study the internal figure of the Earth in hydrostatic equilibrium. The basis is that the equi-density surfaces and the equipotential surfaces of gravity coincide.

This section and the next are preparatory, where spherical harmonic series will be heavily used. For making formulae shorter, we introduce a new notation of surface spherical harmonics

$$\mathcal{H}_n^k(\theta, \lambda) = \begin{cases} P_n^{|k|}(\cos\theta) \cos|k|\lambda \,, \ k \geq 0 \,; \\ P_n^{|k|}(\cos\theta) \sin|k|\lambda \,, \ k < 0 \,. \end{cases} \tag{9.18}$$

With this notation, the degree n surface spherical harmonics can be expressed shortly as

$$Y_n^k(\theta, \lambda) = \sum_{k=0}^{n} (A_n^k \cos k\lambda + B_n^k \sin k\lambda) P_n^k(\cos\theta) = \sum_{k=-n}^{n} A_n^k \mathcal{H}_n^k(\theta, \lambda) \,. \tag{9.19}$$

When $k \geq 0$, the meaning of A_n^k in the two expressions is the same. When $k < 0$, $A_n^k = B_n^{|k|}$. We will need the orthogonality and addition theorem of spherical harmonics expressed using $\mathcal{H}_n^k(\theta, \lambda)$, which can be inferred right away from (3.247) to (3.249) and (3.268) and are written in the following forms for subsequent reference

$$\int_w \mathcal{H}_n^k(\theta', \lambda') \mathcal{H}_m^l(\theta', \lambda') d\omega = \delta_{mn} \delta_{lk} (1 + \delta_k) \frac{2\pi}{2n+1} \frac{(n+|k|)!}{(n-|k|)!} \,, \tag{9.20}$$

$$P_m(\cos\psi) = \sum_{l=-m}^{m} \frac{2}{1+\delta_l} \frac{(m-|l|)!}{(m+|l|)!} \mathcal{H}_m^l(\theta, \lambda) \mathcal{H}_m^l(\theta', \lambda') \,. \tag{9.21}$$

The variables of integration in (9.20) are θ' and λ', i.e., $d\omega = \sin\theta' d\theta' d\lambda'$. These formulae can be easily verified by substituting (9.18) into them, and then compare the result with the original formulae (3.247) to (3.249) and (3.268).

We consider an aspherical homogeneous thin shell, i.e., it has a constant density and a nearly spherical irregular shape. We express its inner surface as (notations are chosen for future convenience)

$$r'(\theta', \lambda') = q \left[1 + \sum_{n=1}^{\infty} \sum_{k=-n}^{n} f_n^k \mathcal{H}_n^k(\theta', \lambda') \right], \qquad (9.22)$$

and its outer surface as

$$r'(\theta', \lambda') + h(\theta', \lambda') = (q + \delta q) \left[1 + \sum_{n=1}^{\infty} \sum_{k=-n}^{n} \left(f_n^k + \delta f_n^k \right) \mathcal{H}_n^k(\theta', \lambda') \right].$$
$$(9.23)$$

These mean that the point (r', θ', λ') is on the inner surface, and the point $(r' + h, \theta', \lambda')$ is on the outer surface. The thickness of the shell is

$$h(\theta', \lambda') = \delta q \left[1 + \sum_{n=1}^{\infty} \sum_{k=-n}^{n} \left(f_n^k + q \frac{\delta f_n^k}{\delta q} \right) \mathcal{H}_n^k(\theta', \lambda') \right]. \qquad (9.24)$$

We denote the density of the shell as ρ. As shown in Fig. 9.3, the gravitational potential of the shell at the point (r, θ, λ) is

$$\delta V(r, \theta, \lambda) = G \int_{\omega} \frac{\rho h}{l} r'^2 d\omega, \qquad (9.25)$$

where the variables of integration are θ' and λ', i.e., $d\omega = \sin \theta' d\theta' d\lambda'$, and

$$l = \left(r^2 + r'^2 - 2rr' \cos \psi \right)^{1/2}. \qquad (9.26)$$

Fig. 9.3 Gravitational potential of an aspherical homogeneous thin shell

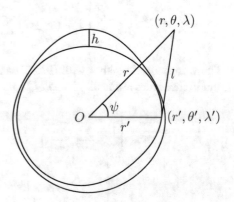

In (9.25), the thin shell is considered as a material surface of density ρh, and $dS = r'^2 d\omega$ is a surface element on the shell at the point (r', θ', λ'). According to the generating equation of the Legendre function, $1/l$ can be developed to a Legendre series. Taking reference to (3.93), we have

$$
\frac{1}{l} =
\begin{cases}
\displaystyle\sum_{m=0}^{\infty} \frac{r'^m}{r^{m+1}} P_m(\cos \psi), & r > r'; \\
\displaystyle\sum_{m=0}^{\infty} \frac{r^m}{r'^{m+1}} P_m(\cos \psi), & r < r'.
\end{cases}
\tag{9.27}
$$

By substituting (9.27) into (9.25), we obtain two equations, one for $r < r'$, and one for $r > r'$. The point (r, θ, λ) is in the interior of the shell when $r < r'$, and we denote δV as δV_I. We have

$$
\delta V_I(r, \theta, \lambda) = G\rho \int_\omega \sum_{m=0}^{\infty} \frac{h r^m}{r'^{m-1}} P_m(\cos \psi) d\omega .
\tag{9.28}
$$

The point (r, θ, λ) is in the exterior of the shell when $r > r_1$, and we denote δV as δV_E. We have

$$
\delta V_E(r, \theta, \lambda) = G\rho \int_\omega \sum_{m=0}^{\infty} \frac{h r'^{m+2}}{r^{m+1}} P_m(\cos \psi) d\omega .
\tag{9.29}
$$

As the shell is assumed to be very close to a sphere, the coefficients f_n^k are small quantities. By neglecting their second and higher order terms, we obtain

$$
\frac{1}{r'^{m-1}} = \frac{1}{q^{m-1}} \left[1 - (m-1) \sum_{n=1}^{\infty} \sum_{k=-n}^{n} f_n^k \mathcal{H}_n^k(\theta', \lambda') \right] ,
\tag{9.30}
$$

$$
r'^{m+2} = q^{m+2} \left[1 + (m+2) \sum_{n=1}^{\infty} \sum_{k=-n}^{n} f_n^k \mathcal{H}_n^k(\theta', \lambda') \right] .
\tag{9.31}
$$

These formulae, together with (9.24), can be used to obtain, by neglecting second and higher order terms of f_n^k and δf_n^k,

$$
\frac{h}{r'^{m-1}} = \frac{\delta q}{q^{m-1}} \left\{ 1 + \sum_{n=1}^{\infty} \sum_{k=-n}^{n} \left[-(m-2) f_n^k + q \frac{\delta f_n^k}{\delta q} \right] \mathcal{H}_n^k(\theta', \lambda') \right\} ,
\tag{9.32}
$$

$$
h r'^{m+2} = q^{m+2} \delta q \left\{ 1 + \sum_{n=1}^{\infty} \sum_{k=-n}^{n} \left[(m+3) f_n^k + q \frac{\delta f_n^k}{\delta q} \right] \mathcal{H}_n^k(\theta', \lambda') \right\} .
\tag{9.33}
$$

The final expression of δV_I can be obtained by substituting (9.32) and (9.21) into (9.28). Due to (9.20), only the terms $m = n$ and $l = k$ are non-vanishing. By writing the term $n = 0$ and $k = 0$ separately, we obtain

$$\delta V_I(r, \theta, \lambda) = G\rho \int_\omega q\,\delta q\,d\omega + G\rho \sum_{n=1}^\infty \sum_{k=-n}^n \frac{r^n \delta q}{q^{n-1}} \left[-(n-2)f_n^k + q\frac{\delta f_n^k}{\delta q} \right]$$

$$\times \frac{2}{1+\delta_k} \frac{(n-|k|)!}{(n+|k|)!} \left\{ \int_\omega \left[\mathcal{H}_n^k(\theta', \lambda') \right]^2 d\omega \right\} \mathcal{H}_n^k(\theta, \lambda).$$

(9.34)

By using (9.20) again, we obtain

$$\delta V_I(r, \theta, \lambda) = 4\pi G\rho q\,\delta q + G\rho \sum_{n=1}^\infty \sum_{k=-n}^n \frac{r^n \delta q}{q^{n-1}} \frac{4\pi}{2n+1} \left[-(n-2)f_n^k + q\frac{\delta f_n^k}{\delta q} \right] \mathcal{H}_n^k(\theta, \lambda)$$

$$= \frac{4\pi G}{3} \left\{ 3\rho q\,\delta q + \sum_{n=1}^\infty \sum_{k=-n}^n \frac{3}{2n+1} r^n \left[-\rho \frac{n-2}{q^{n-1}} f_n^k \delta q \right.\right.$$

$$\left.\left. + \rho\frac{1}{q^{n-2}} \frac{\delta f_n^k}{\delta q} \delta q \right] \mathcal{H}_n^k(\theta, \lambda) \right\}.$$

(9.35)

The final result of δV_E can be inferred just by inspecting the differences between (9.28) and (9.29) and between (9.32) and (9.33). We obtain

$$\delta V_E(r, \theta, \lambda) = \frac{4\pi G}{3} \left\{ \frac{3\rho q^2 \delta q}{r} + \sum_{n=1}^\infty \sum_{k=-n}^n \frac{3}{2n+1} \frac{1}{r^{n+1}} \right.$$

$$\left. \times \left[(n+3)\rho q^{n+2} f_n^k \delta q + \rho q^{n+3} \frac{\delta f_n^k}{\delta q} \delta q \right] \mathcal{H}_n^k(\theta, \lambda) \right\}. \quad (9.36)$$

These last two formulae are the result of this section.

9.3 Gravitational Potential of a Heterogeneous Aspherical Body

We consider an Earth model very close to spherical symmetry. Any of its equi-density surfaces can be expressed as

$$r'(\theta', \lambda') = q \left[1 + \sum_{n=1}^\infty \sum_{k=-n}^n f_n^k \mathcal{H}_n^k(\theta', \lambda') \right], \quad (9.37)$$

where f_n^k is a function of q, and so is the density ρ. Each value of q corresponds to an equi-density surface, where (r', θ', λ') satisfies this equation.

We divide this Earth model to thin shells by equi-density surfaces, and each thin shell can be treated as an aspherical homogeneous thin shell as that in the previous section. We consider a thin shell

$$\delta q = dq \,. \tag{9.38}$$

We then have

$$\frac{\delta f_n^k}{\delta q} \delta q = \delta f_n^k = \frac{d f_n^k}{dq} dq \,. \tag{9.39}$$

The formulae of the gravitational potential of this thin shell can be obtained according to (9.35) and (9.36)

$$\delta V_I(r, \theta, \lambda) = \frac{4\pi G}{3} \left\{ 3\rho q dq + \sum_{n=1}^{\infty} \sum_{k=-n}^{n} \frac{3}{2n+1} r^n \left[\rho d \left(\frac{f_n^k}{q^{n-2}} \right) \right] \mathcal{H}_n^k(\theta, \lambda) \right\} , \tag{9.40}$$

$$\delta V_E(r, \theta, \lambda) = \frac{4\pi G}{3} \left\{ \frac{3\rho q^2 dq}{r} + \sum_{n=1}^{\infty} \sum_{k=-n}^{n} \frac{3}{2n+1} \frac{1}{r^{n+1}} \left[\rho d \left(q^{n+3} f_n^k \right) \right] \mathcal{H}_n^k(\theta, \lambda) \right\} . \tag{9.41}$$

We now derive the formula of the gravitational potential of the aspherical heterogeneous body on the equi-density surface

$$r(\theta, \lambda) = r_0 \left[1 + \sum_{n=1}^{\infty} \sum_{k=-n}^{n} f_n^k \mathcal{H}_n^k(\theta, \lambda) \right] . \tag{9.42}$$

The substitution of this formula into (9.40) and (9.41) yields, by neglecting second and higher order terms of dq,

$$\delta V_I(r, \theta, \lambda) = \frac{4\pi G}{3} \left\{ 3\rho q dq + \sum_{n=1}^{\infty} \sum_{k=-n}^{n} \frac{3}{2n+1} r_0^n \left[\rho d \left(\frac{f_n^k}{q^{n-2}} \right) \right] \mathcal{H}_n^k(\theta, \lambda) \right\} . \tag{9.43}$$

$$\delta V_E(r, \theta, \lambda) = \frac{4\pi G}{3} \left\{ \frac{3\rho q^2 dq}{r_0} \left[1 - \sum_{n=1}^{\infty} \sum_{k=-n}^{n} f_n^k \mathcal{H}_n^k(\theta, \lambda) \right] \right.$$
$$\left. + \sum_{n=1}^{\infty} \sum_{k=-n}^{n} \frac{3}{2n+1} \frac{1}{r_0^{n+1}} \left[\rho d \left(q^{n+3} f_n^k \right) \right] \mathcal{H}_n^k(\theta, \lambda) \right\} . \tag{9.44}$$

We emphasize that V_I is the gravitational potential of a thin shell with the equi-density surface (9.42) inside it. In other words, V_I is the gravitational potential of a thin shell outside the equi-density surface (9.42). Similarly, V_E is the gravitational potential of a thin shell inside the equi-density surface (9.42). Hence, the total gravitational potential on the equi-density surface (9.42) is

$$V = \int_0^{r_0} dV_E + \int_{r_0}^R dV_I , \tag{9.45}$$

where R is the value of q in (9.37) as well as that of r_0 in (9.42) on the surface of the Earth model. Substitute (9.43) and (9.44) into (9.45). We obtain

$$
\begin{aligned}
V =& \frac{4\pi G}{3}\left\{ \frac{1}{r_0}\int_0^{r_0} 3\rho q^2 dq\left[1 - \sum_{n=1}^{\infty}\sum_{k=-n}^{n} f_n^k \mathcal{H}_n^k(\theta,\lambda)\right]\right.\\
&+ \sum_{n=1}^{\infty}\sum_{k=-n}^{n}\frac{3}{2n+1}\frac{1}{r_0^{n+1}}\left[\int_0^{r_0}\rho d\left(q^{n+3} f_n^k\right)\right]\mathcal{H}_n^k(\theta,\lambda)\right\}\\
&+ \frac{4\pi G}{3}\left\{\int_{r_0}^R 3\rho q dq + + \sum_{n=1}^{\infty}\sum_{k=-n}^{n}\frac{3}{2n+1}r_0^n\left[\int_{r_0}^R\rho d\left(\frac{f_n^k}{q^{n-2}}\right)\right]\mathcal{H}_n^k(\theta,\lambda)\right\}\\
=& \frac{4\pi G}{3}\left[\frac{1}{r_0}\int_0^{r_0}3\rho q^2 dq + \int_{r_0}^R 3\rho q dq + \sum_{n=1}^{\infty}\sum_{k=-n}^{n}\left\{-\frac{f_n^k}{r_0}\int_0^{r_0}3\rho q^2 dq\right.\right.\\
&+ \frac{3}{2n+1}\left[\frac{1}{r_0^{n+1}}\int_0^{r_0}\rho d\left(q^{n+3}f_n^k\right) + r_0^n\int_{r_0}^R\rho d\left(\frac{f_n^k}{q^{n-2}}\right)\right]\right\}\mathcal{H}_n^k(\theta,\lambda)\right].
\end{aligned}
\tag{9.46}
$$

This is the result of this section.

9.4 Clairaut's Equation of the Earth's Internal Flattening

The principle for determining the Earth's internal flattening is that the equipotential surfaces of gravity coincide with the equi-density surfaces. We always assume that the Earth is very close to spherical symmetry.

The gravitational potential of such a body has already been formulated in the previous two sections. The centrifugal potential is given in (1.43). On the equi-density surface (9.42), it can be written as

$$Q = \frac{1}{2}\Omega^2 r^2 \sin^2\theta = \frac{1}{3}\Omega^2 r_0^2 - \frac{1}{3}\Omega^2 r_0^2 \mathcal{H}_2^0(\theta,\lambda) , \tag{9.47}$$

where Ω is the rotation rate of the Earth, and by definition,

$$\mathcal{H}_2^0(\theta, \lambda) = P_2(\cos\theta) = \frac{3}{2}\cos^2\theta - \frac{1}{2}. \tag{9.48}$$

As Q itself is a small quantity, which is on the relative order of f_n^k in magnitude as compared to V, we have approximated r by r_0. The gravity potential $W = V + Q$ on the equi-density surface (9.42) is, according to (9.46) and (9.47),

$$W = \frac{4\pi G}{3}\left[\frac{1}{r_0}\int_0^{r_0} 3\rho q^2 dq + \int_{r_0}^R 3\rho q dq + \frac{\Omega^2 r_0^2}{4\pi G} + \sum_{n=1}^\infty \sum_{k=-n}^n \left\{-\frac{f_n^k}{r_0}\int_0^{r_0} 3\rho q^2 dq\right.\right.$$
$$\left.\left. + \frac{3}{2n+1}\left[\frac{1}{r_0^{n+1}}\int_0^{r_0}\rho d\left(q^{n+3} f_n^k\right) + r_0^n \int_{r_0}^R \rho d\left(\frac{f_n^k}{q^{n-2}}\right)\right] - \underbrace{\frac{\Omega^2 r_0^2}{4\pi G}}_{n=2,k=0}\right\}\mathcal{H}_n^k(\theta, \lambda)\right], \tag{9.49}$$

where the term underlined with "$n = 2, k = 0$" exists only in the particular case when $n = 2$ and $k = 0$.

The meaning of (9.42) is that each value of r_0 defines an equi-density surface, i.e., for each value of r_0, the point (r, θ, λ) with r satisfying (9.42) is on an equi-density surface. The coincidence of equi-density surfaces and equipotential surfaces of gravity implies that, for any r_0,

$$W(r, \theta, \lambda) = \text{Constant}. \tag{9.50}$$

Hence, all coefficients of $\mathcal{H}_n^k(\theta, \lambda)$ in (9.49) should vanish except the case $n = k = 0$. We obtain

$$-\frac{f_n^k}{r_0}\int_0^{r_0} 3\rho q^2 dq + \frac{3}{2n+1}\left[\frac{1}{r_0^{n+1}}\int_0^{r_0}\rho d\left(q^{n+3} f_n^k\right) + r_0^n \int_{r_0}^R \rho d\left(\frac{f_n^k}{q^{n-2}}\right)\right]$$
$$-\underbrace{\frac{\Omega^2 r_0^2}{4\pi G}}_{n=2,k=0} = 0. \tag{9.51}$$

This is the equation to determine f_n^k, which can be further simplified. It will be shown that only the coefficient f_2^0 is non-vanishing, which is expected, since the rotation influences this term only. Without rotation, the shape in hydrostatic equilibrium is expected to be spherically symmetrical.

In forthcoming derivations, the Newton–Leibniz formulae

$$\frac{d}{dr_0}\int_0^{r_0} f(q)dq = f(r_0) \quad \text{and} \quad \frac{d}{dr_0}\int_{r_0}^R f(q)dq = -f(r_0) \tag{9.52}$$

will be repeatedly used.

Multiply both sides of (9.51) by r_0^{n+1}, and then take the derivative with respect to r_0. We obtain

$$
\left(-nr_0^{n-1} f_n^k - r_0^n \frac{df_n^k}{dr_0} \right) \int_0^{r_0} 3\rho q^2 dq - 3\rho r_0^{n+2} f_n^k
$$

$$
+ \frac{3}{2n+1} \left[(n+3)\rho r_0^{n+2} f_n^k + \rho r_0^{n+3} \frac{df_n^k}{dr_0} + (2n+1)r_0^{2n} \int_{r_0}^R \rho d\left(\frac{f_n^k}{q^{n-2}} \right) \right.
$$

$$
\left. + (n-2)\rho r_0^{n+2} f_n^k - \rho r_0^{n+3} \frac{df_n^k}{dr_0} \right] - \frac{5\Omega^2 r_0^4}{4\pi G} \Bigg|_{n=2, k=0} = 0, \tag{9.53}
$$

which can be simplified to, by rearranging terms,

$$
\left(-nr_0^{n-1} f_n^k - r_0^n \frac{df_n^k}{dr_0} \right) \int_0^{r_0} 3\rho q^2 dq + 3r_0^{2n} \int_{r_0}^R \rho d\left(\frac{f_n^k}{q^{n-2}} \right) - \frac{5\Omega^2 r_0^4}{4\pi G} \Bigg|_{n=2, k=0} = 0. \tag{9.54}
$$

Divide both sides of this equation by r_0^{2n}, and then take the derivative with respect to r_0. We obtain

$$
\left[\frac{n(n+1)}{r_0^{n+2}} f_n^k - \frac{n}{r_0^{n+1}} \frac{df_n^k}{dr_0} + \frac{n}{r_0^{n+1}} \frac{df_n^k}{dr_0} - \frac{1}{r_0^n} \frac{d^2 f_n^k}{dr_0^2} \right] \int_0^{r_0} 3\rho q^2 dq
$$

$$
+ 3\rho r_0^2 \left(-\frac{n}{r_0^{n+1}} f_n^k - \frac{1}{r_0^n} \frac{df_n^k}{dr_0} \right) + \rho \frac{3(n-2)}{r_0^{n-2}} f_n^k - \frac{3\rho}{r_0^{n-2}} \frac{df_n^k}{dr_0} = 0, \tag{9.55}
$$

which can be written as, by multiplying both sides by r_0^n and then rearranging terms,

$$
\left[\frac{d^2 f_n^k}{dr_0^2} - \frac{n(n+1)}{r_0^2} f_n^k \right] \int_0^{r_0} 3\rho q^2 dq + 6\rho r_0^2 \left(\frac{df_n^k}{dr_0} + \frac{f_n^k}{r_0} \right) = 0. \tag{9.56}
$$

Define

$$
D = \frac{3}{r_0^3} \int_0^{r_0} \rho q^2 dq. \tag{9.57}
$$

We can finally write (9.56) as, by dividing both its sides by r_0^3,

$$
D \left[\frac{d^2 f_n^k}{dr_0^2} - \frac{n(n+1)}{r_0^2} f_n^k \right] + \frac{6\rho}{r_0} \left(\frac{df_n^k}{dr_0} + \frac{f_n^k}{r_0} \right) = 0. \tag{9.58}
$$

Before further studying (9.58), let us show that D is the average density of the mass enclosed within the equi-density surface defined by (9.42). By definition, the mass is

$$M_{r_0} = \int_0^{2\pi} \int_0^{\pi} \int_0^r \rho(r', \theta', \lambda') r'^2 \sin\theta' dr' d\theta' d\lambda', \tag{9.59}$$

where r is related to r_0 according to (9.42). Substitute r' by q according to (9.37). We can write (9.59) as, with $r'^2 dr' = d(r'^3/3)$, and neglecting second and higher order terms of f_n^k,

$$M_{r_0} = \int_0^{2\pi} \int_0^{\pi} \int_0^{r_0} \rho(q) d \left\{ \frac{q^3}{3} \left[1 + 3 \sum_{n=1}^{\infty} \sum_{k=-n}^{n} f_n^k \mathcal{H}_n^k(\theta', \lambda') \right] \right\} \sin\theta' d\theta' d\lambda'$$

$$= \int_0^{2\pi} \int_0^{\pi} \int_0^{r_0} \rho(q) q^2 \left[1 + \sum_{n=1}^{\infty} \sum_{k=-n}^{n} \left(3 f_n^k + q \frac{f_n^k}{q} \right) \mathcal{H}_n^k(\theta', \lambda') \right] dq$$

$$\times \sin\theta' d\theta' d\lambda'$$

$$= \left(\int_0^{2\pi} \int_0^{\pi} \sin\theta' d\theta' d\lambda' \right) \left[\int_0^{r_0} \rho(q) q^2 dq \right]$$

$$+ \sum_{n=1}^{\infty} \sum_{k=-n}^{n} \left[\int_0^{r_0} \rho(q) q^2 \left(3 f_n^k + q \frac{f_n^k}{q} \right) dq \right]$$

$$\times \left[\int_0^{2\pi} \int_0^{\pi} \mathcal{H}_n^k(\theta', \lambda') \sin\theta' d\theta' d\lambda' \right]$$

$$= 4\pi \int_0^{r_0} \rho(q) q^2 dq + \sum_{n=1}^{\infty} \sum_{k=-n}^{n} \left[\int_0^{r_0} \rho(q) q^2 \left(3 f_n^k + q \frac{f_n^k}{q} \right) dq \right]$$

$$\times \left[\int_\omega \mathcal{H}_n^k(\theta', \lambda') d\omega \right]$$

$$= 4\pi \int_0^{r_0} \rho(q) q^2 dq . \tag{9.60}$$

where the orthogonality (9.20) is used to derive the last identity. Setting $\rho = 1$ yields the volume enclosed in the equi-density surface

$$v_{r_0} = 4\pi \int_0^{r_0} q^2 dq = \frac{4}{3} \pi r_0^3 . \tag{9.61}$$

Hence, the average density is

$$D = \frac{M_{r_0}}{v_{r_0}} = \frac{3}{r_0^3} \int_0^{r_0} \rho q^2 dq \,, \tag{9.62}$$

which is identical to (9.57).

We now proceed to show that $f_2^0 \neq 0$, but $f_n^k = 0$ for all other n and k.

First, we show that $|f_n^k|$ increases with r_0 when r_0 is extremely small. We know that ρ decreases with r_0 — under the long term action of gravity, Earth materials flow, and heavier materials sink down with time. Hence, we have $D = \rho$ when $r_0 = 0$, and $D > \rho$ when $r_0 > 0$. When r_0 is extremely small, we can approximately take $D = \rho$, and (9.58) is simplified to

$$\frac{d^2 f_n^k}{dr_0^2} - \frac{n(n+1)}{r_0^2} f_n^k + \frac{6}{r_0} \left(\frac{df_n^k}{dr_0} + \frac{f_n^k}{r_0} \right) = 0 \,. \tag{9.63}$$

The general solution of this second order differential equation is

$$f_n^k = A_n^k r_0^{n-2} + B_n^k r_0^{-n-3} \,, \tag{9.64}$$

where A_n^k and B_n^k are arbitrary coefficients. Since f_n^k must be finite, we have $B_n^k = 0$. When $n > 2$, (9.64) implies that $|f_n^k|$ increases with r_0. The cases when $n = 1$ and $n = 2$ have to be studied particularly. When $n = 1$, it can be easily verified that $f_1^k = A_1^k / r_0$ is in fact an exact solution of (9.58). We now substitute this solution into the equation of equi-density surface (9.42). After further substituting $\mathcal{H}_n^k(\theta, \lambda)$ by its definition (9.18), and then the Legendre functions by their expressions in Table 3.2, we obtain

$$r(\theta, \lambda) = r_0 + A_1^0 \cos \theta + A_1^1 \sin \theta \cos \lambda + A_1^{-1} \sin \theta \sin \lambda \,. \tag{9.65}$$

This is a sphere with the center's Cartesian coordinate to be $(-A_1^1, -A_1^{-1}, -A_1^0)$, implying that this solution is a rigid shift of the whole sphere. In order for $r = 0$ when $r_0 = 0$, we must have $A_1^k = 0$. Hence, the case $n = 1$ is irrelevant, and there is no need to be further considered. When $n = 2$, (9.64) implies that $f_2^k = A$ is a constant. Some refinement is required. We now assume

$$\frac{\rho}{D} = 1 - H r_0^\alpha \,, \qquad f_2^k = A + B r_0^s \,, \tag{9.66}$$

where $H > 0$ and $\alpha > 0$ must hold, as $\rho / D < 0$. The substitution of this assumption into (9.58) yields

$$s(s+5) B r_0^{s-2} - 6H A r_0^{\alpha-2} = 0 \,. \tag{9.67}$$

For this relation to hold, we must have $s = \alpha$, and A and B must have the same sign. These implies that $|f_2^k|$ increases with r_0. In summary, we proved that, for any n, $|f_n^k|$ increases with r_0 when r_0 is extremely small.

We can further show that $|f_n^k|$ increases with r_0 throughout $0 \leq r_0 \leq R$. It is known that $|f_n^k|$ increases with r_0 when r_0 is extremely small. In order for $|f_n^k|$ to start decreasing with r_0, $df_n^k/dr_0 = 0$ must hold at a certain point, where (9.58) becomes

$$\frac{d^2 f_n^k}{dr_0^2} = \left[n(n+1) - 6\frac{\rho}{D}\right]\frac{f_n^k}{r_0^2} . \tag{9.68}$$

When $n \geq 2$, the value within the brackets at the right-hand side is always positive, meaning that $d^2 f_n^k/dr_0^2$ and f_n^k have the same sign. This implies that, according to the Taylor series

$$f_n^k(r_0 + \delta r_0) = f_n^k(r_0) + \frac{df_n^k}{dr_0}\delta r_0 + \frac{1}{2}\frac{d^2 f_n^k}{dr_0^2}\delta r_0^2 , \tag{9.69}$$

$|f_n^k|$ keeps increasing with r_0 after this point. As a result, $|f_n^k|$ increases with r_0 throughout $0 \leq r_0 \leq R$, and f_n^k does not change its sign.

Finally, we examine the value of f_n^k on the Earth's surface, $f_n^k(R)$. By setting $r_0 = R$ in (9.54), and then dividing both sides of it by $R^{n-1}\int_0^R 3\rho q^2 dq$, we obtain

$$nf_n^k(R) + R\left(\frac{df_n^k}{dr_0}\right)_{\substack{r_0=R \\ n=2,k=0}} = -\frac{5\Omega^2}{4\pi G D(R)} , \tag{9.70}$$

where

$$D(R) = \frac{3}{R^3}\int_0^R \rho q^2 dq \tag{9.71}$$

is the average density of the whole Earth. As $|f_n^k|$ increases with r_0 throughout $0 \leq r_0 \leq R$, the two terms at the left-hand side of (9.70) have the same sign, and consequently, they must vanish when the right-hand side of the equation vanishes, which is not the case only when both $n = 2$ and $k = 0$ hold.

Thus we have shown that only f_2^0 is non-vanishing, which is a consequence of the Earth's rotation, as can be seen from (9.70). In this circumstance, the equipotential surface of gravity (9.42) is simplified to

$$r(\theta, \lambda) = r_0\left[1 + f_2^0 P_2(\cos\theta)\right] . \tag{9.72}$$

We now show that this is the equation of an ellipsoid of revolution accurate to the first order of flattening. The equation of an ellipsoid of revolution with semi-major axis a and flattening α is

$$r^2 \left[\frac{\sin^2 \theta}{a^2} + \frac{\cos^2 \theta}{a^2 (1 - \alpha)^2} \right] = 1. \tag{9.73}$$

The equi-volumetric radius is

$$r_0 = a (1 - \alpha)^{1/3} = a \left(1 - \frac{1}{3}\alpha \right). \tag{9.74}$$

We obtain, from (9.73),

$$
\begin{aligned}
r &= a \left[\sin^2 \theta + \frac{\cos^2 \theta}{(1 - \alpha)^2} \right]^{-1/2} \\
&= a \left[\sin^2 \theta + (1 + 2\alpha) \cos^2 \theta \right]^{-1/2} \\
&= a \left(1 + 2\alpha \cos^2 \theta \right)^{-1/2} \\
&= a \left(1 - \alpha \cos^2 \theta \right) \\
&= a \left[1 - \frac{1}{3}\alpha - \frac{2}{3}\alpha P_2(\cos \theta) \right] \\
&= r_0 \left[1 - \frac{2}{3}\alpha P_2(\cos \theta) \right].
\end{aligned} \tag{9.75}
$$

By comparing this formula with (9.72), we obtain

$$f_2^0 = -\frac{2}{3}\alpha. \tag{9.76}$$

The substitution of f_2^0 by α according to (9.76) in the differential equations (9.51), (9.54), and (9.56) yields, with $n = 2$ and $k = 0$,

$$\frac{5}{3} D\alpha - \frac{1}{r_0^5} \int_0^{r_0} \rho d(q^5 \alpha) - \int_{r_0}^{R} \rho d\alpha - \frac{5\Omega^2}{8\pi G} = 0, \tag{9.77}$$

$$\frac{D}{3} \left(2\alpha + r_0 \frac{d\alpha}{dr_0} \right) - \int_{r_0}^{R} \rho d\alpha - \frac{5\Omega^2}{8\pi G} = 0, \tag{9.78}$$

and

$$r_0^2 \frac{d^2 \alpha}{dr_0^2} - 6\alpha + \frac{6\rho}{D} \left(r_0 \frac{d\alpha}{dr_0} + \alpha \right) = 0, \tag{9.79}$$

respectively, where the definition of average density (9.62) has been used. These are all referred to as Clairaut's equation.

Customarily, the second order ordinary differential equation (9.79) is solved to compute the internal flattening. Therefore, two boundary conditions are required to uniquely determine the solution. At the vicinity of the Earth's center, with the approximation $D = \rho$, the general solution of (9.79) is $\alpha = A + B r_0^{-5}$, where A and B are arbitrary constants, just as in the formulation of (9.63) and (9.64). A condition $B = 0$ has to be imposed at the vicinity of the Earth's center in order for α and $d\alpha/dr_0$ to be finite, which is referred to as regularity boundary condition. This condition can be expressed as $(d\alpha/dr_0)_{r_0=0} = 0$ or $[r_0(d\alpha/dr_0)]_{r_0=0} = 0$, both leading to $B = 0$. A condition on the Earth's surface can be obtained from (9.70)

$$2\alpha(R) + R\left(\frac{d\alpha}{dr_0}\right)_{r_0=R} = \frac{15\Omega^2}{8\pi G D(R)} = \frac{5}{2}m, \tag{9.80}$$

where m is defined as

$$m = \frac{3\Omega^2}{4\pi G D(R)} = \frac{\Omega^2 R^3}{G\left[\frac{4}{3}\pi R^3 D(R)\right]} = \frac{\Omega^2 R^3}{GM}, \tag{9.81}$$

M being the mass of the Earth. Actually, this is the same m as that defined in (4.2), which is determined from observations. When ρ has a jump, the value of α must continue across the place of jump, as a cavity or overlap may occur otherwise. It can be inferred from (9.78) that $d\alpha/dr_0$ is also continuous at the place of jump of ρ, as D and $\int_{r_0}^{R} \rho d\alpha$ are continuous functions of r_0.

For readers familiar with numerical integration of ordinary differential equations, the Clairaut's equation (9.79) can be solved by converting it into a system of two first order differential equations. Let

$$\alpha_1 = r_0 \frac{d\alpha}{dr_0}. \tag{9.82}$$

We can write (9.79) as

$$\begin{cases} \dfrac{d\alpha}{dr_0} = \dfrac{1}{r_0}\alpha_1, \\ \dfrac{d\alpha_1}{dr_0} = \dfrac{6}{r_0}\left(1 - \dfrac{\rho}{D}\right)\alpha + \dfrac{1}{r_0}\left(1 - 6\dfrac{\rho}{D}\right)\alpha_1. \end{cases} \tag{9.83}$$

The boundary condition (9.80) becomes

$$2\alpha(R) + \alpha_1(R) = \frac{5}{2}m. \tag{9.84}$$

A conceptual procedure of computation is as follows: (1) We start the integration from the Earth's center. For $r_0 \leq \delta r$, we approximately take $D = \rho$. The solution is known to be $\alpha = A$, and $d\alpha/dr_0 = 0$. Hence, at $r_0 \leq \delta r$, we set $\alpha = A$, and $\alpha_1 = r_0(d\alpha/dr_0) = 0$, where A is a constant arbitrarily chosen. (2) Numerically integrate the system of equation (9.83) from $r_0 = \delta r$ to the Earth's surface $r_0 = R$. At places where ρ has a jump, the integration is continued by keeping both α and α_1 continuous. (3) At the Earth's surface, multiply both α and α_1 by a constant, so that the boundary condition (9.84) is satisfied. (4) Multiply α and α_1 throughout $0 \leq r_0 \leq R$ by the same constant. The result is what we are seeking.

9.5 An Analytic Approximate Solution of Clairaut's Equation

We present a classical approximate analytic solution for the internal flattening of the Earth. Define

$$\eta = \frac{r_0}{\alpha} \frac{d\alpha}{dr_0}, \tag{9.85}$$

which is referred to as Radau's parameter. We can readily obtain

$$\begin{cases} \dfrac{d\alpha}{dr_0} = \dfrac{\eta\alpha}{r_0}, \\[2mm] \dfrac{d^2\alpha}{dr_0^2} = \left(\dfrac{1}{r_0} \dfrac{d\eta}{dr_0} + \dfrac{\eta^2 - \eta}{r_0^2} \right) \alpha. \end{cases} \tag{9.86}$$

Substitute these formulae into Clairaut's equation (9.79). We obtain

$$r_0 \frac{d\eta}{dr_0} + \eta^2 - \eta - 6 + \frac{6\rho}{D}(\eta + 1) = 0. \tag{9.87}$$

Multiply both sides of (9.57) by $r_0^3/3$, and then take the derivative with respect to r_0. We obtain

$$\frac{1}{3} \frac{d}{dr_0}(r_0^3 D) = r_0^2 D + \frac{1}{3} r_0^3 \frac{dD}{dr_0} = \rho r_0^2, \tag{9.88}$$

whence

$$\frac{\rho}{D} = 1 + \frac{1}{3} \frac{r_0}{D} \frac{dD}{dr_0}. \tag{9.89}$$

Substitute this formula into (9.87). We obtain

$$r_0 \frac{d\eta}{dr_0} + \eta^2 + 5\eta + 2 \frac{r_0}{D} \frac{dD}{dr_0}(\eta + 1) = 0. \tag{9.90}$$

By eliminating $d\eta/dr_0$ from the identity

$$\frac{\frac{d}{dr_0}\left(Dr_0^5\sqrt{1+\eta}\right)}{Dr_0^5\sqrt{1+\eta}} = \frac{1}{D}\frac{dD}{dr_0} + \frac{5}{r_0} + \frac{1}{2(1+\eta)}\frac{d\eta}{dr_0} \tag{9.91}$$

and (9.90), we obtain

$$\frac{2(1+\eta)}{Dr_0^4\sqrt{1+\eta}}\frac{d}{dr_0}\left(Dr_0^5\sqrt{1+\eta}\right) - \frac{2r_0}{D}\frac{dD}{dr_0}(1+\eta) - 10(1+\eta)$$

$$+ \eta^2 + 5\eta + 2\frac{r_0}{D}\frac{dD}{dr_0}(1+\eta) = 0. \tag{9.92}$$

This equation can be simplified to

$$\frac{d}{dr_0}\left(Dr_0^5\sqrt{1+\eta}\right) = 5Dr_0^4\psi(\eta), \tag{9.93}$$

where

$$\psi(\eta) = \frac{1 + \frac{1}{2}\eta - \frac{1}{10}\eta^2}{\sqrt{1+\eta}}. \tag{9.94}$$

To this stage, Clairaut's second order differential equation is reduced to a first order differential equation (9.93).

Equation (9.93) can be approximately simplified based on the range of η and $\psi(\eta)$, for which we derive another expression of η. Subtract (9.78) from (9.77). We obtain

$$\frac{5}{3}D\alpha - \frac{1}{r_0^5}\int_0^{r_0}\rho d(q^5\alpha) - \frac{D}{3}\left(2\alpha + r_0\frac{d\alpha}{dr_0}\right) = 0. \tag{9.95}$$

By substituting the first equation of (9.86) into this one, we obtain

$$\frac{5}{3}D\alpha - \frac{1}{r_0^5}\left(\rho r_0^5\alpha - \int_0^{r_0}q^5\alpha d\rho\right) - D\left(\frac{2}{3}\alpha + \frac{1}{3}\alpha r_0^4\eta\right) = 0. \tag{9.96}$$

An expression of η can then be obtained by solving this equation

$$\eta = 3\left(1 - \frac{\rho}{D} + \frac{1}{r_0^5\alpha D}\int_0^{r_0}q^5\alpha\frac{d\rho}{dr_0}dr_0\right). \tag{9.97}$$

For the Earth, we know that $\alpha > 0$ and $d\rho/dr_0 \leq 0$. Hence, the integral in the above formula is always negative, and we obtain

$$\eta \leq 3\left(1 - \frac{\rho}{D}\right) \leq 3. \tag{9.98}$$

The value of η on the Earth's surface can be obtained by substituting the first equation in (9.86) into the boundary condition (9.80)

$$\eta(R) = \frac{5}{2}\frac{m}{\alpha(R)} - 2. \tag{9.99}$$

For the Earth, $m \approx 1/289$ and $\alpha(R) \approx 1/298$. We have

$$\eta(R) \approx 0.57. \tag{9.100}$$

This implies that η varies from 0 at the Earth's center to ~ 0.57 on the Earth's surface. According to

$$\frac{1}{\psi}\frac{d\psi}{d\eta} = \frac{d}{d\eta}\ln\psi = \frac{\eta(1 - 3\eta)}{20(1 + \eta)\left(1 + \frac{1}{2}\eta - \frac{1}{10}\eta^2\right)}, \tag{9.101}$$

we see that $\psi(\eta)$ reaches extreme when $\eta = 0$ or $1/3$. Hence, $\psi(\eta)$ may attain its minimum or maximum at $\eta = 0, 1/3, 0.57$, or 3. We have

$$\begin{cases} \eta = \quad 0, \quad\quad 1/3, \quad\quad 0.57, \quad 3; \\ \psi(\eta) = 1.00000, \ 1000074, \ 0.99961, \ 0.8. \end{cases} \tag{9.102}$$

Therefore, setting $\psi(\eta) = 1$ within the Earth introduces an error smaller than the order of only $1/1000$, which is referred to as Radau's approximation. We can thus simplify (9.93) to an approximate equation for practical computation

$$\frac{d}{dr_0}\left(Dr_0^5\sqrt{1 + \eta}\right) = 5Dr_0^4. \tag{9.103}$$

An analytic formula of the flattening can be derived based on the approximate equation (9.103). Replace the variable r_0 by q in (9.103), and then integrate from 0 to r_0. We obtain

$$\eta = \left(\frac{5}{Dr_0^5}\int_0^{r_0} Dq^4 dq\right)^2 - 1. \tag{9.104}$$

The right-hand side of this formula can be expressed as a function of the parameter

$$y = \frac{2}{Dr_0^5} \int_0^{r_0} \rho q^4 dq ,$$

(9.105)

which is referred to as moment of inertia factor,, and its physical meaning will be made clear in the next section. After replacing r_0 by q in (9.88), we obtain $\rho q^2 dq = (1/3)d(Dq^3)$. Hence, (9.105) can be further developed as, by integrating by part,

$$y = \frac{2}{3Dr_0^5} \int_0^{r_0} q^2 d(Dq^3) = \frac{2}{3Dr_0^5} \left(Dr_0^5 - 2 \int_0^{r_0} Dq^4 dq \right) .$$

(9.106)

The substitution of this formula into (9.104) yields

$$\eta = \left(\frac{5}{2} - \frac{15}{4}y \right)^2 - 1 ,$$

(9.107)

where y is supposed to be computed by numerically evaluating the integrals in (9.105) and (9.57).

The exact value of $\eta(R)$ can be computed using the density profile according to (9.107). The flattening of the Earth's surface, $\alpha(R)$, can be computed by solving (9.99)

$$\alpha(R) = \frac{5m}{2[\eta(R) + 2]} .$$

(9.108)

Some interesting theoretical extreme values of $\alpha(R)$ can be inferred from this formula and (9.97). Newton assumed the Earth to be homogeneous. In this case, we have $\rho(R)/D(R) = 1$ and $d\rho/dq = 0$, and hence, $\eta(R) = 0$ according to (9.97). We obtain $\alpha(R) = 1/230$ from the above formula with $m = 1/289$. This is the same result we expect to obtain by taking the Earth as a Maclaurin ellipsoid. Huygens assumed that the gravitational attraction points toward the Earth's center, which implies that the mass of the whole Earth is concentrated at the Earth's center. In this case, we have $\rho(R)/D(R) = 0$, $d\rho/dq = 0$, $\eta(R) = 3$, and finally $\alpha(R) = 1/577$.

With η and $\alpha(R)$, the profile of α can be computed by solving (9.85), which we write as

$$\frac{d\ln\alpha}{dq} = \frac{\eta}{q} .$$

(9.109)

Integrate it from R to r_0. We obtain the solution

$$(\ln\alpha)_R^{r_0} = \int_R^{r_0} \frac{\eta}{q} dq ,$$

(9.110)

which leads to

$$\alpha(r_0) = \alpha(R)e^{\int_R^{r_0} \frac{\eta}{q} dq} . \tag{9.111}$$

This is the formula for computing α.

It can be seen that the integrals in the expressions of D, y, and α have to be evaluated numerically, for which the simplest way is to use the rectangle or trapezoidal method.

9.6 Earth's Internal Flattening and Gravity

In hydrostatic equilibrium, the equi-value surfaces of density, pressure, and gravity potential coincide. They are shown to be an ellipsoid of revolution accurate to the first order of flattening and are given in (9.75), which is rewritten here for convenience of reference,

$$r = r_0 \left[1 - \frac{2}{3}\alpha P_2(\cos\theta) \right] . \tag{9.112}$$

The result of computation shows that the value of α is $1/414.8$ at the Earth's center, $1/413.0$ at the inner core boundary, $1/392.5$ at the core–mantle boundary, and $1/299.8$ on the Earth's surface. By comparing the theoretical value of $1/299.8$ in hydrostatic equilibrium to the value of $1/298.3$ determined by observations, we see that hydrostatic equilibrium is indeed very close to reality. A profile of the Earth's internal flattening is shown in Fig. 9.4, where the reciprocal of the flattening is plotted for physically more apparent.

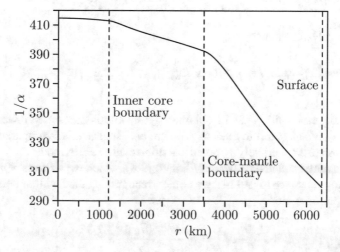

Fig. 9.4 A profile of the Earth's internal flattening. The reciprocal of the flattening, $1/\alpha$, is actually plotted to be physically more apparent

The mass and mean density within the equi-value surface are provided in (9.60) and (9.62). The values of mean density are 5.514, 10.987, and 12.894 g·cm^{-3} for the whole Earth, the whole core, and the inner core, respectively.

The gravity potential on the equi-value surface is given in (9.49), where the coefficient of $\mathcal{H}_n^k(\theta, \lambda)$ vanishes in hydrostatic equilibrium. Hence, we have

$$W = \frac{4\pi G}{3} \left(\frac{1}{r_0} \int_0^{r_0} 3\rho q^2 dq + \int_{r_0}^R 3\rho q dq + \frac{\Omega^2 r_0^2}{4\pi G} \right). \tag{9.113}$$

The gravity on the equi-value surface is

$$\vec{g} = \nabla W = \frac{\partial W}{\partial r} \hat{e}_r + \frac{1}{r} \frac{\partial W}{\partial \theta} \hat{e}_\theta = \frac{dW}{dr_0} \frac{\partial r_0}{\partial r} \hat{e}_r + \frac{1}{r} \frac{dW}{dr_0} \frac{\partial r_0}{\partial \theta} \hat{e}_\theta$$

$$= \frac{dW}{dr_0} \left(\frac{\partial r_0}{\partial r} \hat{e}_r + \frac{1}{r} \frac{\partial r_0}{\partial \theta} \hat{e}_\theta \right). \tag{9.114}$$

To further develop this formula, we write (9.112) as

$$r_0 = r \left[1 + \frac{2}{3} \alpha P_2(\cos\theta) \right] \tag{9.115}$$

to obtain, accurate to first order of α,

$$\frac{\partial r_0}{\partial r} = 1 + \frac{2}{3} \alpha P_2(\cos\theta) + \frac{2}{3} r \frac{d\alpha}{dr_0} \frac{\partial r_0}{\partial r} P_2(\cos\theta) = 1 + \frac{2}{3} \left(\alpha + r_0 \frac{d\alpha}{dr_0} \right) P_2(\cos\theta), \tag{9.116}$$

$$\frac{1}{r} \frac{\partial r_0}{\partial \theta} = \frac{2}{3} \alpha \frac{d}{d\theta} P_2(\cos\theta) + \frac{2}{3} \frac{d\alpha}{dr_0} \frac{\partial r_0}{\partial \theta} P_2(\cos\theta) = \frac{2}{3} \alpha \frac{d}{d\theta} P_2(\cos\theta). \tag{9.117}$$

It can be easily shown that

$$\frac{dW}{dr_0} = -4\pi G \frac{1}{r_0^2} \int_0^{r_0} \rho q^2 dq + \frac{2}{3} \Omega^2 r_0 = -G \frac{M_{r_0}}{r_0^2} + \frac{2}{3} \Omega^2 r_0. \tag{9.118}$$

The final expression of $\vec{g}$ can be obtained by substituting these last three formulae into (9.114), which is left to readers, who are also suggested to compare this result with the case of a spherically symmetrical non-rotating Earth model.

The pressure is related to W according to (9.8), which we write as $\nabla p = \rho \nabla W$ here. Taking reference to (9.114), we obtain, recognizing the fact that both ρ and p are functions of r_0 alone,

$$\frac{dp}{dr_0} \left(\frac{\partial r_0}{\partial r} \hat{e}_r + \frac{1}{r} \frac{\partial r_0}{\partial \theta} \hat{e}_\theta \right) = \rho \frac{dW}{dr_0} \left(\frac{\partial r_0}{\partial r} \hat{e}_r + \frac{1}{r} \frac{\partial r_0}{\partial \theta} \hat{e}_\theta \right). \tag{9.119}$$

The differential equation of pressure is thus

$$\frac{dp}{dr_0} = \rho \frac{dW}{dr_0}.$$
(9.120)

Readers are also suggested to compare this result with the case of spherically symmetrical non-rotating Earth models.

The deviation of gravity and pressure from those of the spherically symmetrical non-rotating Earth model plotted in Fig. 9.2 is on the relative order of magnitude of the flattening.

The moments of inertia are defined as

$$A = \int_v \rho(y'^2 + z'^2)dv, \quad B = \int_v \rho(z'^2 + x'^2)dv, \quad C = \int_v \rho(x'^2 + y'^2)dv.$$
(9.121)

We first compute the average of the three components defined in (9.121) for the mass within the equi-value surface (9.112)

$$I = \frac{1}{3}(A+B+C) = \frac{2}{3}\int_V \rho(x'^2+y'^2+z'^2)dV = \frac{2}{3}\int_0^{2\pi} d\lambda' \int_0^\pi d\theta' \int_0^r \rho r'^4 \sin\theta' dr'.$$
(9.122)

We emphasize that r' is defined according to (9.112), which is written here as

$$r' = q\left[1 - \frac{2}{3}\alpha P_2(\cos\theta')\right].$$
(9.123)

We obtain

$$r'^4 dr' = \frac{1}{5}dr'^5 = \frac{1}{5}d\left\{q^5\left[1 - \frac{10}{3}\alpha P_2(\cos\theta')\right]\right\}$$

$$= q^4\left[1 - \frac{10}{3}\alpha P_2(\cos\theta')\right]dq - \frac{2}{3}q^5 \frac{d\alpha}{dq}P_2(\cos\theta')dq$$

$$= q^4\left[1 - \left(\frac{10}{3}\alpha + \frac{2}{3}q\frac{d\alpha}{dq}\right)P_2(\cos\theta')\right]dq.$$
(9.124)

The substitution of this formula into (9.122) and then the use of the orthogonality relation of spherical harmonics yield

$$I = \frac{8\pi}{3}\int_0^{r_0} \rho q^4 dq.$$
(9.125)

The difference between the axial and equatorial components of moment of inertia for the mass within the equi-value surface (9.112) is

$$
\begin{aligned}
C - A &= \int_V \rho(x'^2 - z'^2) dV \\
&= \int_0^{2\pi} d\lambda' \int_0^\pi d\theta' \int_0^r \rho r'^4 (\sin^2 \theta' \cos^2 \lambda' - \cos^2 \theta') \sin \theta' dr' \\
&= \int_0^{2\pi} d\lambda' \int_0^\pi d\theta' \int_0^r \rho r'^4 \left[\cos^2 \lambda' - (1 + \cos^2 \lambda') \cos^2 \theta' \right] \sin \theta' dr' \\
&= \int_0^{2\pi} d\lambda' \int_0^\pi d\theta' \int_0^r \rho r'^4 \left[\cos^2 \lambda' - \frac{2}{3}(1 + \cos^2 \lambda') \left(\frac{3}{2} \cos^2 \theta' - \frac{1}{2} + \frac{1}{2} \right) \right] \\
&\quad \times \sin \theta' dr' \\
&= \int_0^{2\pi} d\lambda' \int_0^\pi d\theta' \int_0^r \rho r'^4 \left[-\frac{1}{3} + \frac{2}{3} \cos^2 \lambda' - \frac{2}{3}(1 + \cos^2 \lambda') P_2(\cos \theta') \right] \\
&\quad \times \sin \theta' dr' .
\end{aligned}
\tag{9.126}
$$

Substitute (9.124) into this formula, and then use the orthogonality of the spherical harmonics. We obtain

$$
\begin{aligned}
C - A &= \int_0^{2\pi} d\lambda' \int_0^\pi d\theta' \int_0^{r_0} \rho q^4 \left(-\frac{1}{3} + \frac{2}{3} \cos^2 \lambda' \right) \sin \theta' dq \\
&\quad + \int_0^{2\pi} d\lambda' \int_0^\pi d\theta' \int_0^{r_0} \rho q^4 \left(\frac{2}{3} + \frac{2}{3} \cos^2 \lambda' \right) \left(\frac{10}{3} \alpha + \frac{2}{3} q \frac{d\alpha}{dq} \right) \\
&\quad \times \left[P_2(\cos \theta') \right]^2 \sin \theta' dq \\
&= \int_0^{2\pi} d\lambda' \int_0^\pi d\theta' \int_0^{r_0} \frac{1}{3} \rho q^4 \cos 2\lambda' \sin \theta' dq \\
&\quad + \int_0^{2\pi} d\lambda' \int_0^\pi d\theta' \int_0^{r_0} \frac{2}{3} \rho \left(5q^4 \alpha + q^5 \frac{d\alpha}{dq} \right) \left(1 - \frac{1}{3} \cos 2\lambda' \right) \\
&\quad \times \left[P_2(\cos \theta') \right]^2 \sin \theta' dq \\
&= (2\pi) \times \left(\frac{2}{5} \right) \times \int_0^{r_0} \frac{2}{3} \rho \left(5q^4 \alpha + q^5 \frac{d\alpha}{dq} \right) dq \\
&= \frac{8\pi}{15} \int_0^{r_0} \rho \left(5q^4 \alpha + q^5 \frac{d\alpha}{dq} \right) dq \\
&= \frac{8\pi}{15} \int_0^{r_0} \rho \frac{d}{dq} (q^5 \alpha) dq .
\end{aligned}
\tag{9.127}
$$

The values of I are 8.020×10^{37}, 9.125×10^{36}, and 5.858×10^{34} for the whole Earth, the whole core, and the inner core, respectively. The values of $C - A$ are 2.603×10^{35}, 2.323×10^{34}, and 1.418×10^{32} for the whole Earth, the whole core, and the inner core, respectively.

Finally, we mention that the moment of inertia factor y defined in the previous section is in fact the ratio

$$y = \frac{I}{M_{r_0} r_0^2} = \frac{2}{D r_0^3} \int_0^{r_0} \rho q^4 dq , \qquad (9.128)$$

where M_{r_0} is the mass given in (9.62). For the Earth as a whole assuming axial symmetry $A = B$, the moment of inertia factor is computed using the value of the dynamic form factor $J_2 = (C - A)/(M a^2) = 1.0826 \times 10^{-3}$ determined by measuring artificial satellite orbit and that of the dynamical flattening $H = (C - A)/C = 1/305.44$ determined by observing the precession rate of the Earth's axis around the ecliptic pole. With $R = a(1 - \alpha)^{-1/3}$ and $I = (2A + C)/3$, we obtain

$$y(R) = \frac{J_2}{H} \left[1 + \frac{2}{3}(\alpha - H) \right] = 0.3307 . \qquad (9.129)$$

It can be inferred from (9.128) that $y(R) = 0.4$ when the density ρ is a constant. A value smaller than 0.4 indicates that the density is larger in the deep interior.

In Chap. 4, we have constructed a fictitious mass distribution to produce the normal gravitational potential. It is just one of the possibilities of mass distribution that produces the same external gravitational field, but in no mean it signifies any information of mass distribution within the Earth, nor any information of the gravity field within the Earth. The hydrostatic equilibrium figure formulated in this chapter is a physically meaningful approximation of the Earth's interior.

Reference

Dziewonski, A. M., & Anderson, D.L. (1981). Preliminary reference Earth model. *Physics of the Earth Planetary Interior, 25*, 297–356. https://doi.org/10.1016/0031-9201(81)90046-7.

Further Reading

Bullen, K. E. (1975). *The Earth's density* (Chapters 4 and 5). Chapman and Hail.
Jeffreys, H. (1976). *The Earth: Its origin, history and physical constitution* (Chapter 4). 6th edn. Cambridge University Press.
Lanzano, P. (1982). *Deformation of an elastic Earth* (Chapter 2). Academic Press.
Moritz, H. (1990). *The figure of the Earth: Theoretical geodesy and the Earth's interior* (Chapters 2 to 4). Karlsruhe, Germany: Wichmann.

Supplementary Materials

A

A.1 Spherical Trigonometry

Spherical trigonometry provides relations among quantities of a spherical triangle, which is defined by three points on a sphere, each pair being connected by a great arc, i.e., the arc with center on the center of the sphere. The angle between the planes holding two arcs is called an angle, and the angle between two points with respect to the center of the sphere is called a side. Definitions of an angle A and a side a are illustrated in Fig. A.1. Conventionally, as shown in Fig. A.2, A, B, and C are used to represent the angles at the respective vertex, and a, b, and c are used to represent the sides opposite to the respective vertex.

As shown in Fig. A.3, for a spherical triangle ABC, define a point D on the great arc through A and B such that $\widehat{AD} = 90°$ and a point E on the great arc through A and C such that $\widehat{AE} = 90°$. We then draw a great arc to connect D and E. Another two great arcs can be similarly drawn. The three great arcs drawn form a triangle $A'B'C'$. The vertex A of the triangle ABC is the pole of the side $B'C'$ of the triangle $A'B'C'$, i.e., if A is defined as the pole of a spherical coordinate system, the arc $B'C'$ is in the equatorial plane. The same is for the other two vertex–arc combinations. Hence, the triangle ABC is referred to as the polar triangle of $A'B'C'$. It can be readily shown that $\widehat{A'I} = \widehat{A'F} = 90°$ (hint: the angles at vertices I and F are all right angles). The same is for the other two pairs of sides. Therefore, $A'B'C'$ is also a polar triangle of ABC, implying that the relation of a triangle to its polar triangle is reciprocal. We can see that $A = \widehat{DE}$, $\widehat{EB'} = \widehat{DC'} = 90°$, and $A' = \widehat{IF}$, $\widehat{BF} = \widehat{CI} = 90°$. Consequently, we have

$$a' + A = 180°, \qquad A' + a = 180°. \tag{A.1}$$

In the following, we derive the fundamental relations among the angles and sides of a spherical triangle.

© The Author(s), under exclusive license to Springer Nature Switzerland AG 2023
J.-Y. Guo, *Physical Geodesy*, Springer Textbooks in Earth Sciences, Geography
and Environment, https://doi.org/10.1007/978-3-031-23320-3

Fig. A.1 Definition of an
angle and a side

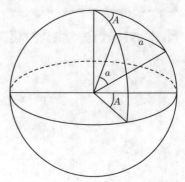

Fig. A.2 Notations of angles
and sides

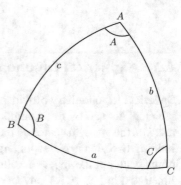

Fig. A.3 Definition of the
polar triangle

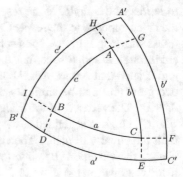

As shown in Fig. A.4, draw a straight line CD perpendicular to the plane OAB, a
straight line DE perpendicular to the line OA, and a straight line DF perpendicular
to the line OB. We then have $A = \angle CED$ and $B = \angle CFD$, and $\angle CEO$ and $\angle CFO$
are right triangles. Therefore,

$$CD = CE \sin A = OC \sin b \sin A, \tag{A.2}$$

$$CD = CF \sin B = OC \sin a \sin B. \tag{A.3}$$

Fig. A.4 Proof of the sine
formula

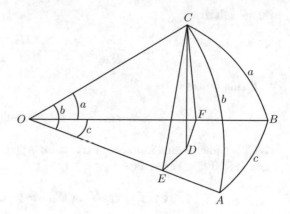

Fig. A.5 Proof of the cosine
formula

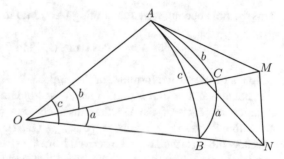

Divide one of them by the other. We obtain

$$\frac{\sin a}{\sin A} = \frac{\sin b}{\sin B}, \tag{A.4}$$

which is called sine formula.

As shown in Fig. A.5, draw a straight line AM that is tangential to the side b at
A, and draw another line AN that is tangential to the side c at A. We have

$$MN^2 = OM^2 + ON^2 - 2OM \cdot ON \cos a, \tag{A.5}$$

$$MN^2 = AM^2 + AN^2 - 2AM \cdot AN \cos A. \tag{A.6}$$

These equations can be solved to obtain

$$2OM \cdot ON \cos a = ON^2 - AN^2 + OM^2 - AM^2 + 2AM \cdot AN \cos A$$

$$= 2OA^2 + 2AM \cdot AN \cos A, \tag{A.7}$$

which leads to

$$\cos a = \frac{OA}{OM} \cdot \frac{OA}{ON} + \frac{AM}{OM} \cdot \frac{AN}{ON} \cos A . \tag{A.8}$$

We finally obtain

$$\cos a = \cos b \cos c + \sin b \sin c \cos A , \tag{A.9}$$

which is called cosine formula for the sides. Apply the above formula to the polar triangle $A'B'C'$. We obtain

$$\cos a' = \cos b' \cos c' + \sin b' \sin c' \cos A' . \tag{A.10}$$

This formula becomes, by substituting (A.1) into it,

$$\cos A = - \cos B \cos C + \sin B \sin C \cos a , \tag{A.11}$$

which is called cosine formula for the angles.

The sine and cosine formulae are most fundamental. We have given only one instance for each of them. Other instances can be obtained by rotating letters according to $(A, a) \rightarrow (B, b) \rightarrow (C, c) \rightarrow (A, a)$.

Many other formulae can be derived from the sine and cosine formulae, and some are listed below. The following are called five-part formulae:

$$\sin a \cos B = \cos b \sin c - \sin b \cos c \cos A , \tag{A.12}$$

$$\sin a \cos C = \cos c \sin b - \sin c \cos b \cos A , \tag{A.13}$$

$$\sin A \cos b = \cos B \sin C + \sin B \cos C \cos a , \tag{A.14}$$

$$\sin A \cos c = \cos C \sin B + \sin C \cos B \cos a . \tag{A.15}$$

The following are called four-part formulae:

$$\cot A \sin C = - \cos C \cos b + \sin b \cot a , \tag{A.16}$$

$$\cot A \sin B = - \cos B \cos c + \sin c \cot a , \tag{A.17}$$

$$\cot a \sin c = \cos c \cos B + \sin B \cot A , \tag{A.18}$$

$$\cot a \sin b = \cos b \cos C + \sin C \cot A . \tag{A.19}$$

Other instances of the five-part and four-part formulae can be obtained by rotating letters according to $(A, a) \rightarrow (B, b) \rightarrow (C, c) \rightarrow (A, a)$.

A.2   Some Elementary Properties of the Reference Earth Ellipsoid

As shown in Fig. A.6, for a point P on the surface of the reference Earth ellipsoid, we derive the formulae of geocentric coordinates θ and r assuming the geodetic latitude B to be known. These formulae will be used in subsequent sections.

The equation of the ellipse in the plane Opz is

$$\frac{p^2}{a^2} + \frac{z^2}{b^2} = 1 .\tag{A.20}$$

The derivative has the following geometrical property:

$$\tan\left(\frac{\pi}{2} + B\right) = \frac{dz}{dp} = -\frac{b^2}{a^2}\frac{p}{z} .\tag{A.21}$$

The geocentric coordinates r and θ as shown in Fig. A.6 are related to p and z in the form of

$$p = r\sin\theta , \quad z = r\cos\theta .\tag{A.22}$$

With these relations, we can simplify (A.21) to

$$\cot B = \frac{b^2}{a^2}\tan\theta .\tag{A.23}$$

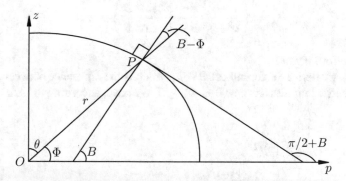

Fig. A.6 Relation between geocentric and geodetic coordinates for a point on the surface of the reference Earth ellipsoid

This is the formula for computing B using θ. The formula for computing θ using B is

$$\tan \theta = \frac{a^2}{b^2} \cot B . \tag{A.24}$$

From this formula, we can further obtain

$$\sin \theta = \frac{1}{(1 + \cot^2 \theta)^{1/2}} = \frac{1}{[1 + (b^4/a^4) \tan B]^{1/2}} = \frac{a^2 \cos B}{(a^4 \cos^2 B + b^4 \sin^2 B)^{1/2}} , \tag{A.25}$$

$$\cos \theta = (1 - \sin^2 \theta)^{1/2} = \frac{b^2 \sin B}{(a^4 \cos^2 B + b^4 \sin^2 B)^{1/2}} , \tag{A.26}$$

where we used the facts that $\sin \theta$ and $\cos B$ have the same sign, and $\cos \theta$ and $\sin B$ have the same sign.

Substitute (A.22) into (A.20), and then solve for r. We obtain, using (A.23),

$$
\begin{aligned}
r &= \left(\frac{a^2 b^2}{a^2 \cos^2 \theta + b^2 \sin^2 \theta} \right)^{1/2} \\
&= \left[\frac{a^2 b^2 (1 + \tan^2 \theta)}{a^2 + b^2 \tan^2 \theta} \right]^{1/2} \\
&= \left\{ \frac{a^2 b^2 [1 + (a^4/b^4) \cot^2 B]}{a^2 + (a^4/b^2) \cot^2 B} \right\}^{1/2} \\
&= \left(\frac{a^4 \cos^2 B + b^4 \sin^2 B}{a^2 \cos^2 B + b^2 \sin^2 B} \right)^{1/2} .
\end{aligned} \tag{A.27}
$$

This is the formula of r.

Finally, we derive the formula of $B - \Phi$, which is the difference between geodetic and geocentric latitudes. As $\cos(B - \Phi) \geq 0$, we have, according to (A.23),

$$
\begin{aligned}
\cos(B - \Phi) &= \sin(B + \theta) \\
&= [1 + \cot^2(B + \theta)]^{-1/2} \\
&= \left[1 + \left(\frac{\cot B \cot \theta - 1}{\cot B + \cot \theta} \right)^2 \right]^{-1/2} \\
&= \left[1 + \left(\frac{\cot B - \tan \theta}{\cot B \tan \theta + 1} \right)^2 \right]^{-1/2}
\end{aligned}
$$

$$= \left\{ 1 + \left[\frac{\cot B - (a^2/b^2) \cot B}{(a^2/b^2) \cot^2 B + 1} \right]^2 \right\}^{-1/2}$$

$$= \left\{ 1 + \left[\frac{(b^2 - a^2) \sin B \cos B}{a^2 \cos^2 B + b^2 \sin^2 B} \right]^2 \right\}^{-1/2}$$

$$= \frac{a^2 \cos^2 B + b^2 \sin^2 B}{(a^4 \cos^2 B + b^4 \sin^2 B)^{1/2}} , \tag{A.28}$$

$$\sin(B - \Phi)$$

$$= [1 - \cos^2(B - \Phi)]^{1/2}$$

$$= \left[1 - \frac{(a^2 \cos^2 B + b^2 \sin^2 B)^2}{a^4 \cos^2 B + b^4 \sin^2 B} \right]^{1/2}$$

$$= \left[\frac{a^4 \cos^2 B + b^4 \sin^2 B - (a^2 \cos^2 B + b^2 \sin^2 B)^2}{a^4 \cos^2 B + b^4 \sin^2 B} \right]^{1/2}$$

$$= \left[\frac{a^4 \cos^2 B + b^4 \sin^2 B - (a^4 \cos^4 B + b^4 \sin^4 B + 2a^2 b^2 \cos^2 B \sin^2 B)}{a^4 \cos^2 B + b^4 \sin^2 B} \right]^{1/2}$$

$$= \left[\frac{a^4 \cos^2 B - a^4 \cos^4 B + b^4 \sin^2 B - b^4 \sin^4 B - 2a^2 b^2 \cos^2 B \sin^2 B}{a^4 \cos^2 B + b^4 \sin^2 B} \right]^{1/2}$$

$$= \left[\frac{a^4 \cos^2 B \sin^2 B + b^4 \sin^2 B \cos^2 B - 2a^2 b^2 \cos^2 B \sin^2 B}{a^4 \cos^2 B + b^4 \sin^2 B} \right]^{1/2}$$

$$= \left[\frac{(a^2 - b^2)^2 \sin^2 B \cos^2 B}{a^4 \cos^2 B + b^4 \sin^2 B} \right]^{1/2}$$

$$= \frac{(a^2 - b^2) \sin B \cos B}{(a^4 \cos^2 B + b^4 \sin^2 B)^{1/2}} , \tag{A.29}$$

where we used the fact that $\sin(B - \Phi)$ has the same sign as $\sin B$. We finally obtain

$$\tan(B - \Phi) = \frac{(a^2 - b^2) \sin B \cos B}{a^2 \cos^2 B + b^2 \sin^2 B} . \tag{A.30}$$

A.3   Radii of Curvature of Meridians and Prime Verticals

In order to derive the formulae of the radii of curvature, we have to formulate the length of an infinitesimal arc within the meridian and prime vertical. According to Figs. A.7 and A.8, the length of an infinitesimal arc in the meridian, $\Delta X = PP'$, and that in the parallel, $\Delta Y = PP''$, are

$$\Delta X = -\frac{r\Delta\theta}{\cos(B - \Phi)} \quad \text{and} \quad \Delta Y = r\sin\theta\,\Delta\lambda\,, \tag{A.31}$$

respectively, where $\Delta\theta = \theta_P - \theta_{P'}$ is negative, so that $\Delta B = B_P - B_{P'}$ is positive. As the prime vertical and the parallel are tangential to each other at P, the length of an infinitesimal arc in the prime vertical is indeed ΔY. We now find their expressions in terms of B and λ. The expressions of r, $\sin\theta$, and $\cos(B - \Phi)$ have already been

Fig. A.7 Radius of curvature of a meridian

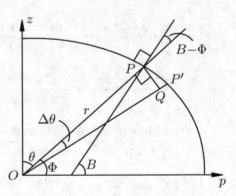

Fig. A.8 Radius of curvature of a prime vertical

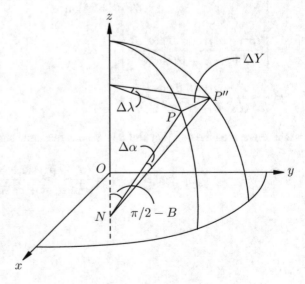

provided in (A.27), (A.25), and (A.28). Here we only need to formulate $\Delta\theta$, which can be done by taking the differential of the relation between θ and B, (A.24),

$$\sec^2\theta\,\Delta\theta = -\frac{a^2}{b^2}\csc^2 B\,\Delta B\,. \tag{A.32}$$

By using (A.24) again, we obtain

$$\begin{aligned}
\Delta\theta &= -\frac{a^2}{b^2\sin^2 B(1+\tan^2\theta)}\Delta B \\
&= -\frac{a^2}{b^2\sin^2 B[1+(a^4/b^4)\cot^2 B]}\Delta B \\
&= -\frac{a^2 b^2}{a^4\cos^2 B + b^4\sin^2 B}\Delta B\,. \tag{A.33}
\end{aligned}$$

Substitute this formula and (A.27), (A.25), and (A.28) into (A.31). We obtain

$$\Delta X = \frac{a^2 b^2}{(a^2\cos^2 B + b^2\sin^2 B)^{3/2}}\Delta B\,, \qquad \Delta Y = \frac{a^2\cos B}{(a^2\cos^2 B + b^2\sin^2 B)^{1/2}}\Delta\lambda\,. \tag{A.34}$$

Within the meridional plane Opz, the angle between the perpendiculars of the ellipse at P and P' is ΔB. Hence, the radius of curvature of the meridian is

$$R_M = \frac{\Delta X}{\Delta B} = \frac{a^2 b^2}{(a^2\cos^2 B + b^2\sin^2 B)^{3/2}}\,. \tag{A.35}$$

As shown in Fig. A.8, draw the perpendiculars to the surface of ellipsoid at P and P'', both intersecting the symmetry axis at N. The angle between these perpendiculars is then $\Delta\alpha = \cos B\,\Delta\lambda$. Hence, the radius of curvature of the prime vertical is

$$R_N = \frac{\Delta Y}{\cos B\,\Delta\lambda} = \frac{a^2}{(a^2\cos^2 B + b^2\sin^2 B)^{1/2}}\,. \tag{A.36}$$

A.4   Transformation Between Geodetic and Cartesian Coordinates

As shown in Fig. A.9, we consider a point P in the plane Opz; its geocentric coordinates are r and θ, and its geodetic coordinates are B and h.

Fig. A.9 Relation between geocentric and geodetic coordinates for a point above the surface of the reference Earth ellipsoid

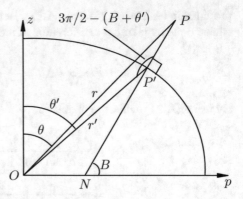

Transformation from Geodetic to Cartesian Coordinates Draw a straight line perpendicular to the surface of the reference Earth ellipsoid from P, which intersects with the surface at the point P'. Then, B is the geodetic latitude of P', and $h = PP'$. We have

$$r = [h^2 + r'^2 + 2hr' \sin(B + \theta')]^{1/2}. \tag{A.37}$$

According to (A.28) and (A.27), we have

$$\sin(B + \theta') = \frac{a^2 \cos^2 B + b^2 \sin^2 B}{(a^4 \cos^2 B + b^4 \sin^2 B)^{1/2}}. \tag{A.38}$$

$$r' = \left(\frac{a^4 \cos^2 B + b^4 \sin^2 B}{a^2 \cos^2 B + b^2 \sin^2 B}\right)^{1/2}. \tag{A.39}$$

Substitute these last two formulae into (A.37). We obtain

$$r = \left[h^2 + \frac{a^4 \cos^2 B + b^4 \sin^2 B}{a^2 \cos^2 B + b^2 \sin^2 B} + 2h \frac{a^2 \cos^2 B + b^2 \sin^2 B}{(a^2 \cos^2 B + b^2 \sin^2 B)^{1/2}} \right]^{1/2}$$

$$= \left\{ \left[h^2 + \frac{a^4}{a^2 \cos^2 B + b^2 \sin^2 B} + 2h \frac{a^2}{(a^2 \cos^2 B + b^2 \sin^2 B)^{1/2}} \right] \cos^2 B \right.$$

$$\left. + \left[h^2 + \frac{b^4}{a^2 \cos^2 B + b^2 \sin^2 B} + 2h \frac{b^2}{(a^2 \cos^2 B + b^2 \sin^2 B)^{1/2}} \right] \sin^2 B \right\}^{1/2}$$

$$= \left\{ \left[h + \frac{a^2}{(a^2 \cos^2 B + b^2 \sin^2 B)^{1/2}} \right]^2 \cos^2 B \right.$$

$$\left. + \left[h + \frac{b^2}{(a^2 \cos^2 B + b^2 \sin^2 B)^{1/2}} \right]^2 \sin^2 B \right\}^{1/2}$$

$$= \left[(h + R_N)^2 \cos^2 B + \left(h + R_N \frac{b^2}{a^2} \right)^2 \sin^2 B \right]^{1/2} . \tag{A.40}$$

This is the formula of r.

To derive the formula of θ, we take reference to the triangle OPN. According to the sine formula of plane trigonometry, we have

$$\frac{h + P'N}{\sin(\pi/2 - \theta)} = \frac{r}{\sin(\pi - B)} . \tag{A.41}$$

Hence,

$$\cos \theta = \frac{h + P'N}{r} \sin B . \tag{A.42}$$

So we have to derive the formula for $P'N$. In the triangle $OP'N$, the sine formula leads to

$$\frac{P'N}{\sin(\pi/2 - \theta')} = \frac{r'}{\sin(\pi - B)} . \tag{A.43}$$

Hence,

$$P'N = r' \frac{\cos \theta'}{\sin B} . \tag{A.44}$$

According to (A.24), we also have

$$\tan \theta' = \frac{a^2}{b^2} \cot B . \tag{A.45}$$

By substituting this formula and (A.39) into (A.44), we obtain the formula for $P'N$

$$P'N = \frac{r'}{\sin B \sec \theta'}$$

$$= \frac{r'}{\sin B (1 + \tan^2 \theta')^{1/2}}$$

$$= \frac{r'}{\sin B[1 + (a^4/b^4)\cot^2 B]^{1/2}}$$

$$= \frac{b^2 r'}{(a^4 \cos^2 B + b^4 \sin^2 B)^{1/2}}$$

$$= \frac{b^2}{(a^2 \cos^2 B + b^2 \sin^2 B)^{1/2}} \cdot \tag{A.46}$$

Substituting this formula into (A.42), we obtain the formula for $\cos\theta$

$$\cos\theta = \frac{1}{r}\left[h + \frac{b^2}{(a^2\cos^2 B + b^2\sin^2 B)^{1/2}}\right]\sin B = \frac{1}{r}\left(h + R_N\frac{b^2}{a^2}\right)\sin B .$$
$$\tag{A.47}$$

The formula for $\sin\theta$ can be obtained using this one and the formula of r, (A.40), according to $\sin\theta = (1 - \cos^2\theta)^{1/2}$,

$$\sin\theta = \frac{1}{r}\left[h + \frac{a^2}{(a^2\cos^2 B + b^2\sin^2 B)^{1/2}}\right]\cos B = \frac{1}{r}(h + R_N)\cos B .$$
$$\tag{A.48}$$

These formulae degenerate to (A.25) and (A.26) when $h = 0$. Divide (A.48) by (A.47). We obtain

$$\tan\theta = \frac{a^2 + h(a^2\cos^2 B + b^2\sin^2 B)^{1/2}}{b^2 + h(a^2\cos^2 B + b^2\sin^2 B)^{1/2}}\cot B = \frac{h + R_N}{h + R_N(b^2/a^2)}\cot B .$$
$$\tag{A.49}$$

This is the formula for θ, which degenerates to (A.24) when $h = 0$.

The relation between the Cartesian coordinate (x, y, z) and the geodetic coordinate (B, λ, h) can be obtained from (A.47) and (A.48) as

$$\begin{cases} x = r\sin\theta\cos\lambda = (h + R_N)\cos B\cos\lambda , \\ y = r\sin\theta\sin\lambda = (h + R_N)\cos B\sin\lambda , \\ z = r\cos\theta = \left(h + R_N\frac{b^2}{a^2}\right)\sin B . \end{cases} \tag{A.50}$$

Transformation from Cartesian to Geodetic Coordinates According to (A.50), we readily have

$$\lambda = \text{atan2}(y, x) , \tag{A.51}$$

where atan2 is the arctan function provided in most programming languages accepting two arguments.

The major task is to derive the formulae for computing B and h. Define

$$d = (x^2 + y^2)^{1/2} ; \qquad (A.52)$$

we can write (A.50) as, taking reference to (A.36),

$$d = \left[h + \frac{a^2}{(a^2 \cos^2 B + b^2 \sin^2 B)^{1/2}} \right] \cos B , \ z = \left[h + \frac{b^2}{(a^2 \cos^2 B + b^2 \sin^2 B)^{1/2}} \right] \sin B . \qquad (A.53)$$

Define an angle Ψ such that

$$\cos \Psi = \frac{a \cos B}{(a^2 \cos^2 B + b^2 \sin^2 B)^{1/2}} , \qquad \sin \Psi = \frac{b \sin B}{(a^2 \cos^2 B + b^2 \sin^2 B)^{1/2}} . \qquad (A.54)$$

It can be verified that this definition satisfies $\cos^2 \Psi + \sin^2 \Psi = 1$, which is a must. We can write (A.53) as

$$d - a \cos \Psi = h \cos B , \qquad z - b \sin \Psi = h \sin B . \qquad (A.55)$$

We can further obtain

$$\frac{d - a \cos \Psi}{z - b \sin \Psi} = \frac{\cos B}{\sin B} = \frac{b \cos \Psi}{a \sin \Psi} , \qquad (A.56)$$

where (A.54) has been used to derive the second identity. This equation can be reduced to

$$ad \sin \Psi - bz \cos \Psi - (a^2 - b^2) \sin \Psi \cos \Psi = 0 . \qquad (A.57)$$

Define an angle Ω such that

$$\cos \Omega = \frac{ad}{[(ad)^2 + (bz)^2]^{1/2}} , \qquad \sin \Omega = \frac{bz}{[(ad)^2 + (bz)^2]^{1/2}} . \qquad (A.58)$$

We then have

$$\Omega = \text{atan2}(bz, ad) , \qquad (A.59)$$

and we can write (A.57) as, by multiplying both its sides by $2/[(ad)^2 + (bz)^2]^{1/2}$,

$$f(\Psi) = 2 \sin(\Psi - \Omega) - c \sin 2\Psi = 0 , \qquad (A.60)$$

where we have used $f(\Psi)$ to represent the function for further reference, and c is defined as

$$c = \frac{a^2 - b^2}{[(ad)^2 + (bz)^2]^{1/2}} . \tag{A.61}$$

Now the procedure to solve for B and h is made clear. We first compute Ψ by solving (A.60), where Ω and c are computed according to (A.59), (A.61), and (A.52). We then compute B using

$$B = \text{atan2}(a \sin \Psi, b \cos \Psi) , \tag{A.62}$$

which is obtained from (A.54), and compute h using

$$h = (d - a \cos \Psi) \cos B + (z - b \sin \Psi) \sin B , \tag{A.63}$$

which is obtained from (A.55).

Lastly, we outline the solution procedure of (A.60) for computing Ψ, which is quite straightforward using Newton's iterative method. A starting value can be obtained by setting $h = 0$ in (A.55)

$$\Psi_0 = \text{atan2}(az, bd) . \tag{A.64}$$

Let the solution be

$$\Psi = \Psi_0 + \delta\Psi . \tag{A.65}$$

The equation $f(\Psi) = 0$ then becomes

$$f(\Psi) = f(\Psi_0) + \left[\frac{df(\Psi)}{d\Psi} \right]_{\Psi=\Psi_0} \delta\Psi = 0 . \tag{A.66}$$

Solve for $\delta\Psi$ from the last equation, and then substitute the result into the preceding one. We obtain

$$\Psi = \Psi_0 - f(\Psi_0) \bigg/ \left[\frac{df(\Psi)}{d\Psi} \right]_{\Psi=\Psi_0} . \tag{A.67}$$

Substituting the explicit expression of $f(\Psi)$ from (A.60) into it, we finally obtain

$$\Psi = \Psi_0 - \frac{2 \sin(\Psi_0 - \Omega) - c \sin 2\Psi_0}{2[\cos(\Psi_0 - \Omega) - c \cos 2\Psi_0]} , \tag{A.68}$$

which can be used iteratively by setting Ψ_0 to be the result obtained in the last iteration. It can be easily inferred that c is on the order of e^2 in magnitude, where e is the first eccentricity. According to (A.60), $\Psi - \Omega$ is also on the order of e^2 in magnitude. Hence, the denominator of the second term of (A.68) is on the order of 1 in magnitude, while the numerator should be extremely small, and therefore, the result would converge rapidly. We mention again that B and h are to be computed according to (A.62) and (A.63).

A.5   Coordinate Transformation Between Cartesian Coordinate Systems

By rotating a coordinate system an angle around a straight line through the origin, we obtain a new coordinate system. The subject of this subsection is to formulate the coordinate transformation between these two coordinate systems following the convention in the community of geodesy.

We first consider a simple case, as shown in Fig. A.10, where a new coordinate system $Ox'y'z'$ is obtained by rotating the coordinate system $Oxyz$ an angle α around the x-axis according to the right-hand rule: point the thumb to the direction of the x-axis, and the rotation is in the direction of the other 4 fingers curved to inside the palm. We formulate the relation between the coordinates of a point P in the two coordinate systems. Evidently, we have $x = x'$, and only the transformation between x, y and y', z' is in need of formulation. The Cartesian coordinates of the point P in the two coordinate systems expressed using polar coordinates are

$$\begin{cases} y = r\cos(\psi + \alpha), & z = r\sin(\psi + \alpha); \\ y' = r\cos\psi, & z' = r\sin\psi. \end{cases} \tag{A.69}$$

Fig. A.10 Transformation of coordinates by rotation

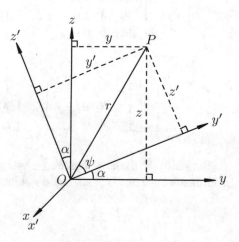

The transformation from y, z to y', z' is

$$\begin{cases} y' = r\cos[(\psi + \alpha) - \alpha] = y\cos\alpha + z\sin\alpha\,; \\ z' = r\sin[(\psi + \alpha) - \alpha] = -y\sin\alpha + z\cos\alpha\,. \end{cases} \tag{A.70}$$

Its inverse, i.e., the transformation from y', z' to y, z, is

$$\begin{cases} y = y'\cos\alpha - z'\sin\alpha\,; \\ z = y'\sin\alpha + z'\cos\alpha\,. \end{cases} \tag{A.71}$$

These coordinate transformations can be written in matrix forms as

$$R_1(\alpha) = \begin{bmatrix} 1 & 0 & 0 \\ 0 & \cos\alpha & \sin\alpha \\ 0 & -\sin\alpha & \cos\alpha \end{bmatrix}, \quad R_1'(\alpha) = \begin{bmatrix} 1 & 0 & 0 \\ 0 & \cos\alpha & -\sin\alpha \\ 0 & \sin\alpha & \cos\alpha \end{bmatrix}; \tag{A.72}$$

$$\begin{bmatrix} x' \\ y' \\ z' \end{bmatrix} = R_1(\alpha) \begin{bmatrix} x \\ y \\ z \end{bmatrix}, \quad \begin{bmatrix} x \\ y \\ z \end{bmatrix} = R_1'(\alpha) \begin{bmatrix} x' \\ y' \\ z' \end{bmatrix}. \tag{A.73}$$

If the new coordinate system $Ox'y'z'$ is obtained by rotating the coordinate system $Oxyz$ an angle β around the y-axis according to the right-hand rule, the coordinate transformation can be obtained by changing the letters x, y, and z to y, z, and x, respectively, and the angle α to β in (A.72) and (A.72). After rearranging the coordinates in sequential order, we write the coordinate transformations in the form

$$R_2(\beta) = \begin{bmatrix} \cos\beta & 0 & -\sin\beta \\ 0 & 1 & 0 \\ \sin\beta & 0 & \cos\beta \end{bmatrix}, \quad R_2'(\beta) = \begin{bmatrix} \cos\beta & 0 & \sin\beta \\ 0 & 1 & 0 \\ -\sin\beta & 0 & \cos\beta \end{bmatrix}; \tag{A.74}$$

$$\begin{bmatrix} x' \\ y' \\ z' \end{bmatrix} = R_2(\beta) \begin{bmatrix} x \\ y \\ z \end{bmatrix}, \quad \begin{bmatrix} x \\ y \\ z \end{bmatrix} = R_2'(\beta) \begin{bmatrix} x' \\ y' \\ z' \end{bmatrix}. \tag{A.75}$$

Similarly, if the new coordinate system $Ox'y'z_3'$ is obtained by rotating the coordinate system $Oxyz$ an angle γ around the z-axis according to the right-hand rule, the coordinate transformation is

$$R_3(\gamma) = \begin{bmatrix} \cos\gamma & \sin\gamma & 0 \\ -\sin\gamma & \cos\gamma & 0 \\ 0 & 0 & 1 \end{bmatrix}, \quad R'_3(\gamma) = \begin{bmatrix} \cos\gamma & -\sin\gamma & 0 \\ \sin\gamma & \cos\gamma & 0 \\ 0 & 0 & 1 \end{bmatrix}; \quad (A.76)$$

$$\begin{bmatrix} x' \\ y' \\ z' \end{bmatrix} = R_3(\gamma) \begin{bmatrix} x \\ y \\ z \end{bmatrix}, \quad \begin{bmatrix} x \\ y \\ z \end{bmatrix} = R'_3(\gamma) \begin{bmatrix} x' \\ y' \\ z' \end{bmatrix}. \quad (A.77)$$

It can be readily verified that the transformation matrices have the following property:

$$\begin{cases} R'_1(\alpha) = [R_1(\alpha)]^{-1} = [R_1(\alpha)]^T = R_1(-\alpha), \\ R'_2(\beta) = [R_2(\beta)]^{-1} = [R_2(\beta)]^T = R_2(-\beta), \\ R'_3(\gamma) = [R_3(\gamma)]^{-1} = [R_3(\gamma)]^T = R_3(-\gamma). \end{cases} \quad (A.78)$$

These formulas imply that R'_i and R_i are the inverse of each other, R_i is orthogonal, and the inverse of a rotation is equivalent to a backward rotation.

The transformation matrices R_i and R'_i can be used sequentially. For example, if $Ox'y'z'$ is obtained by rotating $Oxyz$ an angle α around the x-axis, $Ox''y''z''$ is obtained by rotating $Ox'y'z'$ an angle β around the y'-axis, and $Ox'''y'''z'''$ is obtained by rotating $Ox''y''z''$ an angle γ around the z''-axis, the coordinate transformations between $Oxyz$ and $Ox'''y'''z'''$ are

$$\begin{bmatrix} x''' \\ y''' \\ z''' \end{bmatrix} = R_3(\gamma)R_2(\beta)R_1(\alpha) \begin{bmatrix} x \\ y \\ z \end{bmatrix}, \quad \begin{bmatrix} x \\ y \\ z \end{bmatrix} = R'_1(\alpha)R'_2(\beta)R'_3(\gamma) \begin{bmatrix} x''' \\ y''' \\ z''' \end{bmatrix}. \quad (A.79)$$

In these formulae, the sequence of the matrices R_i and R'_i cannot be altered. The angles α, β and γ are referred to as Euler angles. It is left to readers to formulate the relation between the Euler angles and the angles among the axes of two Cartesian coordinate systems defined in Sect. 2.5.1.

If the Euler angles are very small, so that the approximations $\sin\alpha = \alpha$, $\sin\beta = \beta$, $\sin\gamma = \gamma$, and $\cos\alpha = \cos\beta = \cos\gamma = 1$ can be made, we obtain, by neglecting second and higher order terms of the Euler angles,

$$
R_3(\gamma)R_2(\beta)R_1(\alpha) = \begin{bmatrix} 1 & \gamma & -\beta \\ -\gamma & 1 & \alpha \\ \beta & -\alpha & 1 \end{bmatrix}, \quad R_1'(\alpha)R_2'(\beta)R_3'(\gamma) = \begin{bmatrix} 1 & -\gamma & \beta \\ \gamma & 1 & -\alpha \\ -\beta & \alpha & 1 \end{bmatrix}.
$$
$$(A.80)$$

In these formulae, the results are independent of the sequence of R_i and R_i'.

In geodesy, the transformation from the reference system $O_1x_1y_1z_1$ to $O_2x_2y_2z_2$ is referred to as Helmert transformation, which is written as

$$
\begin{bmatrix} x_2 \\ y_2 \\ z_2 \end{bmatrix} = \begin{bmatrix} \Delta x \\ \Delta y \\ \Delta z \end{bmatrix} + \begin{bmatrix} 1+s & \gamma & -\beta \\ -\gamma & 1+s & \alpha \\ \beta & -\alpha & 1+s \end{bmatrix} \begin{bmatrix} x_1 \\ y_1 \\ z_1 \end{bmatrix},
$$
$$(A.81)$$

where $(\Delta x, \Delta y, \Delta z)$ is the coordinate of the Origin O_1 in the coordinate system $O_2x_2y_2z_2$, α, β, and γ are the Euler angles, $O_1x_1y_1z_1$ rotates so that its axes be parallel to those of $O_2x_2y_2z_2$, and s is a scale factor of length.

General Literature

B

Fang, J. (1975). *Gravimetry and Earth's shape, Vol 2* (In Chinese). China Science Publishing and Media Ltd.

Garland, G. D. (1965). *The Earth's shape and gravity*. Pergamon Press Ltd.

Guan, Z. L., & Ning, J. S. (1981). *Earth's shape and external gravity field, Vols 1 and 2* (In Chinese). Beijing, China: Press of Surveying and Mapping.

Guo, J. Y. (1994). *Introduction to physical geodesy* (In Chinese). Wuhan, China: Press of Wuhan Technical University of Surveying and Mapping.

Heiskanen, W. A., & Moritz, H. (1967). *Physical geodesy*. W.H. Freeman and Company.

Hofmann-Wellenhof, B., & Moritz, H. (2005). *Physical geodesy*. Springer-Verlag.

Jekeli, C. (2000). *Heights, the geopotential, and vertical Datums*. Report 459, Department of Geodetic Science and Surveying, The Ohio State University.

Levallois, J. J. (1970). *Géodésie général, Vol 3: Le champ de la pesanteur* (In French). Paris, France: Eyrolles.

Moritz, H. (1980). *Advanced physical geodesy*. Abacus Press.

Sansò, F., Reguzzoni, M., & Barzaghi, R. (2019). *Geodetic heights*. Springer.

Sansò, F., & Sideris, M. G. (eds.) (2013). *Geoid determination: Theory and methods*. Springer-Verlag.

Sjöberg, L. E. (2015). The development of physical geodesy during 1984–2014—A personal review. *Journal of Geodetic Science, 5*, 1–8. https://doi.org/10.1515/jogs-2015-0003.

Sjöberg, L. E., & Bgherbandi, M. (2017). *Gravity inversion and integration: Theory and applications in geodesy and geophysics*. Springer International Publishing AG.

Torge, W. (1989). *Gravimetry*. de Gruyter.

Veníček, P. (1976). *Physical geodesy*. Department of Surveying Engineering Lecture Notes 43. New Brunswick, Canada: University of New Brunswick, Fredericton.

Vermeer, M., 2020. *Physical geodesy*. Aalto University publication series. Available over the Internet https://users.aalto.fi/~mvermeer/fys-en.pdf.

Index

A

Addition theorem, 131, 133, 147, 155, 212, 225, 228, 452
Adjusted disturbing potential, 362, 370, 426
Adjusted Earth, 300, 305, 306, 322, 362, 410, 426
Adjusted geoid, 300
Adjusted geoidal height, 300, 304, 428
Adjusted gravity disturbance, 362
Adjusted gravity field, 300, 362
Adjusted quasigeoid, 362, 371
Airy-Heiskanen model (isostasy), 264, 265, 267, 270, 271, 293, 298–300, 313
Analytical downward continuation, 348, 357
Angular velocity, 7
Aspherical heterogeneous body, 456
Aspherical homogeneous thin shell, 453, 456
Associated Legendre equation, 86, 101, 105
Associated Legendre function, 103, 104, 109, 142, 410, 414
Astrogeodetic deflection of the vertical, 432
Astro-gravimetric leveling, 443–445
Astronomical latitude, 431
Astronomical leveling, 439, 443, 444
Astronomical longitude, 431
Average curvature, 41
Average density, 460–462, 464
Azimuth, 217, 438, 444
Azimuthal average, 132, 415

B

Bi-linear interpolation, 405
Bjerhammar sphere, 361, 391
Bouguer correction, 255
Bouguer plate correction, 255, 256, 290
Boundary value problem, 60, 322, 390
Brillouin sphere, 132, 134, 273, 354, 392

Bruns' formula (curvature), 42
Bruns' formula (geoidal height), 205, 209, 227, 308
Bruns' formula (height anomaly), 324, 342, 363, 381, 389

C

Cap-shaped shell, 284, 289, 290, 295, 310
Center of mass, 7, 135, 237, 241, 306
Centrifugal acceleration, 6, 189
Centrifugal potential, 14, 160, 204, 457
Chasles' theorem, 57
Clairaut's equation, 464, 465
Clairaut's theorem, 187
Cogeoid, *see* Adjusted geoid
Cogeoidal height, *see* Adjusted geoidal height
Compensation bottom, 262, 263
Compensation depth, 263, 266, 293
Compensation function, 266, 268
Complete Bouguer anomaly, 258, 260, 261, 294, 301
Complete Bouguer correction, 258
Complete Bouguer gravity anomaly, *see* Complete Bouguer anomaly
Convolution, 65
Core-mantle boundary, 451, 469
Coriolis force, 7
Cosine formula for the angles (spherical trigonometry), 478
Cosine formula for the sides (spherical trigonometry), 478
Crustal density, 260, 262, 321–323, 326, 361
Curl, 77
Curvature, 39, 45
Curvature of the meridian, 483
Curvature of the prime vertical, 483

Printed in the United States
by Baker & Taylor Publisher Services